# INTRODUCTORY STATISTICAL ANALYSIS

**Donald L. Harnett**
Indiana University

**James L. Murphy**
University of North Carolina

▲ **ADDISON-WESLEY PUBLISHING COMPANY, INC.**
Reading, Massachusetts
Menlo Park, California · London · Amsterdam · Don Mills, Ontario · Sydney

ISBN 0-201-02749-6
DEFGHIJ-MA-79

*To Jan and Linda*

# Preface

## Purpose and Approach

This text is designed for use in a beginning statistics course for business, economics, or social sciences. It is intended to provide a mixture of intuitive explanation, examples, and mathematical rigor for students who have had some exposure to college mathematics. We do not assume the reader has studied calculus; however, there are occasions when our explanation will indicate how calculus could be used. In many instances we will present, in footnotes, complementary explanations which involve calculus, or more complicated mathematical proofs and derivations. We assume in this book that the reader is familiar with the basic notation of summations, but a brief review of this subject is given in Appendix A.

Readability, organization, and flow of material have been a major concern throughout the writing and the classroom trials of this material. All topics are treated without excess verbiage, since the authors believe that re-explaining the same purpose, definition, or concept several times in slightly different terms has little value to the intelligent reader. Instead, our approach emphasizes a single explanation stated as clearly as possible, followed by examples to reinforce the discussion.

Our approach is less mathematically demanding than some of the rigorous and highly symbolic texts; at the same time, however, by including more depth in probability theory, statistical inference, and regression analysis than is found in most books, we have tried to avoid compromising mathematical concepts. Sampling theory is treated as the link between probability models and statistical decision methods. We have inserted the chapter on statistical decision theory into the general development of inference, rather than adding it on at the end of the book. Material on nonparametric methods and analysis of variance has been incorporated into the chapters on hypothesis testing, regression, and correlation. Another feature of this book is the nontechnical presentation of multiple regression, including an econometric discussion of the underlying assumptions and problems of multicollinearity, autocorrelation, and heteroscedasticity.

Although the authors recommend that the text chapters be covered consecutively, it is possible for instructors to omit certain sections without damage to the general conceptual development. For example, all of the sections covering the chi-square and $F$-distributions in Chapters 5 through 8 can be omitted without loss in continuity. Also, the second chapter on hypothesis testing (Chapter 10), the second chapter on regression and correlation (Chapter 12), and the chapter on time series and index numbers (Chapter 13) can be selected individually. Finally, Sections 9.8 and 9.9 in Chapter 9 can be omitted if a less intensive coverage of statistical decision theory is desired.

The text examples and exercises were selected to be of general interest to undergraduate students, and also to demonstrate the usefulness of statistics in everyday decision-making situations. As we indicate in our section *For the Reader*, the problems at the end of each chapter are divided into Review Problems and Exercises. The student is encouraged to work as many of the former set as possible.

**Acknowledgments**

The authors express their gratitude to Professor Kirk Roberts for helpful reviews and comments on this material, and to the many students who have helped instruct us over the years in better methods of presentation. We also give credit to Ann Durham, Cheryl Capps, and Jeanie Tate for their secretarial and typing work in preparation of the manuscript.

Finally, we express our thanks to our wives and children, some of whom may someday agree with us that the entire effort was worthwhile.

*Bloomington, Indiana*                                                    D.L.H.
*Chapel Hill, North Carolina*                                        J.L.M.
*November 1974*

# For the Reader

Several distinctive pedagogical and organizational features of this book should be noted. First, the reader should take time to read Appendix A in order to review the notation associated with summations. While browsing through that portion of the book, stop a moment to notice that Appendix B contains the eight different statistical tables we will be using in this book, and that after these tables there are answers to selected (odd-numbered) problems.

A second useful feature is that those formulas in the book which are of particularly important use in statistics have been set off in a box. Other important explanations and definitions are italicized. Finally, the reader will notice that the problems at the end of each chapter are divided into two groups: Review Problems and Exercises. The Exercises involve independent work, or they include some extra challenge. The Review problems are similar to the text examples, and hence are designed to test the reader's knowledge of the basic concepts in each chapter. We suggest the student work as many as possible (if not all) of these review problems in preparing for examinations. In general, the Review Problems at the end of each chapter follow the same order as the topics presented in that chapter.

# Contents

# 1

# Introduction and Descriptive Statistics

## 1.1  STATISTICS AND STATISTICAL ANALYSIS

The techniques of statistics and their applications are involved, in one form or another, in almost all advancements in modern science, and in many other phases of human activity as well. As Solomon Fabricant said over 20 years ago, "the whole world now seems to hold that statistics can be useful in understanding, assessing, and controlling the operations of society." For good or bad, progress in our society appears to be measured more by numbers than by quality, and it is the description and manipulation of these numbers with which statistics is concerned.

### Beginning Applications of Statistics

Although the origins of statistics can be traced to studies of games of chance in the 1700's, it has been only in the past fifty years that statistical methods have grown to include applications in almost all fields of science—social, behavioral, physical, etc. Most early applications of statistics consisted primarily of methods for presenting data in the form of tables and charts. Soon, however, these techniques of *descriptive statistics*, as they are called, grew to include a large variety of measures concerned with arranging, summarizing, or somehow conveying the characteristics of a set of numbers. Presently, they represent what is certainly the most visible form of the application of statistics, attested to by the mass of quantitative information collected and printed in our culture every day. Crime rates, births and deaths, divorce rates, price indexes, the Dow–Jones average, and batting averages are but a few of the many "statistics" familiar to all of us.

In addition to conveying the characteristics of quantifiable information, descriptive measures provide an important basis for analysis in almost all academic disciplines, especially in the social and behavioral sciences, where human behavior cannot in general be described with the preciseness generated in the physical sciences. Statistical measures of satisfaction, intelligence, job aptitude, and leadership, for example, serve to expand our knowledge of human motivation and performance; in the same fashion, indexes of prices, productivity, gross national product, employment, free reserves, and net exports serve as the tools of management and government in considering policies directed toward promoting long-term growth and economic stability.

### The Use of Statistics in Decision-making

Despite the growth and increasing importance of descriptive methods over the past several hundred years, these measures now represent only a minor, relatively unimportant portion of the body of statistical literature. The phenomenal growth in statistics since the turn of the century has come in the field of what is called *statistical inference*, or *inductive statistics*. Analysis for this purpose is concerned with the process of making generalizations or predictions, or estimating the relationship between two or more variables. This process is called inferential or inductive analysis because it involves the drawing of conclusions (or "inferences") about the unknown characteristics of some phenomena on the basis of only limited or imperfect information. Generally this involves making conclusions about certain characteristics of a set of data (called a *population*) on the basis of the values observed in a sample drawn from that population. In such a role, statistics and decision-making are closely related, since statistical inference often provides the quantitative information necessary for deciding among alternative courses of action when it is impossible to predict exactly which consequence will follow each alternative action.

The process of drawing conclusions from limited information is one familiar to all of us, for almost every decision we face must be made under conditions of uncertainty about what the consequences of this decision will be. In deciding to watch television tomorrow rather than study you may, at least subconsciously, be inferring that your grades will not suffer by this decision. You probably have given considerable time and thought to your choice of a major in college, but here again your decision must be made on the basis of the limited amount of information that can be provided by aptitude tests, guidance counselors, and the advice of your parents and friends. If you make a poor choice, you may risk the loss of considerable time and money, at best. The business man faces similar problems. Should he introduce a new product? What about plant expansion? How much should be spent on advertising? The economic advisor to policy-makers must choose among various alternative recommendations for preventing unemployment, improving the trade balance, dampening inflation spirals, and increasing production and income. In general the answers to these questions and choices can only be "inferred" from less than perfect information about the optimal strategy. As a result, such decisions often take place under conditions which expose the decision-maker to considerable uncertainty and risk (in terms of the "costs" of making the wrong decision) in deciding on the appropriate action to take. The process of making decisions under these circumstances is usually referred to as *decision-making under uncertainty*, or as *decision-making under risk.**

Statistics, as a decision-making tool, plays an important role in problems of research and development and of guidance and control in a wide variety of fields. Both government and industry, for instance, participate in the process of developing, testing, and certifying new drugs and medicines, a process which often requires a large number of statistical tests (and decisions) concerned with the safety and effectiveness of these drugs for public use. Similarly, the psychologist, the lawyer, or almost any person making decisions involving uncertain quantities, such as human behavior, will often base these decisions on data of a statistical nature. Since complex decision situations almost always involve some form of

---

* Although the two terms "decision-making under uncertainty" and "decision-making under risk" are often given slightly different interpretations in the decision-theory literature, we shall consider them synonymous in this book.

statistical analysis, be it formal or informal, explicit or implicit, it is difficult to overemphasize the importance of the inferential portion of statistics as an aid to the decision-maker. In fact, statistics is often defined as the set of methods for making decisions under uncertainty.

### Meaning of the Statistical Population

In problems of statistical inference, the set of all values under consideration—that is, all pertinent data—is customarily referred to as a *population*, or as a *universe*. In general, any set of quantifiable data can be referred to as a population if that set of data constitutes *all values of interest*. For example, in deciding on the choice of a major, you may want to make inferences about the set of grades you can expect to receive in the courses in this field; or perhaps you may want to estimate the salaries being earned by people with degrees in this field. Similarly, the businessman may want to learn something about the number of customers who might buy a new product he is considering, or he may wish to know the expected increased sales resulting from a particular advertising campaign. A policy-maker may want to estimate the changing demand for food brought about by increased welfare payments to the handicapped or by a reduction of limitations on imported products. A legislator might need to infer the level of state revenue available from a special excise tax on tobacco products, or a sales tax, so that he can prepare an expenditure budget including funding for programs of importance for his constituency. In each of these cases, the set of *all relevant values* constitutes a population. The I.Q.'s of all students in the sophomore class or the I.Q.'s of all students in a university, past levels of advertising expenditures, sales, and profits of all companies, or consumption patterns for all households could represent populations.

In making decisions, we naturally would prefer to have access to as much information as possible about the population or populations relevant to any given decision. Only when *all* the information contained in a population is available can one avoid the possibility of an incorrect inference about this population. Unfortunately, it is usually not possible, or is much too costly, to collect all the information contained in the population associated with a practical problem; so inferences (and the resulting decisions) must be made on the basis of only limited or imperfect information about these values. The function of statistics as an aid to the decision-maker is to help decide (a) what information is needed for a particular type of decision, and (b) how this information can best be collected and analyzed for the decision at hand.

In trying to decide what information about a population is necessary for making a decision, we shall be referring to certain numerical characteristics which serve to describe or distinguish that population. These numerical characteristics, referred to as the *parameters* of the population, describe specific properties of the population such as the central tendency of all values or the variability of these values. For instance, one parameter of the population "executive salaries in the steel industry" would be the "average salary in that industry," since this measure describes the *central tendency* of all salaries in that population. Another parameter would be a measure of the *spread*, or *variability*, of all salaries. A precise definition of these parameters will be given in Sections 1.3 and 1.4. For now, it is important to note that the task of determining the exact value of a population parameter may be quite difficult. For one reason, it may not be convenient or practical to collect these data. If we are interested in the population "executive salaries," for example, it may not even be possible to identify all the executives in a given industry, much less find out their salaries. Suppose we are interested in future executive salaries, say one year from now. The relevant population in this case, "executive salaries one year from now," may be impossible to describe not only because of the reluctance of executives to disclose their salaries, but also because many of these people may not even know what their salaries will be one year from now.

### Use of Samples from a Population

Since it is often impossible or impractical to determine the exact value of the parameters of a population, the characteristics of a given population are normally judged by collecting or observing only a limited or restricted amount of the set of all possible values. Data collected or observed for this purpose are called a *sample*. The individual values contained in a sample are often referred to as "observations," and the population from which they come is sometimes called the "parent" population. We may, for example, take a sample of 100 executives, determine their current salaries, and on the basis of these "observations," make statements about different characteristics (parameters) of the population of interest (such as the average salary, or the variability of salaries in a certain industry, or perhaps the average salary in that industry a year from now). As another example, suppose a quality-control engineer has responsibility for assuring the reliability of electrical components produced in some production process. Testing each and every item may be prohibitively expensive, or even impossible, if inspection destroys the usefulness of the items. Consider the problem of producing a stereo cartridge which should last 1000 hours of playing time. An inspector might test each cartridge for 1000 hours or until it becomes defective; but then, what is left for the manufacturer to sell? The solution to this problem lies in determining the reliability of all items produced (a population parameter) after inspecting only a subset of the items (i.e., on the basis of a sample).

The numerical characteristics of a sample are used to *estimate* the parameters of the parent population from which this sample was drawn. A numerical characteristic used for this purpose is referred to as a *sample statistic*, or usually just a *statistic*. If 100 executives are sampled, their "average salary" is a sample statistic. A measure of the spread of these 100 salaries represents another sample statistic. Figure 1.1 illustrates these important terms in their relationship to each other.

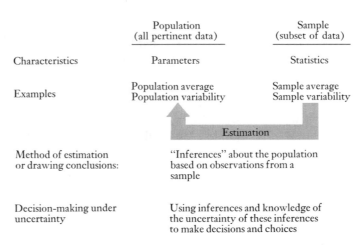

|                 | Population<br>(all pertinent data) | Sample<br>(subset of data) |
|-----------------|------------------------------------|----------------------------|
| Characteristics | Parameters                         | Statistics                 |
| Examples        | Population average<br>Population variability | Sample average<br>Sample variability |

Estimation

| Method of estimation<br>or drawing conclusions: | "Inferences" about the population<br>based on observations from a<br>sample |
|---|---|
| Decision-making under<br>uncertainty | Using inferences and knowledge of<br>the uncertainty of these inferences<br>to make decisions and choices |

**Fig. 1.1.** Statistical terms

The methods of statistical inference and decision-making under uncertainty will be described in greater detail beginning in Chapter 6. Before doing so, however, we will devote considerable attention to discussing a number of concepts and techniques fundamental to these methods. The use of probability measures discussed in Chapters 2 through 5 provides the foundation for almost all statistical tools. The methods of presenting or describing data, and the techniques for measuring characteristics of sets of data, are presented in the remainder of this chapter and in Chapter 6. It is important to realize the major purpose for which they will be used—*to draw inferences about population parameters on the basis of sample statistics and to make decisions based on these inferences.*

## DESCRIPTIVE STATISTICS

### 1.2 GRAPHICAL FORMS

Graphs and charts, the most popular and often the most convenient means for presenting data, are usually employed when a *visual* representation of all or a major portion of the information is desired. Although a large variety of alternative methods exist for presenting data in this form, only a few will be discussed here. The "pie chart," for example, is a familiar device for describing how a given quantity is divided into subsets. In Fig. 1.2 the quantity in question is the 1974 U.S. Federal tax dollar, and the subsets represent where the dollar comes from and where it goes.

Another popular descriptive measure is the chart showing changes in the size of some measure over time, such as the short-term interest rates in Fig. 1.3.

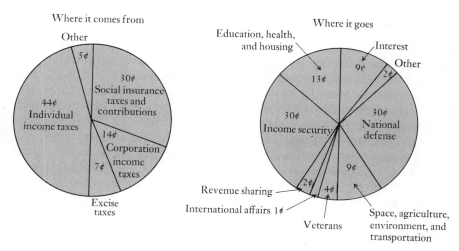

**Fig. 1.2.** The Federal budget dollar. Fiscal year 1974 estimate.

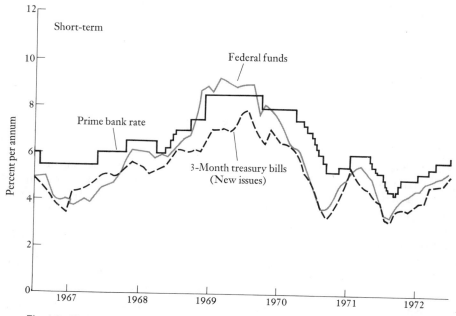

**Fig. 1.3.** Short-term interest rates. (Source: Economic report of the President, 1973.)

**Frequency Distributions**

Although graphs and charts such as those shown in Figs. 1.2 and 1.3 often serve a very useful purpose, these forms are not appropriate for most purposes of statistical analysis and decision-making because they provide only a representation of the actual information and not the data themselves. For statistical purposes, the data usually need to be presented in a form which gives a more precise indication of the information at hand. That is, some method is needed which not only *summarizes* or *describes* large masses of data without producing a loss or distortion in the essential characteristics of the information, but which also makes the data easier to present. One way to do this is to arrange the data into what is called a *frequency distribution*, or a *frequency table*. In constructing a frequency distribution, it is first necessary to divide the data into a limited number of different categories or *classes*, and then record the number of times (the frequency) an observation falls (or is "distributed") into each class. To illustrate the construction of a frequency distribution, suppose the data in Table 1.1 represent the number of pitchers of draft beer sold during "happy hours" at a certain college bar during an eight-week summer school session (40 school days), where these sales have been arranged from the lowest value to the highest.

In trying to summarize this beer-sales data, one has a wide choice of classes or categories he could use. Statisticians have developed certain guidelines for the construction of such categories. These can be listed and illustrated in terms of the data from Table 1.1.

**Table 1.1**

| | | | | | | | | | |
|----|----|----|----|----|----|----|----|----|----|
| 63 | 68 | 71 | 74 | 76 | 78 | 81 | 84 | 85 | 89 |
| 66 | 70 | 73 | 75 | 76 | 79 | 82 | 84 | 85 | 90 |
| 67 | 71 | 73 | 75 | 76 | 79 | 82 | 85 | 86 | 92 |
| 68 | 71 | 74 | 75 | 77 | 79 | 84 | 85 | 86 | 94 |

1) Classes are generally chosen so that the width of each class, called the *class interval*, is the *same for all categories*. Otherwise, interpretation of the frequency distribution may be difficult. For example, grouping the beer quantities into unequal categories such as 60–64, 65–74, 75–89, 90–95 is ill advised. Comparisons of the number of observations between each category would be misleading since the size of the categories differs from 5 units to 15 units. Instead, an equal interval size might be obtained by finding the difference between the largest and smallest value in the data and dividing this difference by the number of classes.

2) The number of classes used should probably be *less than* 20 (for ease of handling and to assure sufficient condensation of the information) and *at least* 6 (to avoid large losses of information due to grouping).

3) Open-ended intervals should be avoided. Too much information is lost if categories such as "values $< 65$" or "values $>90$" are used. Try to enclose all the values specifically. If one or a few extreme values do not conveniently fit in your frequency categories, list them separately and determine the best frequency distribution for the other values excluding these extremes.

4) Overlapping categories should not be used. Two adjacent categories such as 65–70 and 70–75 are ambiguous; rather, one might use classes, 64.5–69.5 and 69.5–74.5 so it is clear which class contains the value of 70.

5) The midpoints of each category should be *representative of the values* assigned to that category. This is important because these midpoints, called *class marks*, will be used in the calculation of summary measures, as proxies for all the values in each respective class. For our data, no obvious problem of misrepresentation by midpoints occurs. However, consider a grouping of sales items in a clothing store where many items are priced $2.98, 3.99, 9.95, etc. To set up classes such as $1.00–1.99, 2.00–2.99, 3.00–3.99, . . . , $9.00–9.99, would show poor judgment. The midpoints of the classes would be $0.50, 1.50, 2.50, . . . , $9.50, whereas most of the values assigned into the categories would fall very close to the upper endpoints of each class. It would be better to set up categories with even dollar values as the midpoints, such as $0.50–1.49, 1.50–2.49, 2.50–3.49, . . . , $9.50–10.49.

Following these basic rules, suppose we break the data in Table 1.1 into seven classes of equal width, 59.5–64.5, 64.5–69.5, . . . , 84.5–89.5, and 89.5–94.5. The boundaries of each of these classes are referred to as their *upper* and *lower class limits*. In this example, the lower class limits are 59.5, 64.5, . . . , 89.5, while the upper class limits are 64.5, 69.5, . . . , 94.5. Table 1.2 shows the frequency distribution of sales in these classes, and in addition gives the relative frequency of each class. *Relative frequency* is determined by dividing the frequency of each class by the total number of observations and expressing the result as a decimal. Thus the first relative frequency in Table 1.2 is $1/40 = 0.025$, the second is $4/40 = 0.100$,

**Table 1.2**

| Class | Frequency distribution | Relative frequency |
|---|---|---|
| 59.5–64.5 | 1 | 0.025 |
| 64.5–69.5 | 4 | 0.100 |
| 69.5–74.5 | 8 | 0.200 |
| 74.5–79.5 | 11 | 0.275 |
| 79.5–84.5 | 6 | 0.150 |
| 84.5–89.5 | 7 | 0.175 |
| 89.5–94.5 | 3 | 0.075 |

and so forth. Note that the sum of all relative frequencies must equal 1.000. It is also more clear now that sales of 74.5–79.5 pitchers of beer occurred most often and that sales above and below this level were spread similarly, but not exactly symmetrically.

This frequency distribution may lead to more statistical analysis in order to help solve some decision problem. Why are not the sales the same each day? Are sales higher on specific days during the week? Perhaps sales are related to the number of advertisements that day on local radio, or perhaps to the temperature of the particular summer day. Could the proprietor increase the overall average sales if he could determine such regularities of relationships? Could he then better plan his own advertising and purchasing of draft beer? Should he buy an outdoor thermometer sign which always reads five degrees too high, or should he hire an extra employee on Fridays?

Although these questions, related to this small operation, do not represent earth-shattering decisions, perhaps it is evident that statistical analysis of available data can be useful in finding answers. If the data represented compounded portfolio yields by the 40 largest (total assets) insurance companies, similar kinds of questions could be very important to the financial management of a single company, or to stockholders, bankers, and brokers in general. The same type of statistical analysis developed in this text for simple illustrative examples could and does apply to a wide range of very important managerial and governmental decision-making problems.

**Visual Representation of Data**

While it is often useful to arrange the values in a data set from smallest to largest (as in Table 1.1), or to classify data into categories and determine frequencies and relative frequencies (as in Table 1.2), many analysts prefer a more pictorial representation. Perhaps the most common means of doing this is by constructing a graph in which the classes are plotted on the horizontal axis and the frequency of each class is plotted on the vertical axis. This type of graph is called a *histogram*, or sometimes a *bar graph*. Figure 1.4 represents the histogram for the frequency distribution in Table 1.2.

Another means of describing the data of Table 1.2 is to construct a *frequency polygon* by drawing a straight line between the midpoints of adjacent class intervals. The *class marks* in this example equal 62, 67, 72, . . . , 92. The frequency polygon for the data in Fig. 1.4, as indicated by the dashed line through the class marks, serves to smooth a set of values. The advantage of using smoothed approximations to discrete data is that working with these approximations is in general much easier. The reader should verify at this point that plotting the *relative* frequencies in Table 1.2 yields the same graph as the frequency distribution except that the vertical scale is different (as shown on the right side of Fig. 1.4).

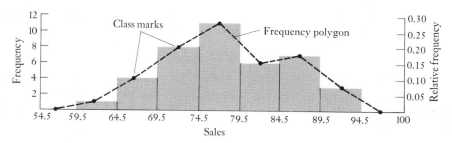

**Fig. 1.4.** Histogram and frequency polygon of data in Table 1.2.

### Cumulative Frequency Distributions

In many circumstances it is desirable to be able to describe data by recording the frequency of values *less than or equal to* some number. For instance, how many times were beer sales in Table 1.2 less than or equal to 79.5 pitchers? Descriptions of this nature are given by a *cumulative frequency distribution*, and a *cumulative relative frequency distribution*. These distributions are obtained by summing the values in a frequency distribution and in a relative frequency distribution, respectively. In Table 1.3, for instance, the cumulative frequency of class 64.5–69.5 is 5 because there are five values less than or equal to 69.5; next, the cumulative frequency for the class 69.5–74.5 is 13, since eight additional values fall between 69.5 and 74.5. Note that the cumulative relative frequency for each of these classes can be derived by dividing the cumulative frequency by the total number of observations, 40. The cumulative relative frequency for the second class is 5/40 = 0.125; and the cumulative relative frequency of the third class is 13/40 = 0.325. Each cumulative relative frequency value in Table 1.3 can also be derived by summing the previous cumulative value and the current relative frequency value. The relative frequency that beer sales were 79.5 or less is obtained by summing the

**Table 1.3**

| Class | Frequency | Cumulative frequency | Cumulative relative frequency |
|---|---|---|---|
| 59.5–64.5 | 1 | 1 | 0.025 |
| 64.5–69.5 | 4 | 5 | 0.125 |
| 69.5–74.5 | 8 | 13 | 0.325 |
| 74.5–79.5 | 11 | 24 | 0.600 |
| 79.5–84.5 | 6 | 30 | 0.750 |
| 84.5–89.5 | 7 | 37 | 0.925 |
| 89.5–94.5 | 3 | 40 | 1.000 |

cumulative relative frequency for 74.5 or less, which is 0.325, and the relative frequency for 74.5–79.5, which is 0.275, for a total of

$$0.325 + 0.275 = 0.600.$$

This value indicates that 60 percent of all sales of pitchers of beer during daily happy hours were less than or equal to 79.5. The cumulative frequency for the highest class, 89.5–94.5 in this case, must equal 1.00, since all values certainly fall within this class or some lower class.

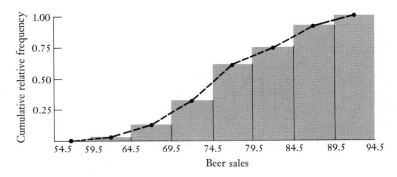

**Fig. 1.5.** Cumulative relative frequency.

Just as a graph of the frequencies of a set of values provides a visual description of the original data, so a graph of the cumulative frequency or the cumulative relative frequency provides visual information about cumulative values. Note in Fig. 1.5 that cumulative relative frequencies can be plotted in the same fashion as relative frequencies, and that this cumulative distribution can be smoothed by a line similar to the frequency polygon used in Fig. 1.4. The concepts of relative frequency and cumulative relative frequency will be important for our discussion in Chapters 2, 3, and 4, where we shall interpret these values as probabilities.

**Summary Measures**

Thus far we have concentrated on describing an entire set of observations, either graphically or by means of a frequency distribution. In many cases, however, it is not convenient to work with all observations, but it is preferable to have one or more descriptive measures which summarize the data in some quantitative form. In particular, interest usually centers on the two measures we mentioned earlier, the *central location* of the data and the *variability* or *spread* of the observations. Such characteristics of a data set are called *summary measures*. As we indicated in Fig. 1.1, when these summary measures apply to an entire population they are called *parameters*; when the data set represents a sample drawn from a population, they are called *statistics*. Their usefulness is obvious if one considers the difficulty of making a logical presentation of the meaning and interpretation of a given data set. Simple intuitive or "naked-eyeball" analysis of the values can be misleading and may easily miss some important implications. Presenting such analysis of large data sets, furthermore, is tedious and boring for the listener or reader. If the important and useful information in a data set can be condensed into a few summary measures, then understanding and comparing various features of different populations or samples becomes much easier. In operations research work (for example, in statistical analysis of weapons systems or the space program), summary presentation of this nature is called *data reduction*. All the information in a data set which is useful for a particular purpose is "reduced" into a single measure, such as a reliability measure or an average velocity.

The two types of summary measures most often used in statistical inference and decision-making concern the *central location* of the data and the *variability* of the data. There are a number of different ways of measuring these two characteristics, as shown in Table 1.4. Some of these terms are perhaps already familiar to you, while others are new, technical terms. Although each one is useful for certain purposes, this text will emphasize the two measures which are most common and useful in statistical inference, namely, the *arithmetic mean* and the *standard deviation*.

**Table 1.4** Common summary measures

| Central location | Variability |
| --- | --- |
| Arithmetic mean | Standard deviation |
| Median | Variance |
| Mode | Range |
| Geometric mean | Interquartile range |

Before presenting the meaning and the methods of computation of these and other summary measures, we will discuss why *both* the characteristic of central location and the characteristic of variability are used in most statistical analysis. Some examples of simple decision problems can illustrate why one characteristic without the other is, in general, insufficient.

### Inadequacy of Central Location Measures Only

Consider the case of a person choosing between two sales jobs each having the same potential earnings, say $7000 a year. One company representative says that the average number of hours worked per week in his company by new sales persons earning $7000 is 30; the other prospective employer says the average for his new employees is 50. One might decide on this basis to work for the first company with the "average" work week of 30 hours. Be careful! The hours worked per week to earn $7000 in the first company may range from 25 to 80 hours; whereas in the second company, the hours may range from 45 to 55 hours. Thus, we see that it is necessary to know more about the *distribution* of the data set, hours worked per week, before this information can be used intelligently in making a decision.

### Inadequacy of Variability Measures Only

Consider the same job choice as above, but suppose an important factor in the decision is the current earnings of employees who five years ago had similar starting positions to the one now being offered. Suppose the data for the first company indicates that the range from the lowest to the highest salary is $5,000; the data for the second company indicates a much larger range of $100,000. The job with the second company might appear preferable because it seems to have a higher potential for rewards. However, suppose it is also learned that the earnings data of these five-year employees in the first company shows an average of $15,000, with a spread from $13,500 to $18,500. Meanwhile, in the second company, the average is $10,000 with a spread from $6,000 to $106,000 (perhaps one of the employees was a niece of the chairman of the board and received all the sales accounts for New York City). We see in this example that a better decision can be made when one has more complete information about the distribution of the relevant data.

Whether analysis is based on an entire population or on a sample, it will be extremely important to make use of all the available information in decision-making. This especially includes the use of measures for both central location and variability. The techniques for calculating such measures and the discussion of their use and interpretation follows.

### 1.3 CENTRAL LOCATION

The single most important measure describing numerical information concerns the location of the center of the data. The term "central location" may refer to any one of a number of different measures including the *mean*, the *median*, and the *mode*. Each of these measures is appropriate for certain descriptive purposes, but completely inappropriate for others.

### The Mode

The *mode* is defined as that value which *occurs most often* or, equivalently, that point (or class mark) corresponding to the value with the *highest frequency*. For the sales shown in Table 1.1, the mode is a score of 85, since 85 occurs more often than any other number. For the grouped data of Fig. 1.4, the mode is at the class-

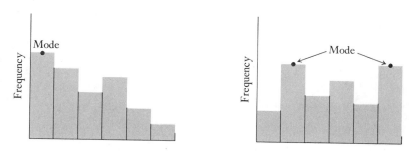

**Fig. 1.6.** The mode as a measure of central location.

mark score of 77, representing the class with the highest frequency. Note that the mode may be a poor measure of central location if the most frequently occurring value does not appear near the center of the data. The mode need not even be unique. Consider the frequency distributions shown in Fig. 1.6. The first distribution has its mode at the lowest class and certainly cannot be considered representative of central location. The second distribution has two modes, neither of which appears to be representative of the central location of the data. For these reasons, the mode has limited use as a measure of central location for decision-making. For descriptive analysis, however, the mode is a useful measure of the most frequently occurring value.

### The Median

Another measure of the central tendency of a data set is given by the *median*, defined to be the *middle value* in a set of numbers arranged according to magnitude. When it is desirable to divide data into two groups, each group containing exactly the same number of values, the median is the appropriate measure. Finding the median of a set of numbers is not difficult when these numbers are arranged according to their magnitude. If the number of values in the data set ($N$) is *odd*, the middle value can be determined by counting off, from either the highest or lowest value, $(N + 1)/2$ numbers; the resulting number divides the data into the two desired groups, and thus represents the median. For example, in a list of five values, the median is found by counting down (or up) $(5 + 1)/2 = 3$ values, while in a list of seven values, the median is found by counting down (or up) $(7 + 1)/2 = 4$ values.

When $N$ is *even*, there are *two* middle values, and the median is usually defined to be the number *halfway between* these two numbers. The median of six values is thus halfway between the third and fourth numbers. In Table 1.1, for instance, there are 40 numbers, so the median must be halfway between the twentieth and the twenty-first; since the twentieth value (from the lowest end) is 77 and the

**Table 1.5**

|            | Data set A           | Data set B         | Data set C      | Data set D       |
|------------|----------------------|--------------------|-----------------|------------------|
| Observations | 2, 2, 3, 4, 8, 10, 13 | −5, 8, 8, 9, 10, 12 | 2, 3, 4, 4, 4, 7 | 2, 3, 4, 4, 4, 19 |
| Median     | 4                    | 8.5                | 4               | 4                |
| Mode       | 2                    | 8                  | 4               | 4                |
| Mean       | 6                    | 7                  | 4               | 6                |

twenty-first value is 78, the median in this case is defined to be 77.5. Note, in the example given in Table 1.5, that no matter how many items of data there are, the median is always located in the numerical ordering so that the *number of values* on each side of the median *is equal*. The only time this rule causes some confusion (but technically, still holds) is when there are several values equal to the median, as in Examples C and D in Table 1.5.

### The Mean

By far the most common measure of central location is the *arithmetic mean*, commonly referred to as the "average." Calculating the *mean* (the word "arithmetic" is usually omitted) of a set of numbers is a relatively straightforward process, one that most of us learned long before knowing anything else about statistics. The mean of a set of numbers is nothing more than the arithmetic sum of all values being considered divided by the total number of values in the data set. As we indicated in Fig. 1.1, it is possible to determine either the mean of a sample or the mean of a population. The symbol $\bar{x}$ (read "x-bar") traditionally denotes the former, the mean of a sample, while $\mu$ (the Greek letter mu) usually represents the latter, the mean of a population.* Perhaps the mean I.Q. of all members of the sophomore class at your college is $\mu = 123.4$. A sample of 25 of the I.Q.'s of these people might yield an average value denoted by $\bar{x} = 116.8$, or $\bar{x} = 125.1$, or any of an infinite number of possible values. In most statistical problems the value of the population parameter $\mu$ is unknown and must be estimated on the basis of a sample statistic such as $\bar{x}$.

In order to develop a formula for the mean of a population, suppose we continue to denote the number of values in the population by the letter $N$, and let $x_1$ equal the first value in the population, $x_2$ equal the second value, and so forth,

---

* In statistics it is traditional to let Greek letters represent the characteristics of a population, and to let letters from the English alphabet stand for the attributes of a sample. The bar symbol over a letter typically denotes the arithmetic mean of all the data represented by the letter; thus $\bar{x}$ is the mean of all values in the data set denoted by $x$.

with $x_N$ equaling the last value. The mean of these $N$ values equals their arithmetic sum divided by $N$, as shown in Formula (1.1).

$$\text{Population mean:} \qquad \mu = \frac{x_1 + x_2 + \cdots + x_N}{N} = \frac{1}{N} \sum_{i=1}^{N} x_i. \qquad (1.1)$$

To illustrate the use of Formula (1.1), suppose we find the mean of the values in Table 1.1. In this case $N = 40$, $x_1 = 63$, $x_2 = 66$, and $x_N = x_{40} = 94$:

$$\mu = \frac{(63 + 66 + 67 + \cdots + 94)}{40} = \frac{3128}{40} = 78.20.$$

Thus, the average number of pitchers of beer sold during happy hours is 78.20.

As additional examples, the mean has been calculated for each data set shown in Table 1.5, to permit a comparison between the three measures of central tendency for these four data sets. Note from Table 1.5 that the mean may be less than, greater than, or equal in value to the median. For the data from Table 1.1, the mean (78.20) is larger than the median (77.5), but smaller than the mode (85). The mean of a data set may be thought of as the "point of balance" of the data, analagous to the *center of gravity* for a distribution of mass in physics.

### Comparison of the Mode, Median, and Mean

The arithmetic mean is the most widely used measure of central location. It has the disadvantage for descriptive purposes of being more affected by extreme values than the median or the mode, because it takes into account the difference among all values, not merely the rank order (as does the median) or their frequency (as does the mode). In comparing data sets D and C in Table 1.5, we see that substituting the value 19 for the value 7 does not change the median and the mode, but the mean is raised so that it exceeds all but one of the numbers in data set D. A recent cartoon illustrated this problem quite well by depicting a small town worker commenting to a reporter that "the average yearly income in this town is $100,000—there's one guy making a million, and ten of us workers making $10,000."

The mean requires and uses more information about the values in the data set than the mode or the median. The mode requires only a frequency count. For example, suppose a study is being made of the frequency of various makes of automobiles (such as Fords, VW's, etc.) passing a certain point. The *mode* would be a good descriptive measure in denoting which automobile make was observed most often. The mean or median value of automobile makes has no meaning here, since there is no way of assigning numbers to the makes in such a way that the ranks or differences among these numbers have any meaning.

The median requires knowledge of not only the *frequency* of the values in a data set, but also their *ranking*, so that these values can be ordered and the middle value obtained. Suppose each of 50 pro-football rookies can be rated I, II, III, IV, or V according to his performance in a tryout camp, as well as on past records. A ranking of I indicates the player is sure all-pro, while a ranking of V indicates the player should try some other career opportunity quickly. The frequencies of each rank can be tabulated and both the mean and median can be calculated. Consider, for example, the data in Table 1.6.

**Table 1.6** Ranking of pro-football rookies

| Rank | I | II | III | IV | V |
|------|---|----|-----|----|----|
| Frequency | 3 | 17 | 14 | 6 | 10 |

The *mode* for this data is rank II, while the median occurs at rank III (since the 25th and 26th values both lie in this rank). Again, it would be inappropriate to calculate a mean, since the difference between ranks is not precisely known, nor can these differences be assumed to be equal.

In contrast to the examples above, economic and business problems generally involve data in which the differences among values are known, such as income measures, output quantities, profits retained, prices, and interest rates. The disadvantage of the mean for frequency and ranked data becomes its special advantages in these cases—that is, it is a more reliable measure of central location because it *uses more of the information* within the data set. The mean uses frequencies, the rank order of the data, and the differences among the values.

In concluding this section, we must point out that the mean, median, and mode are not the only measures of central location. For example, two other types of means, the *geometric mean* and the *harmonic mean* are especially useful in certain types of problems in business and economics. We will not present these measures here, but rather refer the reader to Neiswanger's *Elementary Statistical Methods*, 3rd ed. (New York: The Macmillan Company, 1960).

## 1.4 MEASURES OF DISPERSION

Measures of central location usually do not give one enough information to provide an adequate description of the data, because variability or spread is ignored. The person who knows *just* the mean may be compared to the fellow with his head in a refrigerator and his feet in an oven who declares "on the average, I feel fine." Another measure is needed which indicates how spread out or *dispersed* the data are.

Since there are a number of ways to measure spread, let's consider some logical properties for a *good* measure to have. One desirable property for such a measure is that any index of spread should be *independent of the central location* of the observations—that is, independent of the mean of the data. This property implies, in effect, that if a constant is added to (or subtracted from) each value in a set of observations, this transformation will not influence the measure of spread. Another logical property is that *all* observations should be used to measure spread, rather than just a few selected values, such as the highest and lowest. Finally, a good measure should reflect the typical spread of the data, and should be one which is *convenient to manipulate mathematically.*

**The Range**

One simple example of a measure of spread is the *range*, which is defined to be the *absolute difference* between the highest and lowest values. The range for the data in Table 1.1 is the absolute difference between the highest value, 94, and the lowest value, 63, or

$$|94 - 63| = 31.$$

Similarly, the ranges of the four examples in Table 1.5 are

$$|13 - 2| = 9, \quad |12 - (-5)| = 17, \quad |7 - 2| = 5, \quad \text{and} \quad |19 - 2| = 17,$$

respectively. The range has the advantages of being independent of the measure of central location and being easy to calculate. It has the *disadvantages* of ignoring all but two values of the data set and not necessarily giving a *typical* measure of the dispersion, since a single extreme value changes the range radically.

**Midranges**

To counter this last complaint, it is possible to calculate a measure of variability called a *midrange*. A midrange is defined the same way as the range, except that it includes only a certain proportion of the values in the *middle* of the data set. Since this measure excludes values at both the upper and lower ends of the data set when the numbers are ordered numerically, a few extreme values do not affect the measure of variability. Both an 80-percent midrange and a 50-percent midrange are commonly used.

The 80-percent midrange excludes the upper 10 percent of the values and the lower 10 percent of the values, and finds the range of the remaining middle 80 percent of the original values. Using the 40 values in Table 1.1, this measure is calculated by excluding the *four largest* (10% of 40) and the *four smallest* values. This leaves values spread from 68 to 86, so the 80 percent midrange is 18.

The 50-percent midrange has the same advantages as the 80-percent midrange, and also provides a measure more typical of the spread of individual elements in the data set from the central location measure. To calculate a 50-percent midrange, one excludes the upper 25 percent and lower 25 percent of the values, and then finds the range of the remaining 50 percent. Again using the values from Table 1.1, this measure requires the exclusion of the ten highest (25% of 40) and ten lowest values. The remaining spread is from 73 to 84, representing a 50-percent midrange of 11.

Midranges still have a disadvantage, in that they do not use all the information about variability in the data set. They do not use the individual values or the differences between them, but they are independent of the measure of central location.

### Deviation

When a measure of variability is said to be independent of the central location, this usually implies that it is independent of the arithmetic mean, since the mean is the most commonly used measure of central location. In general, it is very easy to transform one set of numbers into a new data set so that the new set is independent of the mean. This is accomplished by transforming *each* piece of data in the set into a new number, where the *mean of these new numbers equals zero*. The transformation necessary to do this is quite simple: merely subtract from every number a value equal to the mean of the data set. Each resulting number, called a *deviation*, indicates how far and in which direction the original number is from the mean. The sum of these deviations must be zero. Consider, for example, a data set consisting of the five values of $x$ shown in Table 1.7. If we subtract the mean of these five numbers (which is $\mu = 10$) from each value of $x$, the mean of the new set of values, labeled $x - \mu$ must equal zero. Any set of numbers can be transformed in this fashion into a set of deviations with a mean of zero.

To formalize the above process, consider a population with $N$ values $x_i$, $i = 1, 2, \ldots, N$, and having a mean $\mu$. When these $N$ values are transformed into

**Table 1.7**

|        | $x$ | Deviations $x - \mu$ |
|--------|-----|----------------------|
|        | 4   | $-6$                 |
|        | 8   | $-2$                 |
|        | 10  | 0                    |
|        | 13  | $+3$                 |
|        | 15  | $+5$                 |
| Sum    | 50  | 0                    |
| Mean   | 10  | 0                    |

deviations from the mean, the new values are $x_i - \mu$, for $i = 1, 2, \ldots, N$, and the sum of these deviations must be zero. That is,

$$\sum_{i=1}^{N} (x_i - \mu) = 0. \qquad (1.2)$$

Since the transformation $(x_i - \mu)$ gives all sets of data a common central location (i.e., they all have means of zero), measures of dispersion defined in terms of deviations about the mean have the good property that they are independent of central location. Of course, the average deviation of a set of values,

$$\frac{1}{N} \sum_{i=1}^{N} (x_i - \mu),$$

cannot itself be used as a measure of variability because this number always equals zero. Some deviations are positive, some are negative, and by definition they balance each other about the mean.

### Standard Deviation and Variance

Now instead of average deviations about the mean, we consider *average squared deviations* about the mean. This squaring avoids the problem inherent in ordinary deviations about the mean, namely, that their sum always equals zero, since *all squares are positive* in sign. Indeed, this index meets all the properties of a good measure of spread, and is the traditional method for measuring the variability of a data set. Since it uses the deviations about the mean, it is independent of central location. It uses every value in the data set and is reasonably easy to compute mathematically. It is very sensitive to any change in the values—even a single change of one value in a set of 100 would result in a different measure of variability.

This "average squared deviation" measure is called the *variance* and is denoted for a population by the symbol $\sigma^2$, which is the square of the lower case Greek letter *sigma*. It is defined as follows:

$$\text{Population variance:} \quad \sigma^2 = \frac{1}{N} \sum_{i=1}^{N} (x_i - \mu)^2. \qquad (1.3)$$

Very often the square root of the above measure, denoted by $\sigma$ and called the *population standard deviation*, is used in place of (or in conjunction with) the population variance to describe variability. This is because the standard deviation is usually more convenient than the variance for interpreting the variability of a data set, since $\sigma^2$ is in squared units while $\sigma$ is in the same units as the data. The population standard deviation is defined as follows:

$$\text{Population standard deviation:} \quad \sigma = \sqrt{\frac{1}{N} \sum_{i=1}^{N} (x_i - \mu)^2}. \qquad (1.4)$$

To illustrate the calculation of a population variance and standard deviation, assume that the values of $x$ in Table 1.8 represent the number of tape recorders assembled by ten different workers on an assembly line over the past month. That is, worker 1 assembled 115 recorders, worker 2 assembled 122 recorders, etc. The mean number of recorders assembled is seen in the first column to be

$$\mu = \tfrac{1200}{10} = 120.$$

The deviations from the mean are shown in the second column. Note that the sum of these deviations equals zero, as it must. The third column of values gives the *squared* deviations about the mean, of which the sum is 436. Hence, the *average squared deviation* of this population (the variance) is:

$$\sigma^2 = \frac{1}{N} \sum_{i=1}^{N} (x_i - \mu)^2 = \frac{436}{10} = 43.6.$$

While the variance meets the criteria desired for a good measure of dispersion, it does *not* represent the typical size of a deviation from the mean because all

**Table 1.8**

|  | $x$ | $x - \mu$ | $(x - \mu)^2$ |
|---|---|---|---|
|  | 115 | $-5$ | 25 |
|  | 122 | $+2$ | 4 |
|  | 129 | $+9$ | 81 |
|  | 113 | $-7$ | 49 |
|  | 119 | $-1$ | 1 |
|  | 124 | $+4$ | 16 |
|  | 132 | $+12$ | 144 |
|  | 120 | 0 | 0 |
|  | 110 | $-10$ | 100 |
|  | 116 | $-4$ | 16 |
| Sum | 1200 | 0 | 436 |
| Mean | 120 | 0 | 43.6 |

deviations are squared. On the other hand, since the standard deviation is the square root of the variance, this measure is in the same units as the original data, and hence gives a good measure of the spread in the number of recorders produced:

$$\sigma = \sqrt{\text{Variance}} = \sqrt{43.6} = 6.60.$$

In general, a precise interpretation of values of $\sigma$ and $\sigma^2$ is difficult because variability depends so highly on the unit of measurement. For instance, variability of income in the U.S. is certainly larger when measured in dollars than when measured in thousands of dollars. In any case, as the spread of a population increases, the values of $\sigma^2$ (and $\sigma$) will also increase. On the other hand, if there is no variability in the data, then all values must be at a single point ($x_i = \mu$ for all $i$) and $\sigma^2 = \sigma = 0$.

One rule of thumb that often provides a good *approximation* to the spread of a set of observations states that *about 68 percent* of all values will fall within *one* standard deviation to either side of the mean, and *about 95 percent* of all values will fall within *two* standard deviations to either side of the mean.* In other words, the interval from ($\mu - \sigma$) to ($\mu + \sigma$), which we will write as ($\mu \pm \sigma$), will often contain about 68 percent of all the population values. Similarly, the interval ($\mu - 2\sigma$) to ($\mu + 2\sigma$), that is, ($\mu \pm 2\sigma$), will often contain about 95 percent of all the population values. We show below that this rule of thumb provides a fairly good approximation to the spread of the data on tape recorders (Table 1.8). Six of the ten values (60%) in the population are contained in the interval

$$\mu \pm \sigma = 113.40 \quad \text{to} \quad 126.60,$$

and all ten values (100%) are contained in the interval

$$\mu \pm 2\sigma = 106.80 \quad \text{to} \quad 133.20.$$

| Interval | Values within interval | Percent of population |
|---|---|---|
| $\mu \pm \sigma = \quad 120 \pm 6.60$ <br> $= \begin{cases} 113.4 \text{ and} \\ 126.6 \end{cases}$ | 115, 116, 119, 120, 122, 124 | 60% |
| $\mu \pm 2\sigma = \quad 120 \pm 2(6.60)$ <br> $= \begin{cases} 106.80 \text{ and} \\ 133.20 \end{cases}$ | 110, 113, 115, 116, 119, <br> 120, 122, 124, 129, 132 | 100% |

* We will show, in Chapter 5, that this rule of thumb is based on the assumption that the population is symmetrical, with a bell-like shape, and is called the *normal distribution*.

As a final example of the process of calculating and interpreting the variance and standard deviation of a population, consider once again the values of beer sales in Table 1.1. We previously calculated the mean of this population to be $\mu = 78.20$. If we now calculate the forty deviations from this mean, we obtain,

$$
\begin{aligned}
63 - 78.20 &= -15.20 \\
66 - 78.20 &= -12.20 \\
67 - 78.20 &= -11.20 \\
68 - 78.20 &= -10.20 \\
\vdots \qquad &\qquad \vdots \\
92 - 78.20 &= \phantom{-}13.80 \\
94 - 78.20 &= \phantom{-}15.80
\end{aligned}
$$

The mean of these deviations must be zero. The average *squared* deviation (or variance) is

$$
\begin{aligned}
\sigma^2 &= \frac{1}{N} \sum_{i=1}^{N} (x_i - \mu)^2 \\
&= \frac{1}{40} \left[ (-15.20)^2 + (-12.20)^2 + (-11.20)^2 + \cdots + (13.80)^2 + (15.80)^2 \right] \\
&= 53.36;
\end{aligned}
$$

and the standard deviation is

$$
\sigma = \sqrt{53.36} = 7.44.
$$

If the rule of thumb described earlier holds, then $\mu \pm \sigma$ should contain about 68 percent of all beer sales, and $\mu \pm 2\sigma$ should contain about 95 percent of all these values. Checking these intervals against the values in Table 1.1 gives the following results:

| Interval | | | Percent of values |
|---|---|---|---|
| $\mu \pm 1\sigma = 78.20 \pm 7.44$ | $= 70.76$ to $85.64$ | | 70% |
| $\mu \pm 2\sigma = 78.20 \pm 2(7.44)$ | $= 63.32$ to $93.08$ | | 95% |

Hence, had we known only that $\sigma = 7.44$ and $\mu = 78.20$, we could have given, for this particular set of observations, a fairly good description of the variability of the data, as well as its central location.

Considering all the measures of central location and dispersion that we have presented, the two measures most often useful in statistical inference and decision-making are the *mean* and the *standard deviation*. These are common household terms to any statistician, used every day in helping to make decisions based on statistical analysis of data sets. The mean is precisely the balance point of all the values. The standard deviation is the typical (or standard) size of the difference (deviation) between a typical selected value and the mean of all the values. As such, it provides a good insight into the extent of variability in the data set, especially when the above rule of thumb applies. The reader should keep in mind that the variance and the standard deviation do not represent two different ways of measuring the variability of a population. Since $\sigma$ is merely the square root of $\sigma^2$, these two measures reflect the *same information* about variability, but are expressed in different units. The standard deviation is easier to interpret, but it is more difficult to manipulate mathematically than the variance because of the square-root sign.

### The Mean and Variance of Data in a Frequency Distribution

The formulas for the mean and variance presented thus far (Formulas (1.1) and (1.3)), have assumed that each value of the data set is given separately. Often, however, it is much easier to manipulate large amounts of data if this data is first grouped into a *frequency distribution*. An example of such a distribution is shown in columns 1 and 2 of Table 1.9 for the starting monthly salaries (reported to the nearest 100 dollars) by recent college graduates.

**Table 1.9** Illustration of calculating the mean for a frequency distribution of monthly salaries of recent college graduates

| (1) Salary $x$ | (2) Frequency $f$ | (3) $fx$ | (4) Relative frequency $f/N$ | (5) $(f/N)x$ |
|---|---|---|---|---|
| 400 | 8 | 3,200 | 0.032 | 12.8 |
| 500 | 23 | 11,500 | 0.092 | 46.0 |
| 600 | 75 | 45,000 | 0.300 | 180.0 |
| 700 | 90 | 63,000 | 0.360 | 252.0 |
| 800 | 43 | 34,400 | 0.172 | 137.6 |
| 900 | 11 | 9,900 | 0.044 | 39.6 |
| Sum | 250 | 167,000 | 1.000 | 668.0 |

One way to find the mean of this population would be to sum all 250 values separately (8 values of \$400 + 23 values of \$500 + $\cdots$ + 11 values of \$900), and then divide by 250. (This is the procedure presented earlier in Formula (1.1).) But most of us learned long ago that multiplication is easier than repeated addition; hence, we should take advantage of the fact that there are only six different values in Table 1.9, not 250. In other words, instead of adding $400 eight times, we can use the product, 8($400). We can use similar products for every value of $x$, as shown in column (3) of Table 1.9. Summing these products for all six values and then dividing by 250 yields $\mu$:

$$\mu = \$167{,}000/250 = \$668.$$

To be formal about the above process, let $x_i$ represent the $i$th value of $x$, and let $f_i$ equal the frequency of that value. If there are $c$ different values, then $\mu$ is the average product of $x_i$ times $f_i$.

---

*Population mean for frequency distribution:*

$$\mu = \frac{1}{N} \sum_{i=1}^{c} x_i f_i.$$

(1.5)

---

Note that we can rewrite Formula (1.5) in a slightly different (but equivalent) form by placing $N$ inside the sum sign.

---

*Population mean for frequency distribution:*

$$\mu = \sum_{i=1}^{c} x_i \left( \frac{f_i}{N} \right).$$

(1.6)

---

Since $f_i/N$ represents the relative frequency of the $i$th value, we can define the mean to be the sum of each value of $x_i$ multiplied by its relative frequency. For now, the reader might wish to practice using Formula (1.6), by following all the steps required to fill in columns 4 and 5 in Table 1.9, to verify that $\mu = 668$.

A variance for a data set given in a frequency-distribution format is calculated in much the same manner as described above for the mean. In this case the mean of the population ($\mu$) is subtracted from each value of $x_i$ (for $i = 1, 2, \ldots, c$); these deviations are then squared, multiplied by their frequency, and summed. The result is,

---

*Population variance for a frequency distribution:*

$$\sigma^2 = \frac{1}{N} \sum_{i=1}^{c} (x_i - \mu)^2 f_i = \sum_{i=1}^{c} (x_i - \mu)^2 \frac{f_i}{N}.$$

(1.7)

---

The data on monthly salaries is repeated in Table 1.10 with all the computations shown to calculate the variance by Formula (1.7). Using the first form of the formula, and column (4) in Table 1.10, we obtain,

$$\sigma^2 = \tfrac{1}{250}(3{,}004{,}000) = 12{,}016.$$

Using the second form and columns 5 and 6 of Table 1.10, the same value is obtained. The standard deviation is the square root of this value, or $\sigma = 109.6$.

**Table 1.10** Illustrating calculation of the variance for a frequency distribution of monthly salaries of recent college graduates

| (1) Salary $x$ | (2) Frequency $f$ | (3) $x - \mu$ | (4) $f(x - \mu)^2$ | (5) Relative frequency $f/N$ | (6) $(f/N)(x - \mu)^2$ |
|---|---|---|---|---|---|
| 400 | 8 | $-268$ | 574,592 | 0.032 | 2,298.368 |
| 500 | 23 | $-168$ | 649,152 | 0.092 | 2,596.608 |
| 600 | 75 | $-68$ | 346,800 | 0.300 | 1,387.200 |
| 700 | 90 | $+32$ | 92,160 | 0.360 | 368.640 |
| 800 | 43 | $+132$ | 749,232 | 0.172 | 2,996.928 |
| 900 | 11 | $+232$ | 592,064 | 0.044 | 2,368.256 |
| Sum   250 | | 0 | 3,004,000 | 1.000 | 12,016.000 |

## 1.5 OTHER MEASURES

While the mean and the standard deviation are the most common descriptive measures, there are a number of other measures which give additional information about the characteristics of a data set. This section is devoted to describing, rather briefly, a few of these measures.

### Percentiles, Deciles, and Quartiles

At times it may be useful to divide a set of numbers into a specified number of groups, each containing the same number of values. *Percentiles* divide the data into 100 equal parts, each of these 100 parts having one percent of all values. The number corresponding to the 90th percentile, for example, is that value which has 90 percent of all values below it and 10 percent above it. Thus, a student scoring higher than 95 percent of all other students on his college board exams and lower than five percent of the other students is said to have scored in the 95th percentile. Percentiles can be *determined exactly* from a table of cumulative relative frequencies on ungrouped data, and *approximated* from a table on grouped data. From Table 1.3, for example, we can only estimate that the value of the 50th percentile, which is the median, is somewhere between the score of 74.5, representing a cumulative relative frequency of 0.325, and the score of 79.5, representing a cumulative relative frequency of 0.600. Determination of the exact value of the median, shown earlier to be 77.5, requires *ungrouped* cumulative relative frequencies.

Quartiles and deciles are defined in much the same fashion as percentiles: *quartiles* divide the data into four equal groups, while *deciles* divide the data into ten equal parts. The *first* quartile value is that point which exceeds one-fourth of the observations and is exceeded by three-fourths of the data. Only three quartile values are necessary to divide the data into four parts. Likewise, nine decile values divide a set of observations into ten equal parts. The *fifth decile* and the *second quartile* values are equivalent to the median.

**Calculation of Quartiles**

In many cases there may be *no one value* which divides the data into the specified groups. When this happens, there are a number of rather arbitrary methods for specifying a single point. We shall describe one such method, as applied to finding quartiles for ungrouped data. Assume there are $N$ numbers ranked in an array from lowest to highest. Then the *ranks* corresponding to the first, second, and third quartile values are as follows:

$$\text{Rank of the first quartile value} = (1/4)N + \tfrac{1}{4},$$
$$\text{Rank of the second quartile value} = (1/2)N + \tfrac{1}{2} = \text{Median},$$
$$\text{Rank of the third quartile value} = (3/4)N + \tfrac{1}{4}.$$

For example, in the data of Table 1.1 there are 40 observations. The rank of the first quartile value is

$$(\tfrac{1}{4})40 + \tfrac{1}{4} = 10.25;$$

this means that the first quartile value is located at a point one-fourth (0.25) of the distance past the 10th ranked value, i.e., one-quarter of the way between the 10th and 11th values. Since ranks 10 and 11 are both 73, the first quartile value is $q_1 = 73$. The second quartile value corresponds to the $(\tfrac{1}{2})N + \tfrac{1}{2} = 20.50$ rank, or $q_2 = 77.5$, which is the median. Finally, since $(\tfrac{3}{4})N + \tfrac{1}{4} = 30.25$, and the 30th rank is 84, while the 31st rank is 85, the third quartile value is $q_3 = 84.25$.

Once the first and third quartile values have been found, it is easy to calculate the 50-percent midrange, for this value is merely the difference $q_3 - q_1$. In fact, $q_1$ and $q_3$ were defined so that their difference would contain exactly 50 percent of the population:

$$q_3 - q_1 = (\tfrac{3}{4}N + \tfrac{1}{4}) - (\tfrac{1}{4}N + \tfrac{1}{4}) = \tfrac{1}{2}N.$$

The 50-percent midrange is often called the *interquartile* range, since it gives the range between the first and third quartiles. For the beer sales example,

$$q_3 = 84.25 \qquad \text{and} \qquad q_1 = 73.00;$$

hence, the interquartile range is

$$q_3 - q_1 = 11.25.$$

The interquartile range is often a convenient measure to use because it is relatively easy to interpret and not too difficult to calculate. In many circumstances it can also be used to give a rough approximation to the standard deviation of a set of values, since the interquartile range includes 50 percent of all values, whereas plus or minus one standard deviation includes approximately 68 percent of all values. That is, if we multiply the interquartile range by the ratio 68/50, then the result should be a good approximation to the interval between $\pm 1\sigma$. In the beer example, for instance, multiplying $q_1 - q_3 = 11.25$ by 68/50 yields 15.3; this is not a bad estimate of the true interval between $\pm 1\sigma$, which is 2(7.44) = 14.88.

**Shapes of Distributions**

In addition to being able to describe the central location or spread of a set of values, it is often helpful to have a method for describing the *shape* of a frequency distribution. Most of the distributions representing real-world problems are called *unimodal* distributions, implying that they have only one peak, or *mode*. A distribution with two peaks is called a *bimodal* distribution. Often distributions with more than one mode actually reflect the combination of two or more *separate* kinds of data into a single set of values.

Consider, for example, the frequency distribution shown in Fig. 1.7, representing the frequency of sales of television sets for a large department store, in intervals of $100. What Fig. 1.7 actually represents is two unimodal distributions, one reflecting the sales of black-and-white television sets, and the other representing the sales of color television sets. If we make this distinction and plot the resulting frequency distributions, the two distributions in Fig. 1.8 are obtained. Note that the distribution in Fig. 1.8(a) has a fairly long "tail" to the right, a characteristic

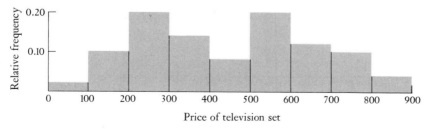

**Fig. 1.7.** Television set sales.

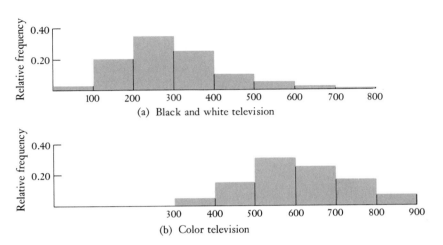

**Fig. 1.8.** Television set sales.

common to many distributions representing data in the behavioral and social sciences, especially income distributions. Figure 1.8(b) represents another commonly occurring distribution, one with a relatively *symmetrical* shape.

Because two distributions may differ in the *shape* of their tails, in the *height* of their peak(s), and in their degree of *symmetry*, even if they are both unimodal distributions with identical means and variances, statistical measures of both the "peakedness" and the symmetry of a distribution have been developed.

### Symmetry and Skewness

A distribution is *symmetric* if it has the same shape on both sides of its median. Imagine folding the picture of a distribution in half at its median. To be symmetric, the two halves must match perfectly—i.e., they must be the "mirror images" of one another. For all symmetric distributions the median equals the mean. The mode will also equal the median if the distribution is unimodal. Figure 1.9 shows three symmetric distributions.

A distribution that is not symmetric, but rather has most of its values either to the right or to the left of the mode, is said to be *skewed*. If most of the values of a distribution fall to the right of the mode (as in Fig. 1.8(a)), this distribution is said to be *skewed to the right*, or *skewed positively*. A distribution with the opposite shape, with most values to the left of the mode, is said to be *skewed to the left*, or *skewed negatively*. Note how, in Fig. 1.10, the lack of symmetry in a distribution affects the relationship between the mean, the median, and the mode. For a completely symmetrical unimodal distribution, such as Fig. 1.9(a), these three values must all be equal. As the distribution becomes *skewed positively*, the mode

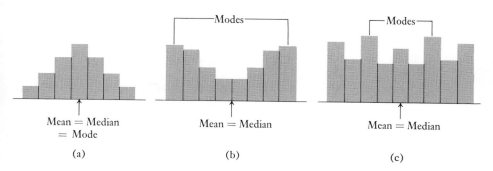

**Fig. 1.9.** Symmetric distributions.

remains at the value representing the highest frequency, but the median and the mean move to the right.

It is easy to remember the direction of skewness and the effect of skewness on the measures of central location in a unimodal distribution by various memory aids. Two useful ones are:

1. The order of magnitude of the central location measures is *alphabetical* in a *negatively* skewed distribution (Mean < Median < Mode) and *reversed* in a distribution with *positive* skewness (such as Fig. 1.10).

2. If one pictures a symmetric distribution with a skewer stuck into one of its tails or sides so that it is stretched sideways, the direction of stretch is the direction of skewness. A distribution with its right side stretched out in the positive (increasing numerical value) direction has positive skewness. A distribution with its left tail stabbed and stretched out in the negative (decreasing numerical value) direction is negatively skewed.

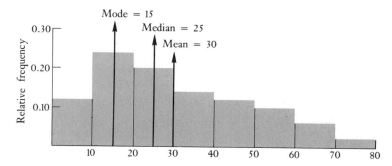

**Fig. 1.10.** The relationship between the mean, the median, and the mode for a distribution with positive skewness.

**Fig. 1.11.** Platykurtic curves have short tails like a platypus, while leptokurtic curves have long tails like kangaroos, noted for "lepping" (after W. S. Gosset).

### Kurtosis

Another descriptive measure of the shape of a distribution relates to its flatness or peakedness. The term applied to this characteristic of shape is *kurtosis*. Prefixes are used to designate whether the distribution is generally a flat variety or a very peaked variety or something in between. As illustrated in Fig. 1.11, a flat distribution with short broad tails is called *platykurtic*. A very peaked distribution with long thin tails is called *leptokurtic*. Although summary measures for kurtosis exist, we will not present any of them in this text.

### 1.6 SUMMARY

In this chapter, the value and use of statistics in making decisions or reaching conclusions was noted. Before we can progress further in discussing these concepts, it is necessary to present some of the fundamental measures necessary for statistical analysis, particularly those relating to *probability*. This chapter has presented many new terms and described the commonly used summary measures of statistics. Of these, the *mean* and *standard deviation* will receive the most attention throughout the book. The next step is to turn to the basics of probability. Then, the summary measures and the probability concepts will be combined, in developing an under-standing of the important methods and interpretation of statistical analysis.

## REVIEW PROBLEMS

1. Define or briefly describe each of the following: decision-making under uncertainty, population, parameter, descriptive statistics, sample statistic.

2. Explain what you understand to be the primary purpose of descriptive statistics. Give an example from a recent newspaper or magazine of some use of descriptive statistics.

3. Suppose a recruiter for a certain company claims that advancement opportunities are great in his company because salaries of five-year employees range from 20 to 200 percent more than their corresponding beginning salaries. Explain why this statistical information might be inadequate for a prospective employee who is trying to decide whether or not to take a job with this company.

4. A $25 sweater is on sale for $15 and a $75 coat is on sale for $60. Find the average per-centage decrease in price for these items.

5. The distribution of the number of long-distance phone calls made per month by 100 students in a certain dormitory is indicated by the following frequencies:

| Number of calls | 0 | 1 | 2 | 3 | 4 |
|---|---|---|---|---|---|
| Frequency $f$ | 47 | 33 | 14 | 5 | 1 |

a) Find the mean number of calls per student.
b) Sketch the relative frequency distribution and the cumulative frequency distribution for the number of calls.

6. Suppose a certain gasoline producer sponsors a mileage economy test involving 30 cars. The miles per gallon, $x$, recorded to the nearest gallon, are given below.

| $x$ | $f$ |
|---|---|
| 15 | 8 |
| 16 | 9 |
| 18 | 7 |
| 20 | 6 |

Find the average miles per gallon and the standard deviation for this distribution.

7. Five instructors report the number of quizzes they will conduct during the semester, namely 2, 8, 1, 4, and 5. Find, for these data,

a) the mean
b) standard deviation,
c) median
d) range
e) Give one reason why the value of part (b) is a better measure of dispersion than the value of part (d).

8. Suppose $x$ takes on the following five values, all with the same frequency:

| $x$ | 2 | 3 | 10 | 5 | 2 |
|-----|---|---|----|---|---|

Find:
a) the mean, median, and mode of $x$;
b) the range and average deviation of $x$;
c) the variance and standard deviation of $x$.

9. Twenty communities provide information on the vacancy rate in local apartments. Find the mean and standard deviation of the following data on vacancy rates, $x$.

| Frequency | Vacancy rate |
|-----------|--------------|
| 10 | 3–7% |
| 6 | 8–12% |
| 4 | 13–17% |

10. Ten student nurses revealed, in a confidential interview, the number of dates they had during their first term in school. The square of the sum of these ten values is 900, and the sum of their squares is 115. Find the mean of this population.
a) Find the variance of this population using the computational formula $\sigma^2 = \Sigma x_i^2 f_i - \mu^2$.
b) A group of college freshman women averaged 10 dates over the same period, with a standard deviation of 4. Compare the two distributions.

11. In the following population frequency distribution, $x$ is the number of students participating in 30 special workshops.

| $x$ | $f$ |
|-----|-----|
| 4 | 5 |
| 5 | 4 |
| 6 | 6 |
| 7 | 9 |
| 8 | 1 |
| 9 | 3 |
| 10 | 1 |
| 11 | 1 |

a) Find the mean and variance of $x$.
b) What percent of the distribution lies within $\mu \pm 1\sigma$? What percent lies within $\mu \pm 2\sigma$?

12. The distribution of coins in a cash drawer is 60 pennies, 20 nickels, 9 dimes, 8 quarters, and 3 half-dollars. Let $x$ be the value in cents of a coin.

a) Sketch the frequency distribution for $x$;
b) Find the mean and variance for $x$;
c) Sketch the cumulative relative frequency distribution for $x$.

13. Given the following frequency distribution:

| Class | Frequency | Class mark | $fx$ | $fx^2$ |
|-------|-----------|------------|------|--------|
| 1–5   | 4         | 3          | 12   | 36     |
| 6–10  | 8         | 8          | —    | —      |
| 11–15 | 3         | —          | —    | —      |
| 16–20 | 5         | —          | 90   | —      |

a) Find the mean of this population by completing the table above.
b) Find the variance using the formula $\sigma^2 = (1/N)\Sigma x^2 f_i - \mu^2$.

14. Given the following set of numbers: 8, 2, 4, 6, 5, 4, 3, 9, 4, 5,

a) Find the mean, the median, and the mode of these observations.
b) Find the variance of these numbers by using Formula (1.7). Check your answer by using the "mean square – square mean" formula shown in Problems 10 and 13.
c) Find the standard deviation of the ten observations. What percent of the observations falls within one standard deviation of the mean? What percent falls within two standard deviations of the mean?
d) Find the first, second, and third quartile values. What is the value of the interquartile range?

15. The following frequency distribution shows the weekly sales of the Snoopy Hat Corporation for last year.

| Sales | Number of weeks |
|-------|-----------------|
| 0–$5,999 | 4 |
| $6,000–11,999 | 6 |
| 12,000–17,999 | 10 |
| 18,000–23,999 | 16 |
| 24,000–29,999 | 12 |
| 30,000–35,999 | 4 |
| | 52 |

a) Draw the histogram for these data.
b) Compute the mean and the mode.
c) Compute the variance.
d) Sketch the cumulative relative-frequency distribution.

16. Five students report the number of miles they live from the center of campus as 2, 8, 4, 11, and 5 miles. Find the (a) mean, (b) standard deviation, (c) median, (d) range, of these values.

17. The number of kitchen employees in local restaurants are given as 3, 7, 10, 4, 1, and 5.

a) Find the mean of the distribution.

b) Find the variance of the distribution.

c) Using some statistical measure, comment on the symmetry or skewness of the distribution.

18. Given the following data of five population values: 3, 2, 5, 8, and 2.

a) Find and compare the mean, mode, and median.

b) Comment on the symmetry or skewness of this distribution.

19. Given the following ten values, find the mean, mode, median, range, and quartile values: 35, 14, 6, 18, 14, 27, 19, 7, 13, and 14.

## EXERCISES

20. What is statistical inference? Why is statistical inference important in the social and behavioral sciences? Give several examples of the use of statistical inference in your major field of study in college.

21. Find an example (from a newspaper, magazine, etc.) of using only *means* to make some conclusion or argumentative point. Examine the argument closely and explain the inadequacy of it, or explain how knowledge of a variability measure could strengthen or change the argument.

22. Acquire 50 observations on a variable of interest to you and your fellow students (e.g., wage rates, apartment rentals, football statistics, anatomical measurements). Construct a frequency distribution for your data and a cumulative frequency distribution. Sketch both distributions.

23. Make up two sets of seven values each with the following characteristics:

a) Same average but different variability;

b) Same range, but different averages;

c) Same average and same range but different 50% midrange.

24. Given the following grade distribution on a statistics quiz, find the mean and variance of $x$. Use the midvalue of each class (the class mark) as the value of $x$.

| Grade | Frequency | Class mark |
|-------|-----------|------------|
| 96–100 | 4 | 98 |
| 91–95 | 5 | 93 |
| 86–90 | 6 | · |
| 81–85 | 5 | · |
| 76–80 | 1 | · |
| 71–75 | 4 | |
| 66–70 | 2 | |
| 61–65 | 5 | |

25. a) Suppose five values for a population are 3, 5, 1, 8, and 3. Show that the mean and standard deviation are 4.0 and 2.37, respectively.

b) Suppose another population has the same values as in part (a) plus the two additional values, 2 and 6. Find the mean, and show that this population has a smaller standard deviation.

c) Suppose another population again has the same values as in part (a) plus the two additional values, −1 and 9. Find the mean, and explain why the standard deviation is larger than in part (a).

26. The following annual starting salaries were offered to 16 students about to receive their master's degrees:

| | | | |
|---|---|---|---|
| $12,500 | $ 9,900 | $11,200 | $11,500 |
| 10,400 | 11,400 | 10,800 | 10,500 |
| 9,600 | 12,600 | 11,800 | 11,000 |
| 10,600 | 11,400 | 9,400 | 9,100 |

a) Find the mean and the median for these 16 observations.

b) Use a class interval of size $500 to form a frequency distribution for the data. Construct a histogram for this distribution, and draw the frequency polygon connecting the class marks (start with the class $9,001–9,500).

c) Form a cumulative relative frequency distribution, and plot this distribution. What percent of the graduates will earn more than $10,500? What percent will earn less than $11,000?

d) Use the frequency distribution you calculated for part (b) to find the mean of the grouped data.

27. For the data given in Exercise 26, compute:

a) The range, and the interquartile range;

b) The sum of the absolute deviations about the mean;

c) The standard deviation;

d) The percent of observations falling within one standard deviation of the mean, and the percent falling within two standard deviations of the mean;

28. Ten residents of Metropolis report the following incomes:

| | | | | |
|---|---|---|---|---|
| $10,500 | $ 4,150 | $ 2,505 | $6,245 | $ 5,570 |
| $ 6,600 | $14,800 | $11,325 | $9,170 | $19,000 |

a) Find the mean, standard deviation, and median of these data.

b) Compare the variability and skewness of this distribution with a regional income distribution that reports a mean of $10,000, standard deviation of $3,000, and median of $8,000.

29. How are the mean, the median, and the mode related in a completely symmetrical frequency distribution? How are they related in a positively skewed distribution and in a negatively skewed distribution? Sketch several distributions to illustrate your answer.

30. A movie producer holds a preview of a new movie and asks viewers for their reaction.

By age groups, he obtains these results:

| | Age Group | | | |
|---|---|---|---|---|
| | Under 20 | 20–39 | 40–59 | 60 and over |
| Liked the movie | 140 | 75 | 50 | 40 |
| Disliked the movie | 60 | 50 | 50 | 20 |

Using some diagrams or summary measures, argue toward which age groups the firm should aim its advertising campaign for the movie.

31. Compute the median and the quartile values for the data in Problem 15. What is the value of the interquartile range?

32. *Shark Loan.* Throckmorton Jones, manager of Shark Loan, Inc., has kept a record of the frequency of the *time between arrivals* of customers at his loan office. These data, shown below, indicate that the time interval between (consecutive) arrivals was between zero and 20 minutes on 50 occasions, that it was between 20 and 40 minutes on 33 different occasions, etc. Since there were 150 customers during this period, there are 150 inter-arrival times (the time to the first customer is counted as one interarrival time).

| Minutes between customers ($t$) | Frequency ($f$) |
|---|---|
| $0 \leqslant t < 20$ | 50 |
| $20 \leqslant t < 40$ | 33 |
| $40 \leqslant t < 60$ | 22 |
| $60 \leqslant t < 80$ | 15 |
| $80 \leqslant t < 100$ | 11 |
| $100 \leqslant t < 120$ | 8 |
| $120 \leqslant t < 140$ | 5 |
| $140 \leqslant t < 160$ | 3 |
| $160 \leqslant t < 180$ | 2 |
| $180 \leqslant t < 200$ | 1 |
| | 150 |

a) Construct a histogram of relative frequencies and a cumulative relative frequency distribution. Draw the polygon for these distributions.

b) Find the mode, the median, the mean, and the standard deviation (use class marks of 10, 30, 50, . . .).

c) What is the value of the interquartile range? What percent of the observations fall in $\mu \pm 1\sigma$? What percent fall in $\mu \pm 2\sigma$?

d) Based on these data, how many customers would you estimate for Shark Loan next week, if the office is open five days a week, ten hours a day?

# Probability Theory:
# Discrete Sample Spaces

## 2.1  INTRODUCTION

As we mentioned in Chapter 1, most problems of statistics involve elements of uncertainty, since it is usually not possible to determine in advance the characteristics of an unknown population, or to foresee the exact consequences which will result from each course of action in a decision-making context. A necessary part of an analytical approach to these problems must thus involve evaluations of *just how likely it is* that certain events have occurred or will occur. "How likely is it that a given sample accurately reflects the characteristics of a certain population?" or "What is the chance that a given consequence will occur following a certain decision?" are examples of inquiries into the probability of an event or set of events of this nature. A probability is a proportional measure between 0 and 1 about the occurrence of an event or set of events. If an event has a probability of zero, then its occurrence is *impossible*; if an event has a probability of 1.0, then its occurrence is *certain*. Finding the value between 0 and 1 which depicts how likely is a certain outcome, a specific event, or a sequence of events, forms a fundamental part of almost all types of statistical analysis; and probability theory provides the foundation for the methods of this analysis.

The origins of probability theory date back to the 1600's when the mathematicians Blaise Pascal and Pierre Fermat became interested in games of chance. Although Pascal and Fermat corresponded regularly about problems involving elements of chance, not until over 100 years later did this new branch of mathematics find many applications beyond the French gambling houses of the seventeenth century. The work of Karl Gauss and Pierre Laplace was especially important during the later 1700's in extending probability theory to problems of the social sciences and actuarial mathematics. Laplace commented that "It is remarkable that a science which began with the consideration of games of chance could have become the most important object of human knowledge." Despite the contributions made in the seventeenth through nineteenth centuries, most modern-day statistics was developed in the past fifty years. R. A. Fisher, J. Neyman, E. S. Pearson, and A. Wald are some of the more prominent researchers who have contributed to the phenomenal growth of statistics in recent years.

### Subjective and Objective Probability

Since there is some disagreement, even among authorities, about the interpretation of probability, it is advisable to begin a discussion of the fundamental concepts of the theory of probability by describing the two major viewpoints, or interpretations. Both of these interpretations are concerned with determining the probability of the occurrence of an event or a set of events. They differ in how this probability is to be determined. The first and more traditional viewpoint defines probability as the relative frequency with which an event occurs over the long run. Probabilities determined by the long-run relative frequency of an event are usually referred to as frequency probabilities, or objective probabilities (since they are determined by "objective evidence" and would have the same value regardless of *who* did the determination). The second interpretation of probability holds that probability represents the decision-maker's subjective estimate of the occurrence of the event based on his knowledge, information, and experience. This interpretation, in which probabilities are referred to as *subjective* probabilities, has recently gained considerable importance in statistical theory, largely because of the influence of such statisticians as L. J. Savage, R. Schlaifer, and H. Raiffa.

Suppose you believe the chances are one in four that you will earn an "A" in statistics, or that the odds are eight to one against the Pittsburgh Pirates' winning the National League pennant next year. These are subjective evaluations, in which your personal opinion about the probability of these events need not agree with those of your statistics professor, or those of the manager of the Pittsburgh Pirates, or with anyone else's, for that matter. You might, for example, believe that the odds of a tail appearing on the flip of a certain coin are less than 50–50, while everyone else thinks the chances of a head and a tail are equal. You are, of course, entitled to your own opinion; presumably, however, if we flipped this coin enough times, your opinion and everyone else's should begin to coincide. In other words, given a sufficient amount of data about the past occurrences of an event we would expect one's subjective opinion to agree fairly closely with the long-run relative frequency of that event.

The problem in the "frequency" approach to estimating probabilities is that, in many real-world problems, there may be little or no historical data available on which to base an estimate of the probability of an event. In such cases, only a *subjective* probability can be determined; and this probability may differ even among experts who have similar technical knowledge and identical information. For example, various space scientists in 1960 gave different estimates of the probability that men would walk on the moon within that decade. Various weathermen observing the same climatic conditions may give different probabilities that rain will occur on a given day. Fortunately, the rules and operations governing probability theory are the same whether the number itself is generated by an objective or subjective approach.

A simple example of the concept of probability as a long-run relative frequency is an experiment in which a coin is tossed over and over again, and where the outcome of each toss is observed and recorded. For the first few tosses the proportion of heads may fluctuate rather wildly. After a sufficiently large number of tosses, however, it should begin to stabilize about one particular value. The more tosses observed, the closer should the proportion of heads approach this long-run, or limiting value. Just such an experiment was performed by the statistician

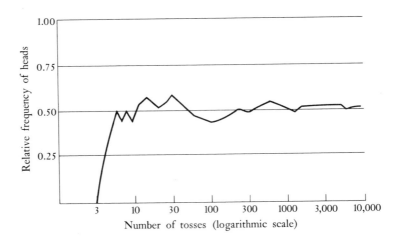

**Fig. 2.1.** Relative frequency of heads in Kerrick experiment.

J. E. Kerrick during his internment in Denmark during World War II; Kerrick tossed a coin 10,000 times and obtained 5067 heads, in the sequence shown in Fig. 2.1.

Note that in Fig. 2.1 the observed relative frequency of heads (to tails) begins to stabilize, after quite a few tosses, about one particular value (which in this case appears to be 0.50). This value represents what we have called long-run relative frequency. One would expect, as the number of tosses becomes even larger, that the observed relative frequency of heads will come closer and closer to this long-run relative frequency until, at the limit (an infinite number of tosses), they are equal. The value at which they are equal is sometimes called the *limit of relative frequency*. It is this concept of the long-run relative frequency of an event which is, in general, implied when referring to an "objective probability" or to "the frequency approach to probability." Thus, a probability of heads equal to 0.50 in the Kerrick experiment means that, in the long run, the expected proportion of heads is 50 percent. In most practical problems, of course, it is seldom possible to observe a sufficient number of observations to precisely determine this limit.

**Expected Relative Frequency**

At times it is convenient to formulate the *expected* or *theoretical* long-run relative frequency of an event. In such cases, the probability of an event depends on assumptions about the conditions underlying the occurrence of this event. For example, if the coin in the experiment by Kerrick had been assumed to be "fair" (i.e., not biased to heads or tails), then the theoretical long-run relative frequency of heads for this coin would have been, by definition, 0.50. The underlying conditions or assumptions about an experiment (such as tossing a coin) represent part of what is called the experimental model. Making assumptions of this nature is useful in determining whether theoretical probability values differ from actual (or observed) probabilities, regardless whether these values are objective or subjective. Note that Kerrick's observed value differs very little from the theoretical value of 0.50.

Two terms used in this discussion, "experiment" and "event," need further elaboration. In the next section we define and discuss the vocabulary and concepts that are important in an introduction to probability theory.

## 2.2 THE PROBABILITY MODEL

Often a person desires to specify a probability associated with some situation. Perhaps you want to know the probability that no automobile accidents will occur in a given location over a certain time period, or the probability that the price of dairy products will decrease, or the probability that candidate X will be elected senator, or the probability that your favorite person will finally phone you right after you go out to get something to eat. To specify the probability associated with a given situation, it is extremely important to define what experiment underlies this situation, and what outcomes can result from this experiment. As we will demonstrate, making probability assessments is simplified by using a probability model as the framework for your mental construction of the problem. Once the nature of the problem is clearly formulated in this way, the solution is routinely obtained by using specified *probability laws and formulas.*

### The Experiment

The first component in the probability model is the definition of the experiment. In the previous section, the word experiment was used to describe the test that was conducted to discover the proportions of heads in coin-flipping. In a probability context an experiment means *any situation capable of replication under essentially stable circumstances.* The repetitions may actually be feasible and be performed (as in coin-flipping), or they may be abstract and theoretically imaginable. Investing $1000 in the stock market could be an experiment, even though it is to be done only once. You could imagine repeating such as investment over and over again many times and theoretically considering the chances that you would make capital gains or losses, or earn a yield of more or less than six percent, etc. Driving a car from New York to Phoenix might be an experiment. You may plan to do it only once, but you might be interested in the theoretical chances of making the trip without tire or engine trouble. Abstractly, you are imagining many replications of the trip under similar circumstances. Even taking an accounting exam might be considered as an experiment. If you imagined many students taking the exam with similar preparation, you might be interested in the chances that more than half will pass the exam, or that you will get an "A" or a "B" on the exam. Clearly, an experiment is very broadly defined to correspond to *any situation involving uncertainty*, whether it in fact actually recurs many times or whether the repetitions are hypothetical.

**Outcomes of an Experiment**

Once an experiment is defined in a situation of uncertainty, it becomes obvious that not all repetitions of the experiment would result in the same outcome. For example, if many persons drive their cars from New York to Phoenix, we can easily imagine that *some* may have no trouble whatsoever, one may have a flat tire, one may have a fan-belt break, some may need minor repairs on the engine, some may have an accident and need major repairs on their engine, etc. The number of different outcomes of this experiment are seemingly endless, but theoretically *countable*. This is quite a contrast from the coin-flipping experiment, in which there are only *two* outcomes, since each repetition results in either "heads" or "tails."

   For another example, suppose one is interested in the number of hours it takes a certain light bulb to burn out if it is left burning continuously. The time it takes the bulb to burn out could be *any* number, of hours, from less than one hour to some large upper bound, say 10,000 hours. Different light bulbs would burn a different number of hours, but clearly, an *infinite number* of outcomes are possible.

   The different outcomes of an experiment are often referred to as *sample points*, and the set of all possible outcomes is called the *sample space*. For example, if we define an experiment to be "flip a coin once," then the sample space is comprised of just two sample points {heads, tails}. If the experiment is to "flip a coin twice," then the sample space consists of the four points

$$\{H, H; \quad H, T; \quad T, H; \quad T, T\}.$$

For these examples, the sample space is said to be *discrete* (because the outcomes can be separated from one another, and counted), and *finite* (the number of outcomes is limited). Suppose, however, we define the experiment to be "flip a coin until the first tail appears." In this case, the number of sample points is infinite, since there is no limit on the number of outcomes. Such a sample space is still discrete, since the number of outcomes can be separated and counted. Thus, a discrete sample space may contain either a finite or an infinite number of outcomes.

   Examples of *discrete sample spaces* which involve a finite number of outcomes include the number of heads in a certain number of flips of a coin, the number of combinations of groups of three companies selected for government contracts from ten companies that submit proposals, or the number of defectives in a shipping lot of a given size. Problems which involve a discrete sample space having an infinite number of outcomes include the quoted dollar value of a common stock (it can vary from 0 to infinity, in units of one-eighths of a dollar), or the number of flaws in material from a textile plant (which can be *any* positive integer).

In contrast to a discrete sample space is the concept of a *continuous sample space*. A sample space is continuous if the number of possible outcomes is infinite and uncountable. Our experiment of the time it takes a light bulb to burn out represents a continuous sample space, because the outcome of the experiment could be *any* real number from zero to the upper bound (such as 10,000 hours). There is obviously an infinite number of outcomes here, and no way to separate and count them. Generally, applications of continuous sample spaces occur when the data involved are obtained by *measurement* rather than by counting. Thus, a set is continuous if *any* value within an interval can occur, such as any value between 0 and 1, or any value between 200 and 1000, or any value from 0 to infinity. For example, the net weight of a box of packaged cereal, the length of a fish caught in a trout stream, the average speed of the winning car at the Indianapolis 500 race, the weight of students in statistics classes, and the distance in miles that cars can be driven on a full tank of gas, are all examples which involve a *continuous sample space*.

Most applications of probability theory involve experiments with a finite number of outcomes, although the number is often large enough so that it makes little difference (for practical purposes) if it is assumed to be infinite. Furthermore, many experiments involving only a discrete set of outcomes can be *approximated* by a continuous set. The advantage of such approximations is that they often *simplify* the derivation of certain statistical results.

We must hasten to add that it is not always clear from the statement of an experiment exactly what outcomes are relevant. For example, in the experiment "take an accounting exam," one student may define the sample space to be all possible *numerical* grades (perhaps a number between 0 and 100). Another student may be interested only in the *letter* grade (A, B, C, D, or F), while a third student might define the sample space as simply {pass, fail}. Similarly, in tossing a coin twice there may be no interest in the *order* of the outcomes, in which case the relevant sample space would be the three points:

{two heads; two tails; one head and one tail}.

When working with continuous sample spaces, it may be advantageous to *group* the sample space into a small number of discrete outcomes. For example, in working with the time it takes a light bulb to burn out, we might group the outcomes into the following three sets: (1) time is less than 100 hours, (2) time is between 100 and 1,000 hours, and (3) time exceeds 1,000 hours. In all probability models it is important that the relevant outcomes be carefully defined.

In defining the sample space of an experiment, *one must be sure* that it is not possible for two or more outcomes to occur in the same replication of the experiment. Outcomes defined in this manner are said to be *mutually exclusive*. For example, the two outcomes {heads, tails} are mutually exclusive for the experiment "toss a coin once," but they are *not* mutually exclusive for the experiment "toss a coin twice" (since two tosses could result in both a head and a tail). Similarly, the student who defines the outcomes of an accounting exam as "A", "Pass," and "Fail" has not specified a mutually exclusive set because both "A" and "Pass" could happen at the same time.

A second requirement, in defining the sample space of an experiment, is that the list of outcomes must be *exhaustive*; that is, no possible outcomes can be omitted. If a flipped coin can stand on its edge, then this outcome must be added to the sample space. If our light bulb could last longer than the specified upper bound (such as 10,000 hours), then this possibility must be added to the sample space. In general, when one is constructing a probability model, it is extremely important that the outcomes associated with the sample space (that is, the sample points) be *both mutually exclusive and exhaustive*.

### Random Variable

Given an experiment and a set of *mutually exclusive* and *exhaustive* outcomes, it is common to consider questions of probability about the occurrence of any one or more of these outcomes by use of the concept of a *random variable*.

*A random variable is a well-defined rule for making the assignment of a numerical value to any outcome of the experiment.*

In many cases, the outcomes of an experiment may meet this definition of a random variable because the outcomes are already well-defined numbers. For example, the measure of hours that a light bulb lasts is a well-defined number; a count of the defectives in a lot of transistors is a well-defined number; and the yield on an investment of $1,000 is a well-defined number. In other cases, the outcomes of an experiment may be qualitative. For example, the outcome of coin-flipping is heads or tails, and the outcome of "taking a course" could be "A", "B", "C", "D", or "F". In these instances the probability model must specify exactly what numerical value corresponds to each qualitative outcome. The registrar does this for grades at many colleges by letting $A = 4, B = 3, C = 2, D = 1$, and $F = 0$. In tossing a coin, one common way to define a random variable is to let heads $= 1$ and tails $= 0$.

In order to define a random variable for the experiment "drive from New York to Phoenix," the sample space would need to be converted to some consistent measure, such as the *number of dollars* required for repair. In working with continuous sample spaces, it is sometimes convenient to reduce the sample space to just a few discrete points. Thus, the yield on a $1000 investment might be classified as falling into one of just a small number of intervals (such as 0 to 2.0, 2.1 to 4.0, etc.); and the dollar amount of repairs on a trip to Phoenix could be classified as either less than $50, between $50 and $100, or over $100. In all these examples we have a random variable only when numerical values are assigned to the outcomes of interest by a well-defined rule.

In making the assignment of numerical values to the outcomes of an experiment, we will denote random variables by letters in *boldface* type, such as $x$, $y$, $z$, or sometimes by subscripted boldface letters such as $x_1$, $x_2$, $x_3$. *Specific* values of such random variables will be denoted by letters in lightface type, such as $x$, $y$, $z$, or perhaps $x_1$, $x_2$, $x_3$. Thus, the designation $x = x$ is read as "the random variable $x$ takes on the value $x$." The following examples will illustrate this notation.

1) *Experiment:* Flip a coin once

   Outcomes: Two discrete outcomes, heads and tails
   Sample space: Discrete and finite
   Random variable: Define $x = 1$ if heads occurs and $x = 0$ if tails occurs.

Although any values might be used to give numerical labels to the outcomes in an experiment such as this one, zero and one are especially convenient mathematically in many situations involving just two outcomes. Since the variable $x$ in this case gives a well-defined rule for assigning numerical values to the experiment, $x$ is a *random variable.*

2) *Experiment:* Taking an exam

   Outcomes: Grades "A," "B," "C," "D," "F"
   Sample space: Discrete and finite
   Random variable: Define $y = 4$ if the grade is "A,"
   $\qquad\qquad\qquad\qquad\quad y = 3$ if the grade is "B,"
   $\qquad\qquad\qquad\qquad\quad y = 2$ if the grade is "C,"
   $\qquad\qquad\qquad\qquad\quad y = 1$ if the grade is "D," and
   $\qquad\qquad\qquad\qquad\quad y = 0$ if the grade is "F."

The familiar four-point grade system is simply an assignment of numbers to a grade measure. Since the variable $y$ is a well-defined rule for assigning numbers to the outcomes of this experiment, $y$ is a random variable.

3) *Experiment:* Driving a car from New York to Phoenix

   Outcomes: Various car troubles which might be encountered on trip
   Sample space: Discrete (infinite but countable)
   Random variable: Define $z =$ nearest number of dollars paid for repairs,
   $\qquad\qquad\qquad z = 0, 1, 2, 3, \ldots$

The random variable $z$ in this case is discrete, and it is also infinite since there is no limit on the amount of repairs. Realistically, however, there is some upper bound to the value of $z$ perhaps equal to the cost of the car if it is a total loss due to an accident. Also, this probability model assumes no negative values for $z$, since we doubt that anyone can find a "Tom Sawyer" mechanic willing to pay for the chance to do the needed repairs.

4)  *Experiment:* Investing $1000 in a common stock

   Outcomes: Values of yield or rate of return
   Sample space: Continuous (always infinite)
   Random variable: Define $x =$ value of yield, $-\infty < x < +\infty$.

A *continuous random variable* is obtained from a continuous sample space whenever a unique value of $x$ is assigned to each outcome in the sample space. Thus, since a yield can be any positive number (or can be negative), $x$ must be continuous.

5)  *Experiment:* Investing $1000 in a common stock

   In this example we simplify Experiment 4 somewhat by grouping the various yields into different classes. For example, we might let one class represent all yields between 0 and 2%, another represent 2.1 to 4.0%, etc. This simplification results in the following probability model.

   Outcomes: Class intervals of yields
   Sample space: Discrete and finite
   Random variable: Define $x =$ the midpoint (or some representative value) of
                  the yields in each class interval (the class marks).

Assigning values of the random variable so they equal the class marks makes it much easier to find the mean and variance of the probability distribution under study. We will describe the process of finding means and variances of probability distributions in Chapter 3. As a final note in this section, we should point out that the numerical value assigned to each outcome in an experiment need not be unique to that outcome. That is, several different outcomes may be assigned the same numerical value. This fact is easily seen in the experiment about driving to Phoenix, for in this case there are certainly many different outcomes (car troubles) which would lead to the same value of the random variable (that is, lead to the same dollar value of cost).

**Probability Distribution**

Once an experiment and its outcomes have been clearly stated, and the random variable of interest has been defined, then the probability of the occurrence of any value of the random variable can be specified. As we shall see shortly, the specification of probabilities is usually easiest when the sample space is discrete and finite. However, as we saw in the probability models labeled 4 and 5 above, a continuous sample space can often be fruitfully transformed to a discrete and finite one in many practical problems by *grouping outcomes.*

For the present discussion let us present some new examples. Suppose 140 students are registered for a certain course, and they are to be divided randomly into four sections. The number assigned to each section is determined by the size of the available classrooms as follows:

| Section | Class size |
|---------|------------|
| 1 | 25 |
| 2 | 45 |
| 3 | 40 |
| 4 | 30 |
| Total | 140 |

If you are one of the students involved in this assignment, you could view this process as an experiment with *four outcomes.* The sample space is discrete and finite; a random variable $x$ may be defined to have values equal to the section number, $x = 1, 2, 3,$ or 4. The probability that you are assigned to each section can be determined, and denoted by the symbol $P(x)$. For example, the probability of your being assigned to section 1 is denoted $P(x = 1)$ or $P(1)$. The probability that you are assigned to Section 1 is simply the proportion of assignments that are made to Section 1, relative to the total number of students, which is

$$P(1) = \tfrac{25}{140} = 0.179.$$

This is the *least probable* outcome of the experiment. What value of $x$ has the highest probability? That is, what section is the most probable outcome? Clearly you have the highest chance of being assigned to Section 2, since this section will have the most students. We find

$$P(2) = \tfrac{45}{140} = 0.321.$$

Continuing in this manner we can find the probability of each possible value for the random variable $x$. When this is done, we have obtained the *probability distribution* for $x$. Table 2.1 and Fig. 2.2 depict the probability distribution for the random variable in this assignment problem.

In this example only four discrete values of $x$ have a positive probability. All other values of $x$ have a probability of occurring equal to zero, indicating that they are impossible. Also, the sum of the probabilities over all values of $x$ is equal to 1.0, indicating that it is a certainty that one of these outcomes occurs every time an assignment is made (that is, they are exhaustive). In other words, we have

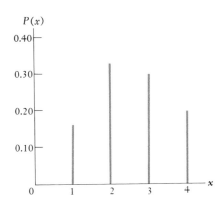

**Fig. 2.2.** Graph of the discrete probability distribution for *x* given in Table 2.1.

**Table 2.1** Probability distribution for *x* in the section assignment example

| Outcome | Value of $x$ | $P(x)$ |
|---------|-------------|--------|
| Section 1 | 1 | $25/140 = 0.179$ |
| Section 2 | 2 | $45/140 = 0.321$ |
| Section 3 | 3 | $40/140 = 0.286$ |
| Section 4 | 4 | $30/140 = 0.214$ |
| Sum | | 1.000 |

tabulated a set of values for the random variable *x* which are both mutually exclusive and exhaustive.

The determination of a probability distribution completes the probability model we began earlier. Figure 2.3 summarizes this model. As shown in this figure, we must clearly state the experiment and recognize its set of *exhaustive* and *mutually exclusive* outcomes. We then define a random variable which assigns a number to each outcome in the sample space. Finally, we determine the probability of each value of the random variable, to obtain the probability distribution.

The construction of the probability model is not always as simple as in the previous example (where the random variable had a unique value for each section). To illustrate a more complicated example, consider the experiment of throwing a pair of "fair" (i.e., perfectly balanced) six-sided dice a single time, and observing the number of dots appearing. The sample space is discrete and finite, as illustrated in Fig. 2.4.

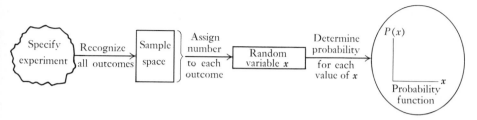

**Fig. 2.3.** The probability model.

Let us define a random variable $x$ to have values equal to *the sum of the dots* appearing on the pair of dice. That is, $x$ can assume any integer value from 2 to 12, depending on the outcome of the roll of the dice. The probability of any value of $x$ is given by the number of sample points for which the sum of the dots equals $x$, divided by the *total number* of sample points. For example, let us find $P(x = 9)$, or $P(9)$. We can observe, from Fig. 2.4, the number of sample points satisfying this value of $x$. Out of all 36 outcomes, those satisfying the condition $x = 9$ are the four ordered pairs

$$(3, 6), \qquad (4, 5), \qquad (5, 4), \qquad \text{and} \qquad (6, 3),$$

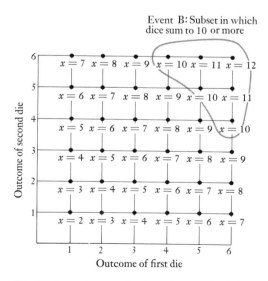

**Fig. 2.4.** Sample space for throwing a pair of dice.

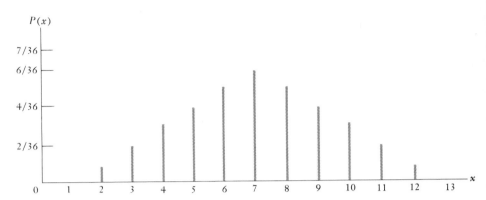

**Fig. 2.5.** Probability distribution of *x* = sum of dots face up after tossing a pair of dice.

where the first entry is the outcome of the first die, and the second entry is the outcome of the second die. Thus,

$$P(9) = \tfrac{4}{36}.$$

Similarly, we could find the probability that the value of *x* would be 3. The ordered pairs (1, 2) and (2, 1) are the only sample points satisfying *x* = 3, and so

$$P(3) = \tfrac{2}{36}.$$

The reader should verify for himself that the probabilities of all values of *x* in this example are those shown in the probability function illustrated in Fig. 2.5. Again, note that only a discrete set of values of *x* have positive probabilities, and the sum of the probabilities over all values of *x* is equal to 1.0. Indeed, these are essential characteristics of all discrete probability distributions. That is, the following two properties are necessary for all probability distributions:

1. $1.0 \geqslant P(x) \geqslant 0$;

2. $\displaystyle\sum_{\text{All } x} P(x) = 1.0.$

## 2.3 EVENTS AND SET NOTATION

### Events

Once a probability model has been built, there are usually some specific questions of interest about the probability of certain subsets of the outcomes of the sample space. Such subsets are called *events*. The smallest event in a sample space is called an *elementary event*, where an elementary event is one which cannot be decomposed any further. In general, the elementary events of an experiment are those involving only a single outcome. In rolling a pair of dice, for example, there are 36 elementary events, corresponding to the 36 different single outcomes of this experiment. If the dice are "fair" (evenly balanced), then all 36 elementary events have an equal chance of occurring on any one throw of the dice. The elementary event "two sixes" will thus occur with probability 1/36 when the dice are fair.

Unfortunately, in most practical problems the events of interest are usually not elementary events, but instead are more complex subsets of the sample space. To determine the probability of such subsets, we will use the concept of a probability distribution (as was defined in Section 2.2). To illustrate this process, suppose that, in rolling a pair of dice, one is interested in the event $A$, where $A$ is the subset of outcomes which correspond to observing a sum equal to nine on the dice. (Events will be denoted by capital letters, such as $A$, $B$, or $E$). This event $A$ is not an elementary event since it is not composed of a single outcome, but rather of four different outcomes. The probability $P(A)$ was found in Section 2.2 by defining a random variable $x$, and then using the probability distribution of $x$. From that discussion we know that

$$P(A) = P(x = 9) = \tfrac{4}{36}.$$

Let's consider another event (set) $B$, where $B$ is defined to be the outcome corresponding to rolling two dice and observing a sum equal to *at least* 10. Event $B$ can be defined in terms of the random variable as $P(B) = P(x \geqslant 10)$. The sample points included in the subset representing event $B$ are encircled in Fig. 2.4. The probability that event $B$ occurs when the dice are rolled can be found by adding together probabilities in Fig. 2.5 for values of $x \geqslant 10$, namely,

$$x = 10, \qquad x = 11, \qquad \text{and} \qquad x = 12.$$

We obtain

$$P(B) = \frac{3 + 2 + 1}{36} = \frac{6}{36}.$$

The event $B$ is composed of six sample points out of the 36 total possible outcomes.

In the following sections, we will develop some probability laws and definitions that are very useful in determining probabilities of more complex events. Their later application to specific decision problems requires a thorough understanding of the components of the probability model. Before continuing on, however, we need to reemphasize the *two basic properties* for defining probabilities of events associated with a given sample space and experiment. If an event $E_i$ is a subset of a discrete sample space denoted by $S$, and $P(E_i)$ is the probability of that event, then the following two basic properties hold:

---

*Property* 1.    $1.0 \geqslant P(E_i) \geqslant 0$   for every subset $E_i$ of $S$.

*Property* 2.    $P(S) = 1.0$.

---

It is easy to recognize that these properties are consistent with our previous results and examples. The first one says that the probability of an event can never be less than zero (which represents an impossibility) nor greater than one (which represents a certainty). The second one says that some event in the sample space ($S$) must *always* take place. If a set of *exhaustive* and *mutually exclusive* events $E_i, E_2, \ldots, E_N$ could be listed (it includes all sample points without any overlapping) then

$$P(S) = \sum_{i=1}^{N} P(E_i) = 1.0.$$

We have observed one way of defining such a set of events which is especially attractive and convenient for further statistical procedures. The method is to use our probability model and define a random variable $x$ over all outcomes of the sample space.

As we indicated above, it is necessary to be especially careful in defining the outcomes of interest in a sample space. One helpful tool in making such definitions is the use of the notation of sets, which we describe in the following section. In studying set notation, the reader should bear in mind that we will be using the terms set and event interchangeably, and that the elements of a set are comparable to the outcomes associated with a particular event. In Section 2.4 we will show how set notation can be used to help determine the probability of certain complex events.

### Set Notation

The objects contained in a set are called the *members* (or *elements*) of that set, and there may be either a finite or an infinite number of these elements. Capital letters such as $A$, $B$, and $C$ are generally used to denote a set, while the elements of a set are designated by separating them by commas and enclosing them in braces. Thus, a set $A$ defined to be *all integers* between 10 and 15 (inclusive) would be written as

$$A = \{10, 11, 12, 13, 14, 15\}.$$

Similarly, a set $B$, defined to be the elements corresponding to a single toss of a coin, would be written as

$$B = \{\text{Heads, Tails}\}.$$

When the number of elements in a set is too large to list, it is necessary to *specify a rule* describing the elements belonging to this set. The traditional convention in specifying such a rule is to let the first symbol (or symbols) inside the brace indicate the variable (or variables) of interest, followed by a vertical bar ($|$), and then to indicate the values of this variable which are elements of the set. The bar separating the variable and the values to be included in the set is read "such that." Thus, the set

$$C = \{x \,|\, 0 \leqslant x \leqslant \infty\}$$

is the set represented by the variable $x$ "such that" the value of $x$ is between zero and infinity. Likewise, the fact that the possible sales ($y$) of a particular product must fall between zero and 300,000 units, may be defined as a set $A$ where

$$A = \{y \,|\, 0 \leqslant y \leqslant 300,000\}.$$

If the sales of this product are also restricted to *integer* values, then this set would be written as

$$A = \{y \,|\, 0 \leqslant y \leqslant 300,000 \quad \text{and} \quad y = \text{integer}\}.$$

Rather than specifying exactly which elements belong to a given set, one may wish to specify merely that the elements of one set are members of other sets. The symbol $\in$ (a variation of the Greek letter for epsilon) usually stands for "is an element of" in representing membership in a set; $x \in A$ is therefore read "$x$ is an element of the set $A$". Similarly,

$$C = \{x \,|\, x \in A \quad \text{and} \quad x \in B\}$$

is the set $C$ such that $x$ is an element of *both* $A$ and $B$. For example, to be eligible to vote in most states in the United States, an individual $x$ must belong to both of the two sets, $A =$ at least 18 years of age and $B =$ United States citizen. Hence, the set "Eligible to vote" $= C$ as defined above would be

$$C = \{x \,|\, x \text{ is at least 18 years of age} \quad \text{and} \quad x \text{ is a U.S. citizen}\}.$$

Some sets have special names. The set of all possible outcomes in a particular experiment, which corresponds to the sample space, is called the *universal set*. The smallest subset in any problem is referred to as the empty or *null set*, which we denote by the symbol $\emptyset$. This subset contains *no* elements or members. For example, an individual cannot be both a male and a female, so the set representing membership in both of these sets must be the null set

$$\emptyset = \{x \,|\, x \text{ is male} \quad \text{and} \quad x \text{ is female}\}.$$

We now turn to the problem of describing the relationship between two or more sets. In this discussion we will be using, for the most part, the term "event" rather than the term "set." We will discuss *complementary events, intersections,* and *unions.*

### Complement of an Event

Perhaps the simplest way of forming a new event from a given event is to form its *complement.* For example, the *complement* of the subset of the sample space *A,* which is denoted by $\bar{A}$ (read *A*-bar), contains all the points in the sample space that are *not part* of *A.* If *S* denotes the total sample space, then

$$\bar{A} = S - A, \quad \text{or} \quad \bar{A} = \{\text{All sample points not in } A\}.$$

To illustrate, suppose the sample space *S* is the set of all students in a given university, and *A* is the subset representing *sophomores*; then the event $\bar{A}$ represents all students in that university who are *not* sophomores. As another example, suppose we define the set *A* to represent companies with sales less than or equal to 300,000 units, written as

$$A = \{x \mid x \leqslant 300{,}000\}.$$

This set is the complement of the set

$$\bar{A} = \{x \mid x > 300{,}000\}.$$

As a final example, the set

$$B = \{y \mid y = \text{Heads}\}$$

is the complement to the set

$$\bar{B} = \{y \mid y = \text{Tails}\}$$

on a single toss of a coin.

Two complementary events must be *exhaustive* because they take into account (or exhaust) all possible events. In addition, complementary events also must always be *mutually exclusive* because the elements in one subset are excluded by definition from the subset of the remaining events. For instance, the outcomes of an investment in the stock market might be classified according to whether the price (= *x*) of the stock a month later is less than \$50, or greater than \$50. The events

$$A = \{x \mid 0 \leqslant x \leqslant 50\} \quad \text{and} \quad \bar{A} = \{x \mid 50 < x \leqslant \infty\}$$

represent *mutually exclusive and exhaustive events.*

### The Intersection of Two Events

The intersection of two events is defined to be those sample points common to both events. Suppose we let the variable *x* represent points of some sample space, and let the two events $A$ and $B$ represent subsets of this space. The intersection of the events $A$ and $B$, which is usually written as $A \cap B$, consists of the set of elements *common to both* the event $A$ *and* the event $B$.

*Intersection of the events $A$ and $B$:*

$$A \cap B = \{x \mid x \in A \quad \text{and} \quad x \in B\}.$$

To graphically illustrate the intersection of two events, suppose in the Venn diagram* shown in Fig. 2.6, that $A$ and $B$ represent subsets of the set $S$. The intersection of $A$ and $B$ equals the shaded portion of these two sets, labeled $A \cap B$.

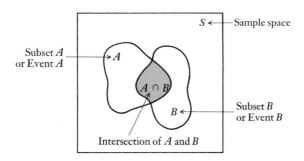

**Fig. 2.6.** The intersection of $A$ and $B$ (shaded portion).

To illustrate the intersection of two events, suppose the set $A$ equals the amount of dollars a college student will have available to spend in a given month, and the set $B$ his expenditures for that month. If, for example,

$$A = \{x \mid 0 \leqslant x \leqslant \$200\} \qquad \text{and} \qquad B = \{x \mid x \geqslant \$200\},$$

then his budget is "balanced" only at the intersection of these two sets,

$$A \cap B = \{x \mid x = \$200\}.$$

Should his income possibilities rise to, say,

$$A = \{x \mid 0 \leqslant x \leqslant \$250\},$$

the income and expenditures could coincide at *any* of the elements *common to both* $A$ and $B$; this is the set which is given by

$$A \cap B = \{x \mid \$200 \leqslant x \leqslant \$250\}.$$

---

* Named after the logician J. Venn (1834–1923).

Note that, if this student insists on living on a higher standard, such as

$$B = \{x \mid x \geqslant \$300\},$$

then he cannot stay within the budget constraint

$$A = \{x \mid x \leqslant \$250\},$$

since the sets $A$ and $B$ are now mutually exclusive—their intersection represents the null set ($A \cap B = \emptyset$).

As another example of the intersection of two sets, let the random variable $x$ represent the number of dots thrown on a single toss of a pair of dice, and define the events $A$ and $B$ as follows:

$$A = \{x \mid 2 \leqslant x \leqslant 8\} \quad \text{and} \quad B = \{x \mid 4 \leqslant x \leqslant 12\}.$$

The elements common to these two sets are $A \cap B = \{x \mid 4 \leqslant x \leqslant 8\}$.

### The Union of Two Events

An important operation involving two (or more) events is called the *union* of these events. The union of such events is defined to be *those sample points satisfying **at least** one of these events*. Thus, the union of the two events $A$ and $B$, written as $A \cup B$, consists of all sample points contained *either* in $A$, *or* in $B$, or in *both* $A$ and $B$.

*Union of the events $A$ and $B$:*

$$A \cup B = \{x \mid x \in A \quad \text{or} \quad x \in B \quad \text{or} \quad x \in A \cap B\}.$$

For example, consider the set of outcomes resulting from a single draw from a deck of cards, and suppose we take the union between the set $A$ = Heart and $B$ = Face card or ace (jack, queen, king, ace). The union $A \cup B$ is the set of 25 cards, each of which is either a heart or a face card or *both* a face card and a heart (13 hearts + 12 nonheart face cards). This type of union is illustrated in Fig. 2.7.

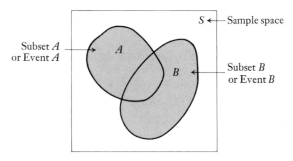

**Fig. 2.7.** The union of $A$ and $B$ (shaded portion).

   The union of two (or more) mutually exclusive and exhaustive sets must be equal to the entire sample space. For example, if the random variable $x$ represents the number of dots showing on a single throw of a pair of dice, and

$$A = \{x \mid 2 \leqslant x \leqslant 5\},$$
$$B = \{x \mid 6 \leqslant x \leqslant 12\},$$

then

$$A \cup B = \{x \mid 2 \leqslant x \leqslant 12\} = S.$$

   The operation of taking the union or the intersection of two events can be extended to three, or to any finite number of events. These operations follow directly from the definitions given above. For example, the intersection of the three events $A$, $B$, and $C$ is the set of elementary events common to *all three* of these events (shown in Fig. 2.8(a)). The union of these three events is the set of elementary events in *either* $A$, $B$, or $C$, in *both* $A$ and $C$, in *both* $B$ and $C$, in *both* $A$ and $B$, or in $A$, $B$, *and* $C$ (shown in Fig. 2.8(b)).

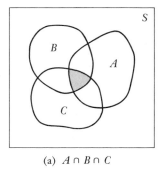

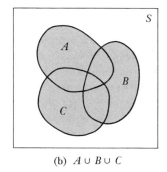

(a) $A \cap B \cap C$                    (b) $A \cup B \cup C$

**Fig. 2.8.** The intersection (a) and union (b) of $A$, $B$, and $C$ (shaded portion).

   The concept of a probability model, and the notation of sets, provide us with the foundation for the analysis of a number of different probability questions. In particular, we now present the rules for determining the probability of the various combined sets described in this section.

### 2.4 COMPLEMENTARY AND CONDITIONAL PROBABILITY

From the basic properties of a probability model, we can determine the probability of the complement of an event. Suppose a sample space contains $N$ sample points and some event $A$ contains $a$ of these points; that is, $P(A) = a/N$. Then $\bar{A}$ must contain $(N - a)$ sample points. Thus, we can write the probability of the event $P(\overline{A})$ as follows:

$$P(\overline{A}) = \frac{(N - a)}{N} = \frac{N}{N} - \frac{a}{N} = 1.0 - \frac{a}{N}.$$

---

*Probability rule for complements:*     $P(\overline{A}) = 1.0 - P(A).$     (2.1)

---

Let us apply this rule to the experiment of throwing dice, as presented previously, where the event $A$ was defined as

$$A = \{x \mid x = 9\}.$$

Hence, $P(\overline{A}) =$ probability that the random variable $x$ takes on values different from nine. This probability is given by

$$P(\overline{A}) = 1.0 - P(A) = 1 - \frac{4}{36} = \frac{32}{36}.$$

By referring to Fig. 2.4, which illustrates this sample space, it is easy to count all outcomes satisfying $\overline{A} = \{x \mid x \neq 9\}$. There are 32 (out of the total number of 36 outcomes) that satisfy $\overline{A}$; we have thus verified the value 32/36 determined above.

As final illustrations of the probability rule for complements, if the probability that the Pittsburgh Pirates will win the National League pennant is 0.15, then the probability that they will *not* win it is

$$1 - 0.15 = 0.85.$$

If the probability that the stock market will advance is 0.68, the probability that it will not advance is $1 - 0.68 = 0.32$.

### Conditional Probability

Suppose we are interested in determining whether some event $A$ occurs "given that" or "on the condition that" some other event $B$ has already taken place (or will take place in the future). Such a conditional event is read as "$A$ given $B$," and is usually written as $(A \mid B)$, where the vertical line is read as "given."

*Conditional event A given B:*

$$(A \mid B) = \{x \mid x \in A \quad \text{given that } x \in B\}.$$

The Venn diagram in Fig. 2.9 can be helpful in illustrating a conditional event. Suppose the sample space under consideration is all 10,000 students in a university, the set $A$ represents the $a = 2000$ students who are seniors, while the set $B$ rep-

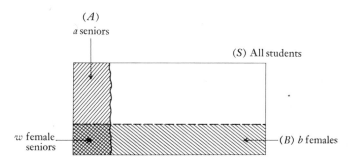

**Fig. 2.9.** Venn diagram of overlapping events.

resents the $b = 3500$ students who are females. Suppose, further, that $w = 800$ of these 3500 females are seniors.

    The conditional event $A \mid B$ represents those students who are seniors selected from those who satisfy the condition of being *female*. The probability of conditional events of this nature is often of interest in specific sampling problems. For example, if a student is to be selected at random, and given that the selected student is female, what is the probability that the student is also a senior? We denote this probability by $P(A \mid B)$, and determine its value by finding the proportion of students satisfying the condition $B$ that are seniors. The answer is

$$P(A \mid B) = \frac{w}{b} = \frac{800}{3500} = 0.228.$$

Similarly, $P(B \mid A)$ can be expressed as the question, "given that a selected student is a senior, what is the probability that the student is female?" The answer is,

$$P(B \mid A) = \frac{w}{a} = \frac{800}{2000} = 0.40.$$

    From the discussion above it should be clear that the intersection of two events $A \cap B$ is composed of the sample points designated by $w$. Since the total number of sample points in the sample space is $N$, we thus know that $P(A \cap B) = w/N$. Similarly, since there are $b$ points in the set $B$, we know that $P(B) = b/N$. Putting these facts together, we see that it is possible to write the probability that a randomly selected student is a senior, *given that* this person is a female, as:

$$P(A \mid B) = P(\text{Senior} \mid \text{Female}) = \frac{P(A \cap B)}{P(B)} = \frac{w/N}{b/N} = \frac{w}{b} = \frac{800}{3500} = 0.228.$$

Similarly, the probability that a randomly selected student is a female *given that* this person is a senior is

$$P(B \mid A) = \frac{P(A \cap B)}{P(A)} = \frac{w/N}{a/N} = \frac{w}{a} = \frac{800}{2000} = 0.40.$$

We can now formalize the definition of a conditional probability.

> *Conditional probability of A, given B:* $P(A \mid B) = \dfrac{P(A \cap B)}{P(B)}.$
>
> *Conditional probability of B, given A:* $P(B \mid A) = \dfrac{P(A \cap B)}{P(A)}.$

(2.2)

    As another example of using the intersection event to obtain a conditional probability, consider the sample space in Fig. 2.4 related to the dice-tossing experiment, where the random variable $x$ is the sum of the dots facing up on a pair of dice. Define the events $A$ and $B$ to be:

$$A = \{x \mid x < 5\} \quad \text{and} \quad B = \{x \mid x \text{ is an odd number}\}.$$

Remember, the number of elements in $S$ for this experiment is $N = 36$. There are $a = 6$ and $b = 18$ sample points satisfying $A$ and $B$ respectively. Included in $A$ are the points

$$(1, 1), \quad (1, 2), \quad (1, 3), \quad (2, 1), \quad (2, 2), \quad \text{and} \quad (3, 1).$$

Also, it is easy to find $w$, which is the number of sample points such that the value of $x$ is both smaller than 5 and *odd*. The intersection $A \cap B$ contains the two outcomes, (1, 2) and (2, 1), and so $w = 2$.

    Now consider the problem of finding $P(A \mid B)$. What proportion of sample points satisfying the condition of being odd numbers are also numbers smaller than 5? By straightforward counting, we can determine that

$$P(A \mid B) = \frac{w}{b} = \frac{2}{18}.$$

By calculation of $P(A \mid B)$ using the definitional formula (2.2), we get

$$P(A \mid B) = \frac{P(A \cap B)}{P(B)} = \frac{w/N}{b/N} = \frac{2/36}{18/36} = \frac{2}{18}.$$

Similarly, the conditional probability, $P(B \mid A)$ for these events is

$$P(B \mid A) = \frac{P(B \cap A)}{P(A)} = \frac{w/N}{a/N} = \frac{w}{a} = \frac{2}{6}.$$

Out of the six sample points satisfying the condition $A$, only two also have sums that are odd numbers.

As a final example, notice that it is not necessary to know all the single probabilities of the events involved, in order to find a conditional probability. Suppose that in a production process, two parts of a particular product are produced simultaneously; and it is known that the probability that both parts were defective is

$$P(D_1 \cap D_2) = 0.05.$$

If the probability that one part is defective is 0.15, then the probability that the second is defective *given that* the first is defective equals

$$P(D_2 \mid D_1) = \frac{P(D_1 \cap D_2)}{P(D_1)} = \frac{0.05}{0.15} = \frac{1}{3}.$$

## 2.5 PROBABILITY OF AN INTERSECTION

In using Formula (2.2), we have assumed that $P(A \cap B)$, $P(B)$, and $P(A)$ are known values, and that a conditional probability, $P(B \mid A)$ or $P(A \mid B)$ is an unknown to be found. But in many problems it is often much more difficult to determine (or to think of) the probability of the intersection of two events than it is to determine the conditional probability of one event, given another. For instance, you might be able to assess quite accurately the conditional probability of making the Dean's List given that you receive an "A" in statistics, but you may not know the probability that *both* these events will occur. Formula (2.2) can be rewritten to handle difficulties of this nature by merely solving this equation for $P(A \cap B)$. The resulting formula is called the *General Rule of Multiplication*, and provides a rule for finding the probability of an intersection.

*General rule of multiplication:*

$$P(A \cap B) = P(A)P(B \mid A) = P(B)P(A \mid B). \qquad (2.3)$$

The first part of Formula (2.3) can be interpreted as follows: The probability that *both* $A$ and $B$ take place is given by two occurrences—first, event $A$ takes place, with probability $P(A)$, and then event $B$ takes place on the condition that $A$ has already occurred, with probability $P(B \mid A)$. The probability that both occurrences take place is the *product* of these two probabilities, or $P(A)P(B \mid A)$. For example, in the production problem just presented, if the probability of receiving one defective component is $P(D_1) = 0.15$, and the probability of receiving a second defective component, given that one has already occurred, is $P(D_2 \mid D_1) = \frac{1}{3}$, then

$$P(D_1 \cap D_2) = \tfrac{1}{3}(0.15) = 0.05.$$

The probability that the intersection of two (or more) events occurs in a given experiment is often referred to as the *joint probability* of these events, where the term "joint probability" implies that the events under consideration take place in the same replication of an experiment. Depending on the nature of the experiment, events which occur jointly do not necessarily take place at identical points in calendar or clock time. For example, the results of drawing first one card, and then another, from a deck of cards can be considered a joint occurrence even though the two cards are not drawn simultaneously. If $C_1$ = club on the first card and $C_2$ = club on the second card, then $P(C_1 \cap C_2)$ is the joint probability that both cards are clubs. The value of $P(C_1)$ obviously equals 13/52. After drawing one club (and not replacing it), 51 cards remain, 12 of which are clubs; hence,

$$P(C_2 \mid C_1) = \tfrac{12}{51},$$

and

$$P(C_1 \cap C_2) = P(C_1)P(C_2 \mid C_1) = \tfrac{13}{52} \times \tfrac{12}{51} = \tfrac{1}{17}.$$

It is important to note that writing $P(E_1 \cap E_2)$ or $P(E_2 \cap E_1)$ does not imply any ordering, over time, of the two events $E_1$ and $E_2$, but only that these events both occur in a single trial of the experiment. Therefore, $P(E_1 \cap E_2)$ must equal $P(E_2 \cap E_1)$. Also, writing $P(E_2 \mid E_1)$ does not imply that $E_1$ precedes $E_2$ in chronological order. For instance, it is just as legitimate to determine the probability that two cars on a two-lane highway will "pass safely" (event $A$) given that one driver is "legally drunk" (event $B$) by $P(A \mid B)$, as it is to determine the probability that one of the drivers is legally drunk given that the two cars pass safely, $P(B \mid A)$.

Although the notation becomes rather cumbersome and the analysis more complex, it is possible to extend the rules for conditional and joint probability to apply to problems involving three or more events. To illustrate the concepts involved, consider a single, relatively simple example, that of determining the probability of receiving a club on *each* of the first three draws (without replacement) from a deck of cards. This probability can be broken down into three occurrences: (1) receiving a club on the first draw, (2) receiving a club on the second draw *given that* a club was received on the first draw, and (3) receiving a club on the third draw *given that* a club was received on *both* the first and second draw. The product of the probability of these three occurrences gives the joint probability of $C_1 \cap C_2 \cap C_3$,

$$P(C_1 \cap C_2 \cap C_3) = P(C_1)P(C_2 \mid C_1)P(C_3 \mid C_1 \cap C_2)$$

$$= \tfrac{13}{52} \times \tfrac{12}{51} \times \tfrac{11}{50} = \tfrac{11}{850}$$

**Special Case of Independence**

Two events are independent if the occurrence (or nonoccurrence) of each event cannot influence (or be influenced by) the occurrence (or nonoccurrence) of the other event. That is, the probability of any given event must be unaffected by the probability of the other event in order for independence to hold. Dependence implies just the opposite, namely, that the occurrence of one event *is* influenced by the occurrence of one or more other events being considered. For example, the probability of making the Dean's List is not independent of the probability of receiving an "A" in statistics. On the other hand, the probability that you receive an "A" in statistics is probably independent of whether the Pirates win the National League pennant.

Similarly, the probability of drawing a club on the second draw from a deck of cards is dependent on the result of the first draw if the first card drawn is not replaced before the second card is drawn. However, if that first card is replaced, then the two events $C_1$ = Club on the first draw, and $C_2$ = Club on the second draw would be independent. Thus, we see again that it is very important that an experiment be carefully defined (does the process include replacement, or not?) before probabilities can be determined.

Formula (2.3) for the probability of an intersection is a general rule which applies to either independent or dependent events. However, if the event $A$ does not influence the occurrence or nonoccurrence of the event $B$, then a simpler formula for the probability of an intersection can be used. If the two events $A$ and $B$ are independent, then the probability that $B$ takes place *given $A$* must equal the *unconditional probability* of $B$. Thus, we can define independent events $A$ and $B$ as follows,

*Definition of independence:*
$$P(B \mid A) = P(B) \quad \text{and} \quad P(A \mid B) = P(A).$$

   (2.4)

Since Formula (2.4) holds true for independent events, the joint probability of two independent events must equal the product of their unconditional probabilities. That is, if $A$ and $B$ are independent, then,

*Special case of multiplication rule (given independent events):*
$$P(A \cap B) = P(A)P(B) = P(B)P(A).$$

   (2.5)

To present a simple example, repeated flips of a coin are usually considered as independent events. Thus, if we denote the probability of receiving a head on both the first and second flips of a fair coin as $P(H_1 \cap H_2)$, then it must be true that

$$P(H_1 \cap H_2) = P(H_1)P(H_2 \mid H_1) = P(H_1)P(H_2) = (\tfrac{1}{2})(\tfrac{1}{2}) = \tfrac{1}{4}.$$

We can calculate $P(H_1 \cap H_2)$ using the special-case formulas for independence since the outcome of the first flip does not influence the outcome of the second flip. That is, the probability of a head on the second flip equals $\tfrac{1}{2}$ regardless of whether a head or tail occurred on the first flip. In terms of Formula (2.4), this can be written,

$$P(H_2 \mid \text{Outcome on first flip}) = P(H_2) = \tfrac{1}{2}.$$

For another example, suppose a soft-drink manufacturer uses two machines in the capping process. The probability of a defective cap from either of these two machines is 1/1000. If the machines work independently, and one bottle is inspected at random from each machine, then the probability that the bottle capped on machine I is defective is not affected by whether or not the bottle capped on machine II is defective. Therefore,

$$P(D_2 \mid D_1) = P(D_2) = \tfrac{1}{1000}.$$

Also, the probability that both capped bottles are defective follows the special case of the multiplication rule, Formula (2.5):

$$P(D_1 \cap D_2) = P(D_1)P(D_2) = (\tfrac{1}{1000})(\tfrac{1}{1000}).$$

There is a chance of one in a million that *both* selected bottles were capped defectively.

The joint probability of more than two independent events is a simple extension of Formula (2.5). If a sequence of events, $E_1, E_2, \ldots, E_k$, are independent, then,

*Special multiplication rule for $k$ independent events:*

$$P(E_1 \cap E_2 \cap \cdots \cap E_k) = P(E_1)P(E_2) \cdots P(E_k) \tag{2.6}$$

Either of the preceding two examples could be extended to illustrate the use of this formula. Consider the experiment of flipping a fair coin $k$ times. Then, the probability of receiving a head on each of the $k$ flips of the coin is

$$P(H_1 \cap H_2 \cap \cdots \cap H_k) = P(H_1)P(H_2) \cdots P(H_k) = (\tfrac{1}{2})^k,$$

since each flip is independent.

## 2.6 PROBABILITY OF A UNION

Since a union, $A \cup B$, is itself a well-defined event, its probability can be found by the simple method of identifying the proportion of all sample points which are included in this union.

Using the same formal notation as earlier, the total number of sample points is $N$. As before, the number satisfying event $A$ is $a$, the number satisfying event $B$ is $b$, and the number satisfying *both* $A$ and $B$ is $w$. With a little intuitive cleverness, we can determine the number of sample points in the union, and then the probability of a union. Figure 2.10 is useful in representing this problem.

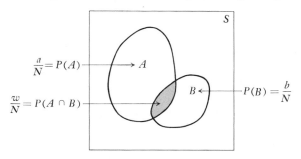

**Fig. 2.10.** Illustration of the general rule of addition:

$$P(A \cup B) = \frac{a + b - w}{N} = \frac{a}{N} + \frac{b}{N} - \frac{w}{N} = P(A) + P(B) - P(A \cap B).$$

The points in $A \cup B$ certainly include all $a$ points in $A$ and all $b$ points in $B$. However, the sum of points $a + b$ is greater than the total number in $A \cup B$ because some points are counted twice. The points being double-counted are all those that are in *both* $A$ and $B$, namely $w$ points. Consequently, to get a true count of all points in $A \cup B$, we must subtract from $(a + b)$ the number of points in the intersection $A \cap B$. The resulting number of points in the union $A \cup B$ is $a + b - w$. Since the total number of points in the sample space $S$ is $N$, then the probability of the union is

$$P(A \cup B) = \frac{(a + b - w)}{N}.$$

Rewriting, we have

$$P(A \cup B) = \frac{a}{N} + \frac{b}{N} - \frac{w}{N}$$

which gives the following rule:

*General rule of addition:*

$$P(A \cup B) = P(A) + P(B) - P(A \cap B).$$

(2.7)

Let us now find the probability of the union of two events using the example experiment of tossing a pair of dice (refer to the sample space given in Fig. 2.4).

Suppose we define the events of interest to be

$$A = \{x \mid x < 5\},$$
$$B = \{x \mid x \text{ is an odd number}\}.$$

We previously determined that $N = 36$, $a = 6$, $b = 18$, and $w = 2$, so that

$$P(A) = \tfrac{6}{36}, \qquad P(B) = \tfrac{18}{36}, \qquad \text{and} \qquad P(A \cap B) = \tfrac{2}{36}.$$

Therefore, the probability of the union of $A$ and $B$, according to Formula (2.7), is,

$$P(A \cup B) = P(A) + P(B) - P(A \cap B) = \frac{6 + 18 - 2}{36} = \frac{22}{36}.$$

You should be able to find exactly 22 different sample points in Fig. 2.4 representing values which are either smaller than five, or odd, or both. The two pairs that are involved in double-counting are (1, 2) and (2, 1).

For another example, consider the experiment of drawing a single card from a deck where the events $A$ and $B$ are

$$A = (\text{Draw an ace}) \qquad \text{and} \qquad B = (\text{Draw a spade}).$$

Here $N = 52$, $a = 4$, $b = 13$, and $w = 1$. Thus,

$$P(A \cup B) = \tfrac{4}{52} + \tfrac{13}{52} - \tfrac{1}{52} = \tfrac{16}{52}.$$

If we had not accounted for the double-counting of the simple event, draw the ace of spades, the probability of the union (Ace $\cup$ Spade) would have been overstated.

### Special Case for Mutually Exclusive Events

In the situation where the events $A$ and $B$ are mutually exclusive, the danger of double counting is eliminated. That is, if $A$ and $B$ are mutually exclusive events, then there is *no overlap* of sample points in both $A$ and $B$ which can be counted twice; and so the intersection $A \cap B$ in this special case is a *null* set, with probability zero. We can thus write the probability of $A \cup B$, when $A$ and $B$ are mutually exclusive, as the sum of $P(A)$ and $P(B)$.

> *Special case of addition rule (given mutually exclusive events):*
> $$P(A \cup B) = P(A) + P(B).$$      (2.8)

The Venn diagram in Fig. 2.11 illustrates the relationship between the sets $A$ and $B$ when these sets are mutually exclusive.

An example of the type of problem which involves Formula (2.8) would be that of determining the probability that *either* the Pirates *or* the Cubs win the National League pennant this year (some people would say this probability is close to zero). Another example would be that of determining the probability of drawing *either* an ace *or* a king on a single draw from a deck of cards. Since $P(\text{Ace}) = 1/13$, and $P(\text{King}) = 1/13$, the probability of the union is

$$P(\text{Ace} \cup \text{King}) = P(\text{Ace}) + P(\text{King}) = \tfrac{1}{13} + \tfrac{1}{13} = \tfrac{2}{13}.$$

**The Relation of Mutually Exclusive Events to Independence**

Beginning students of statistics often become confused between the concept of independent events and that of mutually exclusive events. It is vital for the reader to understand the difference between these concepts, and to understand the implications of both of them for probability theory. If two events $A$ and $B$ are *not* mutually exclusive, then they can be either dependent or independent events. If $A$ and $B$ *are* mutually exclusive, then these two events are always dependent. For example, consider the experiment "draw a single card from a deck of 52." In this example we can show that the event "card is an ace" is independent of the event "card is a spade" by using Formula (2.4)—try it! These events are certainly not mutually exclusive, however, since the draw could be *both* an ace and a spade. We will show in the discussion below that if two events are known to be mutually exclusive, then they cannot be independent.

Suppose two events, $A$ and $B$, are assumed to be independent; then we know that

$$P(A \mid B) = P(A) = \frac{a}{N} \quad \text{and} \quad P(B \mid A) = P(B) = \frac{b}{N}.$$

On the other hand, assume that $A$ and $B$ are *mutually exclusive* events, as shown in Fig. 2.11. What are the conditional probabilities in this case? For the event $(A \mid B)$, the relevant sample space is restricted to only those $b$ points in $B$ (because

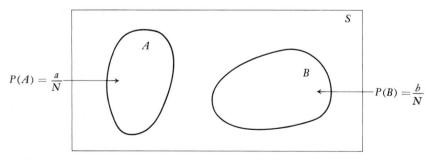

**Fig. 2.11.** Illustration of the special rule of addition for mutually exclusive events:

$$P(A \cap B) = 0, \quad \text{so } P(A \cup B) = P(A) + P(B).$$

event $B$ has occurred). How many of these are also in $A$? Since $A$ and $B$ are mutually exclusive, the answer is *none*; and so the conditional probability, $P(A \mid B)$, equals zero. In other words, it is impossible to satisfy event $A$ given that a point in $B$ is already selected. Similarly, $P(B \mid A) = 0$, since if one is restricted to points in $A$, he cannot find even one point that also satisfies event $B$. By comparing the values of the conditional probabilities under the two different assumptions of independent versus mutually exclusive events, we can conclude that mutually exclusive events can never be independent events since:

$$P(A \mid B) = \frac{a}{N} \quad \text{for independent events.}$$

$$P(A \mid B) = 0 \quad \text{for mutually exclusive events.}$$

### Extensions of the Additional Laws

Both Formulas (2.7) and (2.8) can be extended to three or more events in a union, although the notation and formulas become tedious in the general case. If the events $E_1, E_2, \ldots, E_k$, are *mutually exclusive*, the probability of the union is simply the sum of the probabilities of each event. That is,

$$P(E_1 \cup E_2 \cup \cdots \cup E_k) = \sum_{i=1}^{k} P(E_i).$$

If the events are *not* mutually exclusive, then the task of eliminating the double-counting becomes complex. The reader should refer to Fig. 2.8(b) and satisfy himself that, for the union of three events, the general rule is

$$P(A \cup B \cup C) = P(A) + P(B) + P(C) - P(A \cap B) - P(A \cap C)$$
$$- P(B \cap C) + P(A \cap B \cap C).$$

You must subtract the proportion of sample points that are double-counted but then *add back in* the proportion of sample points that are doubly-omitted. To extend this concept to more than three events becomes mind-boggling and tongue-twisting.

### 2.7 MARGINAL PROBABILITY

In a number of circumstances it is convenient to think of a single event as always occurring jointly with other events, where these other events are assumed to influence the probability of the occurrence of the first event. For instance, it may be helpful not only to identify defective items resulting from a production process, but also to specify exactly *which* machines (or which workers) produced these defectives. Insurance companies are interested in not only the amount of damage associated with each automobile accident but, among other things, the city where the accident took place, and the age and sex of the driver. In such situations, the probability of the event in question (e.g., the probability of producing a defective item across all machines and all workers, or the probability of an accident involving at least $x$ dollars of damage across all cities and all drivers) may not be known directly, but can be calculated by summing its chance of occurrence in combination with the other relevant factors identified in the problem.

To offer a simple illustration, suppose we consider the problem of producing a certain type of battery in three different plants with different equipment and employees. Suppose the weekly average of the number of batteries produced in these three plants, denoted by $E_1$, $E_2$, and $E_3$, is 500, 2000, and 1500, respectively. Further let's assume the probability that a defective ($D$) is produced in *each* of the three plants is:

$$P(D \mid E_1) = 0.020, \qquad P(D \mid E_2) = 0.015, \qquad \text{and} \qquad P(D \mid E_3) = 0.030.$$

Suppose the batteries produced by the three plants supply one automaker. That is, the automaker receives 4000 batteries weekly and the probability that a randomly selected battery would have originated in each plant is

$$P(E_1) = \tfrac{500}{4000}, \qquad P(E_2) = \tfrac{2000}{4000}, \qquad \text{and} \qquad P(E_3) = \tfrac{1500}{4000}.$$

What is the probability that the battery used by the automaker in a randomly selected car is defective? This probability, $P(D)$, is a marginal probability, and its value can be determined by the special rules for marginal probabilities. Figure 2.12 helps in discovering a logical way of finding this probability.

The circle labeled $D$ in Fig. 2.12 represents the proportion of defectives in the sample space $S$. There are three shaded areas in this circle, representing the intersection of $D$ with $E_1$, $E_2$, and $E_3$. The probability of being in these three intersections is $P(D \cap E_1)$, $P(D \cap E_2)$, and $P(D \cap E_3)$. Finally, since there is no overlap among the three intersections, the probability of $D$ is their sum, or $P(D) =$

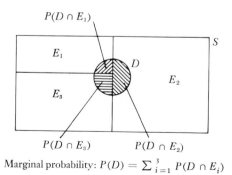

Marginal probability: $P(D) = \sum_{i=1}^{3} P(D \cap E_i)$

**Fig. 2.12.** Illustration of a marginal probability.

$P(D \cap E_1) + P(D \cap E_2) + P(D \cap E_3)$. This relationship can be made general by the following definition.

**Definition.** *If D represents an event such that one of the mutually exclusive events $E_1, E_2, \ldots, E_k$, must always occur jointly with any occurrence of D, then the probability of D is called a marginal (or unconditional) probability, and its value may be determined by the rule*

$$P(D) = \sum_{i=1}^{k} P(D \cap E_i).$$

From the General Rule of Multiplication, Formula (2.3), we know that

$$P(D \cap E_i) = P(E_i)P(D \mid E_i).$$

Thus, the formula for a marginal probability can be written as:

---

*Marginal probability:*

$$P(D) = \sum_{i=1}^{k} P(D \cap E_i) = \sum_{i=1}^{k} P(E_i)P(D \mid E_i). \qquad (2.9)$$

---

Using Formula (2.9) and the probability values from page 71, we can now calculate $P(D)$.

$$P(D) = P(E_1)P(D \mid E_1) + P(E_2)P(D \mid E_2) + P(E_3)P(D \mid E_3)$$
$$= (\tfrac{500}{4000})(0.020) + (\tfrac{2000}{4000})(0.015) + (\tfrac{1500}{4000})(0.030)$$
$$= 0.0025 + 0.0075 + 0.01125 = 0.02125.$$

Slightly more than two percent of the batteries are defective.

As another example of the use of Formula (2.9), consider the problem of estimating the probability that Carl Yastrzemski (Boston Red Sox player) will get a hit on a randomly selected at bat, if it is known that in the past the relative frequency he got a hit off a righthanded pitcher was 0.315, and the relative frequency he got a hit off a lefthanded pitcher was 0.262.* If 0.315 is taken as an estimate of the probability he gets a hit given a righthanded pitcher (i.e., $P(H \mid R) = 0.315$), 0.262 as the probability of a hit given a lefthanded pitcher (i.e., $P(H \mid L) = 0.262$), and we assume the probability that Yastrzemski will face righthanded and lefthanded pitchers to be $P(R) = 0.75$ and $P(L) = 0.25$, then

$$P(\text{Hit}) = P(R)P(H \mid R) + P(L)P(H \mid L)$$
$$= 0.75(0.315) + 0.25(0.262) = 0.302.$$

---

* Data based on Yastrzemski's major-league career record.

## 2.8 APPLICATION OF PROBABILITY THEORY: AN EXAMPLE

Suppose a contractor producing delicate electronic components essential to the manufacturing of certain computer equipment cannot determine whether the component part he produces has been assembled correctly without tearing the component apart, and, in the process, destroying the usefulness of that component. He can, however, purchase a machine which its makers claim will help detect defective components. This machine, which indicates only that the component appears to be good ($+$) or that it appears to be defective ($-$), is not infallible, as it will sometimes indicate positive when the component is defective, or will indicate negative when the component is good. To determine the ability of the machine to distinguish between good and defective items, 400 randomly selected components were first tested by the machine, and then torn apart to see whether they were good or defective. The results of this research are given in Table 2.2. In this case the sample space is discrete and finite (four outcomes). Suppose, for this example, that we define the random variable $x$ as follows:

$$x = \begin{cases} 1 & \text{if the test is positive and the component is good,} \\ 2 & \text{if the test is positive and the component is defective,} \\ -1 & \text{if the test is negative and the component is good,} \\ -2 & \text{if the test is negative and the component is defective.} \end{cases}$$

Using this definition we can now calculate the probability of the intersection and union of several events, as well as a number of marginal probabilities.

### Intersections

The probability of each of the values of $x$ is the probability of the *intersection* of two events, and can be determined directly from the data in Table 2.2. These results, shown in Table 2.3, are seen to satisfy the two properties of a probability

**Table 2.2**

| State of component | Results of test | | Sum across row |
|---|---|---|---|
| | Positive | Negative | |
| Good | 342 | 18 | 360 |
| Defective | 8 | 32 | 40 |
| Sum of column | 350 | 50 | 400 |

distribution (that is, they sum to 1.0, and they are all between zero and one).

**Table 2.3** Probabilities of the four values of the random variable associated with test results in Table 2.2

| State of component | Results of test | | Sum across row |
|---|---|---|---|
| | Positive | Negative | |
| Good | $P(x = 1) = 0.855$ | $P(x = -1) = 0.045$ | 0.900 |
| Defective | $P(x = 2) = 0.020$ | $P(x = -2) = 0.080$ | 0.100 |
| Sum of column | 0.875 | 0.125 | 1.000 |

**Marginals**

The four probabilities in Table 2.3 can now be used to determine various marginal probabilities. For instance, although the marginal (unconditional) probability of a good item can be determined directly (from Table 2.3) to be

$$P(\text{Good}) = \tfrac{360}{400} = 0.90,$$

its value can also be determined via Formula (2.9) as follows:

$$\begin{aligned}
P(\text{Good}) &= P(x = +1) + P(x = -1) \\
&= P(\text{Good} \cap \text{Pos.}) + P(\text{Good} \cap \text{Neg.}) \\
&= 0.855 + 0.045 = 0.90.
\end{aligned}$$

Similarly, the probability that the test reads positive is

$$\begin{aligned}
P(\text{Pos.}) &= P(x = +1) + P(x = +2) \\
&= P(\text{Good} \cap \text{Pos.}) + P(\text{Def.} \cap \text{Pos.}) \\
&= 0.855 + 0.020 = 0.875.
\end{aligned}$$

**Unions**

The probability of the union of two events can be determined by using the General Rule of Addition, Formula (2.7). For example, the probability of *either* a good component *or* a positive test is:

$$\begin{aligned}
P(\text{Good} \cup \text{Pos.}) &= P(\text{Good}) + P(\text{Pos.}) - P(\text{Good} \cap \text{Pos.}) \\
&= 0.90 + 0.875 - 0.855 = 0.92.
\end{aligned}$$

The manufacturer in our problem is primarily interested in the probability that a particular component is good *given the condition* that the machine indicates positive or negative. These values can be determined from Table 2.2 by noting that out of 350 components which tested positive, 342 were good, and of the 50 which tested negative, 32 were defective; therefore,

$$P(\text{Good} \mid \text{Pos.}) = \tfrac{342}{350} = 0.977,$$

and

$$P(\text{Def.} \mid \text{Neg.}) = \tfrac{32}{50} = 0.64.$$

Note that these two probabilities do not add to *one* since they are based on different conditional events. However, the value of $P(\text{Good} \mid \text{Neg.})$ must equal

$$1 - P(\text{Def.} \mid \text{Neg.}) = 1 - 0.64 = 0.36,$$

since these events are complementary.

The results of the machine test in this problem are obviously not independent of the state of the component. If they were independent, $P(\text{Good} \mid \text{Pos.})$ would have to equal $P(\text{Good})$, and $P(\text{Def.} \mid \text{Neg.})$ would have to equal $P(\text{Def.})$, which is not the case. But how much better off is the manufacturer by knowing the results of the test? He is in great shape if the test is positive, as he can then guess the component to be good and be correct 97.7 percent of the time. After a negative indication, however, he will be correct only 64 percent of the time if he guesses the component to be defective. Fortunately, a positive test will occur most often (87.5 percent) and a negative test relatively infrequently (12.5 percent). Multiplying 97.7 by 0.875 and 64.0 by 0.125 gives the total percent of the time he makes a correct assessment, assuming he always accepts the results of the machine test. This weighting is just a common-sense use of the formal rule for adding up all the joint occurrences of those events in which the test indicates the *correct* state of the component,

$$
\begin{aligned}
P[(\text{Good} \cap \text{Pos.}) &\cup (\text{Def.} \cap \text{Neg.})] \\
&= P(\text{Good} \cap \text{Pos.}) + P(\text{Def.} \cap \text{Neg.}) \\
&= P(\text{Pos.})P(\text{Good} \mid \text{Pos.}) + P(\text{Neg.})P(\text{Def.} \mid \text{Neg.}) \\
&= 0.977(0.875) + (0.64)(0.125) \\
&= 0.855 + 0.080 \\
&= 0.935.
\end{aligned}
$$

If we now knew how much this machine costs (to buy and operate), and how much the firm's revenue will increase if it is better able to distinguish between good and defective components, then we could determine whether the machine is worth purchasing. Without this machine, if the manufacturer always presumes that all components are good, he will be correct 90 percent of the time, since

$$P(\text{Good}) = \tfrac{360}{400} = 0.90.$$

## 2.9 BAYES' RULE

One of the most interesting (and controversial) applications of the rules of probability theory involves estimating unknown probabilities and making decisions on
the basis of new (sample) information. Statistical decision theory is a new field of
study which has its foundations in just such problems. Chapter 9 investigates the
area of statistical decision theory in some detail; this section describes one of the
basic formulas of the area, *Bayes' rule.*

An English philosopher, the Reverend Thomas Bayes (1702–1761), was one
of the first to work with rules for revising probabilities in the light of sample
information. Bayes' research, published in 1763, went largely unnoticed for over
a century, and only recently has attracted a great deal of attention. His contribution
consists primarily of a unique method for calculating conditional probabilities.
The so-called "Bayesian" approach to this problem addresses itself to the question
of determining the probability of some event, $E_i$, *given that* another event, $A$, has
been (or will be) observed; i.e., determining the value of $P(E_i \mid A)$. The event $A$
is usually thought of as new information, so that Bayes' rule is concerned with
determining the probability of an event given certain new information, such as
that obtained from a sample, a survey, or a pilot study. For example, a sample
output of 3 defectives in 20 trials (event $A$) might be used to estimate the probability
that a machine is not working correctly (event $E_i$).

Probabilities before revision by Bayes' rule are called *a priori*, or simply *prior*
probabilities, because they are determined before the new information is taken
into account. Prior probabilities may be either objective or subjective values. A
probability which has undergone revision in the light of new information (via
Bayes' rule) is called a *posterior probability*, since it represents a probability calculated *after* this information is taken into account. Posterior probabilities are
always conditional probabilities, the conditional event being the new information.
Thus, by using Bayes' rule, a prior probability, which is an unconditional probability, becomes a posterior probability, which is a conditional probability. In
order to calculate such posterior probabilities, we will first derive Bayes' rule for
the general problem of determining $P(E_i \mid A)$.

Recall that earlier in this chapter the order of events was shown to be im-
material in calculating joint probabilities, which implies that $P(E_i \cap A)$ must be
equivalent to $P(A \cap E_i)$; therefore, the following relationships must hold true
(see Formula (2.3)):

$$P(E_i \cap A) = P(A)P(E_i \mid A),$$
$$P(E_i \cap A) = P(E_i)P(A \mid E_i).$$

If the above two formulas hold, then it must also be true that the two righthand side representations must be equal.

$$P(A)P(E_i \mid A) = P(E_i)P(A \mid E_i).$$

We can now solve for $P(E_i \mid A)$ directly by dividing both sides by $P(A)$:

$$P(E_i \mid A) = \frac{P(E_i)P(A \mid E_i)}{P(A)}. \tag{2.10}$$

The numerator of Formula (2.10) represents the probability that $A$ and $E_i$ will both occur, while the denominator is the probability that $A$ alone will occur. If both these probabilities are calculable, then the conditional probability of the event $E_i$, given some new information $A$, can be determined.

For example, suppose questionnaires are sent out to seek information about the use of the family automobile. Suppose we assume an *a priori* probability that one-half of the questionnaires will be filled out by rural households. Also, we estimate that 0.3 of the questionnaires will be completed by high-income households. Finally, from previous experience with questionnaires sent to rural families, we believe that only 0.2 of those returned will be from high-income households. Based on this description, the following symbols seem useful:

$R$ = rural household;        $H$ = high-income household;
$U$ = nonrural household      $L$ = not a high-income household
      (urban);                      (lower income).

The following probabilities are known from the given information:

$$P(R) = 0.5, \qquad P(H) = 0.3, \qquad P(H \mid R) = 0.2.$$

By the rule of complements, we deduce,

$$P(U) = 0.5, \qquad P(L) = 0.7, \qquad P(L \mid R) = 0.8.$$

Now suppose one questionnaire is received and the location code has been omitted so that it is not known whether it is from a rural or urban household. Our prior probabilities suggest that the probability is 0.5 that it comes from a rural household and 0.5 for an urban household. From analysis of the responses in the questionnaire, suppose the questionnaire was obviously completed by a high-income household. How does this new information affect our probabilistic knowledge of whether it came from a rural or an urban household?

Formula (2.10) can be used to find the revised posterior probability that the questionnaire came from a rural household, given the information that it came from a high-income household, $P(R \mid H)$:

$$P(R \mid H) = \frac{P(R)P(H \mid R)}{P(H)} = \frac{(0.5)(0.2)}{0.3} = \frac{1}{3}.$$

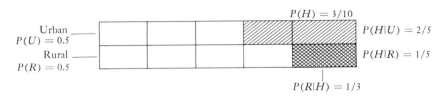

**Fig. 2.13.** Illustration for Bayes' rule.

Figure 2.13 illustrates this result by dividing the universal set into ten equal blocks to represent the probabilities in the problem. The top five represent urban households and the lower five represent rural. The shaded blocks represent the high-income households (3 out of 10). They are located in the urban or rural rows so that $\frac{1}{5}$ of the rural households are high-income, thereby satisfying the given information that

$$P(H \mid R) = 0.2.$$

Clearly, it is logical that if a questionnaire is known to come from a high-income household, (i.e., given a shaded block), then the chances are one out of three that the household would be rural. Hence, $P(R \mid H) = \frac{1}{3}$, as illustrated by the cross-hatched block in Fig. 2.13.

The probability $P(U \mid H)$ can be determined by using either Formula (2.10), or the rule of complements, or directly from the shaded areas in Fig. 2.13. One obtains $P(U \mid H) = \frac{2}{3}$. This is a posterior probability. Since the new information ($H$) is more common among urban families, the posterior probability is larger than the prior, $P(U) = \frac{1}{2}$.

By referring to Formula (2.10), we can recognize that Bayes' rule can be applied to find the posterior conditional probability only when the probabilities are known for the two events and the conditional event on the righthand side. In many cases, the probability of the event described by the new information is not known directly, since it involves other events and is really a marginal probability. Thus, the probability in the denominator of Formula (2.10) is not given explicitly. However, it can be calculated using Formula (2.9) for a marginal probability. A more general form of Bayes' rule is obtained by making this substitution of Formula (2.9) into Formula (2.10) to obtain,

$$\textit{Bayes' rule:} \quad P(E_i \mid A) = \frac{P(E_i)P(A \mid E_i)}{\displaystyle\sum_{j=1}^{k} P(E_j)P(A \mid E_j)}. \tag{2.11}$$

**Example.** The use of Formula (2.11) can be illustrated by considering two types of economic stabilization policy. The first type is fiscal policy, which is controlled by Congress and its tax and revenue legislation. The second is monetary policy which is controlled by the actions of the central bank. Let us presume that these policy decisions are independent of each other, since a legislator responds to varied demands of the populace, while the central bank is relatively autonomous and free from politics. Furthermore, let's assume that the actions of each of these groups are correct (for economic stabilization and growth) 80 percent of the time and are incorrect 20 percent of the time. Then probabilities of these groups acting correctly (for proper economic policy) at a given time are as follows:

$$P(\text{Neither correct}) = P(E_3) = (0.2)(0.2) = 0.04;$$
$$P(\text{Both correct}) = P(E_2) = (0.8)(0.8) = 0.64;$$
$$P(\text{One correct}) = P(E_1) = 1 - [P(E_3) + P(E_2)]$$
$$= 1 - (0.04 + 0.64) = 0.32.$$

Finally, we assume the probabilities that the economy follows a generally stable growth pattern due to (or in spite of) these policy actions are:

$$P\begin{pmatrix}\text{Stable growth given that one of the}\\ \text{bodies is acting correctly}\end{pmatrix} = P(\text{SG} \mid E_1) = 0.7;$$

$$P\begin{pmatrix}\text{Stable growth given that both}\\ \text{are acting correctly}\end{pmatrix} = P(\text{SG} \mid E_2) = 0.99.$$

$$P\begin{pmatrix}\text{Stable growth, given that neither}\\ \text{body is acting correctly}\end{pmatrix} = P(\text{SG} \mid E_3) = 0.4.$$

Given all this information, suppose a situation of stable economic growth is recorded for a particular period. What is the posterior probability that neither stabilizing policy was correct at that time? We need to find $P(E_3 \mid \text{SG})$. By Bayes' rule we have:

$$P(E_3 \mid \text{SG}) = \frac{P(E_3)P(\text{SG} \mid E_3)}{P(\text{SG})} = \frac{P(E_3)P(\text{SG} \mid E_3)}{\sum\limits_{j=1}^{3} P(E_j)P(\text{SG} \mid E_j)}$$

$$= \frac{P(E_3)P(\text{SG} \mid E_3)}{P(E_1)P(\text{SG} \mid E_1) + P(E_2)P(\text{SG} \mid E_2) + P(E_3)P(\text{SG} \mid E_3)}.$$

You should notice that the term in the numerator of Bayes' rule is always one of the terms included in the sum in the denominator. The other similar terms in the denominator refer to the other mutually exclusive events which can occur in conjunction with stable growth. The terms involving $E_1$ and $E_2$ refer to the occasions when one or both stabilizing policies are correct. Substituting the values given above, we find,

$$P(E_3 \mid SG) = \frac{(0.04)(0.4)}{(0.32)(0.7) + (0.64)(0.99) + (0.04)(0.4)}$$

$$= \frac{0.016}{0.224 + 0.6336 + 0.016}$$

$$= \frac{0.016}{0.8736} = 0.0183.$$

The posterior probability that neither fiscal nor monetary policy was correct in this circumstantial problem is 0.0183, as compared to the prior probability of 0.04. The prior was revised downward because the stable growth observed is more compatible with correct policy actions than with the event $E_3$. The posterior probabilities for events $E_1$ and $E_2$ can be determined similarly as follows:

$$P(E_1 \mid SG) = \frac{0.224}{0.8736} = 0.2564, \quad \text{down from } P(E_1) = 0.32;$$

$$P(E_2 \mid SG) = \frac{0.6336}{0.8736} = 0.7253, \quad \text{up from } P(E_2) = 0.64.$$

Note that the *sum* of the posterior probabilities for the three exhaustive and mutually exclusive policy events, given that stable growth was observed, is 1.00, as it must be.

## 2.10  PROBABILITIES OF REPEATED TRIALS

Some of the examples given in this chapter have involved experimental situations in which more than two events can occur and where the events in question are combinations of the occurrence of simple events. Determining probabilities in such problems generally depends on a knowledge of the number of elementary events in such combinations, but this number may not be easily obtained by simply listing all the elements. Determining the number of elementary events (outcomes) for experiments which are repeated many times is a similar problem. Some new mathematical terms and methods are useful in simplifying the calculations in such situations.

### Basic Counting Rule

A basic counting rule is useful for determining the number of distinct outcomes resulting from an experiment involving two or more steps, where each step has several different possible outcomes. If the first step of such an experiment can result in $k$ different outcomes, and there are $m$ different results possible on the second step of the experiment, then the total number of outcomes possible for the two successive steps is $k \cdot m$. The extension of this rule to a three-step process, or more, is done similarly.

For example, suppose a soft-drink manufacturer is contemplating a new marketing scheme involving selection of one of four advertising media (radio, TV, magazine, or store displays), and a choice of one of five new packaging designs (flip-top cans, big-mouth bottles, disposable bottles, quart-size regular bottles, or a powdered instant mix). The basic counting rule can be used to determine the total number of different combinations of advertising and packaging. This number is $k \cdot m = 4 \cdot 5 = 20$.

### Exponential Counting Rule

If a particular step in an experimental process may result in $m$ different outcomes and if this step occurs $k$ times, then the total number of possible outcomes is determined by using the exponential counting rule. This rule is derived by repeating the basic counting rule over and over, as follows:

$$\underbrace{m \times m \times \cdots \times m}_{k \text{ factors}} = m^k.$$

For example, suppose an investor receives a quarterly dividend payment that he may either spend or reinvest. The number of choices he has each quarter is $m = 2$. If we propose that his decisions over two years (8 quarters) be studied, then the number of steps of interest is $k = 8$. Using the exponential counting rule, there would be $m^k = 2^8 = 256$ different possible patterns of decisions to spend or reinvest.

### Permutations

A *permutation* is an arrangement or ordering of outcomes. The number of such arrangements or orderings is often important in determining the probability of some set of particular events. *The number of permutations of m outcomes is the maximum number of different ways that these m outcomes can be arranged or ordered.*

Permutations are involved in probability questions related to the ranking of a set of items, such as investment opportunities, or college football teams. For example, suppose we are interested in the number of ways three candidates (A, B, and C) for election to a particular office might be ranked by the voters. Six different rankings or permutations are possible, as shown in Table 2.4. If these permutations are all equally likely, the probability that any *one* will occur is $\frac{1}{6}$.

**Table 2.4** Permutations example

|       |   | 1 | 2 | 3 | 4 | 5 | 6 |
|-------|---|---|---|---|---|---|---|
|       | 1 | A | A | B | B | C | C |
| Rank  | 2 | B | C | A | C | A | B |
|       | 3 | C | B | C | A | B | A |

If there are four candidates instead of three, we could show (but won't) that there are 24 different permutations. As the number $N$ of candidates gets larger, we soon get tired of listing all possibilities and need an easier way to determine the number of permutations.

Consider the number of different ways each rank can be filled in Table 2.4. There are three ways to fill the first rank (with A, B, or C). Now suppose A is ranked first; then the second rank can be filled in two ways (with either B or C). Once the second rank is filled, there is only *one* way to fill the third. The number of permutations is thus the total ways of filling position one (which is 3 in this example) times the ways of filling position two (2), times the ways of filling position three (1), or

$$(3)(2)(1) = 6.$$

Similarly, with four objects to rank, the number of permutations is

$$(4)(3)(2)(1) = 24.$$

This type of reasoning can be extended to any number of objects. If there are $m$ objects, there are $m$ ways to fill the first position, $(m - 1)$ ways to fill the second, and so forth, the product of these terms being the total number of permutations:

$$m(m - 1)(m - 2) \cdots (1).$$

**Factorials**

Usually the symbol $m!$ is reserved for this type of product, where "!" is read "factorial," and where $m$ can be any integer greater than or equal to zero. For example,

$$5! = (5)(4)(3)(2)(1) = 120,$$

and

$$10! = (10)(9)(8) \cdots (1) = 3,628,800.$$

*By definition*, $0! = 1$. Now, if we have $m$ objects and want to determine the number of permutations of *all* $m$ of these objects (as we did above), this number, which is denoted by the symbol $_mP_m$, is called *the number of permutations of m objects taken m at a time.*

---

*Permutations of m objects taken m at a time:*     $_mP_m = m!$     $^*$     (2.12)

---

Permutation problems often do not involve all $m$ objects at the same time, but rather some subset of these objects. We will denote the number of objects in the subset by $x$. These problems are concerned with the number of permutations of $m$ objects when only $x$ of these objects are considered at any one time (i.e., $x$ positions to fill). In general, there are $x$ terms to multiply when there are $x$ positions to fill. The *last* term in the progression $m(m - 1)(m - 2) \ldots$ must therefore always be $(m - x + 1)$ in order to have only $x$ numbers to multiply (since $x$ integers are included between $m - x + 1$ and $m$). The difference between the number of permutations of $m$ items taken $m$ at a time and $m$ items taken $x$ at a time is that the first sequence of multiplicative integers goes from $m$ down to 1 whereas the latter only goes from $m$ down to $m - x + 1$, and excludes the remaining sequence of multiplicative integers from $m - x$ down to 1. In order to exclude these $(m - x)!$ values from the sequence

$$m! = m(m - 1)(m - 2) \cdots (m - x + 1)(m - x)!,$$

we need to divide $m!$ by $(m - x)!$. Thus, if we let $_mP_x$ represent the number of permutations of $m$ objects considered (or taken) $x$ at a time, then

---

*Permutations of m objects taken x at a time:*     $_mP_x = \dfrac{m!}{(m - x)!}.$     (2.13)

---

To illustrate the use of Formula (2.13), suppose a list of ten investments for a business firm is presented to the board of control, and each member is asked to rank the five projects that he considers to represent the best opportunities. How many conceivable different rankings of 10 items taken 5 at a time exist? In this case, using $m = 10$, and $x = 5$,

$$_mP_x = {_{10}P_5} = \frac{10!}{(10 - 5)!} = 30{,}240.$$

Hopefully, there will be enough consensus among the members of the board of control so that a much smaller number of different rankings will actually be suggested.

---

$^*$ Note that we can also write $m!$ as $m(m - 1)!$. For example, $7! = 7(6!) = 7 \cdot 6 \cdot 5 \cdot 4 \cdot 3 \cdot 2 \cdot 1$.

**Combinations**

The number of permutations of a set of objects depends on how many ways these objects can be ordered. But perhaps one cannot or does not want to be concerned with order. For instance, the order in which voters are surveyed on public issues is generally assumed to be unimportant, as is the order in which cards are received in a bridge hand, or the order in which three companies are selected out of 10 companies submitting engine designs. In these problems interest usually centers on the number of ways a specific *combination* of objects can occur, where two sets of objects are identical if they contain exactly the same elements, no matter how these objects are arranged. *The number of combinations of m outcomes taken x at a time is written as $_mC_x$, and equals the maximum number of different sets which can be collected using x out of m objects.*

There are always *fewer* combinations than permutations for a given $m$ and $x$ since different orderings do not count for combinations, but do count for permutations. It can be shown that $_mP_x$ will always be larger than $_mC_x$ by a factor of $x!$; thus,

$$_mC_x = \frac{_mP_x}{x!},$$

and by substitution for $_mP_x$ from Formula (2.13), we can write:

*Combinations of m objects taken x at a time:*   $_mC_x = \dfrac{m!}{x!(m-x)!}$    (2.14)

Formula (2.14) is often abbreviated as

$$_mC_x = \binom{m}{x},$$

where the term in the parentheses is not a fraction, but merely a different way of denoting the number of combinations of $m$ objects taken $x$ at a time. Suppose we want to select three companies for project-development awards from among ten companies submitting engine designs. In this example, $m = 10$, $x = 3$, and the number of combinations possible is

$$_{10}C_3 = \binom{10}{3} = \frac{10!}{3!(10-3)!} = \frac{10 \cdot 9 \cdot 8 \cdot (7!)}{3 \cdot 2 \cdot 1 \cdot (7!)} = 120.$$

**Probability in More Complicated Situations**

Many problems in statistics use probabilities that are determined, in part, by applying the rule for combinations, Formula (2.14). A few demonstration examples can show where this formula is useful. Reconsider the example of selecting three companies from among ten which submit designs for a low-polluting engine. Without any specific knowledge of the proposals, assume that each company is equally likely to be selected. If you own stock in four of these ten companies, what is the probability that exactly two of the three companies selected are companies whose stock you own?

The probability that a company whose stock you own will be selected first is 4/10. Given this occurrence, the conditional probability that a company whose stock you own will be the second selection is 3/9. Finally, in order for exactly two companies out of three to be selected whose stock you own, the third selection must be filled by a company whose stock you do *not* own. Given the first two selections, there are 8 remaining companies and you do not own stock in 6 of these. Thus, the conditional probability that the third selection satisfies the complicated event in question is 6/8. Using the general rule of multiplication gives the probability of this sequence of three selections:

$$\frac{4}{10} \times \frac{3}{9} \times \frac{6}{8} = \frac{72}{720} = \frac{1}{10}.$$

This probability represents only one possible combination in which two companies whose stock you own is among the three selected. There are other combinations with the same probability of occurring. For example, suppose the first company selected is one in which you do not own stock and the second and third selections are companies whose stock you do own. The probability of such a sequence of three selections will be the same as that for any other sequence;

$$\frac{6}{10} \times \frac{4}{9} \times \frac{3}{8} = \frac{72}{720} = \frac{1}{10}.$$

The question we need to ask ourselves at this point is "how many different ways can three companies be selected, where two are those in which you own stock?" The answer to this question is the number of combinations of three companies taken two (whose stock you own) at a time, which is

$$_3C_2 = \frac{3!}{2!1!} = 3.$$

Thus, since there are three combinations, each having a probability of 1/10, the probability of exactly two companies being selected whose stock you own is $3(1/10) = 3/10$.

Probability questions such as those posed in the stock example above can often be formulated by using the following rule:

$P(Complicated\ event)$

$$= P(Each\ occurrence) \times (Number\ of\ relevant\ occurrences).$$

As another example of the above rule, suppose the probability of a certain basketball player making a free throw is 0.8. If he attempts 7 free throws (with a sufficient time or activity between each so that independence is assumed), what is the probability that he will make exactly four of the seven? To answer this question we must find the probability of a single occurrence satisfying the event (say, making the first four and missing the final three), and then determine the number of such relevant occurrences.

If we denote a miss on the $i$th attempt by $m_i$ and a successful shot by $s_i$, then,

$$P(s_1 \cap s_2 \cap s_3 \cap s_4 \cap m_5 \cap m_6 \cap m_7)$$
$$= 0.8 \times 0.8 \times 0.8 \times 0.8 \times 0.2 \times 0.2 \times 0.2$$
$$= (0.8)^4(0.2)^3.$$

Note from this result that when the same outcome of independent trials is repeated $m$ times, with each trial having a constant probability $p$, then the probability of the *intersection* of such repeated outcomes is $p^m$. In this case, a successful shot is repeated four times (giving $(0.8)^4$) and a miss is repeated three times (giving $(0.2)^3$), thus accounting for all seven trials.

As we indicated above, we have to find all possible ways of making four shots out of seven. It could be that the three misses occur on the first three attempts, or on the first, third, and fifth attempts, or in one of many other sequences. The important fact is that *all* these sequences have the *same* probability of occurring. The number of relevant occurrences is given by the number of combinations of seven attempts with four successful shots in each, or

$$_7C_4 = \binom{7}{4} = \frac{7!}{4!(7-4)!} = \frac{7 \cdot 6 \cdot 5(4!)}{(4!)3 \cdot 2 \cdot 1} = 35.$$

Thus, the probability of exactly four hits and three misses in seven shots is

$$\binom{7}{4}(0.8)^4(0.2)^3 = 35(0.4096)(0.008) = 0.1147.$$

What number of good shots would be more probable than four? Do you think it would be more likely for this basketball player to make 2 shots, or all 7 shots? Such questions can be answered using the same method as in this example. The problem can be formalized by assigning one point for any shot that is good and no points for a miss. Let $x$ be a discrete random variable which has values equal to the number of points made in seven free throw attempts. We formally define $x$ as:

$$\{x \mid 0 \leqslant x \leqslant 7 \quad \text{and} \quad x \text{ is an integer}\}.$$

Any of these eight integers from 0 to 7 inclusive may occur. We calculated above the probability that four shots are good,

$$P(x = 4) = 0.1147.$$

You might try to show that

$$P(x = 2) = 0.0043 \quad \text{and} \quad P(x = 7) = 0.2097.$$

In this example the most probable outcome is six good shots, with $P(x = 6) = 0.3670$.

Finding the probabilities associated with all values of a random variable $x$ in a certain experiment results in the determination of what we have called the *probability distribution for* $x$. Use of the rules for counting and combinations, as well as the use of the probability rules for complements, conditionals, intersections, unions, marginals, and posteriors, is often essential in calculating such probability distributions. The distributions themselves and their means and standard deviations are most often the important key for opening the door to statistical inference or decision-making under uncertainty. We are now ready to leave the foundation work and go on to the next phase in learning the concepts and methods used for these main purposes of statistics. In Chapter 3 and 4, some general properties of probability distributions are presented and some frequently occurring particular distributions relating to discrete random variables are emphasized. The analogy of these results to problems relating to continuous random variables follows in Chapter 5.

## REVIEW PROBLEMS

1. A small boy has five coins in his pocket and selects one at random. The coins are one penny, two nickels, and two dimes. Describe the probability model and find the probability of:

   a) selecting a nickel;
   b) selecting a coin worth less than 10¢.

2. What is meant by the term "random variable"? Give several examples of random variables and describe the sample space for each of these examples.

3. In repeated trials of drawing a single card from a standard deck of cards with replacement after each trial, what is the probability of drawing three consecutive spades?

4. In 900 trials of tossing a fair die, how many times would you expect a number less than three to turn up?

5. A radio repairman wants to replace a defective tube in an old radio. He has seven tubes in the repair kit, but only two of them will work. He selects the tubes at random one after another without replacement. What is the probability that the repairman will have to try exactly four tubes before finding a good one?

6. Consider the experiment of rolling a die, and define $y$ as the number of dots showing face up. Let the events $A$ and $B$ be defined as follows: $A = y$ is an even number, and $B = y < 3$. Find $P(A \cap B)$ and $P(A \cup B)$.

7. State the General Rule of Addition. How is this rule related to the Special Rule of Addition?

8. Suppose that an urn contains five balls, numbered 1 through 5.

   a) Describe the probability model for the experiment "draw two balls without replacement."

   b) What is the probability that the sum of the numbers on the balls will be less than 6, in two draws without replacement?

   c) What is the probability that the two balls will sum to less than 6, given that the first ball was a 2?

9. Repeat parts (a), (b), and (c) of the previous problem, assuming that the balls were drawn with replacement.

10. You have a single die, which is known to be fair. Define an appropriate random variable and probability model, and answer the following questions:

    a) In two throws of this die, what is the probability of rolling either a 6 on the first throw or a 4 on the second throw, or both?

    b) What is the probability, in two throws, that the two numbers rolled are not alike? (*Hint:* Use rule of complements.)

    c) What is the probability of rolling at least one 5 in four throws of the die? (*Hint:* Use rule of complements.)

11. You draw two cards from a deck, with replacement.

    a) What is the probability that the first card is an ace ($A_1$) and the second is a spade ($S_2$)?

    b) What is the probability of drawing either an ace on the first draw or a spade on the second, or both?

    c) Are the events $A_1$ and $S_2$ independent or dependent?

    d) Repeat parts (a), (b), and (c), assuming the draws are without replacement.

12. A study on the probability that a randomly selected person smokes cigarettes divided the U.S. population into three age groups: under 30 ($<30$), between 30 and 50 (30–50), and over 50 ($>50$). Half those under 30 were found to smoke.

    a) If $P(<30) = 1/2$, find the probability a randomly selected person is under 30 and smokes.

    b) If $P(\text{Smokes} \mid <30) = 1/2$, $P(\text{Smokes} \mid >50) = 1/2$, and $P(\text{Smokes} \mid 30\text{–}50) = 1/4$, does this indicate independence or dependence between age and smoking?

    c) If $P(30\text{–}50) = 1/4$ and $P(>50) = 1/4$, find $P(S) = P(\text{Smokes})$.

    d) Replace the probability symbols in the following table with their appropriate values.

|  | $<30$ | 30–50 | $>50$ |  |
|---|---|---|---|---|
| Smokes | $P(S \cap <30)$ | $P(S \cap 30\text{–}50)$ | $P(S \cap >50)$ | $P(S)$ |
| Does not smoke | $P(\bar{S} \cap <30)$ | $P(\bar{S} \cap 30\text{–}50)$ | $P(\bar{S} \cap >50)$ | $P(\bar{S})$ |
|  | $P(<30)$ | $P(30\text{–}50)$ | $P(>50)$ |  |

    e) Find $P(S \cup >50)$.

13. The following data describe certain characteristics of the students enrolled at a university.

|  | Men | Women | Over 21 |
|---|---|---|---|
| Freshmen | 1325 | 1100 | 125 |
| Sophomores | 1200 | 900 | 175 |
| Juniors | 900 | 850 | 325 |
| Seniors | 725 | 775 | 950 |
| Graduates | 1350 | 875 | 2225 |

a) What is the probability that a university student selected at random is a sophomore? What is the joint probability that the student selected is a sophomore and a male? What is the conditional probability that if a male student is selected, this student is a sophomore?

b) Use Formula (2.9) to determine the probability that a randomly selected student is a sophomore.

c) What is the probability that the student selected is over 21? If age and sex are assumed to be independent, what is the probability that the student is a male over 21?

14. Let $x$ and $y$ be the set of real numbers greater than or equal to zero.

a) Graph the set $A = \{(x, y) \mid y \geq x + 1\}$ and the set $B = \{(x, y) \mid y \leq 5 - x\}$.

b) Indicate on your graph for part (a) the intersection and the union of $A$ and $B$.

15. Suppose eight salesmen of a company are cited for outstanding sales records during the past year. Five of these men are married. Suppose four of the eight are selected at random to receive a week's vacation for two in Hawaii. What is the probability that at least three of those selected will be married?

16. Let $A = \{1, 2, 3\}$, $B = \{2, 3, 4, 5, 6, 7\}$ and $C = \{2, 4, 6, 8, 10\}$.

a) List all the subsets of $A$. What is the complement of the set $\{1, 3\}$?

b) Find $A \cup B$ and $A \cap B$; illustrate this union and intersection on a graph.

c) Find $A \cup B \cup C$, $A \cap B \cap C$, and $(A \cup B) \cap C$.

17. Give several examples of sets which:

a) are mutually exclusive and exhaustive,

b) are overlapping.

18. Two boxes sit side by side on a counter. The first box contains 20 balls, 5 of which are white, 4 red, and 11 black. The second box contains 10 balls, 4 of which are white and 6 red. An experiment consists of one person drawing, while blindfolded, one ball from each box.

a) What is the probability that 1 red and 1 white ball will be drawn?

b) What is the probability of drawing 2 red balls, or 1 white ball and 1 black ball?

19. The State U. football coach has just learned from secret sources that his opponent will be using one of three possible quarterbacks, and only one of two possible backfield forma-

tions. His information is limited, but he has managed to determine the following probabilities:

|  | | Quarterback | | |
|---|---|---|---|---|
| | | I | II | III |
| Formation | A | $P(A \cap I) = ?$ | $P(A \cap II) = ?$ | $P(A \cap III) = ?$ | $P(A) = 0.60$ |
| | B | $P(A \cap I) = ?$ | $P(B \cap II) = ?$ | $P(B \cap III) = ?$ | $P(B) = ?$ |
| | | $P(I) = 0.30$ | $P(II) = 0.50$ | $P(III) = 0.20$ |

a) If $P(A \mid I) = 0.20$, find $P(A \cap I)$.
b) Find $P(A \cup I)$.
c) If $P(A \mid II) = 0.8$ and $P(A \mid III) = 0.7$, use Bayes' rule to find $P(II \mid A)$.
d) Show that the choice of a formation and a quarterback are (or are not) independent.

20. The probability that a certain beginner at golf gets a good shot if he uses the correct club is 1/3, and the probability of a good shot with an incorrect club is 1/4. In his bag are five different clubs, only one of which is correct for the shot in question. If he chooses a club at random and takes a shot, what is the probability that he gets a good shot?

21. Out of a group of five summer-school students, three favor a school holiday on July 4. The other two are against the July 4 holiday because it means school lasts one day longer in August. You select at random (without replacement) two of these five leaders for an interview.

   a) What is the probability that:
      1) Neither of the two will favor the holiday?
      2) One will favor and the other oppose?
      3) Both will favor?
   b) The sum of these above three probabilities will (or will not) add to one. Explain.

22. Suppose 2500 freshmen in a university are enrolled in physical education classes. 1000 of these are female, and 1300 weigh over 150 pounds. Also, 300 women and 1400 men are taller than 65 inches. If one of these freshmen is selected at random, what is the probability that the student:

   a) is a male;
   b) weighs over 150;
   c) is taller than 65 inches;
   d) is a male *not* taller than 65 inches.

23. An insurance company will select at random three from among five salesmen to present reports. Suppose two of the salesmen are members of the "MDC" ("Million Dollar Club," whose members are salesmen having insurance sales in one year with over one million dollars' face value). Find the probability that less than two of those selected for reports are MDC members.

24. Given the same situation as in Problem 20, suppose the golfer makes a good shot. What is the probability that he selected one of the incorrect clubs?

25. Suppose I acquire a ski resort. On a given weekend operation, the probability that I make a profit if the weather is "favorable" is 3/4. If the weather is "unfavorable," the probability that I make a profit is 1/8. Suppose the forecast is for a 2/5 chance of "favorable" weather.

a) What is the probability that I will make a profit from the weekend operation?
b) Suppose on Monday I tell you that I made a profit; find the probability that the weather on the preceding weekend was "favorable."

26. A man goes fishing for the first time. He has three types of bait, only one of which is correct for the type of fishing he intends to try. The probability that he will catch a fish if he uses the correct bait is $\frac{1}{3}$. If he uses the wrong bait, his chances of catching a fish are $\frac{1}{5}$.

a) What is the probability that he will catch a fish?
b) Given that the man caught a fish, what is the probability that he used a correct type of bait?

27. In a certain city it is known that one-fourth of the people leave their keys in their cars. The police chief estimates that five percent of the cars with keys left in the ignition will be stolen, but that only one percent of the cars without keys left in the ignition will be stolen. What is the probability that a car stolen in this city had the keys in the ignition?

28. A student recognizes five potential questions he may be asked on a quiz. However, he only has time to study one of them thoroughly, and he selects this one randomly. Suppose the probability that he passes the test if this selected question appears is 0.90, but the probability that he passes the test if one of the other four questions appears is only 0.30. The test contains only one question and it is one of these five.

a) What is the probability that he will pass the test?
b) Suppose you see the student next week and he has passed the test. What is the probability that the question he selected to study was in fact the one on the quiz?

29. For the basketball example at the end of Section 2.10 (where $p$ was 0.8) find the probability of:

a) 2 good shots out of seven;
b) seven good shots out of seven;
c) show that six good shots out of seven is the most probable outcome.

30. a) In how many different ways can six people be seated in a row?
b) How many different ways are there of seating six people in a row when there are ten different people waiting to be seated?
c) How many different combinations of six people can be seated when ten people are waiting to be seated?

31. a) In how many different ways can eight horses finish a race if there are no ties?
b) How many different orderings are possible for the first three positions in an eight-horse race?
c) How many different combinations of three horses can finish in the first three positions in an eight-horse race?

32. Consider the word PANIC.

   a) How many different combinations are possible, using 3 letters at a time?
   b) How many permutations are possible using 3 letters at a time?

33. A club has 14 male members and 10 female members. If a committee of 8 is to be chosen randomly from the membership of this club, what is the probability that one-half of the female members will be on the committee? Be sure to specify the probability model underlying this situation.

## EXERCISES

34. a) Distinguish between objective and subjective probability. Describe what you think to be the advantages and limitations of each of these interpretations of probability.
   b) What is the probability that you will earn at least a "B" in statistics this term? Is this a subjective or objective probability? At what "odds" would you be indifferent between the two sides of a $1 bet, one side saying your grade will be a "B" or better, the other side saying your grade will be a "C" or less? Are these odds consistent with your answer about the probability of a "B" or better? Explain why they are or are not consistent.

35. In defining probability, what is meant by the terms experiment, experimental model, and event? How are these terms related to the limit of relative frequency? How does one go about determining the limit of relative frequency in practical problems?

36. How much should an individual be willing to pay for a raffle ticket when the prize is a $500 color-television set and 4,000 tickets are to be sold?

37. A group of four golfers stumbled across a nest of yellowjackets and two of them were stung. Three of the men broke 100 for the round, and all players either were stung, or broke 100, or both.

   a) What is the probability that a player broke 100 and was stung?
   b) Given that a player has been stung, what is the probability that he broke 100?
   c) Given that a player broke 100, what is the probability that he was stung?

38. Suppose that the outcome of an experiment can be one or more of the three events $E_1$, $E_2$, or $E_3$. Write the set notation for the following statements:

   a) At least one of these three events takes place.
   b) All three events take place.
   c) No more than one of the three events takes place.

39. Define or describe briefly each of the following terms: set, subset, mutually exclusive sets, exhaustive sets, the intersection of two sets, the union of two sets, sample space.

40. Suppose that your instructor in statistics announces that the final exam will consist of five questions, which will be randomly selected from a list of ten questions handed out one week before the exam. In order to pass the exam, a student must be able to answer at least four out of the exam questions selected. What is the probability that a student who can answer only eight of the ten questions will pass the exam?

41. Given the following set of weekly wages in dollars for six employees: 72, 100, 88, 95, 89, and 78. If two of these employees are to be selected at random to serve as labor representatives, what is the probability that at least one will have a wage lower than the average?

42. Suppose that you have three urns, each filled with red and blue marbles. The first urn contains one red and three blue marbles, the second contains two red and two blue, and the third contains three red and one blue. You select an urn at random, and then randomly select a marble from this urn.

   a) If the marble drawn is red, what is the probability that you have drawn from the first urn? What is the probability that you drew from the second urn? From the third urn?
   b) Repeat part (a), assuming that the marble drawn was blue.

43. Given the following table of survey information:

| Salary ($000) | Years of college | |
|---|---|---|
| | At least 2 | None |
| ($A_1$)  5–7.9 | 30 | 50 |
| ($A_2$)  8–10.9 | 50 | 40 |
| ($A_3$)  11–13.9 | 20 | 10 |
| | 100 | 100 |

Relative to a single draw (random selection) of one of the survey respondents;

   a) Explain the meaning of the following symbols
      (i)  $P(A_1 \mid B_2)$
      (ii) $P(A_2 \text{ or } A_3 \cap B_1)$
   b) Illustrate the use of Bayes' theorem by finding the probability of selecting a noncollege respondent, presuming that the one selected is in the highest salary bracket indicated.

44. a) In a poker hand, what is the probability of receiving four of a kind (i.e., four out of five cards with the same face value)?
   b) What is the probability of receiving a full house in a poker hand (i.e., three cards with one face value and two cards with some other face value)?

45. Find and compare the chances in a gambling casino of getting a six when rolling one die four times with the chances of getting a double six when rolling two die 24 times.

46. How many different basketball lineups can be made from a team of ten men if all ten men can play the five positions? How many lineups are possible if the team contains two centers, four guards, and four forwards?

47. In drawing two cards without replacement from a deck of cards, what is the probability of drawing a king on the second draw, given that either a king or an ace was drawn on the first draw?

48. *Shark Loans* (continued from Chapter 1) Shark Loans has recorded both the activity level of the local economy (either Hi, Med, or Low) and the interarrival times (0–20 min, 20–60 min, 60–200 min) for its past 150 customers.

|  | Time Interval Between Customers | State of economy | | |
|---|---|---|---|---|
|  |  | Hi | Med | Low |
| A | 0–20 | 30 | 12 | 8 |
| B | 20–60 | 30 | 21 | 4 |
| C | 60–200 | 30 | 12 | 3 |

Based on these data:

a) Find $P(\text{Hi})$, $P(\text{Med})$, $P(\text{Low})$, $P(A)$, $P(B)$, and $P(C)$.
b) Find $P(A \mid \text{Hi})$, $P(A \mid \text{Med})$, $P(A \mid \text{Low})$.
c) Use Formula (2.9) to find $P(A)$.
d) Are the events $A$, $B$, $C$ independent of the state of the economy?
e) Suppose Shark would like to revise the probabilities $P(\text{Hi})$, $P(\text{Med})$, $P(\text{Low})$ in the light of sample evidence $(S_1)$. Find $P(\text{Hi} \mid S_1)$, $P(\text{Med} \mid S_1)$, and $P(\text{Low} \mid S_1)$ given that $P(S_1 \mid \text{Hi}) = 0.05$, $P(S_1 \mid \text{Med}) = 0.10$, and $P(S_1 \mid \text{Low}) = 0.40$.

# Discrete Random Variables and Expectations

## 3.1 PROBABILITY FUNCTIONS FOR A SINGLE DISCRETE RANDOM VARIABLE

### Probability Mass Function

Now that we have studied the rules for associating a probability value with a single event, or with a combination of events in an experiment, we can formalize the description of a probability distribution for a random variable. Here and in Chapter 4 we focus on the description for *discrete* random variables, while Chapter 5 will present the description for *continuous* random variables.

For discrete random variables, such a description usually involves specifying a function (called a *probability mass function*) which gives the relative frequency of *each* possible value of the random variable.

As we indicated in Chapter 2, the symbol $P(x = x)$ (which is often shortened to $P(x)$) represents the probability that the random variable $x$ assumes the specific value $x$. For example, $P(x = 4)$ or, equivalently, $P(4)$, might represent the probability that the two faces showing on a single throw of a pair of dice sum to 4. The value of $P(x = 4)$ is the probability of just one of the events of an experiment, not an entire probability mass function. One way to specify an entire probability mass function is to list each event and its associated probability. This approach is practical, however, only when the experiment involves a small number of outcomes. For example, we could do this with the dice-throwing experiment, where the dis-

**Table 3.1** Comparison of probability function values and relative frequencies for tossing a pair of dice

| Values of $x$ | Number of outcomes (frequency) | Relative frequency | Values of probability mass function | Cumulative relative frequencies or cumulative probability function |
|---|---|---|---|---|
| 2  | 1 | 1/36 | $P(x = 2) = 1/36$  | $P(x \leqslant 2) = 1/36$ |
| 3  | 2 | 2/36 | $P(x = 3) = 2/36$  | $P(x \leqslant 3) = 3/36$ |
| 4  | 3 | 3/36 | $P(x = 4) = 3/36$  | $P(x \leqslant 4) = 6/36$ |
| 5  | 4 | 4/36 | $P(x = 5) = 4/36$  | $P(x \leqslant 5) = 10/36$ |
| 6  | 5 | 5/36 | $P(x = 6) = 5/36$  | $P(x \leqslant 6) = 15/36$ |
| 7  | 6 | 6/36 | $P(x = 7) = 6/36$  | $P(x \leqslant 7) = 21/36$ |
| 8  | 5 | 5/36 | $P(x = 8) = 5/36$  | $P(x \leqslant 8) = 26/36$ |
| 9  | 4 | 4/36 | $P(x = 9) = 4/36$  | $P(x \leqslant 9) = 30/36$ |
| 10 | 3 | 3/36 | $P(x = 10) = 3/36$ | $P(x \leqslant 10) = 33/36$ |
| 11 | 2 | 2/36 | $P(x = 11) = 2/36$ | $P(x \leqslant 11) = 35/36$ |
| 12 | 1 | 1/36 | $P(x = 12) = 1/36$ | $P(x \leqslant 12) = 1.00$ |
| Sum | 36 | 1.00 | 1.00 | |

crete random variable $x$ = sum of dots for a single throw of a pair of dice. The probability values for this experiment are repeated in Table 3.1.

The number of possible outcomes associated with each value of the random variable is given in the second column of Table 3.1, while the probability mass function is given in column 4. Note the similarity between the concept of relative frequency, as described in Chapter 1, and the concept of a probability mass function. The difference, technically, is that relative frequencies are often only short-run frequencies based on the outcomes of *one or more* replications of an experiment, whereas a probability mass function can be viewed as the theoretical long-run relative frequency for *all conceivable* replications.

**Cumulative Probability Function**

The concepts of *cumulative* relative frequency and the use of graphical representations introduced in Chapter 1 also have their counterparts in the study of probability. A *cumulative probability function* describes how probability accumulates in exactly the same fashion as the cumulative column in Table 1.3 describes how relative frequency accumulates by *summing* over the relative frequency values. The value of the cumulative probability function at any given point $x$ is usually denoted by the symbol $F(x)$, where $F(x)$ is the *sum of all values* of the probability mass function less than or equal to $x$. That is,

> *Cumulative mass function at the point $x = x$:*
>
> $$F(x) = P(x \leqslant x) = \sum_{x \leqslant x} P(x).$$    (3.1)

For example, in Table 3.1 the probability that the sum of the dots on a throw of a pair of dice is three *or less* is

$$F(3) = P(x \leqslant 3)$$
$$= \sum_{x \leqslant 3} P(x) = P(x = 2) + P(x = 3)$$
$$= \frac{3}{36}.$$

The entire cumulative probability function is shown in the final column of Table 3.1.

Graphing a probability mass function or a cumulative probability function also helps describe the probabilities of the outcomes associated with a given random variable. In fact, the name "mass function" derives from the fact that all outcomes associated with a value of a discrete random variable can be represented

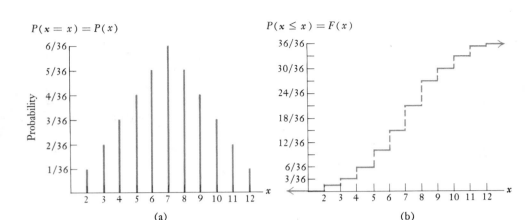

**Fig. 3.1.** Mass and cumulative functions for the random variable *x* = sum of dots on two dice (from Table 3.1).

on a graph by a vertical line whose height (or *mass*) indicates the probability of that value. The standard graphical form for probability mass and cumulative functions is given in Fig. 3.1 for the random variable $x$ = sum of dots on two dice.

### Properties of Probability Functions

At this point we need to be more precise about the properties of $P(x)$ and $F(x)$. First, we know from the discussion in Chapter 1 that a relative frequency must be between 0 and 1.0, and we stated in Chapter 2 that all values of $P(x)$ must lie between these limits.

$$\textit{Property 1:} \quad 0 \leqslant P(x = x) \leqslant 1.$$

Secondly, we know that the sum of the probabilities over all values of the random variable must equal 1.0.

$$\textit{Property 2:} \quad \sum_{\text{All } x} P(x) = 1.0.$$

These two properties also imply several characteristics of the function $F(x)$. Since $P(x)$ can never be negative, the value of $F(x)$ can also never be negative, because $F(x)$ is the *sum* of the values of $P(x)$. Similarly, because all values of $P(x)$ sum to 1.0, the maximum value of $F(x)$ is also 1.0. Note, in the graph in Fig. 3.1(b), that the values of $F(x)$ increase as $x$ gets larger, approaching 1.0 as $x$ goes to its largest value.

Also, note in Fig. 3.1(b) that $F(x)$ is defined for *all* possible values of $x$ from negative infinity $(-\infty)$ to positive infinity $(\infty)$. For example, the cumulative mass function $F(x)$ can be defined for *any* value of $x$ between positive and negative infinity. Suppose we arbitrarily pick a number, say $x = 4.72$. In Fig. 3.1(b) this value is seen to be

$$F(4.72) = \frac{6}{36}.$$

In fact all $x$ values such that $4 \leqslant x < 5$ have a cumulative value for $F(x)$ of $\frac{6}{36}$.

**Examples.** Many other examples of discrete probability functions can be cited. Consider a situation involving oil exploration at one of five possible sites. From past experience, we know that two wells will be dry, two will not be commercially feasible, and one will be a successful well. Thus, the experiment of choosing one site has three different outcomes. Now, suppose we let a random variable $x$ assume values of $-1$, $0$, and $+10$ for the three outcomes, respectively. These values may reflect some management view of the net payoff (or loss) associated with the exploration and drilling at each site. The values of the probability mass function and the cumulative probability function for this random variable are easily determined; they are given in Table 3.2.

**Table 3.2** Mass and cumulative probability functions for the oil exploration example

| | Value of $x$ | $-1$ | $0$ | $+10$ | All other values |
|---|---|---|---|---|---|
| Mass: | $P(x)$ | 2/5 | 2/5 | 1/5 | 0 |
| Cumulative: | $F(x)$ | 2/5 | 4/5 | 1.0 | |

As always, each $0 \leqslant P(x) \leqslant 1.0$, and $F(x)$ is a nondecreasing function of $x$. Only *three* values of $x$ have positive probabilities, so this random variable is discrete. At the largest value, $x = +10$, the cumulative value $F(10) = 1.0$. For any value less than $-1$, the cumulative value would be zero. What is the value, $F(4)$? Be careful! It is true that $P(4) = P(x = 4)$ is zero, but the cumulative value is

$$F(4) = \sum_{x \leqslant 4} P(x),$$

which includes $P(-1)$ and $P(0)$. Thus,

$$F(4) = \tfrac{2}{5} + \tfrac{2}{5} = \tfrac{4}{5},$$

since at the value $x = 4$, there is $\tfrac{4}{5}$ probability accumulated. Indeed, the value of $F(x)$ is $\tfrac{4}{5}$ for all values, $0 \leqslant x < 10$. At the point $x = 10$, the value of $F(x)$ steps up to 1.0. For any discrete random variable, the cumulative function is always a step function (such as that shown in Fig. 3.1(b)).

For another example, consider the size of take-out sales at a donut shop drive-in window. Customers may buy a single donut or packages of 2, 4, 6, or a dozen donuts. In this experiment involving sales to a given customer, there are five discrete outcomes. Suppose we let the random variable $x$ represent the number of donuts sold to a customer, where the values of $P(x)$ and $F(x)$ determined from the relative frequencies of actual sales are given as follows:

| $x$ | 1 | 2 | 4 | 6 | 12 |
|---|---|---|---|---|---|
| $P(x = x)$ | 0.08 | 0.27 | 0.10 | 0.33 | 0.22 |
| $F(x)$ | 0.08 | 0.35 | 0.45 | 0.78 | 1.00 |

We see that for this discrete set of values, $0 \leqslant P(x) \leqslant 1.0$ for each $x$, and

$$\sum_{\text{All } x} P(x) = 1.0,$$

so that these values satisfy the two properties of a probability mass function presented earlier. The definition of a cumulative probability function can be used to find a number of probabilities which might be of interest to the manager of the donut shop.

1) The probability of selling four donuts or less to a given customer is

$$P(x \leqslant 4) = F(4) = 0.45.$$

2) The probability of selling packages of more than two donuts is $P(x > 2)$. This can be easily determined by using the complementary law of probability,

$$P(x > 2) = 1 - P(x \leqslant 2),$$

and since $P(x \leqslant 2) = F(2)$, the solution is

$$P(x > 2) = 1 - F(2) = 1.0 - 0.35 = 0.65.$$

3) The probability of selling packages in which the number of donuts is less than or equal to 6 but more than 1, is $P(1 < x \leqslant 6)$. In this case,

$$P(x \leqslant 6) = F(6) = 0.78.$$

But, since $F(6)$ *includes* the probability $P(x \leqslant 1) = F(1)$, we must *subtract* this value from $F(6)$. Hence,

$$P(1 < x \leqslant 6) = F(6) - F(1) = 0.78 - 0.08 = 0.70.$$

4) Obviously, the probability of selling packages of 12 donuts or *less* is

$$P(x \leqslant 12) = F(12) = 1.0.$$

As you can see from the above examples, the cumulative function is useful in determining probability values of various types of events. Since this function will be used frequently in problems of statistical inference, the student should thoroughly understand this concept and all of the above examples and hints for interpretation before proceeding. A good exercise for the reader would be to sketch the function $F(x)$ for this example of the sales of donuts.

**Some Further Questions**

Perhaps the reader has become curious about what the "average" number of donuts bought per sale might be. Although probabilities of specific outcomes have been calculated for a single sale, presumedly many sales will be made to many different people. This consideration leads to several questions about the theoretical properties of this particular distribution. Recall from Chapter 1 that the important features of any distribution (which we now know include probability distributions) are the measures of *central location* and *variation*. For example, the mean is essential to determine the average or expected number of donuts per sale. This characteristic would be important in determining how many donuts to produce.

A measure of the variation of sales is also important when one is interested in deviations from the expected or theoretical average payoff. For instance, what is the chance that the shop will sell all its donuts because of a succession of sales of one dozen donuts each? For planning purposes, how many packages of each type should be made, so that the chance of selling out after 500 sales is less than 0.05? These and various other questions involving decision-making could be posed. In the next section, the measures of central tendency and variation for a probability distribution are given. To repeat an earlier remark, while the problems and questions used as examples to guide and motivate your study of statistics may not seem worth the effort, remember that the same concepts, measures, formulas, and methods can be and are applied to important decision-making situations involving not pennies but millions of dollars; not tossing dice but choosing portfolio holdings; not flipping coins but controlling production processes; not counting donuts, but making pricing decisions in an oligopoly market, and so on.

## 3.2 EXPECTED VALUE

By means of the discussion presented thus far we can now determine the probability of a single event of an experiment, or describe the probability of the entire set of outcomes associated with a given random variable. This information, however, may not be concise enough for most decision-making contexts. Recall that we had the same problem in Chapter 1, when it was not sufficient merely to present all the data, but in addition several characteristics of these data were given (the most important of which were the *mean* and the *variance*). The same types of measure are also useful in describing probability distributions, but in this case we must speak not of an *observed* mean or an *observed* variance, but of the mean or variance which would be *expected* to result (on the average) for the random variable under consideration. These values are thus given the name *expectations* or *expected values*.

**Expected Value**

> The *expected value of a discrete random variable* $x$ *is found by multiplying each value of the random variable by its probability and then summing over all values of* $x$.

The letter $E$ usually denotes an expected value, and this symbol is followed by brackets enclosing the random variable of interest. Using this notation, we define the expected value of a discrete random variable $x$ as follows:

$$\text{Expected value of } x: \quad E[x] = \sum_{\text{All } x} x\, P(x). \tag{3.2}$$

The expected value of $x$ is the "balancing point" for the probability mass function. That is, it is the *arithmetic mean* of the population of $x$ values:

$$\text{Arithmetic mean:} \quad \mu = E[x] = \sum_{\text{All } x} x P(x).$$

Note that this definition of $\mu = E[x]$ corresponds very closely to our definition in Chapter 1 of the mean of a population for grouped data (Formula 1.6). The major difference is that, in Chapter 1, each value of $x$ was "weighted" by its relative frequency, $f_i/N$, while in this chapter the weight is $P(x)$. As we pointed out previously, $f_i/N$ is the relative frequency for just one (or a few) replications of an experiment, while $P(x)$ is the expected relative frequency for an infinite number of replications of the experiment.

To illustrate the calculation of an expected value, assume that we have (foolishly) agreed to pay you $1 for each dot showing when a pair of dice is thrown (once). Our concern after making the agreement is how much we would lose, *on the average*, if we played the same game many times. That is, if $x$ is the random variable representing the sum of the two numbers thrown, which is the amount we pay out, then we want to determine the *expected value* of $x$ per throw, or $E[x]$.

One method to *approximate* the mean value in any experiment is to replicate the experiment many times and then add up the observed numbers and divide by the number of observations; but such a procedure is often impractical, if not impossible, and gives only an approximation to the desired value. Fortunately, there is no need to replicate an experiment if the probability mass function is known, for we then already have the *expected relative frequency* of each event. For instance, if two dice are thrown 36 times, we would expect to pay the sum of $6 five different times, since

$$P(x = 6) = \tfrac{5}{36}$$

(see Table 3.1). Similarly, we would expect to pay $2 once, $3 twice, $4 three times, and so forth, for a grand total of $252:

$$(\$2 \times 1) + (\$3 \times 2) + (\$4 \times 3) + \cdots + (\$12 \times 1) = \$252.$$

The *average* amount we would *expect* to lose per throw is calculated by dividing $252 by the number of throws, or

$$\frac{\$252}{36} = \$7.$$

Thus, *on the average*, we would expect to lose $7 per throw, since seven is the average number of dots which will be thrown. The value of $7, which is the *expected value* of $x$, can also be determined by dividing both sides of the above equality by 36 and rearranging terms slightly, to give:

$$\$2\left(\frac{1}{36}\right) + \$3\left(\frac{2}{36}\right) + \$4\left(\frac{3}{36}\right) + \cdots + \$12\left(\frac{1}{36}\right) = \frac{\$252}{36} = \$7.$$

Note that each term in the lefthand side of this equality represents one outcome of the throw of the dice (for example, $x = 2, x = 3, \ldots, x = 12$) times the probability of that outcome:

$$[P(x = 2) = \tfrac{1}{36}, \quad P(x = 3) = \tfrac{2}{36}, \quad \ldots, \quad P(x = 12) = \tfrac{1}{36}].$$

This is precisely the procedure outlined in the definition at the beginning of this section. As calculated above, the expected value of $x$, (where $x$ represents the sum of dots facing up after tossing a pair of dice) is $E[x] = 7$. The value 7 is thus the balancing point, or the mean, of this probability mass function.

For another example, consider the donut sales situation with $x$ and $P(x)$ defined as in the previous section. What is the expected value of $x$ in this example? It is the balance point of the probability mass function, and would be the average number of donuts per sale. Using Formula (3.1), we obtain:

$$E[x] = \sum_{\text{All } x} xP(x)$$
$$= 1 \times (0.08) + 2 \times (0.27) + 4 \times (0.10) + 6 \times (0.33) + 12 \times (0.22)$$
$$= 5.64 \text{ donuts.}$$

Over an infinite number of sales, the *average* number of donuts sold would be 5.64. The expected number of donuts sold in a given number of sales, $N$, equals $N \times E[x]$. For example, in 500 sales, the total number of donuts expected to be sold would be

$$500 \times 5.64 = 2820 \text{ donuts.}$$

Thus, if 250 dozen donuts (3000) were produced to cover 500 sales, then the shop could expect a surplus of

$$3000 - 2820 = 180 \text{ donuts.}$$

The number 180 represents only a theoretical value, since one can be sure of this amount of sales only in the limit sense of long-run relative frequencies. Over any finite number of sales, the actual sales may deviate from the 5.64-donut average.

**Expected Value of a Function of a Random Variable**

Not only can we take an expectation of a simple random variable, but we can also take an expectation of *any function* of a random variable. For instance, instead of finding the mean of the random variable $x$, we might be interested in determining the expected value of $x^2$, or of log $x$, or of $e^x$. If $x$ is a random variable, then these functions of $x$ are also random variables, and their expected value can be determined. Suppose we let $g(x)$ represent the random variable whose value is $g(x)$ when the value of $x$ is $x$. The expected value of $g(x)$ is defined as follows:

$$\boxed{\text{Expected value of } g(x): \quad E[g(x)] = \sum_{\text{All } x} g(x)P(x).} \qquad (3.3)$$

The only difference between Formulas (3.3) and (3.2) is that in (3.3) we are weighting each value of $g(x)$ by $P(x)$, rather than weighting each value of $x$ by $P(x)$. The products of $P(x)$ times $g(x)$ are summed to get the balance point of the function $g(x)$. This balance point is the value expected for $g(x)$ for all possible repetitions of the experiment involving the random variable $x$.

To illustrate the use of Formula (3.3), suppose that we had agreed to pay an amount equal to the *square* of the sum of the dots showing on a throw of the dice, instead of just \$1 for each dot showing. Will our losses now be \$49, which is merely the square of the previous expected value, $E[x] = 7$? To answer this question, we need to let $g(x) = x^2$. The expected loss (per throw) is thus $E[g(x)] = E[x^2]$.

$$E[x^2] = \sum_{\text{All } x} x^2 P(x)$$

$$= (\$2)^2(\tfrac{1}{36}) + (\$3)^2(\tfrac{2}{36}) + \cdots + (\$12)^2(\tfrac{1}{36})$$

$$= \$54.83.$$

Therefore, *on the average* we will lose *more* than \$49; \$54.83, to be precise. The result to be emphasized is that it is not necessarily true that $(E[x])^2$ equals $E[x^2]$. Since $E[x^2]$ is often used in statistical formulas, this method of finding it, and its distinction from $(E[x])^2$ are important.

### Variance

Just as the variance of a population is the average squared deviation of the population values from their mean ($\mu$), so can the *variance* of a random variable be defined in terms of the expected squared deviation of the set of outcomes around their expected value $E[x]$. We denote this variance of the random variable $x$ by the symbol $V[x]$, where $V[x]$ is defined as follows:

$$V[x] = \sigma^2 = E[(x - \mu)^2].$$

Since the squared-deviation term within the brackets $[(x - \mu)^2]$ is a function of the random variable $x$, it can be written as $g(x) = (x - \mu)^2$. This means that because we know how to find $E[g(x)]$ from Formula (3.3), we can now find $E[g(x)] = E[(x - \mu)^2]$. Making the substitution for $g(x) = [(x - \mu)^2]$ in Formula (3.3), we obtain:

> *Variance of* $x$:
>
> $$V[x] = \sigma^2 = E[(x - \mu)^2] = \sum_{\text{All } x} (x - \mu)^2 P(x). \qquad (3.4)$$

Formula (3.4) is the traditional way of defining the variance of a discrete random variable. It can also be used to compute a standard deviation, since the standard deviation, denoted by $\sigma$, is always the square root of the variance:

$$\text{Standard deviation of } \boldsymbol{x}: \qquad \sigma = \sqrt{V[\boldsymbol{x}]}. \qquad (3.5)$$

In order to illustrate the use of Formula (3.4), suppose we again use the probability distribution for the donut-sales situation described earlier. The probability mass function for this example is illustrated in Fig. 3.2.

| $x$ | 1 | 2 | 4 | 6 | 12 |
|------|------|------|------|------|------|
| $P(x)$ | 0.08 | 0.27 | 0.10 | 0.33 | 0.22 |

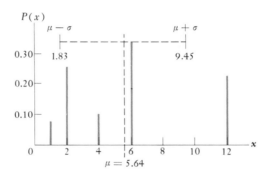

**Fig. 3.2.** Probability mass function for sales of donuts.

Since we know, for this example, that the expected value is

$$E[\boldsymbol{x}] = \mu = 5.64,$$

we substitute this value in Formula (3.4) to calculate $V[\boldsymbol{x}]$:

$$V[\boldsymbol{x}] = \sigma^2 = \sum_{\text{All } x} (x - 5.64)^2 P(x)$$

$$= (1 - 5.64)^2 0.08 + (2 - 5.64)^2 0.27 + (4 - 5.64)^2 0.10$$
$$+ (6 - 5.64)^2 0.33 + (12 - 5.64)^2 0.22,$$

and

$$V[\boldsymbol{x}] = 14.51.$$

The standard deviation is $\sigma = \sqrt{V[\boldsymbol{x}]} = 3.81$ donuts.

The value $\mu = 5.64$ is the center of this probability distribution, in the sense that it is the *expected number of donuts per sale*. The value $\sigma = 3.81$ indicates the size of the spread of the distribution of sales of donuts. According to our rule of thumb, about two-thirds of the probability should be in the interval for $x$ between $\mu + \sigma$ and $\mu - \sigma$. Using the above results, this interval is $5.64 \pm 3.81$, or approximately 1.83 to 9.45, as shown in Fig. 3.2. From the probability distribution, it is clear that this interval includes all donut sales except for sales of single donuts or dozens of donuts. The interval $\mu \pm 1\sigma$ thus contains

$$0.27 + 0.10 + 0.33 = 0.70$$

of the probability.

While Formula (3.4) is theoretically correct, it has a major disadvantage for computational purposes. Specifically, if the mean $\mu$ is not an integer, but has three or more decimals, then subtraction of $\mu$ from each value of $x$ can be tedious. Also, squaring the deviations then gives numbers with at least *six* decimals. Finally, if the number of values taken on by $x$ is quite large, all these subtractions, squares, and the ultimate summation require many steps which are subject to potential computational errors.

Fortunately, a formula equivalent to (3.4) can be derived which only requires calculating the expected value of $x^2$ and the square of the expected value. This formula is:*

> *Equivalent formula for the variance of $x$:*
> $$V[x] = \sigma^2 = E[x^2] - (E[x])^2 \qquad (3.6)$$

Formula (3.6) might be remembered as the *mean of squares minus the squared mean*. This is a general formula which applies to any random variable, discrete or continuous. It will be used again many times throughout this book.

Let us apply Formula (3.6) to the example above to verify that the same result is obtained, namely $V[x] = 14.51$. The first term can be calculated by the expectation Formula (3.2), where $x^2$ is the function $g(x)$ under consideration. Letting $g(x) = x^2$, then

$$E[x^2] = \sum_{\text{All } x} x^2 \, P(x). \qquad (3.7)$$

---

* The equality between Formulas (3.4) and (3.6) is easily demonstrated as follows:

$$
\begin{aligned}
E[(x - \mu)^2] &= \Sigma(x - \mu)^2 P(x) \\
&= \Sigma(x^2 - 2x\mu + \mu^2)P(x) && \text{(by expansion)} \\
&= \Sigma x^2 P(x) - 2\mu \Sigma x P(x) + \mu^2 \Sigma P(x) && \text{(by summing each term)} \\
&= E[x^2] - 2\mu E[x] + \mu^2 && (\Sigma P(x) = 1 \text{ by definition}) \\
&= E[x^2] + (-2 + 1)(E[x])^2 && (\text{since } \mu = E(x)) \\
&= E[x^2] - (E[x])^2;
\end{aligned}
$$

By using Formula (3.7), $E[x^2] = 1^2(0.08) + 2^2(0.27) + 4^2(0.10) + 6^2(0.33) + 12^2(0.22) = 46.32$. The second term in Formula (3.6) is simply the square of the mean, or $\mu^2$. We know that $E[x] = \sum x\,P(x) = 5.64$, and so $(E[x])^2 = (5.64)^2 = 31.81$. Then using Formula (3.6), $V[x] = 46.32 - 31.81 = 14.51$.

The reader might show that the $V[x]$ for the random variable $x$ in the dice-tossing experiment (where we agreed to pay an amount equal to the sum of the dots thrown) is:

$$V[x] = E[x^2] - (E[x])^2$$
$$= 54.83 - (7.00)^2 = 5.83.$$

The standard deviation of $x$ equals $\sqrt{\$5.83} = \$2.41$; thus, according to our rule of thumb from Chapter 1, if we play this game more than once, about 95 percent of our losses should be in the interval

$$\mu \pm 2\sigma = \$7 \pm 2(\$2.41),$$

which is the interval from \$2.18 to \$11.82. We know this result to be reasonable as, on the average, only once in every 36 throws will I pay less than \$2.18 (when two 1's appear) and only once will I pay more than \$11.82 (when two 6's are thrown).

## 3.3 EXPECTATION RULES

The concept of mathematical expectation or expected value will prove so useful in the coming chapters that a prior consideration here of a few of the important properties of expectations should be beneficial. In most cases these rules will be presented without proof, and it is not necessary that they be memorized. They will, however, be referred to in subsequent sections and be useful in solving problems.

*Rule 1:* $E[k] = k$;  *The expected value of a constant is the constant itself.*

*Rule 2:* $V[k] = 0$;  *The variance of a constant is zero.*

*Rule 3:* $E[kx] = kE[x]$;  *The expected value of the product of a constant times a variable is the product of the constant times the expected value of the variable.*

*Rule 4:* $V[kx] = k^2V[x]$;  *The variance of the product of a constant times a variable is the product of the square of the constant times the variance of the variable.*

*Rule 5:* $E[x \pm y] = E[x] \pm E[y]$;  *The expected value of the sum (or difference) of two variables is the sum (or difference) of their expected values.*

If $x$ and $y$ are independent, then

    *Rule 6:* $E[x \times y] = E[x] \times E[y]$;   *The expected value of the product of two **independent** variables is the product of their expected values.*

If $x$ and $y$ are independent, then

    *Rule 7:* $V[x \pm y] = V[x] + V[y]$;   *The variance of the **sum or difference** of two **independent variables** is always the **sum** of their variances.*

    The first four of these rules are illustrated by the use of the probability distributions in Table 3.3. The last three, plus some additional rules for expected values, are illustrated in Section 3.5.

**Table 3.3** Example of rules for expectations

| (1) | (2) | (3) | (4) | (5) | (6) | (7) |
|---|---|---|---|---|---|---|
| $k$ | $x$ | $P(x)$ | $xP(x)$ | $(y = kx)$ | $P(x)$ | $(kx)P(x)$ |
| 2 | 3 | 1/3 | 3/3 | 6 | 1/3 | 6/3 |
| 2 | 2 | 1/3 | 2/3 | 4 | 1/3 | 4/3 |
| 2 | 4 | 1/3 | 4/3 | 8 | 1/3 | 8/3 |
| $E[k] = 2$ | | | $E[x] = 3$ | | | $E[kx] = 6$ |

    Column (1) of Table 3.3 demonstrates Rule 1, that $E[k] = k$. We can also see from this column that Rule 2 holds, $V[k] = 0$. Note that $E[x] = 3$, in column (4). If each value of $x$ is multiplied by 2, the result is column (5), which we have labeled $y$. The expected value of the variable $y = 2x$ is calculated in column (7), and is seen to be $E[2x] = 6$. Thus, we have demonstrated Rule 3, that

$$E[kx] = kE[x].$$

Now let's calculate the variance of the $y$ values in column (5) by using Formula (3.4), and letting $\mu_y = E[2x] = 6$:

$$V[y] = \sum (y - \mu_y)^2 P(y) = (6 - 6)^2(\tfrac{1}{3}) + (4 - 6)^2(\tfrac{1}{3}) + (8 - 6)^2(\tfrac{1}{3})$$
$$= 0 + \tfrac{4}{3} + \tfrac{4}{3} = \tfrac{8}{3}.$$

If Rule 4 holds, $V[y] = V[2x] = 2^2 V[x]$. The variance of $x$ is

$$V[x] = \sum (x - \mu)^2 P(x) = (3 - 3)^2(\tfrac{1}{3}) + (2 - 3)^2(\tfrac{1}{3}) + (4 - 3)^2(\tfrac{1}{3})$$
$$= 0 + \tfrac{1}{3} + \tfrac{1}{3} = \tfrac{2}{3}.$$

Since $V[2x] = \frac{8}{3}$, which equals

$$(2)^2 V[x] = 4(\tfrac{2}{3}) = \tfrac{8}{3},$$

Rule 4 is verified.

### Expectation of Linear Transformations

We will make considerable use of the rules of expectation in the remainder of this book. For now, however, we can demonstrate their use in an especially important area, that of the expectation of a linear transformation. First, we consider the transformation $y = a + bx$ where $a$ and $b$ are constants.

### Expectation of $y = a + bx$

In many statistical problems it is convenient to take a random variable $(x)$, multiply this random variable by some constant $(b)$, and then add another constant $(a)$ to the result $(bx)$. If we let $y = a + bx$, then we would like to determine $E[y]$:

$$\begin{aligned} E[y] &= E[a + bx] \\ &= E[a] + E[bx] \end{aligned} \qquad \text{Rule 5}$$

or

$$E[y] = a + bE[x] \qquad \text{Rules 1 and 3}$$

Note that the above equation can be solved for $E[x]$ as follows:

$$E[x] = \frac{E[y] - a}{b}. \qquad (3.8)$$

One of the advantages of Formula (3.8) is that it can be used to simplify the arithmetic involved in calculating a mean and a variance. To illustrate this simplification, consider the four values of $x$ in Table 3.4, each of which has $P(x) = 0.250$. Finding $E[x]$ for these values is considerably easier if we first multiply each value of $x$ by $b = 1/1000$ (i.e., divide each number by 1000), and then add $a = -2,500$ to the result.* The new values, which are

$$y = a + bx = -2,500 + \frac{x}{1000},$$

are shown in the fourth column of Table 3.4.

Calculating the value of $E[y]$ is a relatively easy task for this data. The values of $P(y)$ must be the same as $P(x)$; hence we see that $E[y] = 2$. The mean of the

---

* We hope the reader understands that many other values of $a$ and $b$ could have been used here. We divided the values of $x$ by 1000 to eliminate the three zeroes in each number, and we subtracted \$2,500 because the resulting numbers were approximately centered about that value.

Table 3.4

| P(x) | x | x/1000 | $y = a + bx$ $y = -2500 + x/1000$ | P(y) |
|------|------|------|------|------|
| 0.250 | 2,483,000 | 2,483 | -17 | 0.250 |
| 0.250 | 2,519,000 | 2,519 | 19 | 0.250 |
| 0.250 | 2,511,000 | 2,511 | 11 | 0.250 |
| 0.250 | 2,495,000 | 2,495 | -5 | 0.250 |
| 1.000 | $E[x] = ?$ | | $E[y] = 2$ | 1.000 |

$x$ values can now be calculated by using Formula (3.8), where $a = -2500$ and $b = 1/1000$:

$$E[x] = \frac{E[y] - a}{b} = \frac{2 - (-\$2,500)}{1/1000} = \$2,502,000.$$

The calculation of $E[x]$ in this manner is usually easier than using the original values of $x$.

Linear transformations of this nature are also useful in determining the variance of a set of values. A formula for $V[y] = V[a + bx]$ can be calculated as follows:

$$V[y] = V[a + bx]$$
$$= V[a] + V[bx] \qquad \left\{ \begin{array}{l} \text{By Rule 7, since} \\ \text{the constant } a \text{ is} \\ \text{independent of } bx \end{array} \right.$$
$$= V[a] + b^2 V[x] \qquad \text{by Rule 4,}$$
$$= b^2 V[x] \qquad \text{by Rule 2.}$$

The fact that $V[a + bx] = b^2 V[x]$ means that adding (or subtracting) a constant ($a$) to a random variable does *not* change its variance. However, changing the scale of a random variable by multiplying it by some constant ($b$) will change the variance by the *square* of that constant. To illustrate how this relationship can be useful, suppose we solve the equation

$$V[a + bx] = b^2 V[x]$$

for $V[x]$, as follows:

$$V(x) = \frac{V[a + bx]}{b^2}. \tag{3.9}$$

To illustrate Formula (3.9), it is not difficult to show, from Table 3.4, that $V[y] = V[a + bx] = 195$. Since we know that $b = 1/1000$, $V[x]$ is:

$$V[x] = \frac{195}{(1/1000)^2} = \$195,000,000.$$

This method of calculating $V[x]$ has saved us the grief of working with the large numbers in the first column of Table 3.4.

**Expectation of $z = (x - \mu)/\sigma$**

Another important linear transformation in statistics involves subtracting from each value of a random variable $(x)$ the mean of that variable $(\mu_x)$, and then dividing each difference $(x - \mu_x)$ by the standard deviation of $x$ $(\sigma_x)$. Such a transformation is often denoted by the letter $z$, where

$$z = (x - \mu_x)/\sigma.$$

The advantage of such a transformation is that the resulting variable $z$ can be shown to always have a mean of zero (that is, $\mu_z = 0$), and will always have a variance of one (that is, $V[z] = 1$). Because of this property, the variable $z$ is usually referred to as a *standardized variable*.

To show how to standardize a variable $x$, consider the data in Table 3.5. As in Table 3.4, assume that all values shown here have equal probabilities. For these values,

$$E[x] = \mu_x = 20,$$

and

$$V[x] = \sigma^2 = 16,$$

as shown at the bottom of columns (1) and (3). The values of $z = (x - \mu)/\sigma$ are given in column (4). These $z$ values are seen to have a mean of

$$E[z] = \mu_x = 0$$

at the bottom of column (4) and a variance of

$$V[z] = \sigma^2 = 1.0$$

at the bottom of column (5).

**Table 3.5**

| (1) $x$ | (2) $x - \mu$ | (3) $(x - \mu)^2$ | (4) $(x - \mu)/\sigma = z$ | (5) $(z - \mu_z)^2$ |
|---|---|---|---|---|
| 13 | $-7$ | 49 | $-7/4 = -1.75$ | $(-1.75)^2 = 3.0625$ |
| 21 | 1 | 1 | $1/4 = 0.25$ | $(0.25)^2 = 0.0625$ |
| 25 | 5 | 25 | $5/4 = 1.25$ | $(1.25)^2 = 1.5625$ |
| 19 | $-1$ | 1 | $-1/4 = -0.25$ | $(-0.25)^2 = 0.0625$ |
| 22 | 2 | 4 | $2/4 = 0.50$ | $(0.50)^2 = 0.2500$ |
| Sum    100 | | 80 | 0 | 5.000 |
| Mean  $\mu_x = 20$ | | $\sigma^2 = 16\,(\sigma = 4)$ | $\mu_z = 0$ | $\sigma_z^2 = 1.000$ |

Thus, we have demonstrated that $E[z] = 0$ and $V[z] = 1.0$, when $z = (x - \mu)/\sigma$. This concept of a standardized variable always having a mean of zero and a variance (or standard deviation) of one, will be especially important in our discussion of probability distributions later in this book.

## *3.4  BIVARIATE PROBABILITY FUNCTIONS

In some situations an experiment may involve outcomes which are related to two (or more) random variables. This section covers the theory of probability functions which involve more than one variable; such functions are called *multivariate* probability functions. In this book only the case of *bivariate* probability functions (two variables) is presented. When a sample space involves two random variables, the function describing their combined probability is called a *joint probability function.*

For an example of a joint probability function, suppose a basketball coach rates the play-making ability of his guard candidates by counting the number of assists and number of turnovers each one makes in scrimmages. By letting $x =$ assists and $y =$ turnovers, this coach can construct a joint probability function that describes the relative frequency of both assists and turnovers.

As another illustration, consider an owner of a taxi company (assuming a nonregulated market) who is trying to decide whether or not to increase his rates. There is uncertainty in this situation about what his competitors intend to do, and the demand of the public. The random variable $x$ might be defined to be the price changes by his competitors, and the random variable $y$ might represent the demand levels for taxi services. If he can assign probabilities to the different potential price changes and demand levels, the owner has a standard statistical decision problem involving combined random variables. The reader can probably begin to see some real potential for complexity in the application of combined random variables in real-life problems. In this beginning study of statistics, we limit our discussion to small examples in order to present the basic concepts.

### The Joint Probability Function

The joint probability function for two discrete random variables $x$ and $y$ is denoted by the symbol $P(x, y)$, where $P(x, y) = P(x = x \text{ and } y = y)$. That is, $P(x, y)$ represents the probability that $x$ assumes the value $x$ while $y$ assumes the value $y$. As we will show shortly, most of the univariate probability rules discussed thus far have comparable rules in the bivariate (or multivariate) case. For example, the two

---

* This section and the next one, (3.5), may both be omitted without loss in continuity. If omitted, they should perhaps be returned to when regression and correlation are covered in Chapter 11.

properties of all probability functions presented in Section 3.1 have direct counter-parts for joint probability functions, as shown below:

$$
\begin{array}{ll}
\textit{Property 1:} & 0 \leqslant P(x, y) \leqslant 1; \\[2mm]
\textit{Property 2:} & \displaystyle\sum_{\text{All } y} \sum_{\text{All } x} P(x, y) = 1.
\end{array}
$$

These and other properties of joint probability functions are discussed throughout this section.

To illustrate a joint probability distribution, consider the results of a study investigating the relationship between the number of jobs a college graduate holds in the first five years after he graduates $(x)$ and the number of increases in respon-sibility he is given (that is, $y$ = number of promotions). Two hundred recent college graduates of comparable age and undergraduate background were surveyed and then classified according to the number of jobs and promotions they received in their first five years out of college. The results of this study are given in Table 3.6.

**Table 3.6** Frequencies for job-promotion study

| | | No. of Promotions $(y)$ | | | | Marginal total |
|---|---|---|---|---|---|---|
| | | 1 | 2 | 3 | 4 | |
| No. of jobs $(x)$ | 1 | 20 | 30 | 24 | 12 | 86 |
| | 2 | 10 | 14 | 20 | 10 | 54 |
| | 3 | 8 | 4 | 28 | 20 | 60 |
| Marginal total | | 38 | 48 | 72 | 42 | 200 |

Now we want to translate this data on frequencies to relative frequencies (or probabilities, as we will interpret them). For example,

$$P(x = 2 \text{ and } y = 3) = P(2, 3)$$

is the probability that one person drawn randomly from this population had two jobs and was promoted three times in the five years. We see that 20 people out of the total of 200 had these characteristics; hence,

$$P(2, 3) = \tfrac{20}{200} = 0.10.$$

The remaining joint probabilities, which are calculated in a similar fashion, are shown in Table 3.7.

**Table 3.7** Probabilities for job-promotion study

|  |  | No. of promotions (y) | | | | Marginal total |
|---|---|---|---|---|---|---|
|  |  | 1 | 2 | 3 | 4 |  |
| No. of jobs (x) | 1 | 0.10 | 0.15 | 0.12 | 0.06 | 0.43 |
|  | 2 | 0.05 | 0.07 | 0.10 | 0.05 | 0.27 |
|  | 3 | 0.04 | 0.02 | 0.14 | 0.10 | 0.30 |
| Marginal total |  | 0.19 | 0.24 | 0.36 | 0.21 | $1.00 = \sum\sum P(x, y)$ |

**Cumulative Joint Probability**

Analogous to a cumulative probability function for a single random variable is the cumulative joint probability function. This function is denoted as $F(x, y)$ and defined as:

*Cumulative joint probability:*

$$F(x, y) = P(x \leqslant x \quad \text{and} \quad y \leqslant y). \tag{3.10}$$

The value of $F(2, 3) = P(x \leqslant 2 \quad \text{and} \quad y \leqslant 3)$ in Table 3.7 can be seen to equal

$$P(1, 1) + P(1, 2) + P(1, 3) + P(2, 1) + P(2, 2) + P(2, 3) = 0.59.$$

Note that $F(x, y)$, like $F(x)$, can assume values only between zero and one.

**Marginal Probability**

The concept of a marginal probability as used here is the same as that discussed in Chapter 2, except that now we must be careful in using abbreviations not to confuse the marginal probability $P(x = x)$ with the marginal probability $P(y = y)$. For instance, it is not clear whether $P(2)$ refers to the former or the latter case. To make this distinction, we will abbreviate $P(x = 2)$ as $P_x(2)$ and abbreviate $P(y = 2)$ as $P_y(2)$. With this notation, and using Formula (2.9) as a reference, we can write the marginal probability of $x$ and $y$ as follows:

*Marginal probability of $x$:*    $P_x(x) = \sum_{\text{All } y} P(x, y);$

*Marginal probability of $y$:*    $P_y(y) = \sum_{\text{All } x} P(x, y).$    (3.11)

The second formula above can be used to find the marginal probability that a person selected randomly from this population held only one job. To find $P_x(1) = P(x = 1)$, the values of $P(x, y)$ in Table 3.7 are summed across all values of $y$ for the particular outcome $x = 1$. Thus,

$$P_x(1) = \sum_{\text{All } y} P(1, y) = P(1, 1) + P(1, 2) + P(1, 3) + P(1, 4)$$
$$= 0.10 + 0.15 + 0.12 + 0.06 = 0.43.$$

Similarly, we can calculate $P_y(3)$, which is the probability that a randomly selected person had exactly three promotions:

$$P_y(3) = \sum_{\text{All } x} P(x, 3) = P(1, 3) + P(2, 3) + P(3, 3)$$
$$= 0.12 + 0.10 + 0.14 = 0.36.$$

**Conditional Probability**

A conditional probability for two random variables $x$ and $y$ is defined in the same manner in which this concept was defined in Chapter 2, except that again we have to be careful about notation. Suppose we let $P_{x|y}(x \mid y)$ denote $P(x = x \mid y = y)$; then the conditional probabilities are defined as follows:

Conditional probability of $x$, given $y$:

$$P_{x|y}(x \mid y) = P(x = x \mid y = y) = \frac{P(x, y)}{P_y(y)};$$

Conditional probability of $y$, given $x$:

(3.12)

$$P_{y|x}(y \mid x) = P(y = y \mid x = x) = \frac{P(x, y)}{P_x(x)}.$$

These formulas can be used to calculate any particular conditional probability of interest. For example, we might determine $P_{x|y}(2 \mid 3)$ for the data in Table 3.7. The formula to be used is,

$$P_{x|y}(2 \mid 3) = P(x = 2 \mid y = 3) = \frac{P(2, 3)}{P_y(3)}.$$

Since $P(2, 3) = 0.10$ and $P_y(3) = 0.36$, we obtain, by substitution,

$$P_{x|y}(2 \mid 3) = \frac{0.10}{0.36} = 0.278.$$

In other words, if it is known that a person had three promotions, the probability that this person had exactly two jobs is 0.278.

### Independence

Just as we were able, in Chapter 2, to determine whether or not two events are independent, we can, in the present context, determine whether or not two random variables are independent. The test for independence in the two cases is very similar. In order for two random variables to be independent, all the joint probability values must equal the product of the corresponding marginal probability values. That is, if $x$ and $y$ are independent, then the following relationship must hold for all values of $x$ and $y$.

> *Joint probability if $x$ and $y$ are independent:*
>
> $$P(x, y) = P_x(x)P_y(y).$$    (3.13)

If the above relationship does not hold for *all* possible combinations of $x$ and $y$, then these values are not independent. Only one violation of Formula (3.13) is necessary to demonstrate *dependence*. For the data in Table 3.7, it is easily shown that *none* of the pairs $x$ and $y$ satisfy Formula (3.13). For example, $P(1, 2) = 0.15$, but this value is not equal to the product

$$P_x(1) P_y(2) = 0.43(0.24) = 0.1032.$$

Thus, we can conclude that the number of jobs and promotions a person has had are *not* independent.

### *3.5 BIVARIATE EXPECTATIONS

We have discussed the important measures of the mean, $\mu = E[x]$, and the variance, $V(x) = \sigma^2$, for a single random variable. These same measures can be used to describe similar concepts in bivariate problems. Suppose we write a function of two random variables $x$ and $y$ as $g(x, y)$. In general, the expectation of such a function is given by a direct extension of Formula (3.3):

> *Expected value of $g(x, y)$:*
>
> $$E[g(x, y)] = \sum_{\text{All } y} \sum_{\text{All } x} g(x, y)P(x, y).$$    (3.14)

---

* This section may be omitted without loss of continuity.

For example, we may want to find the expected value of the product of $x$ times $y$, in which case

$$g(x, y) = x \cdot y, \quad \text{and} \quad E[x \cdot y] = \sum_y \sum_x (x \cdot y)P(x, y).$$

Similarly, if $g(x, y) = x + y$, then

$$E[x + y] = \sum_y \sum_x (x + y)P(x, y).$$

We now investigate these two special cases of Formula (3.14).

### Expectations of $x \cdot y$

To illustrate Formula (3.14) when $g(x, y) = x \cdot y$, consider a lumber-yard which sells plywood paneling in two lengths, 4 ft and 8 ft, and in three different widths, 2 ft, 4 ft, and 6 ft. The lumber-yard is interested in determining the average amount of paneling sold in terms of area (square feet). That is, they want to determine $E[x \cdot y]$, where $x$ = length and $y$ = width. By the basic counting rule of Section 2.7, there are $m = 2$ times $k = 3$, or $m \cdot k = 6$ different arrangements of widths and lengths sold. The distributions $P(x)$, $P(y)$, and $P(x, y)$ for the sale of these six combinations, based on company records, are given in the following matrix.

| $x$ $\diagdown$ $y$ | 2 | 4 | 6 | $P(x)$ |
|---|---|---|---|---|
| 4 | 0.05 | 0.05 | 0.10 | 0.20 |
| 8 | 0.10 | 0.50 | 0.20 | 0.80 |
| $P(y)$ | 0.15 | 0.55 | 0.30 | $1.00 = \sum\sum P(x, y)$ |

The determination of $E[x \cdot y]$ is illustrated for this data in Table 3.8.

**Table 3.8**

| $x$ | $P(x)$ | $xP(x)$ | $y$ | $P(y)$ | $yP(y)$ | $x, y$ | $P(x, y)$ | Area $x \cdot y$ | $E[x \cdot y] =$ $(x \cdot y)P(x, y)$ |
|---|---|---|---|---|---|---|---|---|---|
| 4 | 0.20 | 0.80 | 2 | 0.15 | 0.30 | (4, 2) | 0.05 | 8 | 0.40 |
| 8 | 0.80 | 6.40 | 4 | 0.55 | 2.20 | (4, 4) | 0.05 | 16 | 0.80 |
|   |      |      | 6 | 0.30 | 1.80 | (4, 6) | 0.10 | 24 | 2.40 |
|   |      |      |   |      |      | (8, 2) | 0.10 | 16 | 1.60 |
|   |      |      |   |      |      | (8, 4) | 0.50 | 32 | 16.00 |
|   |      |      |   |      |      | (8, 6) | 0.20 | 48 | 9.60 |
| Sum | 1.00 | 7.20 |   | 1.00 | 4.30 |   | 1.00 |   | 30.80 |

The average number of square feet of paneling sold is thus $\mu = 30.80$, as shown in the last column of Table 3.8.

**Expectation of $x \cdot y$ when $x$ and $y$ are independent**

One special case of the expectation of $x \cdot y$ is worth noting, the case when the variables $x$ and $y$ are independent.

---

*Expectation of $x \cdot y$, assuming independence:*

$$E[x \cdot y] = E[x] \cdot E[y].$$    (3.15)

---

We illustrate Formula (3.15) by considering the responses made by new students regarding university housing and laundry services. A choice is available for single, double, or twin-double (suite) dormitory rooms, as denoted by $x = 1$, 2, or 4 person rooms. A second choice is available between two laundry plans, allowing for \$2 or \$5 worth of laundry per week. The \$2-per-week plan is included in regular dormitory fees. The \$5-per-week plan has an additional fee. The laundry plan selected is denoted by $y = 2$ or 5. The probability distributions of $P(x)$, $P(y)$, and $P(x, y)$, based on the students' responses, are given in the matrix below. They illustrate the phenomenon that dormitory room selections and laundry plan choices are independent (i.e., statistically unrelated decisions) for these new students since $P(x, y) = P_x(x)P_y(y)$ for all $x$ and $y$.

| $x$ \ $y$ | 2 | 5 | $P(x)$ |
|---|---|---|---|
| 1 | 0.24 | 0.16 | 0.40 |
| 2 | 0.12 | 0.08 | 0.20 |
| 4 | 0.24 | 0.16 | 0.40 |
| $P(x)$ | 0.60 | 0.40 | $1.00 = \sum\sum P(x, y)$ |

The fact that $x$ and $y$ are independent in this example can also be shown by using some of the rules of expectation presented earlier in this chapter. Table 3.9 shows a number of expectations that we will use.

From columns (3) and (6) of Table 3.9 we see that

$$E[x] \times E[y] = (2.40)(3.20) = 7.68,$$

**Table 3.9**

| (1) | (2) | (3) | (4) | (5) | (6) | (7) | (8) | (9) $E[x \cdot y] =$ |
|-----|-----|------|-----|------|-------|--------|---------|----------------------|
| $x$ | $P(x)$ | $xP(x)$ | $y$ | $P(y)$ | $yP(y)$ | $(x, y)$ | $P(x, y)$ | $(x \cdot y)P(x, y)$ |
| 1 | 0.40 | 0.40 | 2 | 0.60 | 1.20 | (1, 2) | 0.24 | (2)(.24) = 0.48 |
| 2 | 0.20 | 0.40 | 5 | 0.40 | 2.00 | (1, 5) | 0.16 | (5)(.16) = 0.80 |
| 4 | 0.40 | 1.60 |   |      |      | (2, 2) | 0.12 | (4)(.12) = 0.48 |
|   |      |      |   |      |      | (2, 5) | 0.08 | (10)(.08) = 0.80 |
|   |      |      |   |      |      | (4, 2) | 0.24 | (8)(.24) = 1.92 |
|   |      |      |   |      |      | (4, 5) | 0.16 | (20)(.16) = 3.20 |
| Sum | 1.00 | 2.40 |   | 1.00 | 3.20 |       | 1.00 | 7.68 |

which is the same result as that given in column (9),

$$E[x \cdot y] = 7.68.$$

This result thus verifies that $x$ and $y$ are indeed independent.

**Covariance of $x$ and $y$**

At this point we introduce another measure of variation which is very important in most statistical analysis, that of the *covariance of $x$ and $y$* (which we denote as $C[x, y]$). The covariance of two random variables is a measure of how they vary together (i.e., how they "co-vary"). If we let

$$\mu_x = E[x] \quad \text{and} \quad \mu_y = E[y],$$

then

$$\textit{Covariance of } x \textit{ and } y: \quad C[x, y] = E[(x - \mu_x)(y - \mu_y)]. \quad (3.16)$$

Like a variance, a covariance is somewhat difficult to interpret. If high values of $x$ (high relative to $\mu_x$) tend to be associated with high values of $y$ (relative to $\mu_y$), and low-values associated with low values, then $C[x, y]$ will be a large positive number.* If the covariance is a large negative number, this means that low values of one variable tend to be associated with high values of the other, and vice versa. If two variables are independent, then $C[x, y] = 0$ (i.e., they are not related).

To calculate a covariance, we could use the definition of $E[g(x, y)]$ in Formula (3.14). Rather than do this, we present an equivalent formula which is much more convenient computationally.

---

* This is because when $(x - \mu_x)$ is positive, $(y - \mu_y)$ is also positive, and when $(x - \mu_x)$ is negative, $(y - \mu_y)$ will also be negative; hence the sign of $(x - \mu_x)(y - \mu_y)$ will tend to be positive.

> *Equivalent formula for covariance of* $x$ *and* $y$:
> $$C[x, y] = E[x \cdot y] - E[x]E[y]. \qquad (3.17)$$

The reader may recognize this formula as just a variation of the mean of squares minus squared mean relationship provided in Formula (3.6). Indeed, if the variable $x$ is merely substituted wherever the variable $y$ occurs, the result is a variance formula. $C[x, y]$ becomes $C[x, x]$, which is exactly the same as

$$V[x] = E[x^2] - (E[x])^2.$$

We demonstrate the use of Formula (3.17) to calculate $C[x, y]$ for the data in Table 3.8. From that table,

$$E[x] = 7.20, \qquad E[y] = 4.30, \qquad \text{and} \qquad E[x \cdot y] = 30.80.$$

Thus,

$$
\begin{aligned}
C[x, y] &= E[x \cdot y] - E[x]E[y] \\
&= 30.80 - (7.20)(4.30) = -0.16.
\end{aligned}
$$

The covariance in this case is close to zero, indicating a very small (negative) relationship between the length $(x)$ and width $(y)$ of the lumber sold.

If $x$ and $y$ are independent, we know from Formula (3.15) that

$$E[x \cdot y] = E[x]E[y].$$

Substituting this relationship into Formula (3.17) yields a result that has already been stated; namely, when $x$ and $y$ are independent, then the covariance is zero,

$$C[x, y] = E[x]E[y] - E[x]E[y] = 0.$$

This relationship can be verified for the data in Table 3.9, where $x$ and $y$ are independent.

### Expectations of $x \pm y$

Another special case of importance for $E[g(x, y)]$ occurs when

$$g(x, y) = x + y, \qquad \text{or} \qquad g(x, y) = x - y.$$

Since the formulas for these two functions differ only in a single sign, we can investigate both cases simultaneously by letting $g(x, y) = x \pm y$.

*Expected value of $x + y$ or $x - y$:*

$$E[x \pm y] = \sum_{\text{All } y} \sum_{\text{All } x} (x \pm y)P(x, y) \qquad (3.18)$$

$$= E[x] \pm E[y].$$

To illustrate Formula (3.18), consider a company specializing in sending "care" packages to college students. These packages consist of a fruit box plus a "free gift," which is always either a glass mug or a metal coin bank, which are alternately placed in successive packages. The company is currently trying to analyze the total weight of their packages in order to better control mailing costs. The fruit boxes come in three weights, 2, 5, and 10 pounds. The mug weighs one pound and the coin bank weighs 1/2 pound. The company is interested in determining the average weight of each package, or $E[x + y]$, where $x =$ weight of the fruit and $y =$ weight of the gift. The probability distribution of $P(x)$, $P(y)$, and $P(x, y)$, based on sales records, is given in Table 3.10.

**Table 3.10**

| (1) Weight of fruit $(x)$ | (2) $P(x)$ | (3) $xP(x)$ | (4) Weight of gift $(y)$ | (5) $P(y)$ | (6) $yP(y)$ | (7) $(x, y)$ | (8) $P(x, y)$ | (9) $(x + y)$ | (10) $(x + y)P(x, y)$ |
|---|---|---|---|---|---|---|---|---|---|
| 2 | 0.45 | 0.90 | 1/2 | 0.50 | 0.25 | $(2, \frac{1}{2})$ | 0.20 | 2.5 | 0.500 |
| 5 | 0.35 | 1.75 | 1 | 0.50 | 0.50 | $(2, 1)$ | 0.25 | 3.0 | 0.750 |
| 10 | 0.20 | 2.00 | | | | $(5, \frac{1}{2})$ | 0.15 | 5.5 | 0.825 |
| | | | | | | $(5, 1)$ | 0.20 | 6.0 | 1.200 |
| | | | | | | $(10, \frac{1}{2})$ | 0.15 | 10.5 | 1.575 |
| | | | | | | $(10, 1)$ | 0.05 | 11.0 | 0.550 |
| Sum | 1.00 | 4.65 | | 1.00 | 0.75 | | 1.00 | | 5.400 |

From the final column of Table 3.10 we see that the average weight of the "care" packages is 5.40 pounds. This result is also obtained by summing the totals of columns (3) and (6):

$$E[x] + E[y] = 4.65 + 0.75 = 5.40.$$

Generally, it is easier to find $E[x + y]$ by summing $E[x] + E[y]$. As an exercise, the reader might use the data in Table 3.9 to show that

$$E[x + y] = 5.60.$$

**Variance of $x \pm y$**

Our final expectation in this chapter involves calculating $V[x + y]$. Although the formula given below is presented without proof, its derivation is not difficult, using the concepts of this chapter.

> *Variance of $(x + y)$ or $(x - y)$:*
> $$V[x \pm y] = V[x] + V[y] \pm 2C[x, y].$$    (3.19)

The sign of the final righthand term corresponds to the sign in the lefthand side term. To illustrate the calculation of $V[x + y]$, suppose we use the data in Table 3.8. The values $E[x^2]$ and $E[y^2]$ can be shown to be

$$E[x^2] = 54.4 \quad \text{and} \quad E[y^2] = 20.2.$$

We have already shown that $E[x] = 7.2$ and $E[y] = 4.3$. Thus,

$$V[x] = E[x^2] - (E[x])^2$$
$$= 54.4 - (7.2)^2 = 2.56,$$

and

$$V[y] = E[y^2] - (E[y])^2$$
$$= 20.2 - (4.3)^2 = 1.71.$$

Since we previously calculated $C[x, y] = -0.16$, we can write

$$V[x + y] = V[x] + V[y] + 2C[x, y]$$
$$= 2.56 + 1.71 + 2(-0.16) = 3.95.$$

**Independence**

When $x$ and $y$ are independent, then $C[x, y] = 0$; hence,

*Special case for independence:*

$$V[x \pm y] = V[x] + V[y].$$    (3.20)

Note that the variances are always summed, even if $V[x - y]$ is being found. The reader can convince himself of this result by using the data in Table 3.9, to show that both $V[x + y]$ and $V[x - y]$ are equal to 4.0, because

$$V[x] = 1.84 \quad \text{and} \quad V[y] = 2.16.$$

A summary of the bivariate expectation formulas from this section is shown in Table 3.11.

**Table 3.11**

| Expectation | Symbol | General formula | Special case of independence |
|---|---|---|---|
| 1. $E[x \times y]$ | —— | $\sum\sum(x \times y)P(x, y)$ | $E[x]E[y]$ |
| 2. $E[x \pm y]$ | —— | $E[x] \pm E[y]$ | Same as general formula |
| 3. $E[(x - \mu_x)(y - \mu_y)]$ | $C[x, y]$ | $E[x \cdot y] - E[x]E[y]$ | Zero |
| 4. $E[\{(x \pm y) - E[(x \pm y)]\}^2]$ | $V[x \pm y]$ | $V[x] + V[y] \pm 2C[x, y]$ | $V[x] + V[y]$ |

**REVIEW PROBLEMS**

1. Sketch the cumulative probability function for the random variable $x$ in Section 3.1 (where $x$ is the number of donuts sold to a customer). Compare your sketch to Fig. 3.2, and explain their differences and relationship.

2. Distinguish between a probability mass function and a cumulative probability function for a discrete random variable $x$.

3. a) Compute the expected value of the following five values: 7, 17, 16, 8, 2, by assuming that these values occur with probability $\frac{1}{3}, \frac{1}{4}, \frac{1}{6}, \frac{1}{12}$, and $\frac{1}{6}$, respectively.
   b) Find $E[x^2]$ for these same five values, and then use this result and your answer to part (a) to find $V[x]$.

4. a) Sketch the probability mass function for the population in Problem 3.
   b) Sketch the cumulative function for Problem 3. What probability corresponds to $F(2.0)$, $F(9.5)$, and $F(17.0)$?
   c) Does the mass function graphed in part (a) meet the two conditions of all probability functions described in Section 3.1?

5. A bowl contains five cubes numbered from one to five. Random samples of size two are drawn without replacement from the bowl. A random variable is defined to have values equal to the sum of the numbers on the two cubes drawn in each sample.

   a) Sketch the probability distribution of this random variable.
   b) Find the expected value of this random variable.
   c) Find the standard deviation of this random variable.
   d) Suppose the above experiment is done with replacement of the cube after each draw. Sketch the new probability distribution.

6. A certain gambling device consists of a sphere containing 10 nickels, 10 dimes, 15 quarters, and 15 half-dollars. On each gamble, the sphere is rotated and one of the coins drops out. Each individual coin has the same chance of falling out, but only one falls out each time.

   a) On an individual trial, what is the probability that the coin would be a dime?
   b) What is the probability of getting 50 cents or more on two successive trials?
   c) What is the average payoff expected on the first single trial of this gamble?

7. An entrepreneur is faced with two investment opportunities that each require an initial outlay of $10,000. He estimates that the return on investment $x$ will be either $40,000,

$20,000, or $0, with probability 0.25, 0.50, and 0.25, respectively. For investment $y$ the returns should be $30,000, $20,000, or $10,000, with probability of one-third in each case. Describe the returns from these investments in terms of random variables. Compute the expected value and variance of both $x$ and $y$ by first dividing each variable by 10,000, then subtracting 2.0 from the result, and then using the rules of linear transformations.

8. Given a random variable $x$ with mean 10 and variance 9, find the expected value and variance of the random variable $y$ where $y = 12 + 2x$.

9. Given a random variable $x$ with $E[x] = 5$ and $V[x] = 9$, and another random variable $y$ with $E[y] = 10$ and $V[y] = 25$. The variables $x$ and $y$ are independent.
   a) Find $E[x \cdot y]$, $E[x + 2y]$, and $E[13 - 2x]$.
   b) Find $V[x - y]$, $V[x + 2y]$, and $V[13 - 2x]$.
   c) What is the value of $C[x, y]$?

10. What are the expected value and variance of the sum of the numbers appearing when three dice are thrown simultaneously?

11. a) For the data of Table 3.8, show that $E[x + y] = 5.60 = E[x] + E[y]$.
    b) For the data of Table 3.8, show that $V[x - y] = 4.0 = V[x] + V[y]$.

12. a) Assume that the values of $x$ and $y$ shown below occur with equal probability.

| $x$ | 5 | 2 | 3 | 6 |
|---|---|---|---|---|
| $y$ | 6 | 6 | 6 | 6 |

   1) Find $E[x]$, $E[y]$, $V[x]$, and $V[y]$.
   2) Find $E[x + y]$ and $E[x - y]$.
   3) Are $x$ and $y$ independent or dependent? Explain.
   4) Find $V[x + y]$ and $C[x, y]$.

   b) Repeat the above question for the following data:

| $x$ | 5 | 2 | 3 | 6 |
|---|---|---|---|---|
| $y$ | 8 | 4 | 5 | 7 |

   Explain (intuitively) why $C[x, y]$ is larger for part (b) than for part (a).

## EXERCISES

13. Given the joint probability function with values $f(x, y)$ given in the following table:

| $y$ \ $x$ | 1 | 5 | 10 |
|---|---|---|---|
| 1 | $\frac{1}{20}$ | $\frac{2}{20}$ | $\frac{3}{20}$ |
| 2 | $\frac{4}{20}$ | 0 | $\frac{3}{20}$ |
| 3 | $\frac{2}{20}$ | $\frac{4}{20}$ | $\frac{1}{20}$ |

   a) Find the marginal probability functions for $x$ and for $y$.
   b) Find $E[x]$, $V[x]$, and $E[x + y]$.
   c) Determine whether $x$ and $y$ are independent.

14. Suppose that you are offered the opportunity to participate in an experiment in which a fair coin is tossed until the first tail appears. If a tail appears on the first toss you will be paid $2 for participating. If the first tail appears on the second toss, you will be paid $4 for participating. You will be paid $8 if the first tail appears on the third toss, and $16 if the first tail appears on the fourth toss; in other words, your payment is increased by a factor of 2 for each head that appears, the game ending with the appearance of a tail.

   a) How much would you be willing to pay to participate in this game? (If you said less than $2 you don't understand the game.)
   b) Determine the expected value for this game. Are you willing to pay more or less than the expected value of the game? Explain why.

15. A person is to draw one card at random from a deck of cards with the four aces removed. His payoff, $x$, is $1 if he selects a red card, $2 if he selects a club, $7 if he selects a spade from 2 to 9, and $10 if he selects a 10, jack, queen, or king of spades.

   a) Determine the probability distribution of $x$, and find the fair price to pay to play this game.
   b) What is the standard deviation of the probability distribution of the payoff, $x$?

16. Suppose that an experiment can be described by the following probability mass function:

$$P(x) = \begin{cases} \frac{1}{14} x^2 & \text{for } x = 1, 2, \text{ or } 3, \\ 0 & \text{otherwise.} \end{cases}$$

   a) Graph both the probability mass function and the cumulative probability function for this experiment.
   b) Prove that this function satisfies the two properties described in Section 3.1.
   c) Find the mean and the variance of this distribution.

17. In a Campus Chest bazaar, I run a booth which offers the chance to throw a dart at a balloon. If you break the balloon, you receive a prize equal to the amount hidden behind the balloon. Suppose each balloon is equally likely to be hit and the average chance of a hit is $\frac{1}{2}$ over all expected participants. The awards are distributed as follows:

> 40% have payoff of 5¢,
> 30% have payoff of 10¢,
> 20% have payoff of 25¢,
> 10% have payoff of $1.00.

   If I charge 15¢ a dart, what is my expected return for 500 darts thrown?

18. Let $y$ be a discrete random variable that assumes the value 1 with probability $p$, and the value 0 with probability $(1 - p)$. Find $E[y]$ and $V[y]$.

19. If $y$ is the random variable defined in Exercise 18, determine the values assumed, and probabilities, for the random variable $y^2$. Find $E[y^2]$ and $V[y^2]$.

20. a) Let $x$ be a random variable representing the number of heads that appear when a fair coin is tossed four consecutive times. Let $y = 2x - 4$. Is $y$ a random variable? If so, describe it.

   b) Find $E[x]$ and $V[x]$.

   c) Using the probability distribution for $y$ described in part (a), find $E[y]$ and $V[y]$.

   d) Using the results in part (b), and the expectation rules, find $E[y]$ and $V[y]$. Compare your answer here to your results in part (c).

21. Prove that a random variable $y$ and a constant $k$ are always statistically independent.

22. Given that $x$ is a discrete random variable with values chosen at random from the set $\{1, 2, 3, 4, 5, \text{ and } 6\}$. $y$ is a random variable with values at least as large as $x$ and not greater than 6.

   a) Determine the joint probability function of $x$ and $y$.

   b) Find the conditional probabilities $g(y \mid x = 3)$.

   c) Find the expected value of $y$.

# Discrete
# Probability Distributions

## 4.1 INTRODUCTION

While it is often useful to determine probabilities for a specific discrete random variable or combined random variables, there are many situations in statistical inference and decision-making that involve the same type of probability functions. In such instances, it is important to apply the theory of probability functions from the previous chapters to obtain *general* results about the mean, the variance, independence, and other characteristics of the random variables. Then it is not necessary to derive such results over and over again in each special case using different numbers. It would be quite discouraging to know all these concepts of probability and still have to go through the process of formulating a new probability function and deriving its characteristics every time we are concerned with a slightly different experiment. Fortunately, we can avoid this boredom by recognizing the similarities between certain types, or families, of apparently unique experiments, and then merely matching a given case to the general formulas. Some of these families of discrete probability distributions are discussed in this chapter.

## 4.2 BINOMIAL DISTRIBUTION

Many experiments share the common element that their outcomes can be classified into one of two events. For instance, the experiment "toss a coin" must result in either a head or a tail; the experiment "take an exam" can be considered to result in the outcomes pass or fail; a production process may turn out items which are either good or defective; and the stock market in general goes either up or down. In fact, it is often possible to describe the outcome of many of life's ventures in this fashion merely by distinguishing only two events, "success" and "failure." Experiments involving repeated independent trials, each with just *two* possible outcomes, play an important role in one of the most widely used discrete probability distributions, the *binomial distribution*.

### Bernoulli Trials

Several generations of the Bernoulli family, Swiss mathematicians of the 1700's, usually receive credit as the originators of much of the early research on probability theory, especially that involving problems characterized by the binomial distribution. Therefore, the Bernoulli name has now come to be associated with this class of experiments, and each repetition of an experiment involving only two outcomes (e.g., each toss of a coin) is called a *Bernoulli trial*. For the purposes of probability theory, interest centers not on a single Bernoulli trial but rather on a series of *independent, repeated* Bernoulli trials. That is, we are interested in more than one trial. The fact that these trials must be "independent" means that the results of any one trial cannot influence the results of any other trial. In addition, when a Bernoulli trial is "repeated," it means that the conditions under which each trial is held must be an exact replication of the conditions underlying all other trials. This implies that the probability of the two possible outcomes cannot change from trial to trial. Repeated flips of a coin are independent Bernoulli trials if $P$(Head) and $P$(Tail) remain constant over all flips, and if the outcome of one flip has no bearing on the outcome of any other flip.

In a binomial distribution the probabilities of interest are those of receiving a certain number of successes, $x$, in $n$ *independent trials*, each trial having the *same probability*, $p$, of success.* Note that the two assumptions underlying the binomial distribution, namely independent trials and a constant probability of success, are met by what we have called independent repeated Bernoulli trials and so, by definition, the binomial distribution is appropriate in an experiment involving these trials.

The binomial distribution is completely described by the values of $n$ and $p$, which are referred to as the "parameters" of this distribution. The word "parameter" in this context has the same meaning as it did in Chapter 1—it refers to a characteristic of a population. In the binomial distribution, $n$ is the parameter "number of trials," and $p$ is the parameter "probability of a success on a single trial." Given specific values of $n$ and $p$, one can calculate the probability of any specified number of successes, as well as determine the characteristics of the binomial distribution, such as its mean and variance.

To illustrate a situation where the binomial distribution applies, suppose that a production process is capable of producing a very large number of items; if this process is not working correctly, then there is a *constant* probability, $p$, that a component will be defective. Under this assumption, the number of defectives can range anywhere from zero up to the total number examined. Many probability questions are relevant, such as "What is the probability that a random sample of four will result in one defective?" or "What is the probability that there will be two or fewer defectives in a random sample of 20?" The use of the word "random" in this context implies independence among sample items. The binomial distribution can be used to determine probabilities of this nature.

### Binomial Calculations

It is possible to calculate the probabilities in a binomial situation by using the probability rules developed in Chapter 2. To determine the probability of exactly $x$ successes in $n$ repeated independent Bernoulli trials, each with a constant probability of success equal to $p$, it is necessary to find the probability of *any* one occurrence of this type and then multiply this value by the number of possible

---

* Since it is common in discussing the binomial distribution to refer to the number of trials as the "sample size," we will use a lower-case $n$ to denote this number. Technically, however, the number of trials is a population parameter, and hence could be denoted by capital $N$.

occurrences. Since it makes no difference which ordering we select, let us choose the one in which the $x$ successes come first, followed by the $(n - x)$ failures. Letting $S$ = Success and $F$ = Failure, we can represent this particular ordering as follows:

$$\underbrace{SS \ldots S}_{x \text{ successes}} \underbrace{FF \ldots F}_{\substack{(n - x) \\ \text{failures}}}$$

To determine the joint probability of this particular sequence of successes and failures, recall that all trials are assumed to be independent. This means that the joint probability of these $n$ events equals the product of the probabilities of each event.

$$P(S \cap S \cap \cdots \cap S \cap F \cap F \cap F \cdots \cap F) = P(S)P(S) \cdots P(S)P(F)P(F) \cdots P(F)$$

To simplify the notation, we denote the probability of a failure as $P(F) = q$ where $q = (1 - p)$, and $P(s) = p$.

$$\underbrace{P(S)P(S) \cdots P(S)}_{x \text{ successes}} \underbrace{P(F)P(F) \cdots P(F)}_{(n - x) \text{ failures}} = \underbrace{pp \cdots p}_{x \text{ successes}} \underbrace{qq \cdots q}_{\substack{(n - x) \\ \text{failures}}}$$

Since there are $x$ $p$-values and $(n - x)$ $q$-values, the above joint probability can be written as:

$$P(S \cap S \cap \cdots \cap S \cap F \cap F \cap \cdots \cap F) = p^x q^{n-x}.$$

Let us now relax the assumption that the $x$ successes occur in a specified order. Recall that the joint probability of a series of independent events does not depend on the *order* in which they are arranged; for example,

$$P(E_1 \cap E_2) = P(E_2 \cap E_1).$$

Hence, $p^x q^{n-x}$ represents the probability not only of our one arrangement, but of *any* possible arrangement of $x$ successes and $(n - x)$ failures. The problem which remains is to determine how many occurrences of $x$ successes and $(n - x)$ failures are possible. The answer is simply the number of combinations of $n$ objects, taken $x$ at a time,

$$_nC_x = \binom{n}{x}.$$

Therefore, multiplying the probability of each occurrence, $p^x q^{n-x}$, by the total number of such occurrences, $_nC_x$, gives the probability of $x$ successes in $n$ trials.

### The Binomial Formula

The probability mass function for a random variable $x$ from an experiment involving repeated independent Bernoulli trials, each having the same probability,

$p$, of success, is given by the binomial distribution:

*Binomial distribution:*

$$P(x \text{ successes in } n \text{ trials}) = \begin{cases} \dbinom{n}{x} p^x q^{n-x} & \text{for } \begin{cases} x = 0, 1, 2, \ldots, n, \\ n = 0, 1, 2, \ldots, \end{cases} \\ 0 & \text{otherwise.} \end{cases} \tag{4.1}$$

Note that the binomial probability distribution is a *discrete* distribution, since it has positive probabilities associated with only $(n + 1)$ values of the random variable $x$. Each probability is nonnegative and the sum of these probabilities over all values of $x$ must, of course, equal 1.0—i.e., it satisfies the properties of a probability function.

To illustrate the use of the binomial distribution, let's answer the questions presented earlier in this section about the number of defectives in a sample of four taken from the output of a production process. Define a success in this case to be the identification of a defective item, so that $x$ = the number of defectives. Also, let's denote the number of randomly selected items by $n$, where the fact that these items are *randomly* selected guarantees that each item examined for defectiveness is independent of every other item. Suppose $n = 4$ items are selected and $p = 0.50$ is the probability that an item is defective. In this case, the probability that exactly one item is defective, using Formula (4.1), is:

$$P(x_{\text{ binomial}} = 1) = \binom{n}{x} p^x q^{n-x}$$

$$= P(1) = \binom{4}{1}(0.50)^1(0.50)^3 = 0.2500.$$

The probability that two items in four would be defective is given by,

$$P(2) = \binom{4}{2}(0.50)^2(0.50)^2 = 0.3750.$$

Indeed, the probability of any number of defectives from 0 to 4 may be determined in the same way. Table 4.1 organizes the necessary calculations for all these values, and Fig. 4.1 illustrates the resulting probability function.

We can also use the values in Table 4.1 to calculate the probability of *two or fewer* defective components, when $n = 4$ and $p = 0.50$, as follows:

$$P(x \leqslant 2) = F(2) = P(x = 0) + P(x = 1) + P(x = 2)$$
$$= 0.0625 + 0.2500 + 0.3750 = 0.6875.$$

**Table 4.1** Finding binomial probabilities for $n = 4$, $p = 0.50$

| (1) $x$ | (2) $\binom{n}{x} = {}_nC_x$ | (3) $p^x q^{n-x}$ | (4) = (2) × (3) $P(x) = \binom{n}{x} p^x q^{n-x}$ |
|---|---|---|---|
| 0 | $\binom{4}{0} = \dfrac{4!}{0!\,4!} = 1$ | $(0.5)^0(0.5)^4 = 0.0625$ | $P(0) = 0.0625$ |
| 1 | $\binom{4}{1} = \dfrac{4!}{1!\,3!} = 4$ | $(0.5)^1(0.5)^3 = 0.0625$ | $P(1) = 0.2500$ |
| 2 | $\binom{4}{2} = \dfrac{4!}{2!\,2!} = 6$ | $(0.5)^2(0.5)^2 = 0.0625$ | $P(2) = 0.3750$ |
| 3 | $\binom{4}{3} = \dfrac{4!}{3!\,1!} = 4$ | $(0.5)^3(0.5)^1 = 0.0625$ | $P(3) = 0.2500$ |
| 4 | $\binom{4}{4} = \dfrac{4!}{4!\,0!} = 1$ | $(0.5)^4(0.5)^0 = 0.0625$ | $P(4) = 0.0625$ |
| | | | Sum = 1.0000 |

Several features of the values in Table 4.1 are worth mentioning. First, notice that the number of combinations, ${}_nC_x$, has a symmetrical pattern; both the smallest value of $x$ ($x = 0$) and the largest value of $x$ ($x = n$) give ${}_nC_x = 1$. The largest values of ${}_nC_x$ are in the middle. This symmetry is a characteristic of ${}_nC_x$ no matter what values $n$ and $x$ assume. Secondly, notice that all the values of $p^x q^{n-x}$ are equal. This will happen only when $p = q = 0.50$. Finally, whenever $p = q$, the binomial distribution will always be a symmetrical distribution, with its median equal to

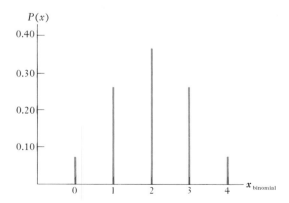

**Fig. 4.1.** Binomial distribution for $n = 4$, $p = 0.50$.

its mean. In the present example, median = mean = mode = 2, as shown in Fig. 4.1.

For most problems, $p$ is not equal to $q$, so the binomial distribution is not generally a symmetric distribution. Table 4.2 and Fig. 4.2 illustrate this situation for the case of $n = 4$ and $p = 0.10$.

The differences between this case and the previous example are obvious. While the number of combinations, $_nC_x$, remains the same, the probability of a given combination changes, so that the final probability distribution has large positive skewness.

**Table 4.2** Finding binomial probabilities for $n = 4$, $p = 0.10$

| (1)<br>$x$ | (2)<br>$\binom{n}{x}$ | (3)<br>$p^x q^{n-x}$ | (4) = (3) × (2)<br>$P(x)$ |
|---|---|---|---|
| 0 | $\binom{4}{0} = 1$ | $(0.1)^0(0.9)^4 = 0.6561$ | 0.6561 |
| 1 | $\binom{4}{1} = 4$ | $(0.1)^1(0.9)^3 = 0.0729$ | 0.2916 |
| 2 | $\binom{4}{2} = 6$ | $(0.1)^2(0.9)^2 = 0.0081$ | 0.0486 |
| 3 | $\binom{4}{3} = 4$ | $(0.1)^3(0.9)^1 = 0.0009$ | 0.0036 |
| 4 | $\binom{4}{4} = 1$ | $(0.1)^4(0.9)^0 = 0.0001$ | 0.0001 |
| | | | Sum = 1.0000 |

**Using a Binomial Table**

The calculations in Tables 4.1 and 4.2 are not very complex but the difficulties could readily become overwhelming if $n$ becomes much larger. Consider a more realistic situation, where $n = 20$ items are sampled and our question posed earlier is now rephrased, "What is the probability of finding four defectives if $n = 20$ and $p = 0.10$?" Using Formula (4.1), we have:

$$P(x_{\text{binomial}} = 4)$$
$$= \binom{n}{x} p^x q^{n-x} = \binom{20}{4} (0.10)^4 (0.90)^{20-4}$$
$$= \frac{20 \cdot 19 \cdot 18 \cdot 17 \cdot (16!)}{4 \cdot 3 \cdot 2 \cdot 1 \cdot (16!)} (0.10)(0.10)(0.10)(0.10) \overbrace{[(0.90)(0.90) \cdots (0.90)]}^{16 \text{ terms}}.$$

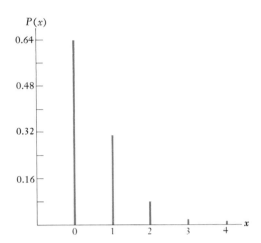

**Fig. 4.2.** Binomial distribution for $n = 4$, $p = 0.10$.

Only the foolish would desire to calculate such a number more than once and only the mighty will do it once. Fortunately, computers can be used to perform such tasks and tabulate the results. The answers to this problem and many others similar to it are readily available in tables similar to Table I at the end of the book. This table gives the probability of $x$ successes for a number of the more commonly referred to values of $n$, and for values of $p$ from 0.01 to 0.99. The probability $P(x = 4)$ for the above problem can be seen to equal 0.0898 by referring to the set of probabilities headed $n = 20$ and finding the values corresponding to $x = 4$ on the lefthand margin and $p = 0.10$ across the top (all decimals are omitted in this table).

The entire probability distribution for values of $x$ when $n = 20$ and $p = 0.10$ can be read from Table I. When the probability values become smaller than 0.00005, the table ends. Thus, if an event cannot be found in the binomial table, such as 18 defectives in 20 items when $p = 0.10$, it should be apparent from the table that this value is very small and for most practical purposes (certainly for any use in this text) can be considered zero, even though theoretically it is *not an impossibility*. If all the probabilities for $p = 0.10$ and $n = 20$ (or any other combination of $n$ and $p$) are summed, the total must equal 1.0000, rounded to four decimals. The reader should use Table I to find the probabilities for all values of $x$ when $n = 4$ and $p = 0.50$, or $p = 0.10$, to match with those calculated in Tables 4.1 and 4.2 for the previous examples.

An important aspect of Table I is its symmetry. Values of $p$ up to 0.50 (found across the top of each set of numbers) are used with the values of $x$ in the lefthand margin, while values of $p$ greater than 0.50 (found across the bottom of each set of numbers) are used with the values of $x$ read from the righthand margin. The symmetry in Table I, which permits values of $p$ from 0.51 to 1.00 to be read from the same set of numbers as those for $p$ in the interval 0.0 to 0.50, results from the fact that the probability of $x$ successes when $p$ is the probability of a success exactly equals the probability of $(n - x)$ failures when the probability of a failure is $(1 - p)$. Thus, in our example about defective components, we could have focused our attention on good items instead of defective items. Since the probability of $(n - x) = 16$ good items when the probability of a good item is 0.90 is the same as $x = 4$ defectives when the probability of a defective is 0.10, we could have let $x =$ number of good components, and obtained the same answer.

Cumulative probabilities can also be obtained from Table I by summing the probabilities of interest. For example, the probability that two or fewer defective components will appear in a sample of $n$ items is written as $P(x \leqslant 2) = F(2)$. If $n = 20$ and $p = 0.10$, this value is:

$$F(2) = P(x \leqslant 2) = P(0) + P(1) + P(2)$$
$$= 0.1216 + 0.2702 + 0.2852$$
$$= 0.6770.$$

Another example of the use of the cumulative probability function is evident in the probability of receiving more than two but no more than five defectives for the situation when $n = 20$ and $p = 0.10$. The answer can be found as follows:

$P(3 \leqslant x \leqslant 5)$
$$= P(x \leqslant 5) - P(x \leqslant 2) = F(5) - F(2)$$
$$= [P(0) + P(1) + P(2) + P(3) + P(4) + P(5)] - [P(0) + P(1) + P(2)]$$
$$= P(x = 3) + P(x = 4) + P(x = 5)$$
$$= 0.1901 + 0.0898 + 0.0319 \qquad \text{(from Table I)}$$
$$= 0.3118.$$

## 4.3 CHARACTERISTICS AND USE OF THE BINOMIAL DISTRIBUTION

Some of the properties of the binomial distribution have already been presented; these properties are now worth summarizing as we present the summary measures of the mean and standard deviation for the binomial variable. The shape of the

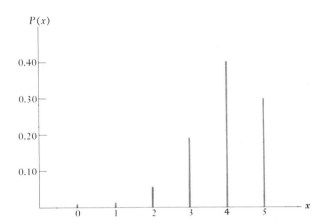

**Fig. 4.3.** Binomial distribution for $n = 5$, $p = 0.80$.

binomial depends on both $n$ and $p$. It will be useful to consider three different combinations of $n$ and $p$:

1. When $n$ is small and $p$ is also small; that is, when $p < \frac{1}{2}$;
2. When $n$ is small and $p$ is large ($p > \frac{1}{2}$);
3. When $p = \frac{1}{2}$ and/or $n$ is large.

**1. Small $n$ and $p < \frac{1}{2}$.** A typical illustration of this case is the binomial distribution for $n = 4$ and $p = 0.10$, shown in Table 4.2 and illustrated in Fig. 4.2. This distribution is skewed to the right, as are all binomial distributions when $n$ is small and $p < 1/2$.

**2. Small $n$ and $p > \frac{1}{2}$.** To illustrate this case, the distribution for $n = 5, p = 0.80$, is shown in Fig. 4.3. As Fig. 4.3 illustrates, binomial distributions when $n$ is small and $p > \frac{1}{2}$ will be skewed to the left, or negatively skewed.

**3. $p = \frac{1}{2}$ and/or large $n$.** When $p = \frac{1}{2}$, the binomial distribution will always be a symmetrical distribution, as we previously demonstrated in the discussion centered about Table 4.1 and Fig. 4.1. An important fact about the binomial is that even when $p \neq \frac{1}{2}$, the shape of the distribution becomes more and more symmetrical the larger the value of $n$. Figure 4.4 illustrates this fact. Note in parts (a) and (b) of this figure that even though $n$ is as small as 20, the distribution if $p = 0.20$ and $p = 0.40$ are fairly symmetrical in appearance. For $n = 100$ and $p = 0.30$, as shown in Part (c) of Fig. 4.4, the distribution is very symmetrical and bell-shaped.

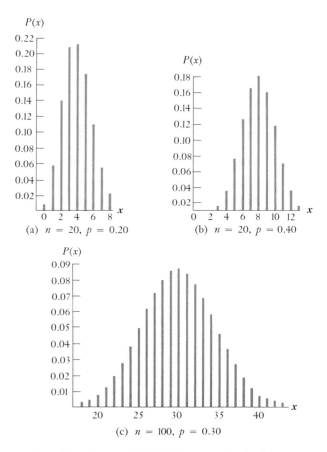

**Fig. 4.4.** The binomial distribution: $P(x) = \binom{n}{x}p^x q^{n-x}$.

## Mean of the Binomial Distribution

Since the binomial distribution is characterized by the value of the two parameters, $n$ and $p$, one might anticipate that the summary measures of the mean and standard deviation also can be determined in terms of $n$ and $p$. It should appear reasonable that the mean number of successes in any given experiment must equal the number of trials ($n$) times the probability of successes on each trial ($p$). If, for example, the probability that a process produces a defective item is $p = 0.10$, then the mean number of defectives in 20 trials is $20(0.10) = 2$; the mean number in 50 trials is

$50(0.10) = 5$, and the mean number in 100 trials is $100(0.10) = 10$. Thus, the mean number of successes in $n$ trials is $np$:*

$$\textit{Binomial mean:} \quad E[x_{binomial}] = \mu = np. \tag{4.2}$$

**Variance of the Binomial Distribution**

The variance (and standard deviation) of the binomial distribution is derived in a similar fashion to the mean and can be shown to be equal to:†

$$\textit{Binomial variance:} \quad V[x_{binomial}] = \sigma^2 = npq;$$
$$\textit{Binomial standard deviation:} \quad \sigma = \sqrt{npq}. \tag{4.3}$$

The variance in 20 Bernoulli trials of a process producing defectives with probability $p = 0.10$ is thus

$$npq = 20(0.10)(0.90) = 1.80;$$

---

* To more formally define the mean of the binomial distribution, we can let the random variable $x$ assume a value of 1 for each success and a value of 0 for each failure, in an experiment consisting of $n$ independent repeated Bernoulli trials. The number of successes in $n$ trials can now be determined merely by summing these $n$ values of $x$. The mean number of successes is given by taking the expectation of this sum, or

$$E[x + x + \cdots + x].$$

By the rules on expectations presented in Section 3.3, this can be simplified as follows:

$$E[\underbrace{x + x + \cdots + x}_{n \text{ values}}] = \underbrace{E[x] + E[x] + \cdots + E[x]}_{n \text{ values}} = nE[x].$$

The expected value $E[x]$ for a single trial equals the sum of the product of each possible value that $x$ can assume ($x = 1, x = 0$) times the probability of each of these values ($P(1) = p$ and $P(0) = q$):

$$E[x] = \sum_{x=0}^{1} xP(x) = (0)q + (1)p = p.$$

Thus, the mean value of $x$ over all $n$ trials, $nE[x]$, is seen to be $nE[x] = np$, which agrees with the value presented in Formula (4.2).

† To derive the variance of the binomial distribution first note that:

$$E[x^2] = \sum_{x=0}^{1} x^2 P(x) = (0^2)q + (1^2)p = p.$$

We can now use the mean square $-$ square mean relationship as follows:

$$\text{Var}(x + x + \cdots + x) = \text{Var}(x) + \text{Var}(x) + \cdots + \text{Var}(x) = n\text{Var}(x),$$
$$n\text{Var}(x) = n\{E[x^2] - (E[x])^2\} = n\{p - p^2\} = np(1 - p) = npq.$$

the standard deviation is

$$\sqrt{npq} = \sqrt{1.80} = 1.34.$$

Note, from the table below, that the intervals $(\mu \pm \sigma)$ and $(\mu \pm 2\sigma)$ for this problem contain slightly more of the total probability obtained from Table I than the rule of thumb given in Chapter 1 (68 and 95 percent) would indicate.

| Interval | Percent of probability |
|---|---|
| $\mu \pm \sigma = 2.00 \pm 1.34 = 0.66$ to $3.34$ | 75% |
| $\mu \pm 2\sigma = 2.00 \pm 2(1.34) = 0$ to $4.68$ | 97% |

The mean and standard deviation of the binomial distribution in Figs. 4.1 and 4.2 can also be found by using Formulas (4.2) and (4.3). For the former example, where $n = 4$ and $p = 0.5$, the mean is

$$np = 4(0.5) = 2.0;$$

the variance is

$$npq = 4(0.5)(0.5) = 1.0;$$

and the standard deviation is

$$\sqrt{V[x]} = \sqrt{1.0} = 1.0.$$

For the latter case of $n = 4$ and $p = 0.10$, the mean is $np = 4(0.10) = 0.4$, the variance is $npq = 4(0.10)(0.90) = 0.36$, and the standard deviation is $\sqrt{V[x]} = \sqrt{0.36} = 0.60$.

### Applications of the Binomial Distribution

The binomial distribution is useful as the basis for investigating a number of special decision problems. One particular problem is that of determining whether the outcomes resulting from repeated trials form a sequence that has a systematic pattern, or whether the different outcomes occur randomly in an unpredictable pattern. For example, in flipping a certain coin, do the results, heads or tails, occur in a random sequence? In a production process, are defectives found at random over the workday, or is there a systematic pattern? One popular belief is that defectives from assembly-line production occur more often on Mondays and Fridays, and immediately before and after rest breaks. One test for randomness in a sequence of outcomes is based on the binomial distribution and is discussed later in Chapter 8. For a more direct example of the use of the binomial, consider the following probability illustration.

**Example 1** Assume that on a given Sunday afternoon in September, television viewers in a certain city may choose between watching a pro football game or a major-league baseball game. Suppose we read in the newspaper two conflicting statements. First a football-league executive claims that 90 percent of the viewers watch the pro football games. Second, an advertising agent states that both games are equally popular among television viewers, implying that 50 percent watch each one. Obviously, we cannot learn the preferences of the entire population of television viewers, but suppose we conduct a small experiment by telephoning at random 25 households and asking what game they are watching. Assume that in 10 of these households a televised game is being viewed, with the following breakdown:

| Viewing football | Viewing baseball | Total |
|:---:|:---:|:---:|
| 7 | 3 | 10 |

Which of the two reports would seem to be more correct? The view of the football executive implies that $p = 0.90$, as opposed to $p = 0.50$ for the advertising agent. Based on $n = 10$, the total number of (presumably) independent trials in our survey, we can calculate the expected value (or mean) of this binomial distribution by using Formula (4.2). If the hypothesized $p$ values are substituted for $p$, the expected number of people viewing pro football would be

$$np = 10(0.90) = 9$$

when $p = 0.90$, and

$$np = 10(0.5) = 5$$

when $p = 0.50$. The observed value of 7 is halfway between these numbers in a numerical sense, but how about in a *probability* sense?

What is the probability that a random survey of 10 viewers would reveal 7 viewers watching pro football, under the two conflicting claims about the true population distribution, namely, $p = 0.90$ or $p = 0.50$? Using Table I for $n = 10$, $x = 7$, we find:

For $p = 0.9$,

$$P(x = 7) = \binom{10}{7}(0.9)^7(0.1)^3 = 0.0574;$$

For $p = 0.5$,

$$P(x = 7) = \binom{10}{7}(0.5)^7(0.5)^3 = 0.1172.$$

The probability of observing seven viewers watching pro football is more than twice as large if $p = 0.50$ as compared to $p = 0.90$. Based on our survey, we would thus tend to agree more with the advertising agent.

In this use of the binomial probability, hypothesized values of the true parameter, $p$, were examined on the basis of probabilistic values for the observed result $x = 7$. It must be pointed out that our conclusion is very sensitive. Suppose the survey had revealed 8 pro football viewers out of 10 instead of 7. Then, for $p = 0.90$,

$$P(x = 8) = 0.1937,$$

as opposed to

$$P(x = 8) = 0.0439$$

for $p = 0.50$. In this case the survey result $x = 8$ is more than 4 times as probable under the claim of the football executive as it is under the claim of the advertising agent. To avoid such sensitive results, larger sample surveys are usually taken. Also, the calculation of the probability in question is often phrased in terms of the set of outcomes which are equal to, or more extreme than, the observed one. For example, in examining the reasonableness of the hypothesis $p = 0.90$, in view of a sample result such as $x = 7$, the usual procedure is to calculate the probability of observing 7 *or less* viewers of pro football. On the other hand, the reasonableness of the hypothesis $p = 0.50$, when $x = 7$, is determined by calculating the probability of 7 *or more* viewers. The reader may wish to compare these two values, namely, $P(x \leqslant 7)$ if $p = 0.90$ and $P(x \geqslant 7)$ if $p = 0.50$, to determine which hypothesis about the value of $p$ appears to be more reasonable.

**Example 2** Another way of using a binomial probability distribution is to reverse the sense of the question, and ask what value for $p$ would be most compatible with the survey result. This process of examining a sample to make statements about a population parameter is the basis of *statistical inference*. Looking at Table I for $n = 10$, we see that the probability of $x = 7$ when $n = 10$ is the greatest when $p = 0.70$. In particular, $P(x = 7)$ for $p = 0.70$ is 0.2668, and $P(x = 7)$ is smaller for all other values of $p$. Thus, we might conclude that a true value of $p = 0.70$ for the population seems most reasonable or likely. Such a conclusion is also very sensitive to the particular sample result. If the survey had found 8 out of 10 watching pro football rather than 7, then the most likely value of $p$ would be $p = 0.8$.

This sensitivity in using probabilities in decision-making or statistical inference can be better understood if we can determine the degree of sensitivity, or more precisely, the chances of error, in our conclusion. Such errors are discussed in detail in Chapters 8 and 9. At this point, we hope merely to give some illustrations, and thus a motivation for studying probability distributions. The following example is an illustration of the use of the binomial distribution in a decision process where the chances of making errors are determined.

**Example 3** Assume that the production process described earlier in this chapter is malfunctioning and will require either a minor or major adjustment. If the defective rate is ten percent ($p = 0.10$), then only a minor adjustment is necessary;

if the number of defectives has jumped to 25.0 percent ($p = 0.25$), then a major adjustment is necessary. The problem at this point is how to decide, on the basis of a random sample of size $n = 20$, whether the process requires a minor or a major adjustment. This decision is not without risks, however, for we assume it to be costly to make the wrong decision—i.e., to make a minor adjustment to a process needing a major adjustment, or to make a major adjustment to a process needing only minor adjustments.

In this circumstance, we need to know how likely it is that $x$ defectives will occur in a sample of $n = 20$ when $p = 0.10$ and when $p = 0.25$. Table 4.3 provides these values.

Suppose, for the moment, that "four defectives" is established as the decision point between a major and minor adjustment: if there are four or more defectives, then major adjustments are made; with three or fewer defectives, minor repairs are made. This decision rule will lead to an *incorrect* decision if $x$ (the number of defectives) is greater than or equal to 4 when $p = 0.10$; that is, the correct decision would be a minor adjustment (since the true $p$ still equals 0.10), but our decision rule leads us to make a major adjustment (since $x \geqslant 4$). Similarly, if $x < 4$ when the true $p$ has really changed to 0.25, then an incorrect decision is also made. We would make a minor adjustment (since $x < 4$) when a major adjustment is really needed (since $p = 0.25$). From Table 4.3 we see that the probability of making these two types of error is as follows:

1.  Making a major adjustment when only minor adjustment is necessary:

$$\text{for } p = 0.10, P(x \geqslant 4) = 0.1331.$$

2.  Making a minor adjustment when a major adjustment is necessary:

$$\text{for } p = 0.25, P(x < 4) = 0.2251.$$

The analysis thus far has been based on using $x = 4$ as a decision point. The probabilities in the analysis will, of course, change if a decision point other than $x = 4$ is used. Suppose the choice between a major and a minor adjustment is set to depend on the critical value of $x = 3$ defectives; then the probability of the two types of error can be calculated from Table 4.3 to be

1.  For $p = 0.10$,

$$P(x \geqslant 3) = 0.3232;$$

2.  For $p = 0.25$,

$$P(x < 3) = 0.0912.$$

For a given value of $n$, one of these types of errors (e.g., making a major adjustment when a minor adjustment is necessary, or vice versa) can be made smaller only if the other is allowed to become larger. Just what decision rule (such as using

**Table 4.3** Using the binomial distribution in decision-making

| $x$ | Decision | If $p = 0.10$ <br> $P(x) = \binom{20}{x}(0.10)^x(0.90)^{20-x}$ | If $p = 0.25$ <br> $P(x) = \binom{20}{x}(0.25)^x(0.75)^{20-x}$ |
|---|---|---|---|
| 0 |  | 0.1216 | 0.0032 ⎫ |
| 1 | Minor | 0.2702 | 0.0211 ⎪ |
| 2 | adjustment | 0.2852 | 0.0669 ⎬ 0.2251 |
| 3 | ↓ | 0.1901 | 0.1339 ⎭ |
| | | ———— Decision (critical) ———— | |
| 4 | ↑ | 0.0898 ⎫    value | 0.1897 |
| 5 | | 0.0319 ⎪ | 0.2023 |
| 6 | | 0.0089 ⎪ | 0.1686 |
| 7 | | 0.0020 ⎪ | 0.1124 |
| 8 | | 0.0004 ⎪ | 0.0609 |
| 9 | Major | 0.0001 ⎬ 0.1331 | 0.0271 |
| 10 | adjustment | 0.0000 ⎪ | 0.0099 |
| 11 | | 0.0000 ⎪ | 0.0030 |
| 12 | | 0.0000 ⎪ | 0.0008 |
| 13 | | 0.0000 ⎪ | 0.0002 |
| 14–20 | ↓ | 0.0000 ⎭ | 0.0000 |
| Sum | | 1.0000 | 1.0000 |

$x = 3$ or $x = 4$ as the critical value) is "best" in a given circumstance depends largely on the costs associated with making these errors. We shall examine this subject in more detail in Chapters 8 and 9.

## 4.4 THE HYPERGEOMETRIC DISTRIBUTION

As we have indicated, the binomial distribution has widespread applications in many different areas, particularly for problems concerned with sampling. For such applications, use of the binomial distribution usually requires the assumption that one is sampling *with replacement*, because $p$ must remain constant from trial to trial. However, many practical sampling problems involve sampling *without replacement*. Fortunately, if $n$ is not too large relative to the population size, the binomial can still be used because it provides a good approximation to the correct answer. However, in cases where *sampling is without replacement and the sample size exceeds five percent of the population size*, this approximation is not sufficiently good and it is necessary to use the *hypergeometric distribution* to determine the precisely correct probability. The hypergeometric distribution applies to problems in which there are only two different types of elements in a finite population, $N_1$ of the first kind and $N_2$ of the second kind. The probability of drawing a sample of $x_1$ of this first type of element and $x_2$ of the second is given by the hypergeometric distribution:

*Hypergeometric distribution:*

$$P(x_1 \text{ out of } N_1 \text{ and } x_2 \text{ out of } N_2) = \frac{\binom{N_1}{x_1}\binom{N_2}{x_2}}{\binom{N_1 + N_2}{x_1 + x_2}}. \qquad (4.4)$$

Formula (4.4) is a direct result of applying a number of probability rules from Chapter 2. The first term in the numerator is the number of combinations of $N_1$ objects taken $x_1$ at a time, while the second term is the number of combinations of $N_2$ objects taken $x_2$ at a time. By the basic counting rule, we multiply

$$\binom{N_1}{x_1} \qquad \text{times} \qquad \binom{N_2}{x_2}$$

to get the total number of combinations of *both* $x_1$ out of $N_1$ *and* $x_2$ out of $N_2$. The denominator is the total number of combinations of $(N_1 + N_2)$ objects taken $(x_1 + x_2)$ at a time.

### Example of the Hypergeometric Distribution

To illustrate the use of the hypergeometric distribution, imagine a production process in which electrical components are produced in lots of 50. If this process is working correctly, there will be no defective items among the 50 produced. Let's assume, however, that at random intervals the process begins to malfunction, so that some of the components produced thereafter are defective. Since inspecting each and every one of the 50 items in a lot is relatively expensive, the hypergeometric distribution can be used to assess the probability that a random sample will detect defective components. Suppose, for example, that the process has, in fact, been malfunctioning, and exactly 5 components in a particular lot are defective. What is the probability that exactly one of these 5 defectives will appear in a sample of 4 randomly selected components? That is, what is the probability that a sample of 4 contains exactly 1 defective and 3 good components? This probability, that the sample contains $x_1 = 1$ of the $N_1 = 5$ defective components and $x_2 = 3$ of the $N_2 = 45$ good components, is given by the hypergeometric distribution:

$$P(1 \text{ defective and 3 good}) = \frac{\binom{5}{1}\binom{45}{3}}{\binom{5+45}{1+3}} = \frac{\binom{5}{1}\binom{45}{3}}{\binom{50}{4}}$$

$$= \frac{\dfrac{5!}{1!4!} \times \dfrac{45!}{3!42!}}{\dfrac{50!}{4!46!}} = 0.308.$$

Similarly, one might want to calculate the probability of receiving two or less defectives in a sample of 4:

$$P(2 \text{ or less defectives}) = P(0 \text{ defectives}) + P(1 \text{ defective}) + P(2 \text{ defectives})$$

$$= \frac{\binom{5}{0}\binom{45}{4}}{\binom{50}{4}} + \frac{\binom{5}{1}\binom{45}{3}}{\binom{50}{4}} + \frac{\binom{5}{2}\binom{45}{2}}{\binom{50}{4}}$$

$$= 0.647 + 0.308 + 0.043 = 0.998.$$

### Comparing the Hypergeometric and the Binomial Distributions

We should emphasize once more that the critical assumption implied by the way probabilities are calculated in the hypergeometric distribution is that sampling takes place from a finite population, without replacement. In terms of the present example, this means that the probability of receiving a defective or good item *changes* after each one of the 4 items is inspected. For instance, the probability that the first item selected is defective, assuming a random sample, is 5/50, or 0.10. If this item is, in fact, defective, then there are only 4 defectives remaining, so that on the second draw the probability of receiving a defective is 4/49, or 0.082. In this way, the experimental situation differs from the binomial situation where the trials are independent. For the binomial, it is necessary that the probabilities, $p$ and $q$, do not change during the trials.

Since calculating hypergeometric probabilities involves boring and tedious arithmetic, sampling experiments are generally devised, whenever possible, so that the binomial distribution can be used instead. That is, experiments involve either infinite populations, or sampling with replacement, or sample sizes of 5 percent or less of the population. In the latter case, it makes little difference whether the binomial or hypergeometric distribution is used, although the hypergeometric always gives exactly the correct answer.

Let's consider a hypergeometric problem in which the size of the population is fairly large, say $N = 100$, and where there are two different kinds of objects, with $N_1 = 40$ and $N_2 = 60$. Now, suppose we are interested in the probability of drawing a sample from this population which contains $x_1 = 2$ out of $N_1 = 40$ and $x_2 = 3$ out of $N_2 = 60$. The total sample size is thus $x_1 + x_2 = 5$. The probability of drawing this sample is given by Formula (4.4) as:

$$P(x_1 = 2 \text{ out of } N_1 = 40, \text{ and } x_2 = 3 \text{ out of } N_2 = 60)$$

$$= \frac{\binom{40}{2}\binom{60}{3}}{\binom{100}{5}} = 0.3545.$$

We can approximate the probability for this same event by using the binomial distribution. First, we denote as $x$ (or a success) the drawing of an object of the first kind; since there are 40 objects of the first kind out of a total of 100 objects, the probability of a success on the first draw is $p = 0.40$. This $p$ will change on every draw after the first, since we assume sampling without replacement. For example, if a success is received on the first draw, then the probability of a success on the second draw is $\frac{39}{99} = 0.3939$. Notice that because the population size is fairly large, the probability doesn't change too much from the first to the second draw. Hence, it is not unreasonable for us to assume, as we must for the binomial, that $p$ is a constant (which in this case is $p = 0.40$). Under this assumption, and where the sample size is $n = 5$ (remember, $x_1 + x_2 = 5$), the probability of exactly two successes $P(x = 2)$ is given by the binomial distribution (Formula (4.1) and Table I) as:

$$P(x = 2 \text{ successes in } n = 5 \text{ trials}) = \binom{5}{2}(0.40)^2(0.60)^3 = 0.3456.$$

The binomial value of 0.3456 is seen to be a fairly good approximation to the correct value from the hypergeometric, 0.3545. Because the binomial probabilities are easier to calculate, and often approximate the hypergeometric quite well, the binomial is more widely used than the hypergeometric distribution in statistical decision-making and inference. Care must be taken not to misuse this binomial approximation, for if the sample size $(x_1 + x_2)$ is large relative to the population size $(N = N_1 + N_2)$, the approximation will not generally be a good one.

**\*4.5  THE POISSON DISTRIBUTION**

Another important discrete distribution, the Poisson distribution, has recently found fairly wide application, especially in the area of operations research. This distribution was named for its originator, the French mathematician S. D. Poisson

---

\* This section may be omitted without loss of continuity.

(1781–1840), who described its use in a paper in 1837. Its rather morbid first applications indicated that the Poisson distribution quite accurately described the probability of deaths in the Prussian army resulting from the kick of a horse, as well as the number of suicides among women and children. More recent and useful applications involve arrivals at a service facility, or requests for service at that facility, as well as the rate at which this service is provided. A few of the many successful applications of the Poisson distribution include problems involving the number of arrivals or requests for service per unit time at toll booths on an expressway, at checkout counters in a supermarket, at teller windows in a bank, at runways in an airport, by maintenance men in a repair shop, or by machines needing repair.

In examples of the above nature, the Poisson distribution can be used to determine the probability of $x$ occurrences (arrivals or service completions) per unit time if four basic assumptions are met. First, it must be possible to divide the time interval being used into a large number of small subintervals in such a manner that the probability of an occurrence in each of these subintervals is very small. Second, the probability of an occurrence in each of the subintervals must remain constant throughout the time period being considered. Third, the probability of two or more occurrences in each subinterval must be small enough to be ignored. Fourth, an occurrence (or nonoccurrence) in one interval must not affect the occurrence (or nonoccurrence) in any other subinterval—i.e., the occurrences must be independent. Consider arrivals at a bank per hour, and suppose we can divide a given hour into intervals of one second, where the *probability* of a customer arriving during any given second is very small and remains constant throughout the one-hour period. Furthermore, assume that only one customer can arrive in a given second (e.g., the door is large enough to admit only one person), and that the number of arrivals in a given time period is independent of the number of arrivals in any *other* time period (e.g., customers don't turn away because of long lines). Under these circumstances, the number of arrivals in the one-hour period follows the Poisson distribution. These four assumptions and how they fit the bank example are summarized below.

| Assumption | Bank example |
| --- | --- |
| 1. Possible to divide time interval of interest into many small subintervals. | 1. Can divide the hour into subintervals of one second each. |
| 2. Probability of an occurrence remains constant throughout the time interval. | 2. The hour is one in which there is a steady flow of customers. |
| 3. Probability of two or more occurrences in a subinterval is small enough to be ignored. | 3. Impossible for two people to enter the bank simultaneously (i.e., in the same second). |
| 4. Independence of occurrences. | 4. Arrivals at the bank are not influenced by the length of the lines. |

Of these four assumptions, numbers (1) and (3) are general enough to apply to almost any setting involving arrivals over time.* The assumptions that occurrences are constant over time, and independent, however, are much less likely to be met in potential applications of the Poisson distribution. Nevertheless, the Poisson does seem to apply in a surprisingly large variety of different situations.

### Parameters of the Poisson Distribution

Examples of the Poisson distribution such as those given above are concerned with the probability of $x$ occurrences (arrivals or service completions) *per unit of time*. The only parameter necessary to characterize a population described by the Poisson distribution is the *mean rate* at which events occur. We shall use the Greek letter lambda, $\lambda$, for this parameter. Lambda can be defined as the mean rate of occurrence for any convenient unit of time, such as one minute, ten minutes, an hour, a day, or even a year. A value of $\lambda = 2.3$, for example, could indicate that there are, on the average, 2.3 requests for service in a particular bank every minute, or perhaps 2.3 customers arriving at a restaurant every 10 minutes. For practical applications the mean rate at which events occur must be determined empirically. That is, $\lambda$ must be known in advance, such as on the basis of a previous study of the situation. Once $\lambda$ is known, the frequency function for the Poisson distribution can be used to determine the probability that exactly $x$ occurrences, or events, take place in the specified interval of time. The value of $\lambda$ must be positive, and $x$ can assume any integer value from 0 to infinity.

$$
\textit{Poisson distribution:}\dagger \\
P(x) = \begin{cases} \dfrac{e^{-\lambda}\lambda^x}{x!}, & \text{for } x = 0, 1, 2, \ldots, \infty, \quad \lambda > 0, \\ 0, & \text{otherwise.} \end{cases} \quad (4.5)
$$

To illustrate the use of Formula (4.5), suppose the bank in our discussion knows from past experience that between 10 A.M. and 11 A.M. of each day the mean arrival rate of customers is $\lambda = 60$ customers per hour. Since arrivals are assumed to be constant over a given time interval, this rate is equivalent to an arrival rate

---

* The Poisson distribution can also be applied to problems involving the number of occurrences of a random variable for a given unit of *area*, such as the number of typographical errors on a page, the number of white blood cells in a blood suspension, the number of flaws in a fabric, or the number of imperfections in a surface of wood, metal, or paint.
† The symbol $e$ in this formula stands for a *constant* whose value is a nonrepeating, nonterminating decimal. We will approximate $e$ with the value 2.71828.

of $\lambda = 1$ customer per minute. Now, suppose the bank wants to determine the probability that exactly two customers will arrive in a given one-minute time interval between 10 and 11 A.M. Substituting $\lambda = 1$ and $x = 2$ into Formula (4.5) yields:

$$P(2 \text{ arrivals}) = \frac{e^{-1}1^2}{2!} = \frac{1}{2e}.$$

Since $e$ (approximately) equals 2.71828, we obtain

$$P(2 \text{ arrivals}) = \frac{1}{2(2.71828)} = 0.1839.$$

Similarly, the bank might want to calculate $P(2 \text{ or less arrivals})$:

$$P(2 \text{ or less arrivals}) = P(0) + P(1) + P(2)$$
$$= \frac{e^{-1}(1)^0}{0!} + \frac{e^{-1}(1)^1}{1!} + \frac{e^{-1}(1)^2}{2!}$$
$$= 0.3679 + 0.3679 + 0.1839 = 0.9197.$$

As was the case for the binomial, Poisson probabilities have been extensively tabulated, so that the task of calculating probabilities using Formula (4.5) can be avoided. Table II gives the probabilities for selected values of $\lambda$ from $\lambda = 0.01$ to $\lambda = 20.0$. The probability values for the above example are shown in Table II under the heading $\lambda = 1.0$. These values are graphed in part (a) of Fig. 4.5.

Each value of the parameter $\lambda$ represents a different member of the family of Poisson distributions. In a supermarket, for example, the mean number of arrivals per minute may be $\lambda = 3.8$. In this case the probability of observing exactly one arrival in a randomly selected minute would be 0.0850 (as determined from Table II, using $x = 1$ and $\lambda = 3.8$) as shown in Fig. 4.5(b). As a final example, the mean number of arrivals per minute might be $\lambda = 10$ at a city subway station during rush hour. Then, the probability of observing two or fewer arrivals in a randomly selected minute would be found in Table II under $\lambda = 10$, by summing the values for $x = 0, 1,$ and 2. This probability is 0.0028, indicating quite a rare event (see part (c) of Fig. 4.5).

Note in Fig. 4.5 that all probabilities are nonzero, and only a discrete number of values of $x$ have positive probabilities. Although all positive integers have some probability of occurring, this probability becomes very small for values of $x$ which differ from $\lambda$ by a factor of $(2\lambda)$ or more. Thus, $x_{\text{Poisson}}$ is an example of a discrete random variable having an infinite number of outcomes. As for all discrete mass functions, the sum of the probabilities over all values of $x$ must be 1.0. The graphs in Fig. 4.5 demonstrate that this distribution has positive skewness, since $x$ cannot

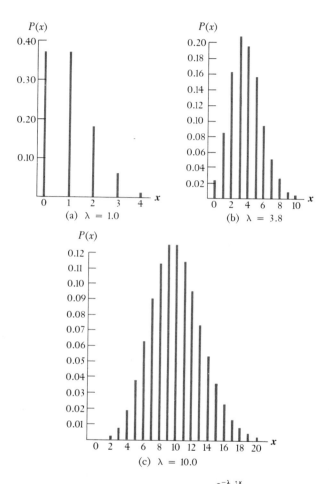

**Fig. 4.5.** The Poisson distribution: $P(x) = \dfrac{e^{-\lambda} \lambda^x}{x!}$ for three values of the parameter $\lambda$.

be lower than zero but may be any positive integer. However, Fig. 4.5(c) also indicates that when $\lambda$ is not too close to zero, the shape of the Poisson distribution appears to be fairly symmetrical.

**Mean and Variance of the Poisson Distribution**

The Poisson distribution has only one parameter ($\lambda$), so the mean and variance of the distribution must be functions of this parameter. Indeed, $\lambda$ is defined as the *mean* number of occurrences of the particular event per time or space unit. There-fore, the *mean of the Poisson distribution is* $\lambda$. What is not obvious at all, but is proven in many statistics books, is that *the variance of the Poisson distribution is also identically equal to $\lambda$.** That is,

$$
\begin{array}{ll}
\textit{Poisson mean:} & \mu = \lambda; \\
\textit{Poisson variance:} & \sigma^2 = \lambda.
\end{array}
\tag{4.6}
$$

Thus, if a value of $\lambda = 1.0$ indicates that customers are arriving at a bank according to a Poisson probability law at an average rate of 1.0 customer every minute, then the variance of these arrivals is also 1.0.

### Determining a Poisson Distribution in a Practical Situation

As another illustration of the Poisson distribution, suppose it has been suggested to the manager of a large supermarket that arrivals at the checkout counters might follow a Poisson distribution during certain periods of time. The manager inves-tigates this probability by first checking the reasonableness of the four basic assumptions listed earlier, especially assumptions number 2 (constant probability) and 4 (independence). After careful consideration, the time interval 4–6 P.M. Monday–Friday, is selected as an appropriate period during which all four as-sumptions seem reasonable.

The next step this manager might take is the collection of observations con-cerning the number of arrivals within the two hours from 4–6 P.M. Rather than count the number of customers arriving for many different *two-hour* time periods, arrivals are recorded in each of 100 randomly selected *one-minute* periods between 4–6 P.M. This procedure is reasonable, since arrivals are assumed to be constant over the two-hour period; hence it makes no difference, theoretically at least, what time unit is used. The number of customers arriving per minute ($x$) in this study ranged from 0 to 10, as shown in Table 4.4. For example, the third line of Table 4.4 indicates that on 19 different occasions there were exactly two arrivals in the one-minute period. The fourth row indicates that there were exactly three arrivals on 23 different occasions.

Recall that, for the Poisson distribution, both $\mu$ and $\sigma^2$ are equal. Hence, in checking to see if the data in Table 4.4 does follow a Poisson distribution, the manager might first wish to see if $\mu = \sigma^2$. Column (4) in this table gives the appro-priate values for calculating $\mu$. Using Formula (1.6), the calculated value of $\mu$ is:

---

* See for example, D. L. Harnett, *Introduction to Statistical Methods*, 2nd edition, Section 4.5 (Reading, Mass.; Addison-Wesley, 1975).

$$\mu = \sum x_i \left(\frac{f_i}{N}\right) = 3.79.$$

The variance of the data in Table 4.4 can be calculated using the mean-square-minus-the-square-mean relationship presented in Formula (3.6). Substituting the appropriate values,

$$E[x] = 3.79 \quad \text{and} \quad E[x^2] = \sum x^2 \frac{f_i}{n} = 18.27$$

(from column 6 of Table 4.4), we obtain

$$V[x] = \sigma^2 = E[x^2] - (E[x])^2$$
$$= 18.27 - (3.79)^2 = 3.91.$$

We see for this data that $\mu$ and $\sigma^2$ are very nearly equal in value (3.79 vs. 3.91). Theoretically, they are supposed to be *exactly* equal but, in a sample of just 100 time periods, we would not expect the observed value of either $\mu$ or $\sigma^2$ to *exactly* equal the population parameter $\lambda$. The closeness of $\mu$ to $\sigma^2$ in this case is encouraging enough for the store manager to make the most important test, namely, to see if the observed relative frequencies correspond to the theoretical frequencies for the Poisson distribution. To make this comparison, suppose we assume that $\lambda = 3.8$. (Note that it is not known from the data in Table 4.4 whether $\lambda = 3.79$ or 3.91, or some other value, and so the choice of $\lambda = 3.8$ is somewhat arbitrary.)

**Table 4.4**

| (1) Arrivals ($x$) | (2) Observed frequency | (3) $\frac{f}{n}$ | (4) $x\frac{f}{n}$ | (5) $x^2$ | (6) $x^2\frac{f}{n}$ |
|---|---|---|---|---|---|
| 0 | 0 | 0.00 | 0.00 | 0 | 0.00 |
| 1 | 8 | 0.08 | 0.08 | 1 | 0.08 |
| 2 | 19 | 0.19 | 0.38 | 4 | 0.76 |
| 3 | 23 | 0.23 | 0.69 | 9 | 2.07 |
| 4 | 17 | 0.17 | 0.68 | 16 | 2.72 |
| 5 | 15 | 0.15 | 0.75 | 25 | 3.75 |
| 6 | 8 | 0.08 | 0.48 | 36 | 2.88 |
| 7 | 3 | 0.03 | 0.21 | 49 | 1.47 |
| 8 | 3 | 0.03 | 0.24 | 64 | 1.92 |
| 9 | 2 | 0.02 | 0.18 | 81 | 1.02 |
| 10 | 1 | 0.01 | 0.10 | 100 | 1.00 |
| Sum | 100 | 1.00 | 3.79 | | 18.27 |

Now, if the number of arrivals in these 100 one-minute periods does follow a Poisson distribution with $\lambda = 3.8$, we would expect the observed frequencies in column 3 of Table 4.4 to closely correspond to the probabilities of the following Poisson distribution

$$P(x) = \frac{e^{-3.8}(3.8)^x}{x!}.$$

These values, from Table II under the heading $\lambda = 3.8$, are reproduced in Table 4.5. We see in Table 4.5 that the observed relative frequencies correspond quite well to the Poisson values for $\lambda = 3.8$. A good exercise for the reader at this point would be to verify, for the probabilities shown in the last column of Table 4.5, that the mean and variance are equal—i.e., that $\mu = \sigma^2 = 3.8$.

**Table 4.5**

| Arrivals | Observed relative frequency | Poisson value $p(x) = e^{-3.8}(3.8)^x/x!$ |
|---|---|---|
| 0 | 0.010 | 0.0224 |
| 1 | 0.080 | 0.0850 |
| 2 | 0.190 | 0.1615 |
| 3 | 0.230 | 0.2046 |
| 4 | 0.170 | 0.1944 |
| 5 | 0.150 | 0.1477 |
| 6 | 0.080 | 0.0936 |
| 7 | 0.030 | 0.0508 |
| 8 | 0.030 | 0.0241 |
| 9 | 0.020 | 0.0102 |
| 10 | 0.010 | 0.0039 |
| 11 | 0.000 | 0.0013 |
| 12 | 0.000 | 0.0004 |
| 13 | 0.000 | 0.0001 |
| Sum | 1.000 | 1.0000 |

The standard deviation of the Poisson distribution when $\lambda = 3.8$ is

$$\sigma = \sqrt{3.80} = 1.94.$$

Using Table 4.5, column (3), the intervals

$$\mu \pm \sigma = 3.8 \pm 1.94 \quad \text{and} \quad \mu \pm 2\sigma = 3.8 \pm 2(1.94)$$

are shown below to contain 71 and 96 percent of the probability, respectively. These are again close to our rule-of-thumb values of 0.68 and 0.95, respectively, for interpreting the meaning of the standard deviation.

| Interval | Probability included |
|---|---|
| $\mu \pm \sigma = 3.80 \pm (1.94) = 1.86$ to $5.74$ | $\sum_{x=2}^{5} P(x) = 0.7082$ |
| $\mu \pm 2\sigma = 3.80 \pm 2(1.94) = 0$ to $7.68$ | $\sum_{x=0}^{7} P(x) = 0.9600$ |

## 4.6  APPROXIMATION OF DISCRETE RANDOM VARIABLES BY CONTINUOUS RANDOM VARIABLES

The use of discrete random variables, especially the binomial random variable, is very important and applies to many experimental situations. Although calculation of the probabilities of discrete random variables is often a quite tedious task, we are fortunate in that such probabilities are already tabulated for many mass functions. A quick perusal of Tables I and II, however, indicates that not all values of $n$ and $p$ in the binomial, or of $\lambda$ in the Poisson, are specified. One might ask what is done in all the cases for which no tabled value exists, such as $n = 26$ or $p = 0.618$ in a binomial, or $\lambda = 12.55$ in a Poisson. The answer is that in these cases it is often advantageous to use a *continuous* approximation to a discrete distribution. This is one of the reasons we study continuous random variables, although not the only one, since there are many instances where the random variable in an experiment may actually assume a continuous form. Our point is that a study of *only* discrete probability distributions is not sufficient to enable us to proceed to more relevant and interesting problems in decision-making and statistical inference.

### REVIEW PROBLEMS

1. Six products are selected at random from a very large group of products of a certain kind. If 40% of the products are defective, what is the probability that no more than four of the six selected will be defective?

2. It is established that a fighter-bomber will hit its target $\frac{3}{4}$ of the time. Suppose military tacticians want a certain crucial target area destroyed. If you assign five planes to strike this same target area one time each, what is the probability that it will be hit two or more times?

3. Suppose that a business submits bids for certain construction projects, and there are always four companies bidding besides this one. Assume that the long-run chances of this firm being the low bidder are one out of five. What is the probability that this business will be awarded the contract as the low bidder in exactly two of the next four projects?

4. If $\frac{2}{3}$ of the students on a certain campus are lower-division students (freshman and sophomores), what is the probability that five students selected at random will include exactly three lower-division students?

5. a) Sketch the binomial mass function and the binomial cumulative function for $n = 8$ and $p = 0.40$. (Use Table I.)
   b) What is the probability $P(x = 7)$ for the above parameters? What is the probability $P(x \geqslant 7)$?
   c) Find the mean and the variance of this distribution.

6. Use Formula (4.1) to determine the binomial mass function for the parameters $n = 5$, $p = 0.50$. Check your answer with Table I. Sketch both the mass function and the cumulative function for these parameters.

7. Past surveys show that 40% of the senior students at a certain college own cars. Suppose six seniors are selected at random (with replacement).

   a) What is the probability that exactly four will own cars?
   b) What is the probability that at least one will own a car?
   c) What is the theoretical mean of the probability distribution under consideration?

8. Find the mean and standard deviation of a binomial random variable, measured in dollars, generated from 12 repeated trials with the probability of success on any given trial being one-third.

9. In 900 trials of tossing a fair die, how many times would you expect a number less than three to turn up?

10. Five products are selected at random from a very large group of products of a certain kind. If 40% of the products are defective, what is the probability that more than 3 of the 5 selected will be defective?

11. In six independent tosses of a fair coin, what is the probability of receiving:

    a) exactly three heads?                    b) at least three heads?
    c) at most three heads?                     d) between two and four heads?
    e) What is the expected value for the number of heads in six tosses? What is the variance?

12. a) If a manufacturing process is working correctly, only 10 percent of the items produced will be defective. You take a random sample of five items. What is the probability that exactly two of these will be defective?
    b) Suppose you selected 25 items at random; what is the expected number of defectives?

13. Suppose that a committee of six is to be chosen from ten men, five of whom are Republicans and five Democrats.

    a) If the committee is to be chosen by random selection, what is the probability that there will be three Republicans and three Democrats on the committee?
    b) What is the probability that a majority of the committee members will be Democrats?

14. A corporation hires five new employees for a training program. Past records indicate that only two out of every three new trainees remain with the company for more than two years. Assuming this trend continues and that the decisions of the new employees are independent,

    a) What is the probability that more than three of the five will continue with the company after two years?
    b) What is the expected number that would remain after two years?

15. In a certain population, 0.3 of the people are lefthanded. If 8 people are selected from this population at random, what is the probability that less than three will be lefthanded?

16. Suppose a large box at a cafe door contains 25 percent coupons for a free beer and 75 percent coupons for a free cola drink. Four customers enter and each draws one coupon. If $x$ is defined as the number of beer coupons that will be drawn,

    a) Find the probability distribution of $x$ and sketch its cumulative probability function.
    b) Find the probability that at least two beer coupons will be drawn among the four.

17. In a certain location, rain falls on one out of three days on the average. If three days are selected at random, what is the probability that rain will fall on one of those dates?

18. How does the hypergeometric distribution differ from the binomial distribution? Under what circumstances is the hypergeometric appropriate? Under what circumstances is the binomial appropriate?

19. Suppose that in a production run of ten units, three are defective. A sample of three units is to be randomly drawn from the ten. What is the probability of receiving *at least* one defective if the samples are drawn: (a) with replacement, (b) without replacement. Should your answer to (a) be higher or lower than to (b)? Why?

20. Use Formula (4.5) to determine the probabilities associated with the Poisson distribution for $\lambda = 2.0$. Sketch both the mass and the cumulative distributions. Check your answers by using Table II of Appendix B. What are the mean and the variance of this distribution?

21. If 10 cards are drawn with replacement from a deck of cards, what is the probability of receiving five aces? (Let $p = 0.08$.) What is the Poisson approximation to this probability? Compare the variance of the binomial distribution for this problem with the variance of the Poisson approximation.

22. In an airport, an average of 8.5 pieces of baggage per minute are handled following a Poisson distribution. Find the probability of 10 pieces of baggage being handled in a selected minute of time.

23. Suppose in a textile manufacturing process, an average of 2 flaws per 10 running yards of material have appeared. What is the probability that a given ten-yard segment will have 0 or 1 defects, if the number of flaws follows a Poisson distribution?

24. Suppose on the average 2.3 telephone calls per minute are made through a central switchboard, according to a Poisson distribution. What is the probability that during a given minute exactly 2 calls will be made?

25. Out of 10 salesmen, seven (call them group $A$) make sales on 20% of their calls. The other three (call them group $B$) make sales on 50% of their calls.

    a) What is the average percentage of sales to calls for all ten salesmen?
    b) Suppose the sales manager selects three of these salesmen at random to assign to a new territory. What is the probability that exactly two of them will be from group $A$?
    c) Suppose we follow one of the salesmen in group $A$ on his next five calls, each of which is considered independent of any of the others. What is the probability that he makes more than one but less than four sales in those five calls?

## EXERCISES

26. It is possible to generalize the binomial distribution to the class of problems where there are more than just two outcomes. Suppose that there are $k$ different (i.e., distinguishable) outcomes, and the probability of occurrence of these outcomes is $p_1, p_2, \ldots, p_k$. Assuming these $k$ outcomes to be mutually exclusive and exhaustive, it must be true that $p_1 + p_2 + \cdots + p_k = 1$. Now, assume that we want to determine the (joint) probability of $n_1$ outcomes of the first kind, $n_2$ outcomes of the second kind, and $n_k$ outcomes of the $k$th kind, where $n = n_1 + n_2 + \cdots + n_k$. An extension of Formula (2.14) gives the number of ways these objects could occur in $n$ trials, where $n = n_1 + n_2 + \cdots + n_k$. Multiplying this formula by the appropriate probabilities results in a probability distribution called the *multinomial distribution*.

*Multinomial distribution:*

$$P(n_1, n_2, \ldots, n_k) = \frac{n!}{n_1! \, n_2! \ldots n_k!} p_1^{n_1} p_2^{n_2} \ldots p_k^{n_k}.$$

   a) Use the multinomial distribution to determine the probability of receiving, in seven draws with replacement from a deck of cards, two black cards, four hearts, and the jack of diamonds.

   b) Suppose that the probability that an individual has Type O blood is 0.45, the probability of Type A blood is 0.40, the probability of Type B blood is 0.10, and the probability of Type AB blood is 0.05. In six randomly selected donors at a blood bank, what is the probability that three people will have Type O blood, two will have Type A blood, and one will have Type AB blood?

27. If a manufacturing process is working correctly, only 10 percent of the items produced will be defective. You take a sample of 7 items.

   a) What is the probability that 3 of these will be defective?

   b) If 3 items were defective would you, as a quality-control inspector, take any action? Why or why not?

28. Suppose that a baseball player has averaged four official at-bats in 150 games and hit 50 home runs. Based on his past performance, what is the probability that in game number 151 this player will hit at least one home run in his first four official times at bat? Are the assumptions of the binomial distribution reasonable for this application? Explain.

29. a) If the two baseball teams playing in the World Series are exactly evenly matched for each game in the series, what is the most likely number of games to be played in a best-out-of-seven series?

   b) What is the most likely number of games if one team is always a $3:1$ favorite?

30. The expected value of $x$ in a series of repeated Bernoulli trials is defined as follows:

$$E[x] = \sum_{x=0}^{n} x \binom{n}{x} p^x q^{n-x}.$$

a) Use the above relationship to prove that the mean of the binomial distribution equals $np$.

b) In the same manner, prove that the variance of the binomial distribution equals $npq$.

31. A classical example of the Poisson distribution resulted from a study of the number of deaths from horse kicks in the Prussian Army from 1875 to 1894. The data for this example are:

| Deaths per corps (per year) | Observed frequency |
|---|---|
| 0 | 144 |
| 1 | 91 |
| 2 | 32 |
| 3 | 11 |
| 4 | 2 |
| 5 and over | 0 |
| Total | 280 |

a) Fit a Poisson distribution to this data. (*Hint*: Note that there were 196 deaths from the 280 observations; hence the mean death rate was $196/280 = 0.70$.) How good does the Poisson approximation appear to be?

b) Do the assumptions of the Poisson distribution seem to be reasonable in this problem? Explain.

32. Use Table I to determine the binomial probabilities for $n = 10$ and $p = 0.20$, and then use Table II to find the Poisson approximation to these probabilities. Comment on how good the Poisson approximation is in this case. Graph both distributions on the same sheet of paper.

33. Using the Poisson distribution in the third column of Table 4.5, show by calculation that its mean and variance are both equal to 3.8.

34. *Shark Loans* (continued)

a) Assume in problem 32 of Chapter 1 that customers arrive at the loan office at a rate of $\lambda = 12$ customers per day. Graph the probability mass function for values of $x$ between 5 and 18.

b) Suppose the people at Shark have noticed that approximately 50 people walk by their office each day. If we assume [from part (a)] that there is a constant probability that each of these people will enter, the binomial distribution can be used to describe the arrival rate.

1) What is the appropriate values of $p$ if we assume 12 of the 50 people enter? What will $\mu$ and $\sigma^2$ be for this binomial distribution?

2) Superimpose on your graph in part (a) the probability mass function for the binomial for $x$ between 5 and 18.

# Probability Theory: Continuous Random Variables

## 5.1 INTRODUCTION

We have examined experiments involving only a discrete set of outcomes, thus limiting ourselves to discrete probability values. As we indicated earlier, however, an outcome set can be continuous as well as discrete, which implies that the random variable in an experiment must be able to assume a continuous form. Fortunately, most probability theory is basically the same for discrete and continuous random variables, and the formulas presented in Chapters 2 and 3 hold for both cases.

Probability functions defined in terms of a continuous random variable are usually referred to as *probability density functions* (abbreviated pdf), or simply as *density functions*. In this chapter, we first discuss the similarities and differences between density functions for continuous random variables and the probability mass functions described in Chapters 3 and 4. Some of the more useful density functions, those used in applications dealing with a wide range of experimental situations and decision problems, are then introduced.

## 5.2 PROBABILITY DENSITY FUNCTIONS

If an experiment can result in an infinite, noncountable number of outcomes, then the random variable defined must be continuous. Typically, whenever the value of a random variable is "measured" rather than "counted," then a continuous random variable is defined. Examples in which outcomes are measured rather than counted might include the water level in a lake, the pressure in a steam boiler, the distance between two points, or even the number of ounces in a cereal box. The value of the random variables in these examples can be any number from zero up to infinity. If we change these examples to be the *errors* in measuring water level, pressure, distances, or ounces (i.e., deviations from the mean), then such a random variable could be any number from minus infinity to plus infinity.

When we say a random variable can be *any* number between two limits, we mean any value is at least *theoretically* possible. For practical purposes, we usually can't measure such variables with a very great accuracy. For example, a swimmer's time in the 100-yard freestyle can, theoretically, be any number from 0 to infinity. But even with electronic timing devices we know that times can be recorded up to only about $\frac{1}{1000}$ of a second. Hence, while such a variable is theoretically continuous, for practical purposes it is discrete. We will see in this chapter that it is often more convenient to manipulate continuous variables than discrete variables.

### A Graphical Representation

Recall that, for a probability mass function, the value of $P(x = x)$ was represented by the *height* of the spike at the point $x = x$. One of the major differences between discrete and continuous probability distributions is that this representation no longer holds. As we will elaborate more thoroughly in this chapter, *probability for a continuous pdf is represented by the area between the x-axis and the density function.* We will show, for all probability density functions, that;

1. The probability $P(a \leqslant x \leqslant b)$ is equivalent to the area between $a$ and $b$ under the density function;

2. The total area under the pdf must be 1.0.

To illustrate the concept of a probability density function, consider a business-man who is trying to estimate the probability of various levels of sales for a new product he is marketing. In estimating sales he decides to assess probabilities using seven different intervals, each of size 1000, ranging from 0 to 7,000 units. Let's assume that Table 5.1 shows his probability assessment, where $x$ = number of units of possible sales.

**Table 5.1** Probabilities of sales

| Interval $(a, b)$ | Midpoint; class-mark | Probability $(a \leqslant x < b)$ |
|---|---|---|
| $0 < x \leqslant 1000$ | 500 | 0.00 |
| $1000 < x \leqslant 2000$ | 1500 | 0.05 |
| $2000 < x \leqslant 3000$ | 2500 | 0.25 |
| $3000 < x \leqslant 4000$ | 3500 | 0.30 |
| $4000 < x \leqslant 5000$ | 4500 | 0.25 |
| $5000 < x \leqslant 6000$ | 5500 | 0.10 |
| $6000 < x \leqslant 7000$ | 6500 | 0.05 |
| Sum | | 1.00 |

This probability distribution can be represented in a histogram, just as we did in Chapter 1, and smoothed with a frequency polygon by connecting the class marks of the intervals. The resulting histogram is shown in Fig. 5.1.

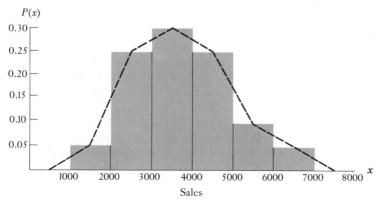

**Fig. 5.1.** Frequency polygon and histogram based on Table 5.1.

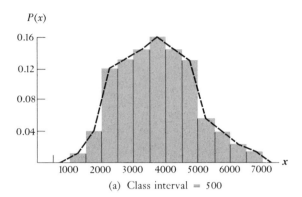

(a) Class interval = 500

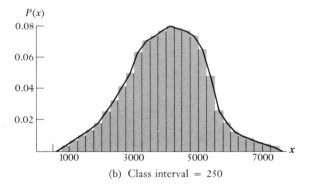

(b) Class interval = 250

**Fig. 5.2.** The frequency polygon for two different class sizes.

The choice of a class interval of size 1000 for this problem was quite arbitrary. Almost any size of interval could have been used. For example, suppose we now decrease the width of the classes, to an interval of, say, 500, or 250, or even to a class interval of 1. Figure 5.2 illustrates how the frequency polygon for this data changes as the width of the interval decreases (and, correspondingly, as the *number* of intervals increases); that is, it begins to look more and more like a *smooth continuous* function. As the width of the class interval becomes closer and closer to zero, the number of classes (or events) under consideration must increase until, at the limit (that is, when the interval size *is* zero) there are an *infinite* number of classes between any two values of *x*. The histogram in this case becomes an *infinite* number of infinitesimally narrow spikes set side by side. According to the rules of probability, the probability that any one of these spikes will occur is $1/\infty$, which is zero. Thus, an important fundamental rule of continuous random variables is:

> *The probability that any one specific value takes place is zero when the random variable is continuous.*

Because the probability of a single point is now zero, we can determine probability values only for intervals, such as $a \leqslant x \leqslant b$.

### Probability as an Area

The histograms in Figs. 5.1 and 5.2 illustrate probability values for class intervals of three different sizes. Let's denote the width of the class interval in a histogram as $\Delta x$ (read "delta-$x$"), where $\Delta x = 1000$ for Fig. 5.1, $\Delta x = 500$ for Fig. 5.2(a), and $\Delta x = 250$ for Fig. 5.2(b). In these three figures, *the height of the histogram indicates the probability that $x$ falls in the interval*. Thus, we can determine a probability, such as $P(a \leqslant x \leqslant b)$, by summing the heights corresponding to each of the events that satisfy $a \leqslant x \leqslant b$.

Note what happens as $\Delta x \to 0$ (compare Figs. 5.1, 5.2, and 5.3). As the size of the intervals gets smaller and smaller, the frequency polygon begins to look more and more like a continuous function. Note particularly in Fig. 5.3 that the sum of the *heights* of the histograms begins to closely approximate the *area* under the frequency polygon. To state this more formally, suppose we denote the height of each rectangle in Figs. 5.1, 5.2, and 5.3 as $P(x)$, and the width as $\Delta x$. The area of each rectangle is thus $P(x)\Delta x$. Now, one approximation to the probability $P(a \leqslant x \leqslant b)$ is the sum of all the area between $a$ and $b$, which is $\sum_{\Delta x = a}^{b} P(x)\Delta x$. The smaller $\Delta x$ is, the better this sum approximates $P(a \leqslant x \leqslant b)$. When $\Delta x = 0$

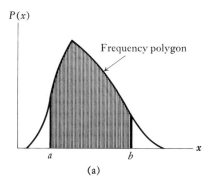
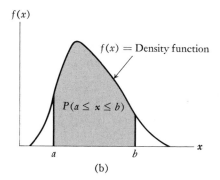

**Fig. 5.3.** The continuous approximation to $P(a \leqslant x \leqslant b)$: (a) Discrete probabilities $P(a \leqslant x \leqslant b) = \sum_{\Delta x=a}^{b} P(x)\,\Delta x$; (b) Continuous approximation $P(a \leqslant x \leqslant b) = \int_{a}^{b} f(x)\,dx$.

the approximation is perfect, but in this case we have a continuous random variable rather than a discrete random variable, and hence we cannot use the concept of a summation to evaluate $P(a \leqslant x \leqslant b)$.

Special symbols are used to denote the evaluation of $P(a \leqslant x \leqslant b)$ in the continuous case. First, at the limit, as $\Delta x \to 0$, the width of the class interval is denoted by the symbol $dx$ instead of $\Delta x$. Secondly, the frequency polygon is now called a density function, and the height of the density function at $x = x$ is denoted as $f(x)$. Finally, the summation sign becomes an integral sign, $\int_a^b$, where this integral sign is interpreted to mean "the limit of summation as $\Delta x \to 0$." Putting all this together gives the following result for continuous random variables:

$$P(a \leqslant x \leqslant b) = \lim_{\Delta x \to 0} \sum_a^b P(x)\Delta x = \int_a^b f(x)dx$$
$$= \text{Area under the curve from } a \text{ to } b.$$

Thus, as we have shown in Fig. 5.3(b), the probability that $x$ falls in the interval $a \leqslant x \leqslant b$ is given by the area under the density function, and *not* the height of the density function itself. The value of $f(x)$ merely represents how high (or "dense") the function is at that point. It cannot represent probability, for we already pointed out that $P(x = x) = 0$ in the continuous case.

Finally, note that for a continuous random variable it makes no difference whether the endpoints $a$ and $b$ are included in the interval or not, since the probability of observing any one specific point, such as $a$ or $b$, equals zero. Thus, for a continuous random variable,

$$P(a < x < b) = P(a \leqslant x < b) = P(a < x \leqslant b) = P(a \leqslant x \leqslant b).$$

**Properties of a Probability Density Function**

A probability density function must satisfy two basic properties that are similar to those for a probability mass function. The major difference between the two is that, while both $f(x)$ and $P(x)$ cannot be negative, the values of $f(x)$ do not necessarily have to be less than or equal to 1.0. On the other hand, it must be true that the total area under $f(x)$, from $-\infty$ to $+\infty$, has to equal 1.0 (for the same reason that $\sum P(x) = 1$).

*Properties of all density functions:*

1. $f(x) \geqslant 0$:          The density function is never negative;
2. $\int_{-\infty}^{\infty} f(x)dx = 1$:     The total area under the density function always equals 1.0.

Given that a function $f(x)$ satisfies these two properties, then we can show that the probability that $x$ falls in the interval $a \leqslant x \leqslant b$ is:

> *For continuous random variables:*
>
> $$P(a \leqslant x \leqslant b) = \text{Area under } f(x) \text{ from } a \text{ to } b. \qquad (5.1)$$

**Example of a pdf**

To illustrate the process of approximating discrete probability values with a continuous function, suppose a mail-order book club is interested in the pattern with which its subscribers pay for the books they order. At the same time that their order is filled, the members of this club are sent a bill on which payment is due within five weeks of the shipping date. In analyzing recent records of this book club, it was found that only about 16 percent of all customers return their payment within the first two weeks; most people wait until weeks four or five to send in their money. Column (2) of Table 5.2 shows the number of payments received in each of the five weeks for the past 100,000 orders; the relative frequency of each value of $x$ is shown in column (3). We will explain column (5) later.

**Table 5.2** Example distribution of book club payments

| (1)<br>Week payment<br>was received<br>$x$ | (2)<br>Number of<br>payments received<br>$f$ | (3)<br>Relative<br>frequency<br>$\frac{f}{n} = P(x)$ | (4)<br>Cumulative<br>relative<br>frequency<br>$F(x)$ | (5)<br>$0.08x - 0.04$ |
|---|---|---|---|---|
| 1 | 3,940 | 0.039 | 0.039 | 0.040 |
| 2 | 12,012 | 0.120 | 0.159 | 0.120 |
| 3 | 20,133 | 0.201 | 0.360 | 0.200 |
| 4 | 27,852 | 0.279 | 0.639 | 0.280 |
| 5 | 36,063 | 0.361 | 1.000 | 0.360 |
| Sum | 100,000 | 1.000 | | 1.000 |

*Discrete case.* Suppose we now plot the mass and cumulative functions describing, for each of the five weeks, the probability that a randomly selected customer will return his payment. Figure 5.4 shows the graph of these functions. The tops of the probability lines in Fig. 5.4(a) form a fairly straight line which has a slope of 0.08 and a vertical intercept of $-0.04$. This relation is shown by the dotted line in the figure. An equation for $P(x)$ represented by a curve connecting the tops of the probability values can thus be written as $f(x) = 0.08x - 0.04$. The fact that this

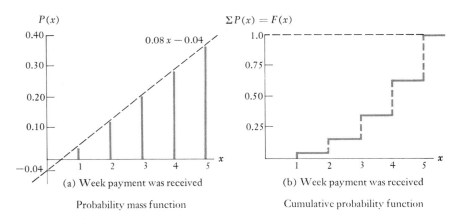

(a) Week payment was received

Probability mass function

(b) Week payment was received

Cumulative probability function

**Fig. 5.4.** Probability functions representing mail-order payments from Table 5.2, columns (3) and (4).

equation quite accurately describes $P(x)$ for the discrete values $x = 1, 2, 3, 4,$ and 5, is shown by comparing columns (3) and (5) in Table 5.2.

*Continuous case.* The probabilities shown in Table 5.2 assume that the book club can distinguish only which week ($x = 1, 2, 3, 4,$ or 5) a customer's payment was received. But suppose we want a continuous approximation which assumes that a payment can be received at *any* value of $x$ between 0 and 5 (that is, $x$ need not be an integer).

To make this approximation we need a function which yields a probability of 0.04 that the payment was received *between* the shipping date and the end of the first week, $P(0 \leqslant x \leqslant 1) = 0.04$, a probability of 0.12 that the payment was received in the second week, $P(1 \leqslant x \leqslant 2) = 0.12$, and so forth, with $P(4 \leqslant x \leqslant 5) = 0.36$. Although we won't present the process, it is not hard to determine that a function giving this approximation is the following:

$$f(x) = \begin{cases} 0.08x & 0 \leqslant x \leqslant 5, \\ 0 & \text{otherwise.} \end{cases}$$

This continuous approximation is graphed in Fig. 5.5.

In the discrete case, the probability $P(x = 1) = 0.039$ represents the probability that a payment was received *during* the first week, meaning from 0 to 1 week. Thus, in the continuous case we can approximate this value by $P(0 \leqslant x \leqslant 1)$. Because we interpret area as probability in the continuous case, $P(0 \leqslant x \leqslant 1)$ is given by the area of the small triangle in the lower lefthand corner of Fig. 5.5 This triangle has a width of $w = 1.0$ and a height of $h = 0.08$; since the area of a triangle is $A = \frac{1}{2}wh$, this area is $\frac{1}{2}(1.0)(0.08) = 0.04$. That is, $P(0 \leqslant x \leqslant 1) = 0.04$, which

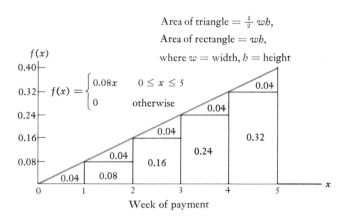

Area of triangle $= \frac{1}{2} wh$,

Area of rectangle $= wh$,

where $w =$ width, $h =$ height

$$f(x) = \begin{cases} 0.08x & 0 \le x \le 5 \\ 0 & \text{otherwise} \end{cases}$$

Week of payment

**Fig. 5.5.** Continuous density function for example of book-club weekly payments.

agrees very closely with the discrete value 0.039. To take another example, consider $P(3 \leqslant x \leqslant 4)$. The total area in this case consists of a triangle, with area $\frac{1}{2}wh = \frac{1}{2}(1.0)(0.08) = 0.04$, plus a rectangle with area $wh = (1.0)(0.16) = 0.160$. The sum of these two areas is $P(3 \leqslant x \leqslant 4) = 0.200$, which again agrees very closely with the discrete value, $P(x = 4) = 0.201$. The remaining probabilities can be determined similarly.

Now, let's see if this density function satisfies the two properties of all probability density functions. First, as shown by Fig. 5.5, the function never goes below the $x$-axis; hence the condition $f(x) \geqslant 0$ is satisfied. Secondly, we can show that the total area under the function and above the $x$-axis equals 1.0, by summing the areas of the individual triangles and rectangles in Fig. 5.5. Or, we can find the area under the large triangle; this triangle has width $w = 5$ and height $h = 0.40$; hence, the total area is $\frac{1}{2}(5.0)(0.40) = 1.0$.

Although we have limited ourselves to integer values in assessing probabilities in this example, it is not necessary to do so. The probability $P(1.7 < x < 3.5)$, for instance, can be determined by the methods of geometry just as we did above. The reader should realize that using geometry is possible in our book club example because the function $f(x)$ is a simple one (a straight line), for which we know how to find areas. For other more complex density functions, geometry can't always be used. In such cases the methods of integral calculus can be used to evaluate the appropriate integral, $\int_a^b f(x)dx$. Fortunately, tables exist for this evaluation for most common density functions in statistics; hence it is not necessary to be able to actually perform such integrations if we want to solve probability problems involving continuous random variables.

**Examples of Other pdf's**

A number of frequently used probability density functions will be investigated later in this chapter. For now, the diagrams in Fig. 5.6 should suffice to give you some insight into different types of density functions. In the first diagram the density function is seen to be a constant, equal to 2.0 for values between $x = 1.0$ and $x = 1.5$, and to zero for all other values of $x$. In the second diagram the function $f(x)$ is a straight line with a slope of two between $x = 0$ and $x = 1$, and equal to zero elsewhere. Finally, in the third diagram, $f(x)$ is a decreasing function for $x$ between zero and infinity, and zero otherwise.*

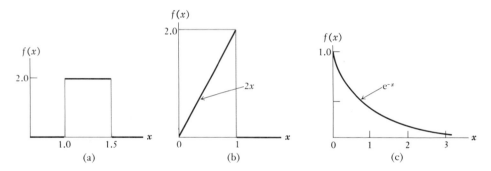

**Fig. 5.6.** Continuous probability density functions.

We should perhaps emphasize again that, although total probability (area) in a given problem must equal one, it is not necessary for $f(x)$ to always assume values less than or equal to one. In other words, the fact that the total *area* under a function equals one does not imply that the *height* of the function is less than one. The first diagram in Fig. 5.6 is a good example of such a case since, by the elementary rules of geometry, the area under this function (which is a rectangle) is seen to equal one; yet the height of $f(x)$ is 2.0 for $1.0 \leqslant x \leqslant 1.5$.

## 5.3  SIMILARITIES BETWEEN PROBABILITY CONCEPTS FOR DISCRETE AND CONTINUOUS RANDOM VARIABLES

As we have indicated, the basic difference between a probability mass function and a probability density function is that, in the former case, probabilities are

---

* In Fig. 5.6(c), the symbol $e$ denotes the same nonrepeating, nonterminating decimal we encountered in Section 4.5 (that is, $e \doteq 2.71828$).

measured by the *height* of the function, while in the latter case, probabilities are measured by *areas* under the function. Most of the probability concepts developed in Chapter 3 are similar for both discrete and continuous random variables. Formal representation of many of the formulas will only *look* different, but not be different in meaning. They look different because, wherever a summation occurs in a formula involving probabilities of a discrete random variable, it is replaced by an integral in the continuous case. That is, rather than summing across values representing heights of a probability function, we take integrals representing sums of areas under a density function. Although we will present, for continuous random variables, the concept of a cumulative distribution function (cdf) as well as the mean and variance of such functions, we will not present the formulas for a joint distribution, marginal distribution, conditional distribution, independence, and covariance. These formulas, however, follow directly from the comparable concepts in Chapter 3.

### Cumulative Distribution Function

As before, the function $F(x)$ represents the probability that the random variable $x$ assumes a value less than or equal to some specified value. To calculate $F(x)$ in the continuous case, it is necessary to *integrate* $f(x)$ over the relevant range, rather than sum discrete probabilities. Hence, if we assume that $f(x)$ is defined for all real values from negative infinity to $x$, then the cumulative function (cdf) is defined as follows:

*Cumulative distribution function:*

$$F(x) = \int_{-\infty}^{x} f(x)dx = \text{Area up to } (x = x). \tag{5.2}$$

Again, we will not be concerned so much with the integration process for finding a cdf, but rather with its concept and interpretation.

It is important to remember that the value of $F(x)$ is a probability. This probability is defined in exactly the same manner as it was in the discrete case:

*Cumulative distribution function:*

$$F(x) = P(x \leqslant x).$$

In the discrete case we said that $F(x)$ could be a list of probabilities, or $F(x)$ could be a formula. In the continuous case, $F(x)$ must be a formula because $f(x)$ must be a formula. To show what types of formulas result from the integration in Formula (5.2), we have calculated $F(x)$ for the three density functions shown in Fig. 5.6. These cumulative functions are shown in Fig. 5.7. For the reader who

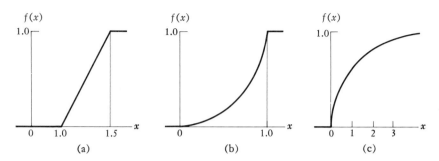

**Fig. 5.7.** Cumulative distribution functions for density functions in Fig. 5.6.

knows calculus, we have completed the integrations necessary for deriving these functions in a footnote.*

The maximum and minimum values which $F(x)$ can assume for these examples are the same as in the discrete case. The maximum value of $F(x)$ must naturally be 1.0; this maximum will always occur at $F(\infty)$ since $F(\infty) = P(x \leq x) = 1.0$. Similarly, $F(x)$ cannot be lower than zero since $f(x) \geq 0$ for all $x$; hence, $F(-\infty) = 0$.

As another example of a cdf, it is not hard to show that the value of $F(x)$ for our book-club example is:

$$F(x) = \begin{cases} 0.04x^2 & 0 \leq x \leq 5, \\ 0 & \text{otherwise.} \end{cases}$$

Let's check to see if this function seems appropriate by verifying that its lowest value, 0, occurs at $x = 0$, and its highest value, 1.0, occurs at $x = 5$. We see that it does, since $F(0) = 0.04(0) = 0$, and $F(5) = 0.04(5^2) = 1.0$.

Instead of working with a cumulative function in the form of a formula, it

---

*

1. $f(x) = \begin{cases} 2 & 1 < x < 1.5, \\ 0 & \text{otherwise.} \end{cases}$

$F(x) = \int_{-\infty}^{x} f(x)\, dx$

$\quad = \int_{1.0}^{x} 2dx$

$\quad = [2x]_1^x$

$\quad = 2x - 2.$

2. $f(x) = \begin{cases} 2x & 0 \leq x \leq 1, \\ 0 & \text{otherwise.} \end{cases}$

$F(x) = \int_{-\infty}^{x} f(x)\, dx$

$\quad = \int_{0}^{x} 2x\, dx$

$\quad = [x^2]_0^x$

$\quad = x^2.$

3. $f(x) = \begin{cases} e^{-x} & x > 0, \\ 0 & \text{otherwise.} \end{cases}$

$F(x) = \int_{-\infty}^{x} f(x)\, dx$

$\quad = \int_{0}^{x} e^{-x}\, dx$

$\quad = [-e^{-x}]_0^x$

$\quad = 1 - e^{-x}.$

is often convenient to list a certain number of the values in a table. For example, in Table 5.3 we list six of the values of $F(x)$ for the book club example.

**Table 5.3** Selected values of $F(x)$ for the example of book club weekly payments, derived from areas in Fig. 5.5

| $x$ | 0 | 1 | 2 | 3 | 4 | 5 |
|---|---|---|---|---|---|---|
| $F(x)$ | 0 | 0.04 | 0.16 | 0.36 | 0.64 | 1.00 |

Once a cumulative distribution function has been tabled (as in Table 5.3), it can be used to find the probabilities of many events of interest. The following formulas, which are illustrated by using Table 5.3, will prove especially useful.

1) $P(x \leqslant b) = F(b)$, by definition of the cumulative distribution function. This gives the probability of observing any value equal to or smaller than a given value $b$. It is represented by the area in a distribution function to the left of the value $b$. Using Table 5.3, $P(x \leqslant 3) = F(3) = 0.36$. You should try to find this area on Fig. 5.5.

2) $P(x \geqslant a) = 1 - P(x < a) = 1 - F(a)$. The area in the upper part of a distribution to the right of a value $a$ may be found using $F(a)$ and the complement rule. For example, $P(x \geqslant 2) = 1 - F(2) = 1 - 0.16 = 0.84$. Find this area on Fig. 5.5.

3) $P(a \leqslant x \leqslant b) = F(b) - F(a)$. The probability that $x$ falls between $a$ and $b$ can be found by subtracting the area to the left of $a$ from the area to the left of $b$, to obtain the amount of area between $a$ and $b$. Using Table 5.3, $P(1 \leqslant x \leqslant 4) = F(4) - F(1) = 0.64 - 0.04 = 0.60$. Find this area on Fig. 5.5 and see that it represents 60 percent of the total area under the probability density function.

**Mean and Variance**

The summary measures of central location and dispersion are as important in describing a density function as they were in describing a mass function. For the most part we will not be concerned in this book with the *process* of calculating such measures, although again it is important for the reader to understand the concepts involved. We therefore present below (but do not elaborate on) the formulas for calculating the mean and the variance for a continuous random variable $x$. The reader should verify that these formulas are the same as those presented in Chapter 3, except that integral signs are substituted for the sum sign and $f(x)dx$ is substituted for $P(x)$.

*Mean of x:*
$$E[x] = \mu_x = \int_{-\infty}^{\infty} xf(x)\,dx;$$

*Expectation of g(x):*
$$E[g(x)] = \int_{-\infty}^{\infty} g(x)f(x)\,dx;$$

$$(5.3)$$

*Expectation of $x^2$:*
$$E[x^2] = \int_{-\infty}^{\infty} x^2 f(x)\,dx;$$

*Variance of x:*
$$V[x] = E[(x - \mu_x)^2] = E[x^2] - (E[x])^2.$$

## 5.4 THE NORMAL DISTRIBUTION

Scientists in the eighteenth century noted a predictable regularity about the frequency with which certain "errors" occur, especially errors of measurement. Suppose, for example, that a machine is supposed to roll a sheet of metal to a width of exactly $\frac{5}{16}$ in.; while this machine produces sheets which are $\frac{5}{16}$ in. wide *on the average*, some sheets are in "error" by being slightly too wide, others by being slightly too narrow. Experiments producing errors of this nature were found to form a symmetrical distribution, which was originally called the "normal curve of errors." The continuous probability distribution which such an experiment approximates is usually referred to as the *normal distribution*, or sometimes the *Gaussian distribution*, after an early researcher, Karl Gauss, (1777–1855).

The normal distribution undoubtedly represents the most widely known and used of all distributions. Because the normal distribution approximates many natural phenomena so well, this distribution has developed into a standard of reference for many probability problems. In addition, under certain conditions the binomial distribution and the Poisson distribution can be approximated by the normal. The normal distribution is so important in the theory of statistics that a considerable portion of the sampling, estimation, and hypothesis-testing theory we will study in the rest of this book is based on the characteristics of this distribution.

### Characteristics of the Normal Distribution

The normal distribution is a continuous distribution in which $x$ can assume any value between minus infinity and plus infinity ($-\infty \leqslant x \leqslant \infty$). Two parameters describe the normal distribution, $\mu$, representing the mean, and $\sigma$, representing the standard deviation.* The normal density function also contains two constants,

---

* Beginning students of statistics are sometimes confused by the fact that for the normal distribution, the general symbols $\mu$ and $\sigma$ are used to represent the specific parameters of this distribution rather than some different ones such as $n$ and $p$ in the binomial or $\lambda$ in the Poisson. This is merely traditional, since the normal is the most commonly used distribution.

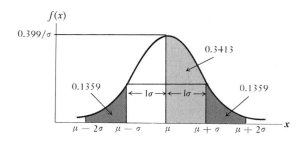

**Fig. 5.8.** Normal distribution with mean $\mu$ and standard deviation $\sigma$.

$\pi$ (where $\pi = 3.14159$ . .) and $e$ (where $e = 2.71828$ . .). The normal pdf is given below. A normal distribution with mean $\mu$ and variance $\sigma^2$ is often denoted by the symbol $N(\mu, \sigma^2)$.

*Normal density function, $N(\mu, \sigma^2)$:*

$$f(x) = \frac{1}{\sigma\sqrt{2\pi}} e^{-(1/2)[(x-\mu)/\sigma]^2} \qquad \text{for } -\infty \leqslant x \leqslant \infty. \qquad (5.4)$$

The curve described by Formula (5.4) is a completely symmetrical, bell-shaped pdf, whose graph is shown in Fig. 5.8.

Note that in Fig. 5.8 the area under the curve from $\mu$ to $\mu + 1\sigma$ is 0.3413 Thus, $P(\mu \leqslant x \leqslant \mu + 1\sigma) = 0.3413$. By symmetry, $P(\mu - 1\sigma \leqslant x \leqslant \mu + 1\sigma) = 2(0.3413) = 0.6826$. We can also see that $P(\mu + 1\sigma \leqslant x \leqslant \mu + 2\sigma) = 0.1359$, and hence

$$P(\mu - 2\sigma \leqslant x \leqslant \mu + 2\sigma) = 0.6826 + 0.1359 + 0.1359 = 0.9544.$$

*The rule of thumb we have been using throughout this book to interpret the size of a standard deviation is now seen to be based on the normal distribution.* The extent to which the intervals we have considered previously have differed from our rule of thumb reflects the fact that these distributions have not been normal distributions.

It is important to remember that all normal distributions have the same bell-shaped curve pictured in Fig. 5.8 regardless of the values of $\mu$ and $\sigma$. The value of $\mu$ merely indicates where the center of the "bell" lies, while $\sigma$ indicates how spread out (or wide) the distribution is. Note in Fig. 5.8 that the height of the density function at the point $x = \mu$ is $0.399/\sigma$. This fact can be derived from Formula (5.4) by substituting $\mu$ for $x$, and then noting that $e^0 = 1$, and $1/\sqrt{2\pi} = 0.399$. Thus, if $\sigma = 1.0$, then $f(\mu) = 0.399/1.0 = 0.399$. When $\sigma = 0.5$, $f(\mu) = 0.399/0.5 = 0.798$, and if $\sigma = 1.5$, $f(\mu) = 0.399/1.5 = 0.266$. The normal distributions corresponding to these three values of $\sigma$ are shown in Fig. 5.9.

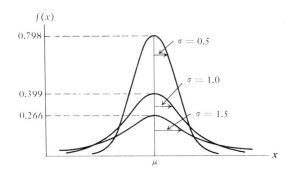

**Fig. 5.9.** Three normal distributions with different standard deviations.

### Examples of Normal Distributions

Even with the limited information we have presented thus far about the normal distribution a number of probability questions can be answered. We will consider two different examples.

**Example 1** (*Bass length*) Suppose it is known that the random variable $x$, representing the length (in inches) of bass in a certain fish breeding pond, is normally distributed with mean, $\mu_x = 10$ inches, and standard deviation $\sigma_x = 0.8$ inches; that is, $N(10, 0.8^2)$. This information is sufficient to completely determine the probability of any event concerning the values of $x$. For example, what is the probability that a bass taken at random will have length greater than 10 inches? Since the normal distribution is symmetrical and 10 is the median as well as the mean, the answer is $\frac{1}{2}$. Now, to move to a slightly more difficult question, what is the probability that the bass will be shorter than 9.2 inches? Using Fig. 5.10, we see that 9.2 is the value exactly one standard deviation below the mean (that

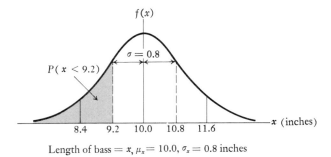

Length of bass $= x$, $\mu_x = 10.0$, $\sigma_x = 0.8$ inches

**Fig. 5.10.** Example of normal distribution.

is, $10.0 - 9.2 = 0.8 = \sigma_x$). The probability $P(x < 9.2)$ is represented by the shaded area under the function $f(x)$.

The probability $P(x < 9.2)$ is thus exactly equivalent to the probability

$$P(x < \mu - 1\sigma),$$

as shown in Fig. 5.8. The problem remains to determine how much area for the normal distribution lies to the left of $\mu - 1\sigma$. To calculate $P(x < \mu - 1\sigma)$ we first recall that, from Fig. 5.8, $P(\mu - 1\sigma \leqslant x \leqslant \mu + 1\sigma) = 2(0.3413) = 0.6826$. The probability we want equals one-half of the complement of this probability, since the area in the interval $x < \mu - 1\sigma$ is half of the area *not* included in the interval $\mu - 1\sigma \leqslant x \leqslant \mu + 1\sigma$. That is,

$$P(x < 9.2) = \tfrac{1}{2}(1 - 0.6826) = \tfrac{1}{2}(0.3174) = 0.1587.$$

Finally, for this same example, suppose we want to determine the probability that the length of a randomly selected bass is between 11.2 and 12 inches (so that it fits nicely into our 12-inch frying pan). This probability is given by the area under the curve between 11.2 and 12.0. To find this area by using calculus, we would have to integrate, from $x = 11.2$ to $x = 12.0$, the normal density function with $\mu = 10$ and $\sigma = 0.8$. Another way to determine this area would be to find a table already calculated for a normal distribution with $\mu = 10$ and $\sigma = 0.8$. We will come back to this problem shortly to reject both these methods and seek some even better way.

**Example 2** (*Tire life*) Suppose the random variable $y$, representing the tread life in miles of a certain new radial tire, is normally distributed with mean, $\mu_y = 40,000$ miles, and standard deviation, $\sigma_y = 3,000$ miles [$N(40,000, 3000^2)$]. This information is sufficient to completely determine the probability of any event concerning the values of $y$. For example, $P(y > 40,000) = \tfrac{1}{2}$, since half of the probability in a normal distribution lies on each side of the mean. Or suppose we calculate the probability that a tire of this model selected at random has a tread life greater than 46,000 miles. Using Fig. 5.11, we see that 46,000 is exactly two standard deviations above the mean ($46,000 - 40,000 = 6,000 = 2\sigma_y$). The probability $P(y > 46,000)$ is represented by the shaded area under the function $f(y)$ in Fig. 5.11. The probability $P(y > 46,000)$ is also equivalent to the $P(x > \mu + 2\sigma)$ in the graph of the normal distribution in Fig. 5.8. Since we already know that $P(\mu - 2\sigma < x < \mu + 2\sigma) = 0.9544$, the value of $P(x > \mu + 2\sigma)$ is easily seen to equal one-half of the complement of 0.9544. Thus

$$P(y > \mu + 2\sigma) = \tfrac{1}{2}(1.0000 - 0.9544) = \tfrac{1}{2}(0.0456) = 0.0228.$$

Finally, for this example, we might ask about the probability that tread life will fall between 44,500 and 47,500 miles. Again, the answer can be determined in two ways. Using calculus, one could integrate the normal pdf with $\mu_y = 40,000$

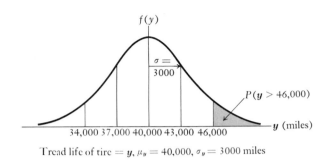

Tread life of tire $= y$, $\mu_y = 40{,}000$, $\sigma_y = 3000$ miles

**Fig. 5.11.** Tire-life example.

and $\sigma_y = 3{,}000$ from 44,500 to 47,500. Or one could use a table of areas already calculated for the precise parameters, $\mu_y = 40{,}000$ and $\sigma_y = 3{,}000$. Both of these methods are unsatisfactory, however, first because it is too tedious to evaluate a new integral every time one investigates a different set of parameters or new $x$-values, and secondly because no such tables exist (there obviously can't be tables listing the infinite number of possible values of $\mu$ and $\sigma$).

## 5.5 STANDARDIZED NORMAL

Values of $x$ for the normal distribution are usually described in terms of how many standard deviations they are away from the mean. The value $x = 200$, for example, has little meaning unless we know in what units $x$ was measured (e.g., feet, miles, pounds). On the other hand, the statement that $x$ is one standard deviation larger (or smaller) than the mean can be given a very precise interpretation, as it is always meaningful to talk of $x$ being a certain number of standard deviations above (or below) the mean, no matter what value $\sigma$ assumes or on what scale the variable $x$ is measured. Now, if $x$ is measured in terms of standard deviations about the mean, it is natural to describe probability values in the same terms— that is, by specifying the probability that $x$ will fall within so many standard deviations of the mean. There are three commonly encountered intervals, the first two of which we have referred to often in the last two chapters: $\mu \pm \sigma$, $\mu \pm 2\sigma$, and $\mu \pm 3\sigma$. In fact, we used the first two of these in the previous section for finding probabilities for normally distributed random variables.

Treating the values of $x$ in a normal distribution in terms of standard deviations about the mean has the advantage of permitting all normal distributions to be compared to one common or standard form. In this standard form, different values of $\mu$ and $\sigma$ no longer generate completely different curves, since $x$ is measured only about $\mu$ and all distances away from $\mu$ are in terms of multiples of $\sigma$. In other words, it is easier to compare normal distributions having different values of $\mu$

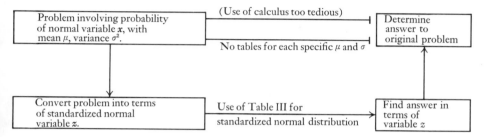

**Fig. 5.12.** Problem-solving tactic using the standardized normal.

and $\sigma$ if these curves are transformed to one common form, which is called the *standardized normal*. The standardized normal, by definition, has a mean of zero ($\mu = 0$) and a standard deviation of one ($\sigma = 1$). Note that if the standard deviation is one, the variance must also be one since $\sigma^2 = 1.0$ when $\sigma = 1.0$.

This process of standardization gives a hint toward the best method of attack in answering questions concerning a normal probability distribution. Instead of trying to directly solve a probability problem involving a normally distributed random variable $x$ with mean $\mu$ and standard deviation $\sigma^2$, an indirect approach is used. We first convert the problem to an equivalent one dealing with a normal variable measured in standard deviation units, called a *standardized normal variable*. A table of standardized normal values (Table III) can then be used to obtain an answer in terms of the converted problem. Finally, by converting back to the original units of measurement for $x$, we can obtain the answer to the original problem. Figure 5.12 is a schematic outline of this method of solving probability problems.

Recall that we discussed in Chapter 3 the process of transforming a random variable $x$ (whether normally distributed or not) with mean $\mu$ and standard deviation $\sigma$ into a standardized measure with mean zero and standard deviation one. The appropriate transformation was shown to be $z = (x - \mu)/\sigma$. The mean of the variable $z$ was shown in Section 3.3 to be $E[z] = 0$ and the variance was shown to be $V[z] = 1.0$.

We now add the additional fact that if the original random variable ($x$) is normally distributed, then the standardized variable $z$ will also be normally distributed. That is, if $x$ is $N(\mu, \sigma^2)$, then $z = (x - \mu)/\sigma$ is $N(0, 1)$.

> *Standardized normal random variable:*
>
> $$z = \frac{x - \mu}{\sigma} \text{ is } N(0, 1).$$

(5.5)

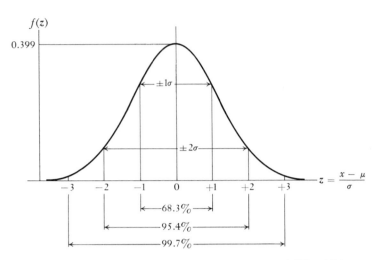

**Fig. 5.13.** Standardized normal distribution: $f(z) = (1/\sqrt{2\pi})e^{-(1/2)z^2}$.

The density function for $z$ can be derived by substituting $z$ for $(x - \mu)/\sigma$, and $\sigma = 1$, into Formula (5.4). The resulting pdf is denoted as $f(z)$.

*Standardized normal density function:*

$$f(z) = \frac{1}{\sqrt{2\pi}} e^{-(1/2)z^2} \qquad \text{for } -\infty \leqslant z \leqslant \infty. \qquad (5.6)$$

This function is shown graphically in Fig. 5.13. The reader should compare Fig. 5.13 with 5.8 to verify that the former is merely a special case of the latter where $\mu = 0$ and $\sigma = 1$.

The interpretation of $z$-values is relatively simple. Since $\sigma = 1$, a value of $x$ is two standard deviations away from the mean whenever $z = \pm 2$; likewise, if $z = \pm 1.56$, the corresponding $x$-value is exactly 1.56 standard deviations away from the mean (or $|x - \mu| = 1.56\sigma$).

**Using the Standardized Normal**

Consider the two examples in the previous section involving normally distributed random variables, where

$$x = \text{length of bass in inches with } \mu_x = 10, \sigma_x = 0.8;$$

and

$$y = \text{tread life of tire in miles with } \mu_y = 40,000, \sigma_y = 3,000.$$

Some probability questions were suggested there that we did not completely answer. To repeat, what is $P(11.2 < x < 12.0)$ and what is $P(44,500 < y < 47,500)$? The answer to each question is shown by the shaded areas under the normal curves in Figs. 5.14(a) and (b) respectively. Since we rejected the two proposed direct ways of answering these questions, we now need to proceed with the indirect approach, using the standard normal distribution.

In transforming the probability $P(11.2 \leqslant x \leqslant 12.0)$ into an equivalent one in standardized normal form, we must apply the transformation $z = (x - \mu)/\sigma$ to each part of the expression in parentheses. For example, the value 11.2 is transformed into its equivalent form by subtracting $\mu = 10.0$ from it, and then dividing by $\sigma_x = 0.8$; the value 12.0 is transformed into its equivalent standardized form in exactly the same manner. Finally, we can think of the variable $x$ being transformed in the same way, since the new variable, $z$, equals $(x - \mu)/\sigma$. Thus,

$$P(11.2 \leqslant x \leqslant 12.0) = P\left(\frac{11.2 - 10.0}{0.8} \leqslant \frac{x - \mu}{\sigma} \leqslant \frac{12.0 - 10.0}{0.8}\right)$$
$$= P(1.5 \leqslant z \leqslant 2.5).$$

We follow the same process in transforming the probability

$$P(44,500 \leqslant y \leqslant 47,500)$$

into standardized normal form, except that in this case $\mu_y = 40,000$ and $\sigma_y = 3,000$:

$$P(44,500 \leqslant y \leqslant 47,500) = P\left(\frac{44,500 - 40,000}{3,000} \leqslant \frac{y - \mu_y}{\sigma_y} \leqslant \frac{47,500 - 40,000}{3,000}\right)$$
$$= P(1.5 \leqslant z \leqslant 2.5).$$

It is now obvious that we constructed our two unanswered probability questions to show how two diverse problems such as bass length and tire life can both reduce to the identical question in terms of the standardized normal. The fact that these probabilities are equivalent can be seen by comparing Figs. 5.14(a) and 5.14(b) with 5.14(c).

### Standardized Normal Values

The problem at this point is: How does one evaluate probabilities in standardized form, such as $P(1.5 \leqslant z \leqslant 2.5)$? As we indicated previously, there are tables of standardized normal values (called $z$-values) for this purpose. Table III is one such table.

Before we describe the use of Table III, let's investigate $F(z)$, the cumulative distribution function for the variable $z$. This function, shown in Fig. 5.15, gives $P(z < z)$.

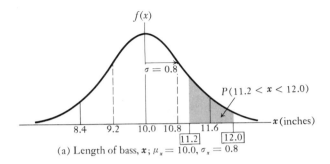

(a) Length of bass, $x$; $\mu_x = 10.0$, $\sigma_x = 0.8$

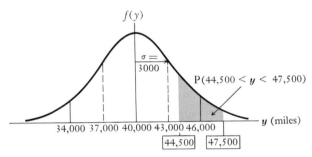

(b) Tread life of tire, $y$; $\mu_y = 40{,}000$, $\sigma_y = 3000$

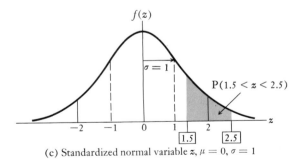

(c) Standardized normal variable $z$, $\mu = 0$, $\sigma = 1$

**Figure 5.14**

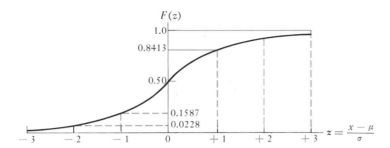

**Fig. 5.15.** Cumulative *z* values.

Note that we have plotted $z$-values only from $-3$ to $+3$ since very little area lies beyond these limits. At $z = 0$ (the mean of $z$), the value of $F(z)$ must be 0.50 since $z = 0$ represents the median of the $z$-values. Most of the other values in Fig. 5.15 should be familiar to you by now. For example, $F(-1) = 0.1587$. This value agrees with the one calculated in the example on bass length, as we saw then that $P(x < \mu - 1\sigma) = 0.1587$. The value $F(-2) = 0.0228$ should also appear familiar, as this is the same value we calculated in the tire-life problem for $P(y > \mu + 2\sigma)$. Because of the symmetry of the normal distribution, $F(-2) = 1 - F(2) = P(y > \mu + 2\sigma)$.

Since the normal distribution is completely symmetrical, tables of $z$-values usually include only positive values of $z$. Thus, the lowest value in Table III is $z = 0$, and the cumulative probability at this point is $F(0) = P(z < 0) = 0.50$. Table III gives other values of $z$, to two decimal points, up to the point $z = 3.49$. The values of $z$ to one decimal are read from the left margin in Table III, while the second decimal is read across the top. The body of the table gives the values of $F(z)$.

To illustrate the use of Table III, we will consider three basic rules. The reader should try to understand (visualize) these rules, not necessarily memorize them.

*Rule* 1: $P(z < a)$ *is given by* $F(a)$ *when* $a$ *is positive* (see figure at top of page 184). We can illustrate this rule by determining the probability $P(y < 45,000)$ in the tire-life problem:

$$P(y < 45,000) = P\left(\frac{y - \mu}{\sigma} \leqslant \frac{45,000 - 40,000}{3,000}\right)$$
$$= P(z < 1.66) = F(1.66).$$

The value $z = 1.66$ in Table III yields a cumulative probability of $F(1.66) = 0.9515 = P(y < 45,000)$.

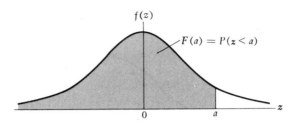

*Rule 2*: $P(z < -a)$ *is given by* $1 - F(a)$.

For this case, let's return to the bass length problem, where

$$P(x < 9.2) = P(z < -1.00).$$

Since $F(1.00) = 0.8413$, $P(x < 9.2) = 1 - 0.8413 = 0.1587$, which agrees with our earlier result.

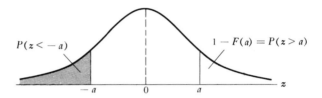

*Rule 3*: $P(z > a)$ *is given by the complement rule as* $1 - F(a)$

For our tire problem we calculated $P(y > \mu + 2\sigma) = 0.0228$. This value is equivalent to $1 - F(2.00)$, and since $F(2.00) = 0.9772$,

$$P(z > 2.00) = 1 - 0.9772 = 0.0228.$$

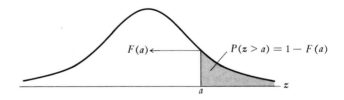

*Rule 4*: $P(a < z < b)$ *is given by* $F(b) - F(a)$

Using this rule we can solve the problem with which we started this section.

$$P(1.5 \leqslant z \leqslant 2.5).$$

From Table III, $F(2.5) = 0.9938$ and $F(1.5) = 0.9332$. Hence,

$$P(1.5 < z < 2.5) = 0.9938 - 0.9332 = 0.0606.$$

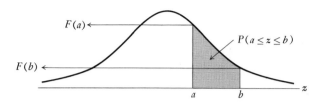

There are many other types of problem that can be solved using the standardized normal distribution. To illustrate these types of problem, we have included the following two additional examples.

**Example 1** Suppose we want to find a value $b$ such that the probability is 0.025 that the tread life of a tire will be at least as large as $b$; that is, $P(y \geq b) = 0.025$. For this example, we know probability values for the problem, but don't know the value of $b$. To find $b$, however, we use the same procedure as before—namely, transforming everything in parentheses into standardized form:

$$P(y \geq b) = P\left(z \geq \frac{b - 40{,}000}{3{,}000}\right) = 0.025.$$

The probability given in this case is shown by the shaded area in Fig. 5.16. If $P(y \geq b) = 0.025$, then $P(y < b) = 1.0 - 0.025 = 0.975$. The value of $z$ from Table III which gives $F(z) = 0.975$ is $z = 1.960$. In this type of problem, the closest probability value to that desired is located within the table, and then the corresponding $z$-value is read from the top and left margins. Now to convert back to the original problem, we solve the formula $z = (b - \mu_y)/\sigma_y$ for $b$, as follows:

$$b = \mu_y + z\sigma_y.$$

By substitution, $b = 40{,}000 + 1.96(3{,}000) = 45{,}880$. The probability is thus 0.025 that a tire lasts over 45,880 miles.

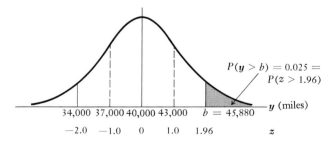

**Fig. 5.16.** Illustration of example problem.

**Example 2** Our final example in this section is to solve a probability problem involving an interval $a \leqslant x \leqslant b$ where both $a$ and $b$ are unknown. For example, in the tire-life problem one might wish to determine two values $a$ and $b$ such that the probability is 0.95 that a randomly selected tire will fall between these values, $P(a \leqslant x \leqslant b) = 0.95$. It should be readily apparent that there is an infinitely large number of such intervals, depending on how the values of $a$ and $b$ are selected. If our interval is to include 95 percent of the area under the curve and exclude 5 percent, we could exclude all of the 5 percent above $b$, or exclude all of the 5 percent below $a$, or exclude part of it below $a$ and part above $b$. In many cases, the best way to split the percent to be excluded among the two tails will be specified in the problem. If it is not specified, then *it is generally agreed that the best way to split the percent to be excluded is in a manner which makes the interval from a to b as small as possible. The smallest interval is obtained by excluding equal areas in both the upper and lower tail of the distribution.*

This result, known as the Neyman-Pearson theorem, is proved in more advanced books on statistics. It tells us that the best way to split our 5 percent is to have 2.5 percent in each tail. Let's now use this result to solve the tire-life problem, $P(a \leqslant y \leqslant b) = 0.95$. Since 2.5 percent is to be excluded above $b$, we first need to find $P(y > b) = 0.025$. Using the standardized transformation,

$$P(y > b) = P\left(\frac{y - \mu_y}{\sigma_y} > \frac{b - 40{,}000}{3{,}000}\right) = 0.025$$

$$= P\left(z \geqslant \frac{b - 40{,}000}{3{,}000}\right) = 0.025.$$

From Table III, $F(1.96) = 0.9750$. Thus, $P(y > 1.96) = 1 - 0.9750 = 0.025$, so that $b = 1.96$ is the appropriate value. Now, by symmetry, if $b = 1.96$ cuts off 2.5 percent from the upper tail, $a = -1.96$ cuts off 2.5 percent from the lower tail; hence, the smallest interval is:

$$P(-1.96 \leqslant z \leqslant 1.96) = 0.95.$$

We can now translate these $z$ values back into the units in our tire-life problem by merely substituting $\mu_y = 40{,}000$, $\sigma_y = 3{,}000$ and $z = 1.96$ and $z = -1.96$ into the standardized transformation $z = (y - \mu_y)/\sigma_y$, and solving for $y$:

$$z = \frac{y - \mu_y}{\sigma_y}.$$

If $z = -1.96$,

$$-1.96 = \frac{y - 40{,}000}{3{,}000},$$

$$y = 34{,}120;$$

and if $z = +1.96$,

$$+1.96 = \frac{y - 40{,}000}{3{,}000},$$

$$y = 45{,}880.$$

The appropriate interval is thus $P(34{,}120 \leqslant y \leqslant 45{,}880) = 0.95$. The reader should verify that 34,120 to 45,800 is the smallest interval possible by trying several other possible splits of the 5 percent to be excluded. Figure 5.17 shows the normal distribution for this example.

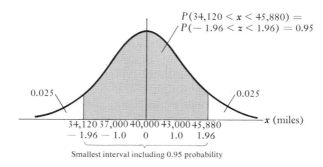

**Figure 5.17**

     Many natural phenomena tend to result in normal distributions, such as length, height, and thickness of animals or plants; medical counts of sugar, white-blood cells, incidence of inner-ear disease; and behavioral, emotional, or psychological measures of human actions, aptitudes, or abilities. Also, the distribution of measured errors or degree of perfection in production processes of many kinds tends to be normal, such as errors from a specified standard in diameters of pistons, cylinders, or gun barrels, weight of packaged products, and even lengths of yardsticks.

     However, the family of probability problems where the standardized normal $z$ is useful extends even beyond these many instances. Many other distributions applying to other types of problems tend to be normal distributions under certain conditions. One example discussed in the next section is that of the binomial distribution. Finally, as we will see in Chapters 7 and 8, the standardized normal distribution is of primary importance in problems of statistical inference dealing with means of samples from a population whose distribution is unknown.

## 5.6 NORMAL APPROXIMATION TO THE BINOMIAL

In studying the binomial distribution we saw that, when the number of trials $n$ is large, this distribution can be tedious to calculate. Fortunately, it is often possible, in this situation, to use the normal distribution to approximate the binomial. Remember that in Fig. 4.4 we saw that when $n$ is no larger than 20, the binomial has a fairly symmetrical (bell-shaped) appearance, even when $p$ is not very close to $\frac{1}{2}$. In general, the larger the value of $n$, and the closer $p$ is to $\frac{1}{2}$, the better the normal will approximate the binomial. Just how large $n$ needs to be depends on how close $p$ is to $\frac{1}{2}$, and on the precision desired, although fairly good results are usually obtained when $npq > 3$.

**Using the Normal $z$ Approximation**

To use a normal distribution to approximate the binomial it is necessary to let the mean of the normal have the same value as the mean of the binomial, $\mu = np$, and let the variance of the normal equal the variance of the binomial, $\sigma^2 = npq$. Then probabilities for any values of $x$ can be approximated by converting to the standardized normal variable $z$ as follows,

$$
\text{Normal approximation to binomial when } npq > 3:
$$

$$
z = \frac{x - \mu}{\sigma} = \frac{x - np}{\sqrt{npq}}. \tag{5.7}
$$

Substituting $np$ for $\mu$ and $\sqrt{npq}$ for $\sigma$ does not assure that the values of $z$ resulting from Formula 5.7 will correspond to a standardized normal distribution with $\mu = 0$ and $\sigma^2 = 1$. It can be shown, however, that as $n$ gets larger and larger, the ratio $(x - np)/\sqrt{npq}$ does, in fact, become a better and better approximation to the standardized normal.*

One additional factor must be considered in using the normal to approximate the binomial, namely that a discrete distribution involving only integer values (the binomial) is being approximated by a continuous distribution (the normal) in which $x$ can take on any value between negative and positive infinity. The problem that can arise in this situation is illustrated in Fig. 5.18.

For the binomial "spikes," we see that $P(x \leqslant a)$ and $P(x \geqslant a + 1)$ will sum to 1.0 whenever $a$ is an integer. But if we sum the area under the normal curve corresponding to $P(x \leqslant a)$ and $P(x \geqslant a + 1)$, this area does *not* sum to 1.0 because the area from $a$ to $(a + 1)$ is missing.

The usual way to handle this problem is to *associate one-half of the interval with each adjacent integer.* The continuous approximation to the probability $P(x \leqslant a)$ would thus be $P(x \leqslant a + \frac{1}{2})$, while the continuous approximation to $P(x \geqslant a + 1)$ would be $P(x \geqslant a + \frac{1}{2})$. This adjustment is called a *correction for continuity.*

---

*In other words, it can be shown that $\lim_{n \to \infty} (x - np)/\sqrt{npq} = N(0, 1)$.

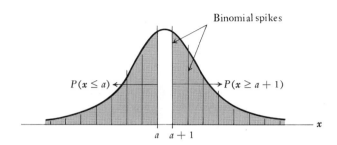

**Figure 5.18**

### A Sample Calculation

Suppose we want to calculate the probability that, in a random sample of 100 people, 60 or more will indicate that they favor a certain legislative proposal when the true population proportion of those favoring this proposal is $p = 0.64$. The parameters necessary for determining the binomial probability $P(60 \leqslant x \leqslant 100)$ are $n = 100$ and $p = 0.64$. For the normal approximation, correcting for continuity, the appropriate probability is $P(59.5 \leqslant x \leqslant 100.5)$, with parameters $\mu = np = 64$ and $\sigma = \sqrt{npq} = \sqrt{100(0.64)(0.36)} = 10(0.8)(0.6) = 4.8$. The values of these two probabilities are:

*Binomial probability:*

$$P(60 \leqslant x \leqslant 100) = \sum_{x=60}^{100} \binom{100}{x} (0.64)^x (0.36)^{100-x}$$
$$= 0.8263 \qquad \text{(from Table I)};$$

*Normal approximation:*

$$P(59.5 \leqslant x \leqslant 100.5) = P\left(\frac{59.5 - np}{\sqrt{npq}} \leqslant z \leqslant \frac{100.5 - np}{\sqrt{npq}}\right),$$

$$P\left(\frac{59.5 - 64}{4.8} \leqslant z \leqslant \frac{100.5 - 64}{4.8}\right) = P(-0.94 \leqslant z \leqslant 7.6)$$
$$= F(7.6) - F(-0.94)$$
$$= 1.000 - [1.00 - F(+0.94)]$$
$$= F(0.94) = 0.8264 \qquad \text{(from Table III)}.$$

The approximation in this case is very good, as in this case $npq = 23.04$, which is much larger than the suggested condition $npq > 3$.

Consider one additional example, the probability of a basketball player missing exactly 4 shots out of 12 *independent* attempts if his probability of missing any individual shot is assumed to be constantly equal to 0.3. With $n = 12$ and $p = 0.3$, the appropriate binomial probability is:

*Binomial probability:*

$$P(x = 4) = \binom{12}{4} (0.30)^4 (0.70)^8 = 0.2310.$$

Using the normal approximation, we can approximate $P(x = 4)$ as

$$P(3.5 \leqslant x \leqslant 4.5).$$

In this case $\mu = np = 12(0.3) = 3.6$, and $\sigma = \sqrt{npq} = \sqrt{2.52} = 1.59$.

*Normal approximation:*

$$
\begin{aligned}
P(3.5 \leqslant x \leqslant 4.5) &= P\left(\frac{3.5 - np}{\sqrt{npq}} < z < \frac{4.5 - np}{\sqrt{npq}}\right) \\
&= P\left(\frac{3.5 - 3.6}{1.59} < z < \frac{4.5 - 3.6}{1.59}\right) \\
&= P(-0.06 < z < 0.57) \\
&= F(0.57) - F(-0.06) \qquad \text{(Using the cumulative} \\
&\qquad\qquad\qquad\qquad\qquad \text{distribution function)} \\
&= 0.7157 - [1.0 - F(+0.06)] \qquad \text{Using the property} \\
&\qquad\qquad\qquad\qquad\qquad\qquad \text{of symmetry and} \\
&\qquad\qquad\qquad\qquad\qquad\qquad \text{Table III} \\
&= 0.7157 - 1.0 + 0.5239 = 0.2396.
\end{aligned}
$$

The normal approximation to the binomial is quite good in this case, even though $npq = 12(0.3)(0.7) = 2.52$ is smaller than 3. Figure 5.19 gives the histogram for the binomial distribution with $n = 12$, $p = 0.30$; superimposed on this histogram is a normal distribution with $\mu = 3.6$ and $\sigma = 1.59$. Notice how good the fit is despite the small sample size.

### *5.7  THE CHI-SQUARE DISTRIBUTION

A number of important probability distributions are closely related to the normal distribution. One of the most widely used of these related distributions is a continuous pdf called the *chi-square* distribution. The chi-square distribution gets its name because it involves the *square* of normally distributed random variables, as we will explain below.

Up to this point we have used examples involving just a single normal variable

---

* This section may be skipped if Sections 6.9 and 7.8 are also to be omitted.

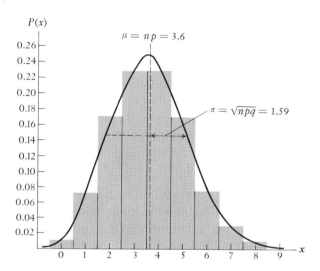

**Fig. 5.19.** Normal approximation to the binomial ($n = 12$, $p = 0.30$; $\mu = 3.6$, $\sigma = 1.59$).

$x$; and we saw how this variable can be transformed into an equivalent $z$-variable in standardized normal form by letting $z = (x - \mu)/\sigma$. Now, let's assume we want to investigate the combined properties of more than just one standardized normal variable, where these variables are assumed to be independent of one another. You might imagine a number of machines (or workers) each producing (independently) a product whose length is normally distributed with mean $\mu$ and variance $\sigma^2$ (that is, $N(\mu, \sigma^2)$). Furthermore, let's assume these lengths have been standardized by the transformation $z = (x - \mu)/\sigma$, so that they form a set of independent standardized normal variables $z_1, z_2, \ldots$ It will be convenient to denote the number of variables in this situation (i.e., the number of machines or workers) by the letter $v$, which is the Greek letter nu. Thus, we have $v$ independent standardized normal variables $z_1, z_2, \ldots, z_v$.

You may recall from Chapter 3 that we can form a new random variable by combining two or more random variables. For the present case, this *new* random variable is defined to be the *sum of the squares* of the variables $z_1, z_2, \ldots, z_v$, and denoted by the letter $\chi^2$ (which is the square of the Greek letter chi). Usually a subscript is added to the $\chi^2$ symbol to denote the value of $v$. That is,

$$\chi_v^2 = z_1^2 + z_2^2 + \cdots + z_v^2.$$

The chi-square distribution is thus the distribution of the random variable $\chi^2$. Such a variable is important in a number of different contexts. For example, if $z_i$ represents a deviation about the mean of the length of a certain product (in standardized units), then $z_i^2$ is the square of this deviation. And $\chi^2 = \sum_{i=1}^{v} z_i^2$ is the sum of the deviations of the $v$ different machines producing this product. Since a variance is also defined in terms of the sum of a set of squared deviations about a mean, it shouldn't surprise you that one of the most important applications of the $\chi^2$ distribution involves variances. We will describe, more completely, this application of the $\chi^2$ distribution in Chapter 6.

**Properties of the $\chi^2$ Distribution**

The chi-square probability density function has only one parameter, which is $v$, and which is called the *number of degrees of freedom.* As shown in Fig. 5.20, the number of degrees of freedom completely determines what the shape of $f(\chi^2)$ will be. When $v$ is small, the shape of the density function is highly skewed to the right. As $v$ gets larger, however, the distribution becomes more and more symmetrical in appearance. Since only squared numbers are involved in calculating $\chi^2$, we know that this variable can never assume a value below zero, but it may take on values up to positive infinity.

The density function for the $\chi^2$ distribution is not of primary importance for our discussion; hence we will not present its formula here, but merely concentrate on its characteristics. First, the mean and the variance of the chi-square distribution are both related to $v$, as follows:

$$Mean = E[\chi_v^2] = v;$$
$$Variance = E[(\chi_v^2 - v)^2] = 2v.$$

Thus, if we have a chi-square variable involving the sum of the squares of four

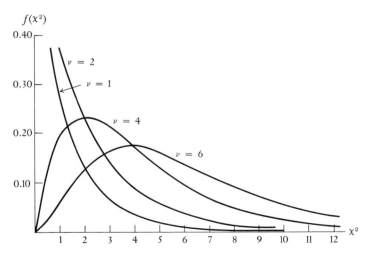

**Fig. 5.20.** The chi-square distribution for various values of $v$.

independent $z$-variables, $\chi_4^2 = z_1^2 + z_2^2 + z_3^2 + z_4^2$, then $v = 4$, and $E[\chi^2] = 4$ and $V[\chi^2] = 8$.

### Chi-square Examples

Table IV in this book gives values of the cumulative $\chi^2$ distribution function for selected values of $v$, and gives (at the bottom) a formula for the normal approximation of $\chi^2$ which can be used for large values of $v$. To illustrate the use of the chi-square table, consider two problems. In the first, we are given a value of a $\chi^2$ random variable and wish to find a probability. In the second, we are given a probability and wish to find an interval of values for $\chi^2$.

**First example** Suppose we want to determine the probability that the value of a chi-square random variable $\chi^2$ is greater than 16 when $v = 10$; i.e., find $P(\chi^2 > 16)$ for 10 degrees of freedom. First, we know that $P(\chi^2 > 16) = 1.0 - P(\chi^2 \leqslant 16) = 1.0 - F(16)$, from the definition of a cumulative distribution function. Then we look for $v = 10$, in the left margin of Table IV, which gives the cumulative $\chi^2$ values. Values of $\chi^2$ are given *in* the table, and selected values of the function $F(\chi^2)$ are given across the top. We need to find the entry from the row with $v = 10$, which is closest in value to 16 (or to interpolate between values) and, then read the value of $F(16)$ at the top. For $v = 10$, $F(15.99) \doteq F(16) = 0.900$. Therefore $P(\chi^2 > 16)$ when $v = 10$ is $1 - 0.900 \doteq 0.10$. Thus, ten percent of the values in the chi-square distribution with ten degrees of freedom will be larger than 16. These values are shown in Fig. 5.21.

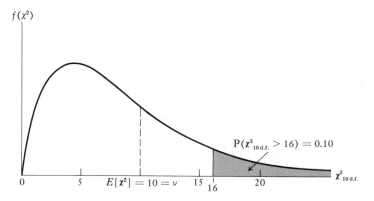

**Fig. 5.21.** $\chi^2$-distribution for $v = 10$.

**Second example** Find the upper and lower endpoints of the narrowest interval so that the probability of $\chi^2$ falling within the interval is 0.95. This time, assume that the number of degrees of freedom is 18. The problem is to find values $a$ and

$b$ such that $P(a < \chi^2_{18\,\text{d.f.}} < b) = 0.95$. To get the narrowest interval, we exclude an equal probability on the left and the right ends of the distribution, namely, 0.025 excluded on each end. In terms of the cumulative distribution function, we want values of $a$ and $b$ to satisfy:

$F(a) = 0.025$ giving 2.5 percent area under $f(\chi^2)$ to the left of $a$, and

$F(b) = 0.975$ giving 2.5 percent area under $f(\chi^2)$ to the right of $b$.

Using the row in Table IV for $v = 18$, we find $F(a) = 0.025$ if $a = 8.23$ and $F(b) = 0.975$ if $b = 31.53$. Therefore, $P(8.23 < \chi^2_{18\,\text{d.f.}} < 31.53) = 0.95$, and the range of the interval is as narrow as possible while still including 95 percent of the area under $f(\chi^2_{18\,\text{d.f.}})$. Note that the distance from each endpoint to the mean is not equal, since the chi-square distribution is not symmetric. It is the amount of area under $f(\chi^2)$ outside each endpoint (0.025) which is made equal in order to solve the problem. Figure 5.22 shows the $\chi^2$ distribution for this example.

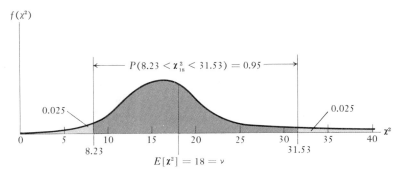

Figure 5.22

In Figs. 5.21 and 5.22, the chi-square distributions have their means greater than the median or mode, because the distribution is positively skewed. The mean is equal to the degrees of freedom, $v$.

## *5.8 EXPONENTIAL DISTRIBUTION

Another important continuous distribution, the *exponential distribution*, is closely related to a discrete distribution discussed previously, the Poisson. Both the Poisson and the exponential distribution have found many applications in operations research, especially in studies of queueing (waiting-line) theory. These two distributions are related in such applications by the fact that if events (e.g., requests for service, or arrivals) are assumed to occur according to a Poisson probability

---

* This section may be omitted without loss in continuity.

law, then the exponential distribution can be used to determine the probability distribution of the time which elapses *between* such events. For example, if customers arrive at a bank in accordance with a Poisson distribution, the exponential may be used to determine the probability distribution of the time between these arrivals. The time it takes to be serviced (called the service time) in these models is another application of the exponential distribution.

The exponential distribution is a continuous function which has the same parameter, $\lambda$, as the Poisson. Lambda, as before, represents the mean rate at which events (arrivals or service completions) occur. Thus a value of $\lambda = 3.0$ might imply that service completions occur, on the average, at the rate of 3.0 per minute (or any other time unit). If a telephone line can handle an average of 20 customers per hour, then $\lambda$, defined as the mean number of customers being served by the telephone facilities, is $\lambda = 20$ (per hour) or $\lambda = \frac{1}{3}$ (per minute). Similarly, $\lambda = 3.8$ might imply, as it did in Section 4.5, that on the average 3.8 customers arrive at a checkout counter in a supermarket every minute.

One major assumption necessary for applying the exponential distribution to applications involving service facilities is that the time between arrivals (if $\lambda = $ the arrival rate) or the time for completing the service (if $\lambda = $ the service rate) is usually relatively short. The longer the time interval becomes, the *less* likely it is that the service completion (or the next arrival) will take that long or longer. Suppose we let the random variable $T$ represent the amount of time between service completions, or between arrivals. As $T$ becomes larger and larger, the value of $f(T)$ for the exponential becomes smaller and smaller. In fact, as can be seen in the graph of the exponential distribution in Fig. 5.23, $f(T)$ approaches zero as $T$ goes to infinity.

Note that the exponential distribution, similar to the Poisson, assumes a value other than zero only when $T$ is greater than or equal to zero and when $\lambda$ is greater than zero. The vertical intercept of the function shown in Fig. 5.23 is seen to equal

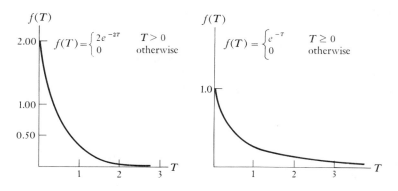

**Fig. 5.23.** The exponential distribution: $f(T) = \lambda e^{-\lambda T}$.

$\lambda$, which means that $f(0) = \lambda$. These relationships characterize the exponential distribution, whose density function is:

> *Exponential distribution:*
>
> $$f(T) = \begin{cases} \lambda e^{-\lambda T} & \text{for } 0 \leqslant T \leqslant \infty, \quad \lambda > 0, \\ 0 & \text{otherwise} \end{cases}$$
>
> (5.8)

**Mean and Variance of the Exponential**

Remember that if we interpret $\lambda$ as the mean arrival (or service) rate, then the exponential gives the probability distribution of the time *between* arrivals (or between service completions). It should thus not be very surprising to learn that the mean of the exponential distribution is $1/\lambda$. For example, suppose the mean service rate of a cashier in a bank equals one-half customer every minute, or $\lambda = \frac{1}{2}$. Since it takes, on the average, one minute to serve half a customer, it takes two minutes to serve one customer; hence, the mean time between service completions is $1/\lambda = 2$. Similarly, if the mean number of arrivals at a check-out counter in a supermarket is $\lambda = 3.8$ per minute, then the mean time between arrivals will be $1/\lambda = 1/3.8 = 0.263$ minutes (or one customer approximately every 16 seconds). As was true for the Poisson, the mean and the variance of the exponential are both functions of $\lambda$. In this case we could show (theoretically) that the mean and variance of the exponential are:

> *Exponential mean:* $\quad \mu = \dfrac{1}{\lambda};$
>
> (5.9)
>
> *Exponential variance:* $\quad \sigma^2 = \dfrac{1}{\lambda^2}.$

Since the exponential and the Poisson distributions can both be applied to problems of arrivals at a service facility, suppose we reconsider the example in Section 4.5 using the exponential function to describe the time between arrivals at a supermarket where the mean arrival rate is $\lambda = 3.8$. By substituting $\lambda = 3.8$ in Formula (5.8) we get

$$f(T) = 3.8e^{-3.8T}.$$

Since $\mu = \lambda = 3.8$ for this example, the mean time between arrivals is $1/3.8 = 0.263$ minutes; the variance of this time between arrivals is $1/\lambda^2 = 1/(3.8)^2 = 0.069$ minutes.

Now, suppose we wanted to calculate the probability that the time between arrivals is greater than one minute, $P(T > 1)$. One way to find such a probability would be to integrate the exponential function to find the area under the curve over the interval in question. Fortunately, such integrations are not necessary since tables have been prepared which permit evaluation of the exponential function directly. Table V in this book gives values of the cumulative exponential distribution $F(T)$ associated with selected values of $\lambda T$.

Thus, to find $P(T > 1)$ when $\lambda = 3.8$, we can use the cumulative exponential distribution table (Table V) to find $F(1)$ associated with $\lambda T = 3.8(1) = 3.8$. Since for $\lambda = 3.8$ the value of $F(1) = 0.978$, we can write $P(T > 1)$ as follows:

$$P(T > 1) = 1.0 - P(T < 1) = 1.0 - F(1)$$
$$= 1.0 - 0.978 = 0.022.$$

Similarly, if one wants to determine the probability that an arrival will occur between one-half and one minute when $\lambda = 3.8$, Table V can be used to find $P(\frac{1}{2} < T < 1) = F(1.0) - F(0.5)$. We already know that $F(1.0) = 0.978$. The value for $F(0.5)$ is found in Table V under $\lambda(0.5) = 3.8(0.5) = 1.9$, and equals $F(0.5) = 0.850$. Thus, $F(1.0) - F(0.5) = 0.978 - 0.850 = 0.128$. This example is illustrated in Fig. 5.24.

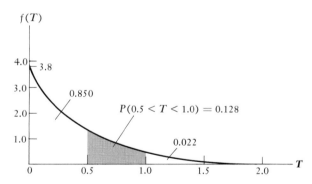

**Fig. 5.24.** Example of exponential probability problem with $\lambda = 3.8$.

Calculations of the above nature can be especially useful in the queueing problems mentioned earlier. If, in addition to studying the pattern of arrivals at our supermarket check-out counter, we had also investigated the service time of the cashiers, then we could develop a relationship which indicates the probability that our cashiers will not be busy for a period of $T$ minutes and the probability that the customers will have to wait more (or less) than $T$ minutes. Ideally, such an investigation could lead to an analysis of the benefits of keeping a customer waiting versus the cost of hiring (or firing) another cashier, the result being a staffing policy designed to balance these costs and benefits. (Too often, it seems, the arrival rate *exceeds* the service rate for extended periods of time in many supermarkets.)

### 5.9 PROBABILITY DISTRIBUTIONS—SUMMARY

The probability distributions presented thus far are summarized in Table 5.4. The first two are discrete distributions and the latter four are continuous distributions. Although many practical problems in statistical inference and decision-making can be handled with a knowledge of these distributions, these are not the only distributions used by statisticians. In Chapter 6 we will study an additional continuous distribution, the *t*-distribution.

### REVIEW PROBLEMS

1. A probability density function for a continuous random variable $x$ is defined to be:

$$f(x) = \begin{cases} 2x & \text{for } 0 \leqslant x \leqslant 1, \\ 0 & \text{elsewhere.} \end{cases}$$

   a) Sketch this pdf.
   b) Show that the total area under $f(x)$ equals 1.0.
   c) Find $P(\frac{1}{2} < x \leqslant 1.0)$.

2. a) Write the formula for the normal density function for $\mu = 10$, $\sigma = 1$; for $\mu = 15$, $\sigma = 10$.
   b) Sketch the probability density function for each of the preceding cases. Sketch the cumulative probability function for both cases.
   c) If $\mu = 15$ and $\sigma = 10$, what is the probability that $5 \leqslant x \leqslant 25$? What is the probability that $-5 \leqslant x \leqslant 35$? What is $P(-10 \leqslant x \leqslant 35)$?
   d) Use your answer to part (c) to determine the following: $P(x \geqslant 35)$, $P(x \leqslant -5)$, and $P(x \leqslant -5 \text{ or } x \geqslant 35)$.

3. a) Find the value of $z$ such that $P(z > z) = 0.01$. Find $P(z < -z) = 0.025$.
   b) Find the value of $z$ such that $P(z > -z) = 0.01$. Find $P(z < -z) = 0.02$.

4. Suppose the number of hours per week of lost work due to illness in a certain automobile assembly plant is approximately normally distributed, with mean of 60 hours and standard deviation of 15 hours. For a given week, selected at random, what is the probability that:
   a) The number of lost work hours will exceed 85 hours?
   b) The number of lost work hours will be between 45 and 55 hours?
   c) The number of lost work hours will be exactly 60?

5. The average age of state congressmen in 1935 was 49.65 with a standard deviation of 5 years. Assume their ages follow a normal distribution; what is the probability that a congressman selected at random would be younger than 38 years old?

**Table 5.4** Summary of probability distributions

| Probability distribution | Parameters | Characteristics | Mass or density function given in formula | Mean | Variance | Reference sections | Probability Table |
|---|---|---|---|---|---|---|---|
| **Discrete** | | | | | | | |
| Binomial | $0 \leqslant p \leqslant 1$ $n = 0, 1, 2, \ldots$ | Skewed unless $p = 0.5$, family of distributions | (4.1) | $np$ | $npq$ | 4.2, 4.3 | $P(\chi_{\text{binomial}})$ in Table I |
| Poisson | $\lambda > 0$ | Skewed positively, family of distributions | (4.5) | $\lambda$ | $\lambda$ | 4.5 | $P(\chi_{\text{Poisson}})$ in Table II |
| **Continuous** | | | | | | | |
| Normal | $-\infty < \mu < +\infty$ $\sigma > 0$ | Symmetrical, family of distributions | (5.4) | $\mu$ | $\sigma^2$ | 5.4 | — |
| Standardized normal $z$ | — | Symmetrical, single distribution | (5.6) | 0 | 1 | 5.5 | $F(z)$ in Table III |
| Chi-square $\chi^2$ | $\nu = 1, 2, \ldots$ | Skewed positively, family of distributions | — | $\nu$ | $2\nu$ | 5.7 | $F(\chi^2)$ in Table IV |
| Exponential | $\lambda > 0$ | Skewed positively, family of distributions. | (5.8) | $1/\lambda$ | $1/\lambda^2$ | 5.8 | $F(T_{\text{exponential}})$ in Table V |

6. Suppose the heights of adult African pygmies has a normal distribution with mean 40 inches and standard deviation 5 inches. What is the probability that the tribe's witch doctor (selected at random) is between 45 and 55 inches tall?

7. Over the years, the total number of points scored by both teams in high-school basketball games follows a normal distribution with mean 150 and standard deviation 15. What is the probability that the number of points scored in a randomly selected game will be between 165 and 180 points?

8. If $z$ is a standardized normal variable, what is the probability of obtaining a value of $z$ between $-1.28$ and $+1.65$?

9. For each of the following, find the specified numerical value and illustrate the interval and area involved on a normal distribution sketch (the symbol $\sim$ means "distributed as").

   a) Assuming $z$ is $N(0, 1)$, find $P(1.5 < z < 2.23)$.
   b) Assuming $z$ is $N(0, 1)$, find $P(-1.34 < z < +0.62)$.
   c) For $z \sim N(0, 1)$, find the value of $b$ so that $P(-0.05 < z < b) = 0.40$.
   d) For $z \sim N(0, 1)$, find the value of $b$ so that $P(z > |b|) = 0.12$.
   e) For $x \sim N(45, 81)$, find $P(33 < x < 51)$.
   f) For $x \sim N(15, 3.24^2)$, find $P(x > 20)$.
   g) For $x \sim N(100, 400)$, find the value of $a$ so that $P(x < a) = 0.95$.

10. If the income in a community is normally distributed, with a mean of $9000 and a standard deviation of $2000, what minimum income does a member of this community have to earn in order to be in the top 10 percent? What is the maximum income one can have and still be in the middle 50 percent?

11. Suppose the sales invoices of a certain company have a normal distribution with mean $32 and standard deviation $8. Also, the service life of telephone poles used by public utility companies is normally distributed with average 15 years and variance 25 years.

    a) Which of these normal distributions has the largest range? (Careful!)
    b) What is the probability that a utility pole selected at random from this latter distribution will have a service life greater than 15 years?
    c) Suppose I observe one utility pole which lasts only 3 years and also, I observe an invoice of $50. Which of these occurrences is the *more* unusual?

12. a) Use the binomial probabilities in Table I to determine the probability of receiving at least 55 but no more than 60 heads in 100 tosses of a fair coin.
    b) Use the normal distribution to approximate your answer to part (a).

13. How many questions would a professor have to put on a multiple-choice exam, where there are four choices for each question, in order for him to be 99.9 percent sure that a student who makes a random guess on each question misses at least one-half of the questions? (Use the normal approximation to the binomial without correcting for continuity.)

14. a) What parameter characterizes the chi-square distribution?
    b) What is the mean and the variance of a chi-square distribution?
    c) Sketch the chi-square distribution for $v = 2$ and $v = 5$. Indicate on your sketch the region cutting off 5 percent of the righthand tail of this distribution.

15. Given the exponential distribution whose parameter, $\lambda$, equals 3.0.

   a) Graph the probability density function and the cumulative probability distribution.
   b) What is the mean and the variance of this distribution?
   c) What percent of the area of this distribution lies within $\pm$ one standard deviation of the mean? Within $\pm$ two standard deviations of the mean?

16. Describe how the Poisson and the exponential distributions are related. What assumptions underlie these distributions?

17. Suppose that a bank can service its customers, on the average, at the rate of four customers per six-minute period. Assume the number of customers serviced to be Poisson-distributed.

   a) What is the probability that this bank will be able to service six or more customers in a six-minute period?
   b) What is the probability that a customer will take longer than three minutes?
   c) What is the probability that a customer will take between two and four minutes?

18. A barbershop has on the average ten customers between 8:00 and 9:00 each morning that it is open. Customers arrive according to the Poisson distribution.

   a) What is the probability that the barbershop will have exactly ten customers between these hours on a given morning?
   b) What is the probability that the barbershop will have more than twelve customers?
   c) What is the probability that the barbershop will have fewer than six customers?

## EXERCISES

19. Find the mean and variance of the distribution defined in Problem 1 of this chapter.

20. The *uniform* or *rectangular* distribution can be defined in terms of its cumulative distribution function, as follows:

$$F(x) = \begin{cases} \dfrac{x-a}{b-a} & \text{for } a \leqslant x \leqslant b, \\ 0 & \text{for } x < a, \\ 1 & \text{for } x > b. \end{cases}$$

   a) Derive the uniform density function for $a \leqslant x \leqslant b$ by differentiating $F(x)$ with respect to $x$.
   b) Sketch both the density function and the cumulative function for the uniform distribution.
   c) Show that the uniform distribution satisfies the two properties of all probability functions.
   d) Find the mean and the variance of this distribution.

21. Given the probability density function

$$f(x) = \begin{cases} \frac{1}{2}x & 0 \leqslant x \leqslant 2, \\ 0 & \text{otherwise.} \end{cases}$$

   a) Determine the cumulative probability function for this problem, and then sketch both the density and the cumulative functions.

    b)  Show that this function possesses the two properties of a probability density function.

    c)  What is the probability that $\frac{1}{2} \leqslant x \leqslant \frac{3}{2}$? Evaluate $F(x)$ at $x = 1$. What probability does $F(1)$ represent?

    d)  Find the mean and the variance of this function.

22.  Should a visitor to a gambling casino complain of loaded dice if he rolls a 7 or an 11 on 17 rolls out of 49? (Use a normal approximation and let $P(7 \text{ or } 11) = \frac{2}{9}$).

23.  In doing some product-control work, I check samples of size 100 from each shipping lot of 500 gross. The company policy is to reject any lot when I am 96% sure that it is more than 10% defective. If I find 15 defectives in a sample, should I reject the lot? What should I do if I find 16 defectives? From these results, can you state some simple rule for me to follow? Assume sampling is with replacement, and use the binomial distribution.

24.  Suppose that you observe the following service times by a teller in a bank: $\frac{1}{2}$, 1, $\frac{1}{2}$, 6, 1, 3, where all times are in minutes. Assume that service times are exponentially distributed.

    a)  What is the mean service time, given these observations? What is the variance?

    b)  Graph the exponential distribution for this application.

    c)  What is the probability that service will take longer than two minutes? What is the probability that service will take less than one minute? What is $P(1 \leqslant x \leqslant 3)$?

25.  The number of automobiles arriving at a certain turnpike tollbooth is considered to be Poisson-distributed, with a mean arrival rate of four cars per minute.

    a)  What is the probability of more than five arrivals in any given minute? What is the probability of fewer than two arrivals in a minute?

    b)  What is the probability that the time between two successive arrivals will exceed two minutes? What is the probability that the time between arrivals will be less than 30 seconds?

    c)  Suppose that the number of customers this particular tollbooth can serve in a minute is also Poisson-distributed, with a mean service rate of three cars per minute. Under these conditions what will happen to the queue of cars waiting to pay their toll?

**6**

# Sampling and Sampling Distributions

## 6.1 INTRODUCTION

This chapter begins the task of relating the probability concepts studied in the past four chapters to the objective of statistics stated in Chapter 1: to draw inferences about population parameters on the basis of sample information.

Basically, there are four questions which need to be asked about samples and the process of inference:

1. What are the methods for collecting samples which best assure that they are representative of the parent population, yet still as inexpensive as possible?

2. What is the best way to describe sample information usefully and clearly?

3. How does one go about drawing conclusions and making inferences about the population?

4. How reliable are the inferences and conclusions drawn from sample information?

We turn first to the question of how sample information can be collected most efficiently by describing various *sample designs*.

## 6.2 SAMPLE DESIGNS

*A sample design is a procedure or plan specified before any data are collected, to obtain a sample from a given population.*

The primary requisite for a "good" sample is that it be representative of the population one is trying to describe. There are, of course, many ways of collecting a "poor" sample. One obvious source of errors of misrepresentation arises when the *wrong population is inadvertently sampled.* The 1936 presidential election poll conducted by the now defunct *Literary Digest* remains a classic example of this problem. The *Literary Digest* predicted, on the basis of a sample of over two million names selected from telephone directories and automobile registrations, that Landon would win an overwhelming victory in the election that year. Instead, Roosevelt won by a substantial margin. The sample collected by the *Digest* apparently represented the population of predominantly middle- and upper-class people who owned cars and telephones; it misrepresented the general electorate, however, and Roosevelt's support came from the lower-income classes, whose opinion was not reflected in the poll.

Another potential source of error in sampling, especially in surveys of public opinion, comes from *response bias.* Poorly worded questionnaires or improper interview techniques may elicit responses which do not reflect true opinions. Kinsey's research on sex practices, for example, received widespread criticism for reporting responses to questions to which most people are fairly sensitive. Such responses are, therefore, not unlikely to be distorted from the truth. Similarly, it is amazing how the economic well-being of certain college alumni can vary between reunion meetings and the time for the annual fund-raising drive.

These types of error are called *nonsampling errors*. Nonsampling errors include all kinds of "human errors" such as mistakes in collecting, analyzing, or reporting data, as well as sampling from the wrong population, and response bias. If the researcher incorrectly adds a column of numbers, this represents a nonsampling error just as much as does the failure of a respondent to provide truthful information on a questionnaire.

In addition, even in well designed and well executed samples, there are bound to be cases where the sample does not provide a good representation of the population under study because samples represent only a portion of a population. In such cases the information contained in the sample may lead one to incorrect inferences about the parent population; that is, an "error" might be made in estimating the population characteristics from the sample information. Errors of this nature, representing the differences which can occur between a sample statistic and the population parameter being estimated, are called *sampling errors*. Sampling errors obviously can occur in all data-collection procedures except a complete enumeration of the population (a census).

One primary objective in sample design is to minimize both sampling and nonsampling errors. Errors are costly, not only in terms of the time and money spent in collecting a sample, but in terms of the potential loss in making a wrong decision on the basis of an incorrect inference from the data. An incorrect public opinion survey, for instance, could cost the politician his job if he based his campaign on inferences from these data. Similarly, investment in real estate or stocks might cost the investor a considerable amount of money if the (sample) information that induced him to make a particular investment proved incorrect.

Note that it is the *decisions* resulting from incorrect inferences that may be costly, not the incorrect inferences themselves; hence, it is customary to refer to the objective of sampling as one of *minimizing the cost* of making an incorrect decision (or an error). But reducing the costs of making an incorrect decision usually comes by increasing the cost of designing and/or collecting the sample. For example, additional effort (or money) devoted to designing a questionnaire, identifying the correct population, or collecting a larger sample usually results in a more representative sample. We can therefore state the primary objective in sample design as one of *balancing the costs of making an error against the costs of sampling*.

Designing an optimal sampling procedure may not be easy. For one reason, the elements of a given population may be extremely difficult to locate, gain access to, or even identify. For example, it may be impractical, if not impossible to identify the population elements of "color television owners" in a particular United States city. Another obvious difficulty already mentioned is cost; budget constraints, for example, may force one to collect fewer data, or to be less careful about collecting these data, than ideal designs would dictate. Also, the costs of making an incorrect decision may be very hard to specify. A discussion of all the problems inherent in sample design, especially those concerned with nonsampling errors and the costs of making an incorrect decision, falls outside the scope of this book. We shall therefore concentrate our attention on the problem of determining sample designs which most effectively minimize sampling errors.

**Types of Sample Designs**

Two distinctions require elaboration before we proceed. First, classification of sampling designs usually depends on whether the methods employed represent *probabilistic or nonprobabilistic* sampling procedures. Probabilistic sampling refers to a process in which the laws of probability determine which elements of the population to include in the sample. In nonprobabilistic sampling, some criterion other than the laws of probability are used to select the items of the sample, for example, the accessibility of the elements, the opinion of experts, or convenience to the researcher. This distinction between probabilistic and nonprobabilistic sampling is an important one in statistics, because most of the theory of sampling depends largely on assumptions concerned with the probability that a given element in the population will be included in the sample.

A second important distinction in sampling theory concerns the difference between sampling from a *finite* population and sampling from an *infinite* population. This distinction becomes crucial when determining the probability that an element of the population is contained in the sample. When sampling from an infinite population, the probability that the sample includes a particular item does not change as more and more elements are selected. The same is true for sampling from a finite population with replacement. Without replacement, however, the size of a finite population *decreases* with each item drawn, and the probability of selecting the remaining items must change accordingly. In Chapter 4 we saw this distinction to be important when the hypergeometric distribution was used to determine probability values for problems in which $p$ changed with each item sampled, and the binomial distribution was used for problems where $p$ remained constant. As we pointed out then, as long as the sample size $n$ is small relative to the population size $N$ (less than about five percent), this distinction may not be too important.* When the distinction is important, adjustments for finite populations are used. It will be convenient, in most of the discussion below, to assume that all samples are drawn either from a population large enough to be considered infinite or from a finite population with replacement.

---

* Remember, we are denoting the sample size with a lower case $n$, and a population size with $N$.

### Probabilistic Sampling

Often the first criterion for a good sample is that each item in the population under investigation has an *equal and independent chance* to be part of the sample; also, it is often advantageous for each set of $n$ items to have an equal probability of being included. Samples in which every possible sample of size $n$ (i.e., every combination of $n$ items from the $N$ in the population) is equally likely are referred to as *simple random samples*. These are the types of sample we have used implicitly throughout our discussion of probability theory for presenting examples of the use of formulas and probability functions.

### Using a Random Number Table

Designing a sample in which each item in the population has an equal probability of being selected usually requires carefully controlled sampling procedures. The stereotype of drawing slips of paper from a goldfish bowl may satisfy the requirements of a simple random sample, but there is no practical way to tell. The 1969 Selective Service draft lottery was highly criticized for using essentially a "goldfish bowl" technique without adequate mixing. A more systematic approach to assuring randomness is to select a sample with the aid of a table of random numbers. In such a table each digit between 0 and 9 is called a *random digit*, where the word random implies that all of these digits have the same long-run relative frequency (i.e., the same probability of occurring), and the occurrence or nonoccurrence of any number is independent of the occurrence or nonoccurrence of all other numbers, or of all sets of $n$ other numbers. In a table of random numbers, random digits are usually combined to form numbers of more than one digit. For example, random digits taken in pairs will result in a set of 100 different numbers (00 to 99), each with a probability of $1/100$ of occurring independent of all other pairs. Likewise, in a table of random numbers consisting of groups of three random digits, each of the 1000 numbers between 000 and 999 will have a probability of $1/1000$ of occurring and will be independent of the remaining numbers. Table VI is a page of random numbers from a book (published by the Rand Corporation) containing one million random digits.

With a finite population of size $N$, a table of random numbers can be used to select a simple random sample of $n$ items in the following manner. First, a unique number between 0 and $N$ must be assigned to each of the $N$ items in the population. The table of random numbers is then consulted. The first $n$ numbers encountered (starting at *any* point in the table and moving systematically across rows or down columns) which are less than $N$ constitute a set of $n$ random numbers. The $n$ elements corresponding to these $n$ numbers form the random sample.

To illustrate this procedure, suppose we use Table VI to select a sample of four items from a population of 75 elements. A random selection of a starting point in the table is customary; let us arbitrarily start with the number which is tenth from the bottom in the first column, 09237. Since our population has less than 100 elements, we need to look at only two digits of each number. Suppose we use the first two digits. Then the first item in our sample becomes item number 09. Reading down, the next three items are 11, 60, and 71 (we skip 79 because we have only 75 elements in the population). Note that if our sample were to contain five items, the number 09 would have occurred again. This duplication could cause problems, as it may not be possible to sample the same item twice. Its usefulness may be destroyed by the first sample, as, for example, in testing the tread life of a tire or measuring the yield from a new seed variety in an agricultural testing station. For a sample size which is small relative to the population size, it will not seriously distort the usefulness of this method to discard the duplicate item, letting the next element on the list take its place (e.g., item 63). In discarding an item it must be recognized that sampling is now taking place without replacement, rather than with replacement.

In simple random sampling one must have access to all items in the population. With a small population of elements which are easy to identify and sample, this procedure normally gives the best results. With a large population, however, simple random sampling may be difficult, perhaps even impossible to implement; at best, it may be quite costly. In this case, a procedure must be designed which, although more restrictive, is also more practical.

In another popular sampling plan, called *systematic sampling*, a random starting point in the population is selected and then every $k$th element encountered thereafter becomes an item in the sample. For example, every 200th name in a telephone directory might be called in order to survey public opinion. This method is *not* equivalent to simple random sampling because every set of $n$ names does not have an equal probability of being selected. A bias such as that in the *Literary Digest* survey, however, would probably not be present in this decade when almost everyone has a telephone. Bias *will* result under systematic sampling, however, if there is a periodicity to the elements of the population. For instance, sampling sales in a supermarket every seventh day will certainly result in a sample that represents only the sales of a single day, say Monday, rather than the weekly pattern.

**Sampling with Prior Knowledge**

Two important random sampling plans depend on prior knowledge about the population: stratified sampling and cluster sampling. *Stratified sampling* assumes that one can divide a population into homogeneous classes or groups, called *strata*, and then sample according to certain criteria within each strata. The advantage of this procedure follows from the fact that if homogeneous subsets of the population can be identified, then only a relatively small number of observations is needed to determine the characteristics of each subset. It can be shown that *the optimal method of selecting strata is to find groups with a large variability between strata, but with only a small variability within strata.*

We illustrate stratified sampling by considering the task of determining the majority political preference in a given city. Assume that it is known, from previous surveys and elections, that political preferences in this town tend to correspond to various income levels. For instance, upper-income families tend to have similar opinions, as do middle-income and lower-income families. Assume, further, that, in this particular city, it is well known that the upper-income families and the lower-income families will have less variability of opinion within their respective groups than will the middle-income group. It may be that upper-income families will in general favor a conservative candidate, lower-income families will favor a labor candidate, and middle-income families are less predictable.

A *proportional* stratified sampling plan selects items from each strata in proportion to the size of that strata. This procedure ensures that each strata is "weighted" in the sample by the number of elements it contains. If the category upper-income families includes ten percent of the voting population, then a proportional stratified sampling plan would select, randomly, ten percent of the sample from this group. Many times, however, a more efficient procedure is to select a *disproportionate* stratified sample. A plan of this nature collects more than a proportionate amount of observations in those strata with the most variability; e.g., the middle-income group in the above example. In other words, by allocating a disproportionate amount of effort (time, money, etc.) to those groups whose opinions are most in doubt, one often obtains a maximum amount of information for a given cost. Similarly, if it is more costly to sample from a particular stratum, fewer items may be taken from that stratum.

*Cluster sampling* represents a second important sampling plan in which the population is subdivided into groups in an attempt to design an efficient sample. The subdivisions or classes of the population in this case are called clusters, where each cluster, ideally, has the same characteristics as the parent population. If each cluster is assumed to be representative of the population, then the characteristics of this population can be estimated by (randomly) picking a cluster and then randomly sampling elements from within this cluster. Sampling within a cluster may take any of the forms already discussed and may even involve sampling from clusters within a cluster (called two-stage cluster sampling). The criterion for the selection of optimal clusters is exactly opposite to that for strata: *there should be little variability between clusters, but a high variability (e.g., representation of the population) within each cluster.*

Cluster sampling can be illustrated by extending our previous example. Assume that we now want to sample political preferences in all United States cities, rather than just one. A simple random sample of all the people in United States cities would probably be very difficult and expensive, if not impossible, to collect; instead, it may be that a number of cities adequately represent the population of all cities, and it would then be sufficient to sample from just one of these cities. Within the chosen city one could use simple random sampling, stratified sampling, or systematic sampling, or one could break the city into smaller clusters. In cluster sampling there is always the danger that a cluster is not truly representative of the population; a geographical bias, for example, may exist when using one city to represent the entire United States.

### Double, Multiple, and Sequential Sampling

One of the most important decisions in any sampling design involves selecting the *size* of the sample. Usually size is determined in advance of any data collection, but in some circumstances this may not be the most efficient procedure. Consider a problem of determining whether a shipment of 5000 items meets certain specified standards. It would be too expensive to check all 5000 items for their quality, so a sample is drawn and each item in the sample is tested for quality. Rather than take one large sample, of perhaps 100 items, a preliminary random sample of 25 items could be drawn and inspected. It may often be unnecessary to examine the remaining 75 items, for perhaps the entire lot can be judged on the basis of these 25 items. If a high percentage of the 25 components is defective, the decision would probably be that the quality of the entire lot may not be acceptable. A low percentage of defectives may lead to accepting the lot. Values other than these extremes may also lead to acceptance or rejection of the entire lot. Nevertheless, there will usually be a range in which there is doubt about the quality of the entire lot. For example, it may be normal to have one or two defectives in a sample of 25; more than three, however, may lead one to suspect the entire lot, but not necessarily reject it. An additional sample, perhaps the remaining 75 items, could then be taken, and the lot judged on the basis of the entire 100 items.

Samples in which the items are drawn in two different stages, such as in the sequential fashion described above or from a cluster within a cluster, represent a process referred to as *two-stage sampling*. Virtually all important samples represent one form or another of multiple-stage sampling, and the sample design is usually not simple to plan. Surely all large-scale surveys represent some form of multiple-stage cluster sampling. The major advantage of double, multiple, and sequential sampling procedures obviously depends on the savings which result when fewer items than usual need to be observed. These procedures are especially appropriate when sampling is expensive, as when inspection destroys the usefulness of a valuable item.

### Nonprobabilistic Sampling

In some sense all nonprobabilistic sampling procedures represent judgment samples, in that they involve the selection of the items in a sample on the basis of the judgment or opinion of one or more persons. Judgment sampling is usually employed when a random sample cannot be taken or is not practical. It may be that there is not enough time or money to collect a random sample; or perhaps the sample represents an exploratory study where randomness is not too important. When the number of population elements is small, on the other hand, the judgment of an expert may be better than random methods in picking a truly representative sample.

In *quota sampling*, each person gathering observations is given a specified number of elements to sample. Often this technique is used in public-opinion surveys, and the interviewer is allocated a certain number of people to interview. The decision as to exactly whom to interview is usually left to the individual doing the interview, although certain guidelines are almost always established. With well trained and trustworthy interviewers, this procedure can be quite effective and can be carried out at a relatively low cost. A great danger exists, however, that procedures left to the interviewers' judgment and convenience may contain many unknown biases not conducive to a representative sample.

The least representative sampling procedure selects observations on the basis of convenience to the researcher; i.e., a *convenience sample*. Street-corner surveys, in which the interviewer questions people as they go by, used to be a favorite method for collecting public opinions. This method obviously cannot be considered very likely to yield a representative sample; more often, the results are biased and quite unsatisfactory. Convenience sampling is not widely used in circumstances other than preliminary or exploratory studies, or where representativeness is not a crucial factor.

## 6.3  SAMPLE STATISTICS

As we have indicated previously, the usual purpose of sampling is to learn something about the population being sampled. In selecting a sampling design, one of the primary considerations is the importance of the information to be gathered, and the accuracy desired in learning about the population. For these purposes, it is important that we structure the problem of taking a sample and analyzing the sample results in terms of the concepts of probability presented in Chapters 2 through 5.

Assume that we are planning on taking a sample of $n$ observations in order to determine the characteristics of some random variable $x$. The process of taking a sample from this population can be viewed as an experiment, and the observations which might occur in such an experiment make up the sample space. Suppose we let the random variables $x_1, x_2, \ldots, x_n$ represent the observations in this sample. For example, the random variable $x_1$ represents the observation which occurs first in a sample of $n$ observations, $x_2$ represents the second observation, and so forth. In simple random sampling, every item in the population has an equal chance of being the observation which occurs first, so in this case the sample space for $x_1$ would be the entire population of $x$-values. It is important to remember that $x_1, x_2, \ldots, x_n$ are all *random variables*, and each of these variables has a theoretical probability distribution. Under simple random sampling, the probability distribution of each of the random variables $x_1, x_2, \ldots, x_n$ will be identical to the distribution of the population random variable $x$ (since the sample space for each one is the entire population of $x$-values).

Once we have collected a random sample of $n$ observations we have one value of $x_1$, one value of $x_2$, and so forth, with one value of $x_n$. We now need to discover how to learn more about the characteristics of the random variable $x$ (the population) by making use of the sample values of $x_1, x_2, \ldots, x_n$. In general, the population parameters of interest are usually those described in Chapter 1— i.e., the summary measures such as central location, dispersion, skewness, or kurtosis. It is intuitively appealing (and mathematically provable) that the best estimate of a population parameter is given by a comparable sample measure (a sample *statistic*). For example, we will see that the best estimate of the central location of a population is a measure of the central location of a sample; and the best estimate of the population dispersion is a measure of the dispersion of the sample. *Thus, a sample statistic is used as an estimate of a population parameter.*

A sample statistic can be defined as a function of some (or all) of the $n$ random variables $x_1, x_2, \ldots, x_n$. That is, a sample statistic is a random variable which is formed out of the sample values of $x_1, x_2, \ldots, x_n$. This means that there is a theoretical probability distribution associated with every sample statistic.* For example, suppose we let $R = x_{\max} - x_{\min}$ be the *range* of the values in a sample. In this case, the sample statistic $R$ is a random variable which is a function of only two values in each sample, the largest value ($x_{\max}$) and the smallest value ($x_{\min}$). Since $R$ is a random variable, it has a theoretical probability distribution which we could develop if this statistic were of interest.

The sample mean is another sample statistic. In this case the statistic is a function of *all n* values in a sample, as we will see in Formula (6.1). There are many different sample statistics which can be used to estimate the population parameters of interest. Generally, the population parameters of most interest are the mean and variance, since these two measures are so useful in describing a distribution and so necessary in decision-making. Consequently, the most useful sample statistics are those which provide the best information about these two parameters,

---

* The method of sampling used will affect the distribution of any sample statistic. For example, if *random* sampling is used, then the random variables $x_1, x_2, \ldots, x_n$ will all be independent, which means (from Chapter 3) that the joint density function $f(x_1, x_2, \ldots, x_n)$ equals $f_1(x_1)f_2(x_2) \cdots f_n(x_n)$. In addition, if *simple* random sampling is used, then the marginal pdf's of each random variable $x_1, x_2, \ldots, x_n$ will all have the same form (such as the normal distribution).

namely the sample mean and the sample variance. In the development to follow, the reader should bear in mind that a similar development could be presented for sample statistics other than the mean and variance.

**Sample Mean and Variance**

The mean of a set of observations representing a sample is calculated in the same manner as we did in Chapter 1 for the mean of a population (Formula (1.5)): by summing the product of each value of $x$ times its relative frequency. In this case, we assume that the sample consists of $n$ observations which can be labeled $x_1$, $x_2, \ldots, x_n$. Furthermore, we will assume that these $n$ observations can be grouped into $c$ different classes, where the frequency of each class is denoted by $f_i$ (for $i = 1, 2, \ldots, c$, where $c$ is the number of classes). The mean of these observations is denoted by the symbol $\bar{x}$ (read $x$-bar), where $\bar{x}$ is defined as follows:

$$\text{Sample mean:} \qquad \bar{x} = \frac{1}{n} \sum_{i=1}^{c} x_i f_i. \tag{6.1}$$

The reader might wish to verify for himself the similarity between Formulas (6.1) and (1.5). Also, note that if the frequency of each sample observation is 1, then Formula (6.1) becomes $\bar{x} = (1/n)\sum x_i$ (comparable to Formula (1.1), which was $\mu = (1/N)\sum x_i$).

Before illustrating the use of Formula (6.1), we present the formula for the variance of a sample. In this case, the sample variance is *not* calculated in the same manner as a population variance, for two reasons. First, we can't take the sum of squared deviations about $\mu$, for in most sampling problems $\mu$ is an unknown. Instead, we will take the sum of squared deviations about $\bar{x}$, or $\sum_{i=1}^{c} (x_i - \bar{x})^2$. As was the case for the population variance, each squared deviation must be multiplied by its frequency, so we now have $\sum_{i=1}^{c} (x_i - \bar{x})^2 f_i$. The second difference is that, in the present case, we divide the sum of squared deviations not by the number of observations, but by $(n - 1)$. The reason for dividing by $(n - 1)$ is that the resulting measure of variability can be shown to provide the *best* measure for making estimates about the (unknown) population variance. This fact will be explained in more detail in Chapter 7, when we discuss estimation procedures. For now, we merely demonstrate the use of this measure of variability, which is denoted by the symbol $s^2$.

$$\text{Sample variance:} \qquad s^2 = \frac{1}{n-1} \sum_{i=1}^{c} (x_i - \bar{x})^2 f_i. \tag{6.2}$$

Recall that we developed a computational formula for variances in Chapter 3. The same type of computational formula can be developed for (6.2), the only difference being that because our divisor is now $(n - 1)$, the old "mean square — square mean" relationship has to be modified slightly, as follows:

*Sample variance (computational form):*

$$s^2 = \left( \frac{1}{n-1} \sum_{i}^{c} x_i^2 f_i \right) - \frac{n}{n-1} \bar{x}^2. \tag{6.3}$$

The *sample standard deviation* is denoted by $s$, and always equals the *square root* of the sample variance.

To demonstrate the use of these formulas, consider the case of a United States Senate committee investigating the number of Federal grants awarded for local projects (such as HUD and TOPICS Programs) in cities with populations ranging from 50,000 to 200,000. In an attempt to measure the characteristics of the population in this case (all United States cities of that size), ten randomly selected cities were surveyed. Table 6.1 shows the results of this survey (where $x$ = number of grants during the past year, $f$ is the frequency of each value of $x$ and $c$ = 5 classes). The sample mean can be calculated using the information in columns (2) and (3):

$$\bar{x} = \frac{1}{n} \sum_{i=1}^{5} x_i f_i = \frac{1}{10}(28) = 2.8.$$

The average number of grants in the ten cities sampled is thus 2.8.

**Table 6.1** Number of Federal grants to cities of size 50,000–200,000

| (1)<br>$x$ | (2)<br>$f$ | (3)<br>$xf$ | (4)<br>$(x - \bar{x})$ | (5)<br>$(x - \bar{x})^2$ | (6)<br>$(x - \bar{x})^2 f$ | (7)<br>$x^2$ | (8)<br>$x^2 f$ |
|---|---|---|---|---|---|---|---|
| 1 | 2 | 2 | −1.8 | 3.24 | 6.48 | 1 | 2 |
| 2 | 3 | 6 | −0.8 | 0.64 | 1.92 | 4 | 12 |
| 3 | 1 | 3 | 0.2 | 0.04 | 0.04 | 9 | 9 |
| 4 | 3 | 12 | 1.2 | 1.44 | 4.32 | 16 | 48 |
| 5 | 1 | 5 | 2.2 | 4.84 | 4.84 | 25 | 25 |
|   | 10 | 28 |   |   | 17.60 |   | 96 |

To measure the variance of these data, we can use either Formula (6.2) or (6.3). Using the product of the frequencies times the squared deviations in column (6), we can calculate $s^2$ as follows:

$$s^2 = \frac{1}{n-1} \sum_{i=1}^{c} (x_i - \bar{x})^2 f_i = \frac{1}{9}(17.60) = 1.955.$$

If the computational Formula (6.3) had been used, only columns (3) and (8) need to be used:

$$\left(\frac{1}{n-1}\sum_{i=1}^{c}x_i^2 f_i\right) - \frac{n}{n-1}\bar{x}^2 = \left[\frac{96}{9} - \frac{10}{9}(2.8)^2\right] = 1.955.$$

The sample standard deviation is $s = \sqrt{1.955} = 1.398$. Again, one way to check to see if the result $s = 1.398$ is reasonable is by our old rule of thumb, which, for samples, says that about 68 percent of the sample values should fall within one standard deviation of the mean, $\bar{x} \pm 1s$. In this case, the result appears reasonable, since seven of the ten observations lie in the interval $\bar{x} \pm 1s = 2.8 \pm 1.398 = 1.402$ to $4.198$.

Perhaps at this point we should reiterate that one of our objectives in calculating sample means and variances is to be able to make statements about the population mean and variance. Since different samples from the same population may have different means and variances, the only way to determine the true population parameters is to enumerate every item in the population (a census). But a census is usually too costly and time-consuming, so we must usually be content to use sample statistics to estimate the population parameters, and then to make statements about how reliable or accurate such a sample statistic is in describing the population parameter of interest.

To establish the reliability or accuracy with which a sample statistic describes a population parameter, one must know how likely it is that specific values of this statistic will occur (1) for every possible value of the population parameter and (2) for every possible sample size. To begin our discussion of the reliability and accuracy of a sample, let's suppose we could take a large number of random samples, all of size $n$, and then calculate $\bar{x}$ for each of these samples. These values of $\bar{x}$ could be put in the form of a frequency distribution. This frequency distribution will have a certain shape as well as a mean and a variance. Now, if we took *all possible* samples of size $n$ (and the number of samples may be infinite), the resulting distribution is the *probability* distribution of all possible values of $\bar{x}$.

The probability distribution of $\bar{x}$ is called a sampling distribution. We could also calculate a sampling distribution for $s^2$ by taking all possible values of $s^2$ from samples of a given size $n$. Such sampling distributions, which are necessary for making probability statements about the reliability and accuracy of sample statistics, will be discussed in detail in the remaining sections of this chapter.

### 6.4 SAMPLING DISTRIBUTION OF $\bar{x}$ ($\sigma$ KNOWN)

As we indicated above, the sampling distribution of $\bar{x}$ is the probability distribution of all values of $\bar{x}$ that could be generated by random samples of a given size $n$. Since the sampling distribution of $\bar{x}$ is itself a population of values, it has its own parameters, such as a mean and a variance. The symbol $\mu_{\bar{x}}$, which traditionally denotes the mean of this population, is thus the mean of all possible sample means. Using the notation of expected values, we can thus write $\mu_{\bar{x}} = E[\bar{x}]$. The variance of this population, which is denoted as $V[\bar{x}] = \sigma_{\bar{x}}^2$, will be discussed shortly.

### Mean of $\bar{x}$, or $\mu_{\bar{x}}$

We now want to determine the relationship between the mean of some parent population, $\mu_x$, and the mean of all sample means of size $n$ drawn from this population, $\mu_{\bar{x}}$. Determining this relationship is not difficult using the rules of expected values, since we can prove *that the expected value* of all possible values of $\bar{x}$ from samples of size $n$ equals the mean of the parent *population*, $\mu_x$.* That is,

$$E[\bar{x}] = \mu_{\bar{x}} = \mu_x. \qquad (6.4)$$

To illustrate this relationship let's consider a population which has only three values, $(1, 2, 3)$, where these values occur with equal probability. You might imagine a statistics instructor who gives only three grades: $B(=3)$, $C(=2)$, and $D(=1)$, each with a probability of $\frac{1}{3}$. The mean of this population is easily seen to be $\mu_x = 2.0$; i.e., the instructor gives out an average grade of $C$. Now let's assume you don't know that he gives out only these three grades, so you decide to take a random sample of two of his former students to see what grade they received. Your sample of $n = 2$ could be any one of the nine samples shown in the lefthand column of Table 6.2. If you calculate the mean of your sample, you could get one of the nine values shown in the righthand column of Table 6.2.

---

* Let $x_1, x_2, \ldots, x_n$ represent independent random variables corresponding to the $n$ observations in a sample from a population with mean $\mu_x$ (i.e., $E[x_i] = \mu$). Now, since $\bar{x} = (1/n)(x_1 + x_2 + \cdots + x_n)$, we can apply the rules of expectation from Section 3.3 as follows:

$$
\begin{aligned}
E[\bar{x}] &= E\left[\frac{1}{n}(x_1 + x_2 + \cdots + x_n)\right] \\
&= \frac{1}{n} E[(x_1 + x_2 + \cdots + x_n)] &&\text{By Rule 3} \\
&= \frac{1}{n} \{E[x_1] + E[x_2] + \cdots + E[x_n]\} &&\text{By Rule 5} \\
&= \frac{1}{n}(\mu + \mu + \cdots + \mu) = \frac{1}{n}(n\mu) &&\text{Since } E[x_i] = \mu, \\
\mu_{\bar{x}} &= \mu.
\end{aligned}
$$

**Table 6.2** All samples of size $n = 2$ from (1, 2, 3)

| Sample | Sample mean |
|--------|-------------|
| (1, 1) | 1.0 |
| (1, 2) | 1.5 |
| (2, 1) | 1.5 |
| (1, 3) | 2.0 |
| (3, 1) | 2.0 |
| (2, 2) | 2.0 |
| (2, 3) | 2.5 |
| (3, 2) | 2.5 |
| (3, 3) | 3.0 |
| Sum | 18.0 |

Because you plan to take only *one* sample, of size $n = 2$, you could get a sample mean as low as $\bar{x} = 1.0$ or as high as $\bar{x} = 3.0$. A logical question for a researcher in this situation might be to ask what would be the *expected* value of $\bar{x}$ in such a situation, or what is $E[\bar{x}] = \mu_{\bar{x}}$? That is, we want to find the average of the sample means shown in the righthand column of Table 6.2. There are nine sample means in this column, and the sum of these means is 18. Hence $E[\bar{x}] = \mu_{\bar{x}} = (1/9)(18) = 2.0$, which is exactly the same value as $\mu_x$. Thus, we have shown that for this population $\mu_{\bar{x}} = \mu_x = 2.0$.

### The Variance of $\bar{x}$

In addition to knowing the mean of the sampling distribution of $\bar{x}$ for a given sample size $n$, we also need to know its variance. The variance of the values of $\bar{x}$ is denoted by either $V[\bar{x}]$ or $\sigma_{\bar{x}}^2$. This variance is defined in the same manner as we previously defined a variance, namely as the expected value of the squared deviations of the variable ($\bar{x}$ in this case) about its mean ($\mu_x$ in this case). Thus,

$$\sigma_{\bar{x}}^2 = V[\bar{x}] = E[(\bar{x} - \mu_x)^2].$$

Although evaluating $E[(\bar{x} - \mu_x)^2]$ directly is difficult or practically impossible for most populations, we are fortunate in statistics never to have to attempt such a calculation if we know the variance of the population from which the samples are drawn. This is because the variance of all possible values of $\bar{x}$ is related to the variance of the parent population ($\sigma_x^2$) and to the sample size ($n$) by a very simple formula (which we will present shortly).

Intuitively it should appear reasonable that the variance of $\bar{x}$ will always be less than the variance of the parent population (except when $n = 1$), because there is less chance that a sample mean will take on an extreme value than there is that a single value of the parent population will take on this value. In order for a sample mean to have an extremely large value, most or all of the sample items would have to be extremely large values. From our knowledge of probability, we know that the probability of a single extremely large value on a single draw is greater than the probability of extremely large values on $n$ repeated draws. It would be very unusual not to draw some middle values or some extremely low values in the set of $n$ draws. Such values would balance out the extremely large values and give a less extreme sample mean. Indeed, this intuitive logic is correct. Not only is the variance of $\bar{x}$ always less than or equal to the variance of the parent population, but it can be shown that $\sigma_x^2$ and $\sigma_{\bar{x}}^2$ are very precisely related. *The variance of the mean of a sample of n independent observations is 1/n times the variance of the parent population.**

$$\sigma_{\bar{x}}^2 = \frac{1}{n}\sigma_x^2.$$

(6.5)

When $n = 1$, all samples contain only one observation, and the distribution of $x$ and $\bar{x}$ are identical. That is, $\sigma_{\bar{x}}^2 = \sigma_x^2/1 = \sigma_x^2$. As $n$ becomes larger $(n \to \infty)$ it is reasonable to expect $\sigma_{\bar{x}}^2$ to become smaller and smaller because the sample means will tend to deviate less and less from the population mean $\mu_x$. When $n = \infty$ (or for finite populations, when $n = N$), there will be no variance of the $\bar{x}$'s, and all sample means will equal the population mean. To illustrate the relationship described by Formula (6.5), let's return to our example involving the population (1, 2, 3). The variance of this population is:

$$\sigma_x^2 = \frac{1}{N}\sum(x - \mu_x)^2 = \frac{1}{3}\left[(1-2)^2 + (2-2)^2 + (3-2)^2\right] = \frac{2}{3}.$$

---

* Let $x_1, x_2, \ldots, x_n$ be independent random variables, each having the same variance (that is $V[x_i] = \sigma_x^2$):

$$V[\bar{x}] = V\left[\frac{1}{n}(x_1 + x_2 + \cdots + x_n)\right] \qquad \text{By definition}$$

$$= \left(\frac{1}{n}\right)^2 V[x_1 + x_2 + \cdots + x_n] \qquad \text{By Rule 4}$$

$$= \left(\frac{1}{n}\right)^2 \left\{V[x_1] + V[x_2] + \cdots + V[x_n]\right\} \qquad \text{By Rule 7}$$

$$\sigma_{\bar{x}}^2 = \frac{1}{n^2}(\sigma_x^2) = \frac{1}{n}\sigma_x^2 \qquad \text{Since } V[x_i] = \sigma_x^2.$$

Since we know the variance of this population is $\frac{2}{3}$ and the sample size $n = 2$, we can calculate $V[\bar{x}]$ using Formula (6.5):

$$\sigma_{\bar{x}}^2 = \frac{1}{n}\sigma_x^2 = \frac{1}{2}\left(\frac{2}{3}\right) = \frac{1}{3}.$$

The appropriate data for verifying, empirically, that $\sigma_{\bar{x}}^2 = \frac{1}{3}$ are shown in Table 6.3.

**Table 6.3** Calculations for $V[\bar{x}]$ for data in Table 6.2 (where $\mu_{\bar{x}} = 2$)

| $(x_1, x_2)$ | $\bar{x}$ | $(\bar{x} - \mu_{\bar{x}})$ | $(\bar{x} - \mu_{\bar{x}})^2$ |
|---|---|---|---|
| (1, 1) | 1.0 | −1.00 | 1.00 |
| (1, 2) | 1.5 | −0.50 | 0.25 |
| (2, 1) | 1.5 | −0.50 | 0.25 |
| (1, 3) | 2.0 | 0.00 | 0.00 |
| (3, 1) | 2.0 | 0.00 | 0.00 |
| (2, 2) | 2.0 | 0.00 | 0.00 |
| (2, 3) | 2.5 | 0.50 | 0.25 |
| (3, 2) | 2.5 | 0.50 | 0.25 |
| (3, 3) | 3.0 | 1.00 | 1.00 |
| Sum | 18.0 | 0.00 | 3.00 |

Dividing the sum of squared deviations, $\sum(\bar{x} - \mu_{\bar{x}})^2 = 3.0$, by the number of values in the population (9) gives the variance of this population of $\bar{x}$'s:

$$V[\bar{x}] = \frac{1}{9}(3.0) = \frac{1}{3}.$$

Thus, we have verified that $\sigma_{\bar{x}}^2 = \frac{1}{n}\sigma_x^2 = \frac{1}{3}$.

It is customary to call the standard deviation of the $\bar{x}$'s (which is the square root of $V[\bar{x}]$ and which is denoted as $\sigma_{\bar{x}}$) the *standard error of the mean*. The word "error" in this context obviously refers to sampling error, as $\sigma_{\bar{x}}$ is a measure of the "standard" (or expected) error when using the sample mean to obtain information or make conclusions about the unknown population mean.

*Standard error of the mean:*    $\sigma_{\bar{x}} = \sqrt{\dfrac{\sigma_x^2}{n}} = \dfrac{\sigma_x}{\sqrt{n}}.$      (6.6)

In the example above, the standard error of the mean is

$$\sigma_{\bar{x}} = \sqrt{\sigma_x^2/n} = \sqrt{(\tfrac{2}{3})/2} = \sqrt{\tfrac{1}{3}} = 0.577.$$

We must emphasize at this point that $\mu_{\bar{x}}$ and $\sigma_{\bar{x}}$ are parameters of the population of all conceivable samples of size $n$, and these population parameters are *unknown* quantities. In fact, the values of $\mu$, $\mu_{\bar{x}}$, $\sigma_x$, and $\sigma_{\bar{x}}$ are usually *all* unknown quantities, which means that the relationship $\mu_{\bar{x}} = \mu$ and $\sigma_{\bar{x}} = \sigma/\sqrt{n}$ cannot be used to solve for the value of one of these quantities. However, knowledge of the fact that such relationships exist is important in determining how far a sample mean can be expected to deviate from the population mean.

In order to emphasize the fact that all possible values of $\bar{x}$ represent a probability distribution, it may be useful to summarize the sample means in Table 6.2 in terms of a probability mass function. For example, out of the nine sample means in Table 6.2, three have a value of 2.0; hence $P(\bar{x} = 2.0) = \frac{3}{9}$. The remaining probabilities from Table 6.2 are listed in Table 6.4, and shown graphically in Fig. 6.1.

Although we have derived the mean and variance of the $\bar{x}$'s, nothing has been said about the *shape* of the distribution. Recall from Chapter 1 that distributions with the same mean and variance may have distinctly different shapes. It is necessary, therefore, to be more specific about the entire distribution of $\bar{x}$'s. To do so, we will first assume that the parent population is normal, and then later drop this assumption.

**Table 6.4**

| $\bar{x}$ | $P(\bar{x})$ |
|-----|-----|
| 1.0 | $\frac{1}{9}$ |
| 1.5 | $\frac{2}{9}$ |
| 2.0 | $\frac{3}{9}$ |
| 2.5 | $\frac{2}{9}$ |
| 3.0 | $\frac{1}{9}$ |
|  | 1.0 |

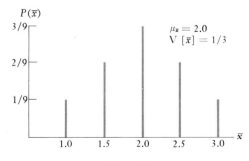

**Figure 6.1**

## 6.5 SAMPLING DISTRIBUTION OF $\bar{x}$, NORMAL POPULATION ($\sigma$ KNOWN)

Suppose samples are drawn from a normal population. Since a mean is a linear transformation of the sample values, the means of samples drawn from a normal population are themselves normally distributed. That is, if we plotted the means of all conceivable samples of size $n$ from a normal population with mean $\mu$ and variance $\sigma^2$ $[N(\mu, \sigma^2)]$, these sample means would themselves form a normal distribution. We already know from Formulas (6.4) and (6.5) that the mean and variance of this normal distribution for $\bar{x}$ would be $\mu$ and $\sigma^2/n$. Thus,

$$\text{Distribution of } \bar{x} \text{ (normal population) } = N(\mu, \sigma^2/n).$$

To illustrate this relationship, suppose a very large number of samples of size $n = 20$ are drawn from a normal population with mean 50 and variance 80. For each of these samples a mean, $\bar{x}$, is calculated. The mean of these sample means (which is sometimes denoted $\bar{\bar{x}}$ and called the *grand mean*) can be calculated and should very nearly equal $E[\bar{x}] = \mu_{\bar{x}} = \mu (= 50$ in this case). Similarly, the variance of the sample means should very nearly equal $V[\bar{x}] = \sigma_{\bar{x}}^2 = \sigma^2/n = 80/20 = 4$; and if the variance equals 4, the standard deviation must equal 2. Further, since we have indicated that the random variable $\bar{x}$ is normally distributed, 68.27 percent of the sample means will fall within plus-or-minus one standard deviation of the mean, $\mu \pm \sigma_{\bar{x}} = 50 \pm 2 = 48$ to 52; 95.44 percent will fall within plus-or-minus two standard deviations of the mean, $\mu \pm 2\sigma_{\bar{x}} = 50 \pm 2(2) = 46$ and 54; and 99.7% of all sample means will fall within plus-or-minus three standard deviations of the mean, $\mu \pm 3\sigma_{\bar{x}} = 50 \pm 3(2) = 44$ to 56. Figure 6.2 shows the sampling

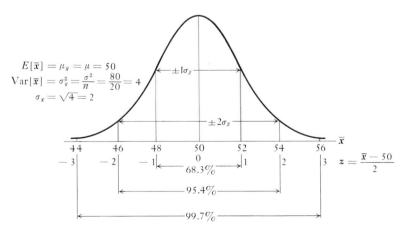

**Fig. 6.2.** Sampling distribution of $\bar{x}$ for sample of $n = 20$ taken from a population with distribution $N(50, 80)$.

distribution of $\bar{x}$ for all possible samples of size 20 taken from a population with the distribution, $N(50, 80)$.

### The Standardized Form of the Random Variable $\bar{x}$

In Chapter 5 we saw that it is usually easier to work with the standard normal form of a variable rather than leave it in its original units. The same type of transformation that was made on the random variable $x$ at that point can now be made on the random variable $\bar{x}$. Recall that, in Chapter 3, the variable $x$ was transformed to its standard normal form by subtracting the mean from each value, and then dividing by the standard deviation. The resulting variable, $z = (x - \mu)/\sigma$ was shown to have a mean of zero and a variance of one. Although now we are interested in transforming the variable $\bar{x}$ instead of $x$ and the standard deviation of this variable is $\sigma/\sqrt{n}$ instead of $\sigma$, the transformation is accomplished in exactly the same fashion. One must just take care to always subtract the mean and divide by the standard deviation corresponding to the variable being standardized. The mean and variance of the resulting variable will always be 0 and 1, respectively. Hence, the random variable

$$z = \frac{\bar{x} - \mu_{\bar{x}}}{\sigma_{\bar{x}}} = \frac{\bar{x} - \mu}{\sigma/\sqrt{n}} \tag{6.7}$$

has a mean of zero and a variance of one. Since we have said that the distribution of the random variable $\bar{x}$ is normal, it follows that the random variable $z$ must also be normally distributed. Thus:

> *When sampling from a normal parent population, the distribution of $z = (\bar{x} - \mu)/(\sigma/\sqrt{n})$ will be normal with mean 0 and variance 1. That is: $z = (\bar{x} - \mu)/(\sigma/\sqrt{n})$ is $N(0, 1)$.*

The standardized normal form of the variable $\bar{x}$ is shown on the $z$-scale in Fig. 6.2.

The limitations of the preceding discussion should be apparent, for although the normal distribution approximates the probability distribution of many real-world problems, one cannot *always* assume that the parent population is normal. What, for example, will be the shape of the distribution of $\bar{x}$'s when sampling from a highly skewed distribution? We consider this situation in the next section.

### 6.6 SAMPLING DISTRIBUTION OF $\bar{x}$, POPULATION DISTRIBUTION UNKNOWN, $\sigma$ KNOWN

When sampling is not from a normal parent population, the size of the sample plays a critical role. When $n$ is small, the shape of the distribution will depend mostly on the shape of the parent population. As $n$ gets large, however, one of the most important theorems in statistical inference says that the shape of the sampling distribution will become more and more like a *normal distribution, no matter what the shape of the parent population.* This theorem, called the central-limit theorem, is stated in formal terms below:

---

**The Central Limit Theorem**
*The distribution of the means of random samples taken from a population having mean μ and finite variance $\sigma^2$ approaches the normal distribution with mean μ and variance $\sigma^2/n$ as n goes to infinity.*

---

We will not prove this theorem, but merely show, in Fig. 6.3, graphical evidence of its validity. The first row of diagrams in Fig. 6.3 shows four different parent populations. The next three rows show the sampling distribution of $\bar{x}$ for all possible repeated samples of size $n = 2$, $n = 5$, and $n = 30$, respectively, drawn from the population shown in the first row. Note in the first column that when the parent population is normal, all the sampling distributions are also normal. This agrees with what we said in Section 6.5.

The second column of values in Fig. 6.3 represents what is called a uniform (or rectangular) distribution. We see here that the sampling distribution of $\bar{x}$ is already symmetrical when $n = 2$, and it is quite normal in appearance when $n = 5$. Moving to the third column of figures, the distribution is now a bimodal distribution with discrete values of $x$ (the central limit theorem applies whether $x$ is discrete or continuous). Again, by $n = 2$ the distribution is symmetrical, and by $n = 5$ its shape is quite bell-shaped. The final parent population is a highly skewed pdf called the exponential distribution. Here we see that for $n = 2$ and $n = 5$ the distribution is still fairly skewed, although becoming more symmetrical as $n$ increases. When $n = 30$, however, even such a skewed parent population results in a symmetrical, bell-shaped distribution, which, by the central-limit theorem, we know to be approximately normally distributed.

In general, just how large $n$ needs to be for the sampling distribution of $\bar{x}$ to be a good approximation to the normal depends, as we saw in Fig. 6.3, on the shape of the parent population. Usually the approximation will be quite good if $n \geq 30$, although the third row of Fig. 6.3 demonstrates that satisfactory results are often obtained when $n$ is much smaller.

### Example of Use of Central-limit Theorem

Now that the sampling distribution of $\bar{x}$ has been specified, at least for large samples where $\sigma$ is known, we can demonstrate the use of the central-limit theorem. Consider the case of a midwestern telephone company which recently asked for a rate increase for all residential telephones, including a 25-percent increase in

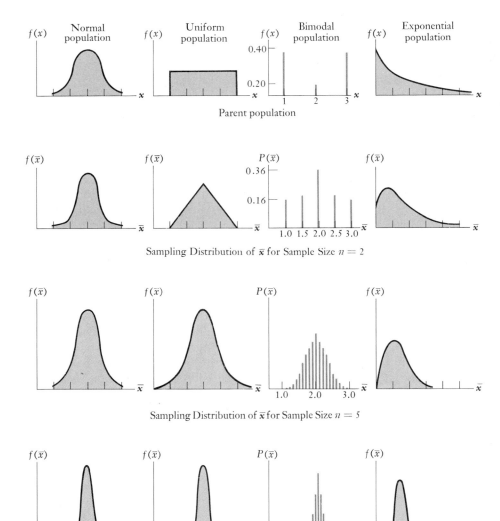

**Fig. 6.3.** Sampling distribution of $\bar{x}$ for various population distributions when $n$ = 2, 5, and 30.

student phones. To oppose this increase, a group of students decided to investigate the typical phone costs incurred in their town. The information provided by the telephone company was that, for the city as a whole, the average monthly bill was $15.30, with a standard deviation of $4.10. The students, however were curious as to whether or not dormitory phones incur the same type of costs, since no information was available on this matter. The students decided to take a random sample of $n = 36$, in an attempt to gain further information.

Now, if the dormitory phone bills come from the same population reported by the telephone company, then we know that the mean of all possible sample means will be $E[\bar{x}] = \mu_{\bar{x}} = \$15.30$, and the standard error of the mean will be $\sigma_{\bar{x}} = \sigma/\sqrt{n} = \$4.10/\sqrt{36} = 0.683$. Furthermore, since the sample size is fairly large ($n = 36$), the central-limit theorem tells us that the sampling distribution of $\bar{x}$ will be normal [i.e., $N(\$15.30, 0.683^2)$]. Use of this distribution, shown in Fig. 6.4, allows us to answer a number of different probability questions in the telephone example.

*Question 1.* Suppose that in the random sample of 36 dormitory residents (with phones), we find the average phone bill is $14.00. A typical question we might ask at this point is: What is the probability that a random sample of $n = 36$ will result in an average bill of $14.00, or less, when $\mu_x = \$15.30$ and $\sigma_{\bar{x}} = 0.683$? Making use of the standard normal transformation, we know that:

$$P(\bar{x} \leqslant \$14.00) = P\left(\frac{\bar{x} - \mu_{\bar{x}}}{\sigma_{\bar{x}}} \leqslant \frac{14.00 - 15.30}{0.683}\right) = P(z \leqslant -1.91).$$

By the central-limit theorem, $z$ is approximately normally distributed; hence (from Table III),

$$P(z \leqslant -1.91) = F(-1.91) = 1.0 - F(1.91)$$
$$= 1.0 - 0.9719 = 0.0281.$$

The probability that a sample of 36 gives an average no bigger than $14.00 is thus 0.0281. The reader should sketch this area on Fig. 6.4.

On the basis of the probability value calculated above, does it appear as if dormitory phone bills are part of the same population as residential phone bills

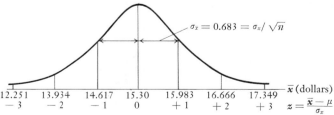

$\sigma_{\bar{x}} = 0.683 = \sigma_x/\sqrt{n}$

| 12.251 | 13.934 | 14.617 | 15.30 | 15.983 | 16.666 | 17.349 | $\bar{x}$ (dollars) |
| $-3$ | $-2$ | $-1$ | $0$ | $+1$ | $+2$ | $+3$ | $z = \frac{\bar{x} - \mu}{\sigma_x}$ |

**Figure 6.4**

in general? We leave such questions concerned with drawing conclusions from sample data to the following chapters, but the reader should keep them in mind to retain a feeling of where we are headed.

*Question 2.* Rather than finding the probability of a specific value of $\bar{x}$, one often wants to determine values $a$ and $b$, such that the probability is 0.99 (or some other probability) that the sample mean will fall between these values. Recall that we did a similar calculation in our discussion of the normal distribution in Chapter 5. For the present case we know that when $n = 36$, $\bar{x}$ will be approximately normally distributed, with mean $\mu_{\bar{x}} = 15.30$ and $\sigma_{\bar{x}} = \$4.10/\sqrt{36} = 0.683$.

As before, we want the smallest interval including 99 percent, which means we need to exclude half of the remaining probability (or $\frac{1}{2}(0.01) = 0.005$) in each tail of the distribution of $\bar{x}$. To do this, let's first see what values of the standardized normal distribution exclude 0.005 in each tail. From Table III, $F(z) = 0.995$ if $z = 2.576$; by symmetry, $F(-2.576) = 1.0 - F(2.576) = 0.005$. Thus,

$$P(-2.576 \leqslant z \leqslant 2.576) = 0.99.$$

Finally, we now have to transform the interval $P(-2.576 \leqslant z \leqslant 2.576) = 0.99$ into the original units (dollars) by finding values $a$ and $b$ to satisfy $P(a \leqslant \bar{x} \leqslant b) = 0.99$. By standardizing the value of $a, \bar{x}$, and $b$, we get:

$$P(a \leqslant \bar{x} \leqslant b) = P\left(\frac{a - 15.30}{0.683} \leqslant z \leqslant \frac{b - 15.30}{0.683}\right).$$

Thus, $(a - 15.30)/0.683 = -2.576$ and $(b - 15.30)/0.683 = 2.576$. Solving these two equations yields $a = \$13.54$ and $b = \$17.06$; this means that the appropriate interval is $P(\$13.54 \leqslant \bar{x} \leqslant \$17.06)$. A diagram of the values in this example are shown in Fig. 6.5.

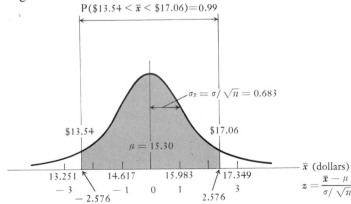

**Fig. 6.5.** Standardized normal form of the sampling distribution of $\bar{x}$, sample mean of monthly phone charges, $\mu_x = \$15.30$, $\sigma_x = \$4.10$, $n = 36$.

## 6.7 FINITE POPULATION CORRECTION FACTOR

The formulas developed for $\sigma_{\bar{x}}$ thus far have assumed that sampling occurred either from an infinite population, or from a finite population with replacement. If samples are drawn without replacement from a finite population of size $N$, then these formulas are not correct, although they may provide a good approximation to the correct values if the population is "large" relative to the sample size. *When the population is finite*, the standard error of the mean is no longer equal to $\sigma/\sqrt{n}$ but is smaller than this amount. That $\sigma_{\bar{x}}$ will be smaller than $\sigma/\sqrt{n}$ should be evident from the logical fact that with a finite population $\sigma_{\bar{x}}$ must approach zero as $n$ approaches $N$. But mathematically $\sigma/\sqrt{n}$ approaches zero as $n$ goes to infinity; hence $\sigma_{\bar{x}}$ will approach zero faster than $\sigma/\sqrt{n}$, unless $N = \infty$.

If the sample size is small relative to the finite population size, say five percent or less, then formulas based on the fact that $\sigma_{\bar{x}} = \sigma/\sqrt{n}$ will be approximately correct. *When the sample size is greater than five percent of the population size, then a correction factor is included in these formulas to adjust $\sigma/\sqrt{n}$ so that it equals $\sigma_{\bar{x}}$.* This factor, called the *finite population multiplier*, is:

$$\text{Finite population multiplier:} \quad \sqrt{(N - n)/(N - 1)}. \tag{6.8}$$

Note that the correction factor is always less than one, except when $n = 1$. To correct for a finite population size, $\sigma/\sqrt{n}$ must be multiplied by this factor as follows:

$$\text{Variance of } \bar{x}: \qquad \sigma_{\bar{x}}^2 = \frac{1}{n}\sigma^2\left(\frac{N - n}{N - 1}\right);$$

$$\text{Standard error of the mean:} \qquad \sigma_{\bar{x}} = \frac{\sigma}{\sqrt{n}}\sqrt{\frac{N - n}{N - 1}}. \tag{6.9}$$

We can illustrate the finite population correction factor by assuming in our previous example that the number of dormitory residents with private phones is 300. Since the sampling of 36 residents was without replacement, and a sample of size $n = 36$ equals 12 percent of this population ($36/300 = 0.12$), the finite population correction factor should be used.

We correct $\sigma_{\bar{x}}$ as follows:

$$\sigma_{\bar{x}} = \frac{\sigma}{\sqrt{n}}\sqrt{\frac{N - n}{N - 1}} = \frac{4.10}{\sqrt{36}}\sqrt{\frac{300 - 36}{300 - 1}}$$

$$= 0.683\sqrt{\frac{264}{299}} = 0.642.$$

The corrected standard error is smaller than the previous value of 0.683. This means that the 99 percent interval we calculated using the latter figure is too big. The reader should verify that when $N = 300$, the appropriate values are

$$P(13.65 \leqslant \bar{x} \leqslant 16.95) = 0.99.$$

## 6.8 SAMPLING DISTRIBUTION OF $\bar{x}$, NORMAL POPULATION, $\sigma$ UNKNOWN

In our discussions thus far about the sampling distribution of $\bar{x}$ we have assumed that the population variance, $\sigma^2$, is known. In most practical problems, however, the value of $\sigma^2$ is not known, but rather is a population parameter about which one wishes to obtain information. As you might suspect, the sample statistic $s^2$ is used as an estimate of the population variance, $\sigma^2$. This is reasonable since we can show that the expected value of $s^2$ is indeed $\sigma^2$.* Thus, on the average over many sample calculations of $s^2$, the idea of substituting $s^2$ for $\sigma^2$ would be perfect. For a single sample, of course, $s^2$ may not equal $\sigma^2$; hence we will have to take this fact into account in substituting $s$ for $\sigma$ in the formula $z = (\bar{x} - \mu_{\bar{x}})/(\sigma/\sqrt{n})$. That is, we are interested in the sampling distribution of the statistic $(\bar{x} - \mu_{\bar{x}})/(s/\sqrt{n})$ as contrasted to the distribution of $(\bar{x} - \mu_{\bar{x}})/(\sigma/\sqrt{n})$ [which in this section is $N(0, 1)$ for all sample sizes, since we are assuming that the parent population is normal].

Intuitively, you can probably guess at what happens when $s$ is substituted for $\sigma$. The mean of the distribution of $(\bar{x} - \mu_{\bar{x}})/(s/\sqrt{n})$ should still be zero, since the

---

* To prove that $E[s^2] = \sigma^2$, we use the rules of expectation in Section 3.3.

$$E[s^2] = E\left[\frac{1}{n-1}\sum(x_i - \bar{x})^2\right]$$

$$= \frac{1}{n-1}E\left[\sum\{(x_i - \mu) - (\bar{x} - \mu)\}^2\right] \qquad \begin{array}{l}\text{By Rule 3 and because}\\ (x - \bar{x}) = (x - \mu) - (\bar{x} - \mu)\end{array}$$

$$= \frac{1}{n-1}\left\{\sum E[(x_i - \mu)^2] - 2E[n(\bar{x} - \mu)(\bar{x} - \mu)]\right. \qquad \left.\begin{array}{l}\text{By expansion of square}\\ \text{term, by Rule 5, and}\\ \Sigma(x_i - \mu)(\bar{x} - \mu) =\\ n(\bar{x} - \mu)(\bar{x} - \mu).\end{array}\right.$$
$$\left. + \sum E[(\bar{x} - \mu)^2]\right\}$$

$$= \frac{1}{n-1}\left(\sum \sigma^2 - (2n\sigma^2/n) + \sum \sigma^2/n\right) \qquad \begin{array}{l}\text{Since } \sigma^2 = E[(x - \mu)]^2 \text{ and}\\ \sigma^2/n = E[(\bar{x} - \mu)^2]\end{array}$$

$$= \frac{1}{n-1}(n\sigma^2 - 2\sigma^2 + n\sigma^2/n) \qquad \text{Since } \sum_{i=1}^{n}(\text{constant}) = n(\text{constant})$$

$$= \frac{1}{n-1}\sigma^2(n-1) \qquad\qquad\qquad\qquad \text{Collecting terms,}$$

$$E[s^2] = \frac{n-1}{n-1}\sigma^2 = \sigma^2.$$

numerator has not been affected by the substitution. In terms of the variance, we should expect the variance of $(\bar{x} - \mu_x)/(s/\sqrt{n})$ to be larger than the variance of $(\bar{x} - \mu_x)/(\sigma/\sqrt{n})$ since one more element of uncertainty (the sample variance) has been added to the ratio. Finally, we would expect the distribution of $(\bar{x} - \mu_x)/(s/\sqrt{n})$ to be symmetrical, since there is no reason to believe that substituting $s$ for $\sigma$ will make this distribution skewed either positively or negatively.

Several additional aspects of the distribution of $(\bar{x} - \mu_x)/(s/\sqrt{n})$ are worth noting. First, it should be apparent that the variability of this distribution depends on the size of $n$, for the sample size affects the reliability with which $s$ estimates $\sigma$. When $n$ is large, $s$ will be a good approximation to $\sigma$; but when $n$ is small, $s$ may not be very close to $\sigma$. Hence, the distribution of $(\bar{x} - \mu_x)/(s/\sqrt{n})$ is a *family* of distributions whose variability depends on $n$.

It is clear from the discussion above that the distribution of $(\bar{x} - \mu_x)/(s/\sqrt{n})$ is not normal, but is more spread out than the normal. The distribution of this statistic is called the "*t*-distribution," and its random variable is denoted as $t = (\bar{x} - \mu_x)/(s/\sqrt{n})$. The variable $t$ is a continuous random variable. One of the first researchers to work on determining the exact distribution of this random variable was W. S. Gossett, an Irish statistician. However, the brewery for which Gossett worked did not allow its employees to publish their research; hence, Gosset wrote under the penname "Student." In honor of Gosset's research, published in 1908, the *t*-distribution is often referred to as the "Student's *t*-distribution."

**Student's *t*-distribution**

Since the density function for the *t*-distribution is fairly complex and not of primary importance at this point, we will not present it, but merely begin by describing the characteristics of this distribution.* As we indicated previously, the *t*-distribution depends on the size of the sample. It is customary to describe the characteristics of the *t*-distribution in terms of the sample size minus one, or $(n - 1)$, as this quantity has special significance. The value of $(n - 1)$ is called the number of *degrees of freedom* (abbreviated d.f.), and represents a measure of the number of observations in the sample that can be used to estimate the standard deviation of the parent population. For example, when $n = 1$, there is no way to estimate the population standard deviation; hence there are *no* degrees of freedom $(n - 1 = 0)$. There is one degree of freedom in a sample of $n = 2$, since one observation is now "free" to vary away from the other, and the amount it varies determines our estimate of the population standard deviation. Each additional observation adds one more degree of freedom, so that, in a sample of size $n$, there

---

* Mathematically, the random variable $t$ is defined as a standardized normal variable $z$ divided by the square root of an independently distributed chi-square variable which has been divided by its degrees of freedom; that is, $t = z/\sqrt{\chi^2/\nu}$.

are $(n-1)$ observations "free" to vary, and hence $(n-1)$ degrees of freedom. The Greek letter $v$ (nu) is often used to denote degrees of freedom, where $v = n - 1$.

A $t$-distribution is completely described by its one parameter, $v =$ degrees of freedom. As we said above, the mean of the $t$-distribution is zero, $E[t] = 0$. The variance of the $t$-distribution, when $v \geq 3$, is $V[t] = v/(v-2)$.

Note that this implies that $V[t] \geq 1.0$ for all sample sizes, in contrast to $V[z]$ which is 1.0 no matter what the sample size. For example, when $v = 3$ the variance of the $t$-distribution is $3/(3-2) = 3.0$. This distribution and the standardized normal are contrasted in Fig. 6.6.

For small sample sizes, the $t$-distribution is seen to be considerably more spread out than the normal. When $v$ is as large as about 30, then $V[t] = 30/(30-2) = 1.07$, which is not much different from $V[z] = 1.0$. In the limit as $n \to \infty$, the $t$- and $z$-distributions are identical. Tables of $t$-values are usually only completely enumerated for $v < 30$, because for larger samples the normal gives a very good approximation, and is easier to use. For this reason it is customary to speak of the $t$-distribution as applying to "small sample sizes," *even though this distribution holds for any size n.*

Probability questions involving a $t$-distributed random variable can be answered by using the cumulative distribution function, $F(t)$ in Table VII. This table gives the values of $t$ for selected values of the cumulative probability ($F(t) = P(t < t)$) given across the top of the table and for degrees of freedom ($v$) given at the left margin. The use of Table VII can be illustrated by several examples.

First, let's determine a probability comparable to one posed earlier for the normal distribution, namely to find an interval $(a, b)$ such that $P(a \leq t \leq b) = 0.95$. Assume $n - 1 = v = 8$ degrees of freedom. As before, the smallest interval is found by putting half the excluded area, $\frac{1}{2}(0.05) = 0.025$, in each tail of the distribution. For example, we want $P(t > b) = 0.025$, which means that, in terms of the cumulative function, we want to find a value $t$ such that $F(t) = 0.975$. From

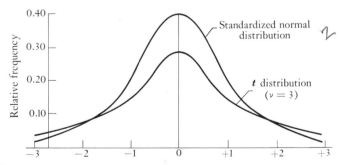

**Fig. 6.6.** The standardized normal and the $t$-distributions compared.

Table VII for $v = 8$, we see that $F(2.306) = 0.975$; hence, $b = 2.306$. Now since the $t$-distribution is symmetrical, the appropriate value for $a$ is merely the negative of the value of $b$, or $a = -2.306$. Thus,

$$P(-2.306 < t < 2.306) = 0.95.$$

Recall, from our previous examples using the standardized normal distribution, that $P(-1.96 < z < +1.96) = 0.95$. The critical values for $z$ which exclude 0.025 probability in the upper and lower tails are $\pm 1.96$, as opposed to $\pm 2.306$ for the $t$-distribution. The difference reflects the fact that the $t$-distribution is more spread out than the $z$-distribution. Note in Table VII that by *increasing* the value of $v$ from 8 to 10, then 20, then 60, then 120, and moving down the column for $F(t) = 0.975$, the critical values for $t$ *decrease* from 2.306 to 2.228, then 2.086, then 2.000, then 1.98 respectively. For larger values of $v$, the spread of the $t$-distribution closes in to match the spread of the $z$-distribution. Indeed, when $n = \infty$, the critical value for $t$ exactly equals the value for $z$ (1.96), as shown by the bottom row in Table VII.

For another example of the $t$-distribution, consider the widely publicized claims of a certain "well-known eastern university" that its students have I.Q.'s which are normally distributed with a mean $\mu = 130$. Skeptical students from a rival university have secured a random sample of the I.Q.'s of 25 students. This random sample yields a mean of $\bar{x} = 126.8$ and a standard deviation of $s = 6$. The question of interest is the probability of receiving a sample mean of 126.8, or lower, if $\mu_x = 130$ (that is, $P(\bar{x} \leqslant 126.8)$). Since the parent population is normal and $\sigma$ unknown, the $t$-distribution applies. First, we translate $P(\bar{x} \leqslant 126.8)$ into standardized form using the relationship $t = (\bar{x} - \mu_x)/(s/\sqrt{n})$.

$$P(\bar{x} \leqslant 126.8) = P\left(\frac{\bar{x} - \mu_x}{s/\sqrt{n}} \leqslant \frac{126.8 - 130}{6/\sqrt{25}}\right) = P\left(t < \frac{-3.2}{1.2}\right) = P(t \leqslant -2.667).$$

Since the $t$-distribution is symmetrical, the probability we want, $P(t < -2.667) = F(-2.667)$ is equivalent to $1 - F(2.667)$; that is, the area under the density function $f(t)$ for $t < -2.667$ equals the area in the upper tail for $t > 2.667$. From Table VII, using the row corresponding to $v = 24$ (since $n - 1 = v = 24$), we see that $F(2.492) = 0.99$, and $F(2.797) = 0.995$, but there is no $F(2.667)$. In this situation we could go to a more extensive table of $t$-values, hopefully to find $F(2.667)$. Rather than doing this, it is common practice, in using the $t$-table, to merely report the two values which we know $F(-2.667)$ lies between. From Table VII we know that $F(2.667)$ lies between 0.99 and 0.995; hence, $1 - F(2.667) = F(-2.667) = P(t < -2.667)$ must be between 0.01 and 0.005. Thus, we can state that $0.01 > P(t < -2.667) > 0.005$. These points on the $t$-distribution for $v = 24$ are shown in Fig. 6.7.

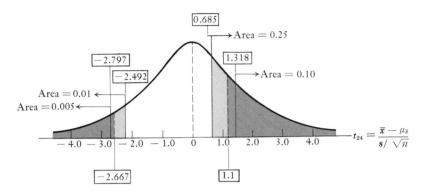

**Fig. 6.7.** The $t$-distribution for $v = 24$.

As a final illustration of the use of the $t$-distribution, suppose the sample by the students in our example ($n = 25$) resulted in $\bar{x} = 131.1$ and $s = 6$. What is the probability of receiving a sample mean this high, or higher, if $\mu = 130$?

$$P(\bar{x} \geqslant 131.1) = P\left(\frac{\bar{x} - \mu_{\bar{x}}}{s/\sqrt{n}} \geqslant \frac{131.1 - 130.0}{6/\sqrt{25}}\right)$$

$$= P\left(t > \frac{1.1}{1.2}\right) = P(t > 0.917).$$

From Table VII, $F(0.685) = 0.75$ and $F(1.318) = 0.90$ when $v = 24$. Hence, $P(t > 0.917) = 1 - F(0.917)$ must lie between $1 - 0.75 = 0.25$ and $1 - 0.90 = 0.10$. That is,

$$0.25 > P(\bar{x} > 131.1) = P(t > 0.917) > 0.10.$$

We see from this result that a sample mean of 131.1, or higher, is not too unlikely when $\mu = 130$ and $n = 25$. These values are also shown in Fig. 6.7.

### Use of the $t$-distribution when the Population is not Normal

It must be emphasized at this point that the $t$-distribution, as well as the chi-square distribution (discussed in the following section), assumes that samples are drawn from a parent population which is normally distributed. In practical problems involving these distributions, the question therefore arises as to just how critical the assumption is that the parent population be exactly normally distributed. Often there is no way to determine the exact distribution of the parent population. Fortunately, the assumption of normality can be relaxed without significantly changing the sampling distribution of the $t$- or the chi-square distributions. Because of this fact these distributions are said to be quite "robust," implying that their usefulness holds up under conditions which do not exactly conform to the original assumptions. The $t$-distribution is much more robust than the chi-square.

At this point perhaps we should emphasize several important aspects of the distribution of $\bar{x}$ when $n$ is "large" (usually $n > 30$). First, we know that $\bar{x}$ will be approximately normally distributed because of the central-limit theorem. And if $\bar{x}$ is normal, this means that the distribution of $(\bar{x} - \mu)/(s/\sqrt{n})$ follows the $t$-distribution.

But we also know when $n$ is large that $s$ will usually be a good approximation to $\sigma$, so that the distribution of $t = (\bar{x} - \mu)/(s/\sqrt{n})$ and that of $z = (\bar{x} - \mu)/(\sigma/\sqrt{n})$ will be approximately the same. In other words, *when n is large*, we can say that $(\bar{x} - \mu)/(s/\sqrt{n})$ is *approximately* $N(0, 1)$. Thus, if a sample of size 100 in our I.Q. example had resulted in $\bar{x} = 129.4$ and $s^2 = 5.2$, the probability $P(\bar{x} \leqslant 129.4)$ can be approximated by using the standardized normal:

$$P(\bar{x} \leqslant 129.4) = P\left(\frac{\bar{x} - \mu_x}{s/\sqrt{n}} \leqslant \frac{129.4 - 130}{5.2/\sqrt{100}}\right)$$

$$\doteq P\left(z \leqslant \frac{-0.60}{0.52}\right) = P(z \leqslant -1.15);$$

$$P(z \leqslant -1.15) = 1 - F(1.15) = 0.1251 \qquad \text{(from Table III)}.$$

Determining the *exact* answer to this problem would involve first finding a $t$-table having values for $v = 100 - 1 = 99$ degrees of freedom, and then determining $P(t \leqslant -1.15)$. Using the normal distribution to approximate the answer is much easier.

## *6.9 THE SAMPLING DISTRIBUTION OF $s^2$, NORMAL POPULATION

The only sampling distribution considered thus far has been for $\bar{x}$, the sample mean. But in many practical problems we need to know about the distribution of the sample variance, $s^2$. That is, we want to investigate the distribution which consists of all possible values of $s^2$ calculated from samples of size $n$. The sampling distribution of $s^2$ is particularly important in problems where one is concerned about the variability in a random sample. For example, the telephone company might be just as interested in the variance in length of calls in a random sample as they are in the mean length. Or a manufacturer of steel beams may want to learn just as much about the variability in tensile strength of his product as he does about its mean strength.

The same statistician who first worked with the $t$-distribution, W. S. Gosset, was also one of the first to describe the sampling distribution of $s^2$. First, note that because $s^2$ must always be positive, the distribution of $s^2$ cannot be a normal

---

* In this section, which can be omitted without loss in continuity, we assume the reader has read Section 5.7, covering the chi-square distribution.

distribution. Rather, the distribution of $s^2$ is a unimodal distribution which is skewed to the right (since $s^2$ can take on any value up to infinity), comparable to a chi-square distribution. In fact, *when the parent population is normal*, with variance $\sigma^2$, the random variables for these two distributions can be shown to be related as follows:

$$s^2 = \frac{\chi^2 \sigma^2}{v} = \frac{\chi^2 \sigma^2}{n-1}. \tag{6.10}$$

In other words, we have to multiply a $\chi^2$ variable by the constant $(\sigma^2/(n-1))$ to get the $s^2$ variable. And since we already know something about the characteristics of the $\chi^2$ distribution from Chapter 5, we can use this information to determine comparable characteristics about the distribution of $s^2$. Recall that $E[\chi^2] = v = n - 1$. Hence,

$$E[s^2] = E\left[\frac{\chi^2 \sigma^2}{v}\right]$$

$$= \frac{\sigma^2}{v} E[\chi^2] \qquad \text{Since } \frac{\sigma^2}{v} \text{ is a constant}$$

$$= \frac{\sigma^2}{v}(v) \qquad \text{Since } E[\chi^2] = v$$

$$= \sigma^2.$$

This agrees with the result we proved in a footnote at the beginning of Section 6.8.

The variance of $s^2$ can be derived in a similar manner, remembering that $V[\chi^2] = 2v$:

$$V[s^2] = V\left[\chi^2 \frac{\sigma^2}{v}\right] = \frac{(\sigma^2)^2}{v^2} V[\chi^2] \qquad \text{By Rule 4}$$

$$= \frac{\sigma^4}{v^2}(2v) \qquad \text{Since } V[\chi^2] = 2v$$

$$= 2\sigma^4/v.$$

We can illustrate the pdf of $s^2$ by assuming that repeated samples of size $n = 21$ are drawn from a population in which $\sigma^2 = 200$. We know for the distribution of $s^2$ that

$$E[s^2] = \sigma^2 = 200,$$

and

$$V[s^2] = \frac{2\sigma^4}{v} = \frac{2(200)^2}{20} = 4000.$$

This distribution is pictured in Fig. 6.8.

For most problems involving the distribution of $s^2$, it is more convenient to

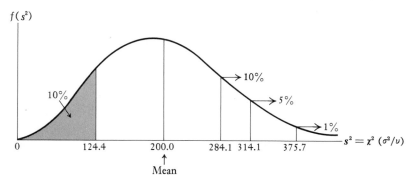

**Fig. 6.8.** The sampling distribution of $s^2$ for $n = 21$, $\sigma^2 = 200$.

work with Formula (6.10) if we solve this formula for the random variable $\chi^2$:

$$\chi^2 = \frac{vs^2}{\sigma^2} = \frac{(n-1)s^2}{\sigma^2}. \tag{6.11}$$

In words, this formula says the following:

*If $s^2$ is the variance of random samples of size n taken from a normal population having a variance of $\sigma^2$, then the variable $(n-1)s^2/\sigma^2$ has the same distribution as a $\chi^2$-variable with $(n-1)$ degrees of freedom.*

Although Gosset was unable to prove Formula (6.11) mathematically, he did demonstrate this relationship in his empirical work. Gosset took the heights of 3000 criminals, calculated the value of $\sigma^2$ for these heights, and then grouped these heights into 750 random samples of 4. For each of these 750 samples Gosset, in effect, calculated a value of $s^2$, multiplied $s^2$ by $(n-1) = 3$, and then divided this number by $\sigma^2$, with the results as plotted in the histogram shown in Fig. 6.9. Note that Gosset's histogram and the chi-square distribution (for $v = n - 1 = 3$) superimposed on it are not in perfect agreement, a fact which Gosset attributed to the particular grouping of heights which he used.

In problems involving variances we are often interested in determining the probability that a normal parent population with variance $\sigma^2$ will yield a sample variance greater than or equal to $s^2$, when the sample size is $n$. Probability problems of this nature are usually determined by taking the *ratio* of the sample value $s^2$ to the population value $\sigma^2$, or $s^2/\sigma^2$. For example, suppose we want to determine the probability that $s^2 \geqslant 400$ when $n = 4$ and the population is normally distributed

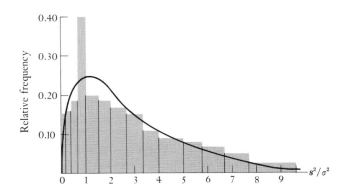

**Fig. 6.9.** Chi-square approximation to Gosset's data on the heights of criminals.

with $\sigma^2 = 100$. Since $s^2/\sigma^2 = 400/100 = 4$, this probability is equivalent to the probability that a sample variance will be at least four times larger than the population variance, or $P(s^2/\sigma^2 \geqslant 4)$.

This problem can't be solved directly because we don't know how $s^2/\sigma^2$ is distributed. But suppose we rewrite our probability problem as $P(vs^2/\sigma^2 \geqslant 3 \times 4)$. This probability is equivalent to $P(s^2/\sigma^2 \geqslant 4)$ since we assumed the sample size was $n = 4$ (which means $v = n - 1 = 3$), and all we have done is multiply the left side of $s^2/\sigma^2$ by $v$ and the righthand side by 3. The advantage of doing this is that we know the distribution of $vs^2/\sigma^2$ follows a $\chi^2$-distribution with $v$ degrees of freedom. We can thus calculate the answer to our original problem $P(s^2/\sigma^2 \geqslant 4)$ as follows:

$$P(s^2/\sigma^2 \geqslant 4) = P(vs^2/\sigma^2 \geqslant 3 \times 4) = P(\chi^2 \geqslant 12).$$

From Table IV for $v = 3$,

$$P(\chi^2 \geqslant 11.34) = 0.01 \quad \text{and} \quad P(\chi^2 \geqslant 12.84) = 0.005.$$

Thus, we can write

$$0.01 > P(s^2/\sigma^2 \geqslant 4) = P(\chi^2 \geqslant 12) > 0.005.$$

In other words, for $v = 3$, a sample variance at least four times as large as the population variance will occur between 0.5 and 1.0 percent of the time. If we had a more detailed table of $\chi^2$ values we could be more precise about the probability $P(s^2/\sigma^2 \geqslant 4)$.

## 6.10  SUMMARY TABLE

Table 6.5 provides a summary of the sampling distributions discussed in this chapter.

**Table 6.5** Summary of sampling distributions

| Random variable | Situation | Reference section | Resulting distribution for problem-solving | Mean | Variance |
|---|---|---|---|---|---|
| (1) $\bar{x}$ | Population normal, $\sigma$ known, sample size $n$ | 6.5 | $z = \dfrac{\bar{x} - \mu}{\sigma/\sqrt{n}}$ | 0 | 1 |
| (2) $\bar{x}$ | Population normal, $\sigma$ unknown, sample size $n$* | 6.8 | $t = \dfrac{\bar{x} - \mu_{\bar{x}}}{s/\sqrt{n}}$ | 0 | $v/(v - 2)$ (where $v = n - 1$) |
| (3) $\bar{x}$ | Population unknown, $\sigma$ known, $n > 30$ | 6.6 | $z = \dfrac{\bar{x} - \mu}{\sigma/\sqrt{n}}$ | 0 | 1 |
| (4) $s^2$ | Population normal, sample size $n$ | 6.9 | $\chi^2 = \dfrac{(n - 1)s^2}{\sigma^2}$ | $\sigma^2$ | $\dfrac{2\sigma^4}{v}$ (where $v = n - 1$) |

* If $n > 30$, $\bar{x}$ will be approximately normally distributed and $s$ should be close to $\sigma$; hence $(\bar{x} - \mu)/(s/\sqrt{n})$ is approximately $N(0, 1)$.

## REVIEW PROBLEMS

1. Define or describe briefly the following terms: sample design, sampling errors, non-sampling errors, probabilistic sampling, nonprobabilistic sampling.

2. Under what conditions is nonprobabilistic sampling more appropriate than probabilistic sampling? Give several examples.

3. Distinguish between:
   a) systematic sampling and simple random sampling,
   b) stratified and cluster sampling,
   c) single-stage sampling and multiple-stage sampling,
   d) judgment, quota, and convenience sampling.

4. In designing a sample survey, what factors are most important in establishing the strata in stratified sampling? The clusters in cluster sampling? How will the cost of sampling affect these decisions?

5. A company packages sunflower seeds. Obviously not all the seeds in a given lot will germinate and produce satisfactory flowers. However, the company does not want to package and present for consumer purchase an excessive number of bad seeds. In the long run, about $\frac{3}{4}$ of the sunflower seeds produce while $\frac{1}{4}$ do not. Before packaging a new lot, the company would like to be sure that at least $\frac{3}{4}$ of the new seeds will germinate.

One clever(?) student says they could test a random sample of only 4 seeds, and if 3 of them grew, they could assume that $\frac{3}{4}$ of all the seeds were good.

a) Do you think the direct relationship between the probability from the sample and the likelihood of good seeds in the entire lot is perfect, as the student suggests? Explain briefly.

b) Suppose exactly $\frac{3}{4}$ of all the seeds were good; what is the probability that exactly 3 out of 4 in a random sample would be good? (Although sampling is without replacement, assume that the number of seeds is very large so that the probability of selecting a good seed, $\frac{3}{4}$, remains the same.)

6. Given the following sample of seven values of total yardage gained by winning teams in college football games: 412, 207, 314, 224, 158, 286, and 317. Find the sample mean and standard deviation.

7. Given the following sample of five values of starting monthly salary for business majors who recently graduated: $675, 690, 725, 770, and 640. Find the sample mean and standard deviation.

8. On a certain Wednesday evening, a check was made of five different television rooms in campus residence units. The number of students watching television in each unit was 22, 16, 34, 27, and 21 respectively.

a) Find the average number of viewers per room.
b) Find the variance of this sample distribution.

9. Given the following sample distribution:

| $x$ | Frequency |
|---|---|
| 1 | 2 |
| 2 | 7 |
| 3 | 10 |
| 4 | 1 |

find $\bar{x}$ and $s_x$.

10. The distribution of the number of defects per square yard of a certain textile is given as follows:

| No. of defects $x$ | Frequency $f$ |
|---|---|
| 0 | 47 |
| 1 | 33 |
| 2 | 14 |
| 3 | 5 |
| 4 | 1 |
| 5 | 0 |
| Total | 100 |

Find the mean and the standard deviation of this distribution.

11. a) Given a normally distributed random variable with mean $\mu = 50$ and standard deviation $\sigma = 8$, find:

$$P(x \geqslant 40), \qquad P(x \leqslant 54), \qquad \text{and} \qquad P(44 \leqslant x \leqslant 56).$$

 b) If a random sample of size $n = 64$ is drawn from this population, find:

$$P(\bar{x} \leqslant 53), \qquad P(\bar{x} \geqslant 49), \qquad \text{and} \qquad P(48 \leqslant \bar{x} \leqslant 52).$$

 c) Sketch the distribution of $x$ and the distribution of $\bar{x}$.

12. What is a "sampling distribution"? Why is the knowledge of the sampling distribution of a statistic important to statistical inference?

13. State the central-limit theorem. Why do you think this theorem is so important to statistical inference?

14. A telephone company randomly selected 121 long distance calls and found that the average length of these calls was 5 minutes, with a standard deviation of 45 seconds.

 a) What is the probability of a sample mean as large as, or larger than, $\bar{x} = 5$ when the true population mean is $\mu = 4\frac{5}{6}$ minutes? What is the probability of a value as small as, or smaller than, $\bar{x} = 5$ when $\mu = 5\frac{1}{5}$?

 b) Do your answers to part (a) depend on any assumptions about the distribution of the parent population?

15. If the mean of all shoe sizes in the United States is 9, and the variance of these sizes is 1, what percent of the population wears a shoe of size 11 or larger? What is $P(\bar{x} > 11)$ if $n = 16$? Assume a normal parent population.

16. Suppose that a random sample is being drawn from a population of housewives known to have a mean age of 30, with a standard deviation of 3 years. The population is normally distributed.

 a) What is the probability that a randomly selected housewife will be over 35 years of age? What is the probability that she will be between 25 and 35?

 b) What is the probability, in a sample of 36 housewives, that the mean age will exceed 31? What is the probability that the mean age will be less than 30.5? What is $P(29 \leqslant \bar{x} \leqslant 31)$?

 c) Does your answer to part (a) of this question depend on the assumption that the parent population is normally distributed? What about your answer to part (b)? Explain.

17. a) What is the "finite population multiplier," and when is it necessary to apply this factor?

 b) If $\sigma = 50$, $n = 25$, and $N = 100$, what is the standard error of the mean?

18. a) Describe the difference between the standardized normal distribution and the $t$-distribution. Under what conditions can each be used?

 b) Is the probability $P(z \geqslant 2.0) = 0.0228$ greater than the probability $P(t \geqslant 2.0)$ for all sample sizes? How do you explain this fact?

19. Find the values of $t_\alpha$ such that $P(t > t_\alpha) = \alpha$ in each of the following:

 a) $\alpha = 0.05$, d.f. $= 10$

 b) $\alpha = 0.01$, d.f. $= 10$

c) $\alpha = 0.05$, d.f. $= 20$    d) $\alpha = 0.01$, d.f. $= 20$
e) $\alpha = 0.01$, d.f. $= 3$.

20. Explain why the standardized normal distribution does not provide an accurate description of the sampling distribution of $\bar{x}$ for small samples drawn from a normal population when $\sigma$ is unknown. What distribution is appropriate in this circumstance?

21. Given a normally distributed random variable with mean $\mu = 50$ and variance unknown:

   a) Suppose that a sample of nine observations yields a sample variance of $s^2 = 36$. What is the probability that $\bar{x}$ is larger than 54 if $\mu = 50$? What is the probability that $\bar{x}$ is less than 44 if $\mu = 50$? What is the probability that $\bar{x}$ lies between 45 and 55?
   b) How would your answers to the above problem change if $n = 36$?

22. What is the "sampling distribution of $s^2$"? How is this distribution related to the chi-square distribution?

23. Suppose that you are drawing samples of size $n = 3$ from a normal population with a variance of 8.25. What is the probability that $(n - 1)s^2/\sigma^2$ will exceed 5.99? What is the probability that $s^2$ will be more than three times as large as $\sigma^2$?

24. Suppose that a random variable is known to be chi-square-distributed with parameter $v = 24$.

   a) What is the mean and the variance of the chi-square distribution for this parameter?
   b) What is the probability that the value of $\chi^2$ will exceed 42.98? What is $P(\chi^2 \geqslant 33.20)$? What is $P(\chi^2 \geqslant 9.89)$?
   c) Use your answers to parts (a) and (b) to draw a rough sketch of the chi-square distribution for $v = 24$.
   d) Superimpose on your sketch for part (c) a graph of the normal distribution for $\mu = 24$ and $\sigma^2 = 48$. How closely do the two distributions agree?

## EXERCISES

25. Suppose that you have a population of 500 elements, numbered from 000 to 499. Use Table VI to determine a random sample of 15 elements.

26. Suppose that you wish to conduct a survey of student opinion in a university. Describe how you might go about this task if you decide to use:

   a) simple random sampling,
   b) systematic sampling,
   c) stratified sampling,
   d) cluster sampling,
   e) judgment, quota, or convenience sampling,
   f) some combination of the above methods.

27. Assume that you have been commissioned to design a survey of the age, income, and occupation of the customers who patronize a nationwide chain of stores. Describe how you would proceed with such a study.

28. Given the following sample distribution by class intervals, find $\bar{x}$ and $s_x$.

| Class | Frequency |
|-------|-----------|
| 190–204 | 3 |
| 205–219 | 5 |
| 220–234 | 9 |
| 235–249 | 6 |
| 250–264 | 7 |

29. The average weekly wage for all workers in a certain industry is \$120. In a sample of 100 workers I find $\bar{x} = \$135$. If the weekly wages of these 100 workers are grouped into classes of size 25, the following distribution results:

| Wages | $f$ |
|-------|-----|
| 38–62 | 4 |
| 63–87 | 15 |
| 88–112 | 20 |
| 113–137 | 30 |
| 138–162 | 15 |
| 163–187 | 10 |
| 188–212 | 6 |
| Total | 100 |

a) Find $\bar{x}$ for the grouped data, using class marks 50, 75, 100, ... , 200.
b) Explain the differences between the three measures of average weekly wages.

30. A manufacturer of razor blades claims that his product will, on the average, give 15 good shaves. Suppose you have five friends who try using one of these razor blades each. The number of shaves reported by your friends are 12, 16, 8, 14, and 10.

a) Find the mean and the standard deviation of this sample.
b) Suggest how you might use this sample evidence to dispute or support the advertiser's claim.

31. A fresh produce distributor has received complaints that his bananas have been arriving spoiled at the retail sale store. He is suspicious of the complaint, since his average delivery time is only 4 days (96 hours), and the bananas are fresh at the time of shipment. He decides to simulate the appropriate conditions and make a test to see how long it takes before the bananas become spoiled. He selects a sample of four crates of bananas at random and measures the number of hours before spoilage occurs. The results for number of hours, $x$, are given as 106, 102, 104, and 108.

a) Find the mean hours before spoilage.
b) Find the standard deviation for $x$.
c) On the basis of these measures, do you think that many of the bananas may indeed be arriving spoiled, or are you also suspicious of the complaints? Explain.
d) Find the range of the sample of $x$ values.

e) Give one reason why the answer of part (b) is better than that of part (d), as an estimate of dispersion for the entire population of bananas from which the sample was taken.

32. Suppose weekly sales of bubble gum are normally distributed with a mean of 684,500 and standard deviation 12,650.
   a) Find the probability that sales in a given week will be greater than 700,692.
   b) One can be 99% confident that the weekly sales will be at least _____ (how much?).

33. a) Given the probability distribution below:

| $x$ | 4 | 5 | 9 | 10 |
|---|---|---|---|---|
| $P(x)$ | $\frac{1}{2}$ | $\frac{1}{6}$ | $\frac{1}{6}$ | $\frac{1}{6}$ |

   Sketch the probability function and find the mean and standard deviation.
   b) Suppose 115 repeated samples of size $n = 5$ are drawn randomly from this probability distribution. The sample means are calculated and their frequency distribution is given below.

| $\bar{x}$ | Frequency | $\bar{x}$ | Frequency |
|---|---|---|---|
| 4.0 | 2 | 6.4 | 15 |
| 4.2 | 6 | 6.6 | 6 |
| 4.4 | 2 | 6.8 | 2 |
| 4.6 | 1 | 7.0 | 3 |
| 5.0 | 2 | 7.2 | 5 |
| 5.2 | 15 | 7.4 | 7 |
| 5.4 | 12 | 7.6 | 8 |
| 5.6 | 5 | 7.8 | 1 |
| 6.0 | 3 | 8.2 | 2 |
| 6.2 | 17 | 8.4 | 1 |
| Total | | | 115 |

   Find the mean and standard deviation of these values. Compare these values to those defined by the Central-Limit Theorem for the distribution of all sample means for samples of size 5.
   c) Sketch the distribution of the 115 sample means in a frequency distribution, using class intervals of size 1.0, beginning with 3.5–4.49, 4.5–5.49, etc. Comment on its shape compared to that defined by the Central-Limit Theorem.

34. Use Table VI to collect a random sample of three observations (with replacement) from the population consisting of the digits 0 through 9. Repeat 4 times.
   a) Calculate $\bar{x}$ for each of your five samples of three observations, and then calculate $\bar{\bar{x}}$, the grand mean.
   b) What is the expected value in the population from which your samples in part (a) were drawn? Is $\bar{\bar{x}}$ reasonably close to $\mu_{\bar{x}}$?

c) Calculate the standard deviation of your five sample means about the grand mean. What is the standard error of the mean in the population from which your samples were drawn? Are the two values reasonably close?

35. The five samples you drew in Problem 34 represent observations from a population with mean $\mu = 4.5$ and a variance of $\sigma^2 = 8.25$ (i.e., the digits 0, 1, 2, . . . , 9). This population is not normally distributed.

a) Calculate, for each of your five sample means in Exercise 34, the value of

$$z = (\bar{x} - \mu)/(\sigma/\sqrt{n}).$$

b) Would you expect the distribution of $z$-values in part (a) to follow a normal distribution? Will these $z$-values have a mean of zero and a variance of one? If not, explain why.

36. a) If you repeat Problem 34, collecting samples of size 100 rather than of size 3, how will your answers to part (b) of that question change?

b) What is the standard error of the mean for samples of size 100?

c) Will your answer to part (b) in Exercise 35 change for $n = 100$?

d) If you plotted the values of $z = (\bar{x} - \mu)/(\sigma/\sqrt{n})$ for these five samples of size $n = 100$, would you expect the distribution of these $z$-values to be normally distributed? Will they have a mean of zero and a variance of one? Explain.

37. Design a simple example of your own to illustrate the use of the finite population multiplier by listing four values of some population, finding $\sigma$, and then finding the standard deviation of all possible samples of size 3 drawn without replacement. Does the standard deviation of your samples equal $\sigma/\sqrt{n}$ multiplied by the population correction factor?

38. Describe whether the $t$-distribution or the standardized normal distribution (or neither) is appropriate in the following circumstances:

a) a small sample from a normal population with known standard deviation,

b) a small sample from a nonnormal population with known standard deviation.

c) a small sample from a normal population with unknown standard deviation.

d) a small sample from a nonnormal population with unknown standard deviation,

e) a large sample from a normal population with unknown standard deviation,

f) a large sample from a nonnormal population with unknown standard deviation.

39. Suppose that you collect the following sample of four observations, drawn randomly from a normal population: 99, 115, 91, 79.

a) What is the probability of obtaining $\bar{x}$ this small or smaller from a population with mean $\mu = 110$, and unknown variance? What is the probability that it is this large or larger if drawn from a population with mean $\mu = 80$, and unknown variance?

b) What is the probability of obtaining $\bar{x}$ this small or smaller from a population with mean $\mu = 110$ and a standard deviation of $\sigma = 14$? What is the probability that it was drawn from a population with mean $\mu = 100$ and $\sigma = 10$?

40. From Formula (6.10), plot the sampling distribution of $s^2$ for samples of size $n = 11$ drawn from a normal population with a variance of $\sigma^2 = 20.0$. What is the mean of this distribution? What is the variance?

# 7

# Estimation

## 7.1  INTRODUCTION

In most statistical studies the population parameters are unknown and must be estimated from a sample because it is impossible or just too much trouble (in terms of time or expense) to look at the entire population. Methods for estimating as accurately as possible the value of population parameters thus assume an important role in statistical analysis. A firm manufacturing electrical components, for instance, may wish to know the average number of defective units being shipped in each batch of 1000 items without inspecting each and every component before shipment. Similarly, the psychologist who wants to determine the mean I.Q. of all college undergraduates will undoubtedly have to rely on sample information. In these cases, the value of a sample statistic such as the mean must be used as an estimate of the population parameter. If the degree of dispersion of defective electrical components from batch to batch or the variability of I.Q.'s is of interest, then this parameter also must be estimated from the sample data. Faced with estimation problems of this nature, our objective in this chapter is twofold: first, to present criteria for judging how well a given sample statistic estimates the population parameter, and second, to investigate several of the most popular methods for actually estimating these parameters.

In conventional use the random variables used to estimate population parameters are called *estimators*, while specific values of these variables are referred to as *estimates* of the population parameters. The random variables $\bar{x}$ and $s^2$ are thus estimators of the population parameters $\mu$ and $\sigma^2$. A specific value of $\bar{x}$, such as $\bar{x} = 120$, is an estimate of $\mu$, just as the specific value $s^2 = 237.1$ is an estimate of $\sigma^2$.

It is not necessary that an estimate of a population parameter be one single value; instead the estimate could be a range of values. Estimates which specify a single value of the population are called *point estimates*, while estimates which specify a range of values are called *interval estimates*. A point estimate for the average I.Q. of college undergraduates might be 120, implying that our best estimate of the population mean is 120. An interval estimate would specify a range of values, say 115 to 125, indicating that we think the mean I.Q. in the population lies in this interval.

The choice of an appropriate point estimator in a given circumstance usually depends on how well the estimator satisfies certain criteria for "good" estimators. The next section of this chapter is devoted to describing these criteria—namely, that an estimator should:

1. On the average equal the value of the parameter being estimated (property of *unbiasedness*);
2. Have a relatively small variance (property of *efficiency*);

3. Use as much as possible of the information available from the sample (property of *sufficiency*);

4. Approach the value of the population parameter as the sample size increases (property of *consistency*).

While these properties are certainly all quite reasonable, one may wish to stress other considerations. We pointed out previously that the inferences (or estimates) made from samples serve as an aid to the process of making decisions, and that samples should be drawn with the objective of minimizing the cost of making an incorrect decision (balanced against the cost of sampling). Since one primary purpose in collecting a sample involves estimating parameters, an estimation procedure should be chosen which will minimize the cost (or loss) of making an incorrect estimate from the sample information. This objective is not necessarily incompatible with any of the above properties of good estimators; in fact, in many cases when these properties are satisfied, the estimator will also minimize the cost of making an error.

In our development of point and interval estimates, we need to use some standard notation. Suppose that we call the population parameter being estimated $\theta$ (the Greek letter for theta), where $\theta$ may be $\mu$ or $\sigma^2$, or any population characteristic. An estimator of $\theta$ is generally designated as $\hat{\theta}$ (read "theta hat"); that is, if $\theta = \mu$, then $\hat{\theta} = \bar{x}$ represents one possible means of estimating $\theta$. The use of this notation will greatly simplify the task of defining the properties of a "good" estimator.

## 7.2 PROPERTIES OF ESTIMATORS

As we indicated above, four properties of estimators that are often used to select the best among a group of alternative estimators are the properties of unbiasedness, efficiency, sufficiency, and consistency. Each of these properties is discussed in this section. Prior to that presentation, it should be emphasized that other properties of estimators have also been defined and are useful in selecting the best estimator possible. The reader may consult other statistics books for further discussion of theoretical properties of estimators. In general, the estimators that we shall emphasize throughout this book are those which have been selected as the best estimators for a particular purpose. We will not dwell here on possible alternative estimators or on demonstrations and proofs that the ones we present are "better" than these alternatives.

### Unbiasedness

An estimator is always a function of sample values, meaning it is a random variable resulting from a sampling experiment. As a random variable, it has a probability distribution with a specific shape, expected value, and variance. Analysis of these characteristics of the distribution of an estimator permit us to specify desirable properties of the estimator. For example, one criterion for a "good" estimator concerns how close the expected value of the estimator comes to the population parameter being estimated. Normally, it is preferable to have the expected value of $\hat{\theta}$ exactly equal or fall close to the true value of $\theta$. When the expected value of $\hat{\theta}$ is not equal to $\theta$, the estimator is said to contain a "bias," and $\hat{\theta}$ is called a *biased* estimator of $\theta$. The magnitude of this bias is defined to be the difference between the true value of $\theta$ and the mean value of the estimator $\hat{\theta}$. Under ideal conditions it would be preferable to have an estimator with a bias of zero, in which case the estimator is said to be *unbiased*. The property of unbiasedness can thus be stated as follows:

> *An estimator is said to be unbiased if the expected value of the estimator is equal to the parameter being estimated,* or
>
> $$E[\hat{\theta}] = \theta.$$

In determining a point estimate for the population mean, it is certainly not difficult to construct examples of a biased estimator. Simply using the largest observation in a sample of size $n > 1$ to estimate $\mu$, and ignoring the rest of the observations, will yield an estimate whose expected value is larger than $\mu$. This is obviously a poor choice for estimating the population mean, especially when there is a much more appealing choice, that of using $\bar{x}$. The sample mean is the most widely used estimator of all, for one of its major advantages is that it provides an unbiased estimate of $\mu$. The fact that $E[\bar{x}] = \mu$ was shown to hold in Section 6.4. The parameter other than $\mu$ most often estimated is $\sigma^2$, the population variance. An unbiased estimator for $\sigma^2$ is $s^2$, since $E[s^2] = \sigma^2$, as we proved in Section 6.8.

As a final example of the property of unbiasedness, consider the problem of estimating $p$, the population proportion of successes in a binomial distribution. It is not difficult to show that if a sample yields $x$ successes in $n$ trials, then the ratio $x/n$ is an unbiased estimate of $p$:

$$E[x/n] = \frac{1}{n} E[x] = \frac{1}{n}(np) = p.$$

The above result implies that if, in a random sample of 100 voters, 60 people indicate they intend to vote for Candidate A, then $60/100 = 0.60$ is an unbiased estimate of the population proportion of people who would say they intend to vote for Candidate A.

One weakness of the property of unbiasedness lies in the fact that the criterion

$E[\hat{\theta}] = \theta$ requires only that the *average* value of $\hat{\theta}$ equal $\theta$. It does not require that most or even any of the values of $\hat{\theta}$ be reasonably close to $\theta$, as would seem desirable in a "good" estimator. To illustrate, suppose $\theta = 200$ and consider the following procedure: Flip a fair coin, letting $\hat{\theta} = 100$ if the coin turns up heads and $\hat{\theta} = 300$ if the coin comes up tails. The expected value of $\hat{\theta}$ equals $\frac{1}{2}(100) + \frac{1}{2}(300) = 200$, which means that $\hat{\theta}$ is an unbiased estimator of $\theta = 200$. Despite being unbiased, however, this estimation procedure never yields (on any one flip of the coin) an estimate closer than 100 units to $\theta$, hardly a "close" approximation to the population parameter of 200.

### Efficiency

For given repeated samples of size $n$, it is desirable that an estimator have values that are close to each other. That is, it would be comforting in estimating an unknown parameter to realize that the value you computed based on a particular random sample would not be much different from the value you or anyone else would compute based on another random sample of the same size. In terms of the estimator viewed as a random variable with a probability distribution, this desired property implies that the variance of the estimator should be small. The distribution of $\theta$ should be quite concentrated, rather than spread out over a wide range of values.

However, just as an unbiased estimator with large variance can frequently give estimates that are quite wrong, so also can an estimator which has a small variance but an unknown bias give estimates which are not very useful or desirable. An extreme case depicting this type of estimator is illustrated by choosing $\hat{\theta}$ equal to a constant. Suppose that, in estimating the population mean, we select a constant value as an estimator, say $\hat{\theta} = 100$. Then $\hat{\theta}$ has zero variance but it is biased unless $\mu$ also happens to equal 100. The use of this estimator merely because it has a small variance makes little sense, since it does not consider the information available in the sample. In general, a small variance is desirable, but so is unbiasedness. We prefer an estimator $\hat{\theta}$ which has a small variance about the true parameter value $\theta$, not merely a small variance about its own expected value.

The property of efficiency of an estimator is defined by comparing its variance to the variance of all other unbiased estimators,

> *The most efficient estimator among a group of unbiased estimators is the one with the smallest variance.*

A most efficient estimator is also called the *best unbiased* estimator, where "best" implies minimum variance. Figure 7.1 illustrates the distributions of three

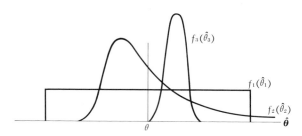

**Fig. 7.1.** Illustration of properties of unbiasedness and efficiency for estimation: $\hat{\boldsymbol{\theta}}_1$, $\hat{\boldsymbol{\theta}}_2$, and $\hat{\boldsymbol{\theta}}_3$.

different estimators (which we label as $\hat{\boldsymbol{\theta}}_1$, $\hat{\boldsymbol{\theta}}_2$, and $\hat{\boldsymbol{\theta}}_3$) based on samples of size $n$. Of the three distributions, $\hat{\boldsymbol{\theta}}_1$ and $\hat{\boldsymbol{\theta}}_2$ both have expected values equal to the true parameter $\theta$, whereas $\hat{\boldsymbol{\theta}}_3$ has a positive bias since $E[\hat{\boldsymbol{\theta}}_3] > \theta$. The variance of the estimators decreases in size from the first to the third, $V[\hat{\boldsymbol{\theta}}_1] > V[\hat{\boldsymbol{\theta}}_2] > V[\hat{\boldsymbol{\theta}}_3]$, so that $\hat{\boldsymbol{\theta}}_3$ has the smallest variance. However, $\hat{\boldsymbol{\theta}}_3$ is not the most efficient estimator among this group because it is not unbiased. Our definition of efficiency requires that the estimator be *unbiased* and have smaller variance than any other unbiased estimator of $\theta$. Thus, $\hat{\boldsymbol{\theta}}_2$ is the most efficient among those illustrated. Whether it is the most efficient among *all possible* unbiased estimators has not been shown and is often difficult to prove.

*Relative Efficiency.* Since it is generally quite difficult to prove that an estimator is the best among all biased ones, the most common approach in measuring efficiency is to determine the *relative efficiency* of two estimators. If $\hat{\boldsymbol{\theta}}_1$ is an unbiased estimator of $\theta$, and $\hat{\boldsymbol{\theta}}_2$ represents another unbiased estimator of $\theta$, then the relative efficiency of $\hat{\boldsymbol{\theta}}_1$ to $\hat{\boldsymbol{\theta}}_2$ is given by the following ratio:

$$Relative\ efficiency: \qquad \frac{V[\hat{\boldsymbol{\theta}}_2]}{V[\hat{\boldsymbol{\theta}}_1]}.$$

As an illustration of the use of relative efficiency, consider the sample mean vs. the sample median as estimators of the mean of a normal population. Both estimators are unbiased when sampling from a normal population, since the normal is symmetric. From Section 6.4 we know that the variance of $\bar{x}$ equals $\sigma^2/n$. It is also possible to find the variance of an estimator which uses the sample median to estimate the population mean; this variance is $\pi\sigma^2/2n$. The ratio of these quantities gives their relative efficiency:

$$\frac{V(\text{median})}{V(\bar{x})} = \frac{\pi\sigma^2/2n}{\sigma^2/n} = \frac{\pi}{2} = 1.57.$$

The ratio 1.57 implies that the mean is 1.57 times as efficient as the median in estimating $\mu$. In other words, an estimate based on the mean of a sample of 100 observations has approximately the same reliability as an estimate based on the median of a sample of 157 observations, assuming a normal parent population.

In some problems it is possible to determine precisely the most efficient estimator, that is, the unbiased estimator with the lowest variance. In estimating the mean of a normal population, for example, it can be shown that the variance of any estimator must be greater than or equal to $\sigma^2/n$. Since the variance of $\bar{x}$ in this case exactly equals $\sigma^2/n$, the sample mean must be the most efficient estimator of $\mu$. In a design of a sample, the most efficient estimator may not always be the best choice because of other factors, such as the time available to collect the sample or the accessibility of the observations. That is, statistical efficiency may have to be sacrificed in order to obtain an estimate in the allowed time; or some other estimator may be less costly to obtain or more meaningful than the most efficient one. Many high schools, for example, publish statements about a student's academic performance only in terms of his *rank* relative to the rest of the class (e.g., "he ranks tenth in his class"). Trying to determine a most efficient estimator in this case (for the average or mean academic performance) may be difficult and, perhaps, the estimator may not be as meaningful as a measure based on the ranked performance.

### Sufficiency

The properties of unbiasedness and efficiency are desirable properties for an estimator, particularly as related to small samples. Another property sometimes used is the property of *sufficiency*, where an estimator is sufficient if it uses all the information about the population parameter that the sample can provide. For example, we certainly want an estimator to use all the sample observations, as well as all the information provided by these observations. The sample median, however, uses only the *ranking* of the observations and not their precise numerical values; hence the median is not a sufficient estimator. A primary importance of the property of sufficiency is that it is a necessary condition for efficiency.

### Consistency

Since the distribution of an estimator will in general, change as the sample size changes, the properties of estimators in the limit (as $n \to \infty$) become important. Recall, from Fig. 6.3, the sampling distribution for $\bar{x}$ when the population has a uniform distribution. For samples of size $n = 1$, the distribution of $\bar{x}$ is the same as the distribution of $x$, a uniform distribution. For $n = 2$, the distribution of $\bar{x}$ has a triangular shape. For $n = 5$ and $n = 30$, the distribution of $\bar{x}$ becomes more like a normal distribution. The central-limit theorem of Section 6.6 states that in the limit as $n$ approaches a very large size, the distribution of $\bar{x}$ approaches the normal distribution.

Properties of estimators based on limits as $n \to \infty$ are called *asymptotic properties*, as compared to finite or small-sample properties. The most important of these asymptotic properties is the property of *consistency*, which involves the variability of the estimator $\hat{\theta}$ about the true value $\theta$ as the size of $n$ increases:

> *An estimator is said to be consistent if it yields estimates which approach the population parameter being estimated as n becomes larger.*

To say that an estimator must approach the parameter being estimated implies that, when considering larger and larger samples, the variance of this estimator about the population parameter (i.e., the variance of $\hat{\theta}$ about $\theta$) must become smaller and smaller. More specifically, if $E[(\hat{\theta} - \theta)^2]$ approaches zero as $n$ goes to infinity, the estimator $\hat{\theta}$ is said to be consistent. Note that this requirement is not equivalent to specifying that $V[\hat{\theta}]$, goes to zero as $n$ goes to infinity *unless* $\hat{\theta}$ is an unbiased estimator of $\theta$. If $\hat{\theta}$ is not unbiased, $V[\hat{\theta}] = E[(\hat{\theta} - E[\hat{\theta}])^2]$ may get smaller and smaller as $n$ increases, but $\hat{\theta}$ will not necessarily get any closer to $\theta$. For instance $\hat{\theta}$ might get closer and closer to 100 as $n$ increases, but unless the true value of the population parameter is also equal to 100, this estimator is not necessarily improving as $n$ becomes large. The following two conditions are thus sufficient to define consistency.

*Property of consistency:*

1. $V[\hat{\theta}] \to 0$ as $n \to \infty$;
2. $\hat{\theta}$ is unbiased ($E[\hat{\theta}] = \theta$).

Previously, we showed that $\bar{x}$ is an unbiased estimator of the population mean, and that $x/n$ is an unbiased estimator of the population proportion. It is not difficult to also show that both these estimators are consistent as well as unbiased. We shall prove that $x/n$ is a consistent estimator of $p$, leaving it to the reader to prove that $\bar{x}$ is also consistent.

$$V[x/n] = \frac{1}{n^2} V[x] \qquad \text{(From Section 3.3, Rule 4)}$$

$$= \frac{1}{n^2} (npq) \qquad \text{(Since } x \text{ is binomially distributed)}$$

$$= \frac{pq}{n}.$$

Since the value of $pq/n$ goes to zero as $n$ goes to infinity, and since $x/n$ is an unbiased estimator of $p$, then $x/n$ must be a consistent estimator of $p$.

## 7.3 ESTIMATING UNKNOWN PARAMETERS

In the 1920's R. A. Fisher developed the method of maximum likelihood as a means of finding estimators which satisfy some (but not necessarily all) of the criteria discussed previously. This method is popular because maximum-likelihood estimators are usually relatively easy to obtain and the resulting estimates are often efficient and approximately normally distributed about $\theta$ for large samples. One disadvantage of the method is that maximum-likelihood estimates are usually not unbiased.

The method of maximum-likelihood estimates the value of a population para-meter by selecting the most likely sample space from which a given sample could have been drawn. In other words, the sample space is selected which would yield the observed sample more frequently than any other sample space. The population parameter corresponding to this space is called the maximum-likelihood estimate of $\theta$, and the name *maximum likelihood* is derived from this process of selecting the most likely sample space.

As an illustration of the process of determining a maximum-likelihood esti-mate, consider the problem of estimating the binomial parameter $p$. Suppose that, in a sample of 5 trials, 3 successes are observed. What sample space (i.e., what binomial population) is most likely to give this particular result; or equivalently, what is the value of $p$ most likely to give the observed sample? The most likely population parameter can be determined by calculating the probability of ob-taining exactly 3 successes in 5 trials for all possible values of the population parameter and selecting that value yielding the highest probability. Table 7.1 examines 9 possible values of $p$, indicating for each value the probability of 3 successes in 5 trials (i.e., if $p = 1/10$, the appropriate probability is

$$\binom{5}{3}\left(\frac{1}{10}\right)^3\left(\frac{9}{10}\right)^2 = 0.0081).$$

**Table 7.1**

|  | Probability of |
| Value of $p$ | three successes |
| --- | --- |
| 0.10 | 0.0081 |
| 0.20 | 0.0512 |
| 0.30 | 0.1323 |
| 0.40 | 0.2304 |
| 0.50 | 0.3125 |
| 0.60 | 0.3456 |
| 0.70 | 0.3087 |
| 0.80 | 0.2048 |
| 0.90 | 0.0729 |

The value of $p$ most likely to yield a sample of 3 successes in 5 trials, as given by Table 7.1, is $p = 0.60$, where the associated probability is 0.3456. The reader will note that this estimate exactly equals the sample proportion $x/n = 3/5 = 0.60$. It is often true that the most likely value for a population parameter is the intuitively appealing one, the corresponding measure of the sample. For example, it can be shown that the maximum-likelihood estimator of a population mean is the sample mean. That is, the value of $\bar{x}$ is the most likely value of $\mu$ that can be found based on a sample of size $n$.

The choice of a particular value as most likely for a population parameter is called finding a *point estimate*. We know that it would be an exceptional coincidence for this estimate to be identical to the population parameter (because of sampling error). Thus, even though the best possible value is used as the point estimate, we have very small confidence that this value is exactly correct. One of the major weaknesses of a point estimate is that it does not permit the expression of any degree of uncertainty about the estimate. The most common way to express uncertainty about an estimate is to define an interval, or a range of values, and to assert with a certain degree of confidence that the population parameter will fall within this interval. This process is known as *interval estimation*.

### Confidence Intervals

You will recall that on a number of occasions thus far we have determined values $a$ and $b$ so that $P(a \leqslant \bar{x} \leqslant b)$ equals some predetermined value. The values $a$ and $b$ were determined from a knowledge about the parent population and its parameters. The interval $(a, b)$ is called a probability interval for $\bar{x}$. For example, if we calculated $P(a \leqslant \bar{x} \leqslant b) = 0.90$, based on a random sample of size $n$ drawn from a population with mean $\mu$, we know the random variable $\bar{x}$ will fall in the probability interval $(a, b)$ 90 percent of the time.

Although it is important to be able to construct confidence intervals for $\bar{x}$ based on a knowledge of the population parameter $\mu$, for most statistical purposes it is $\mu$ which is the unknown, and we want to construct a confidence interval for $\mu$ based on $\bar{x}$. For example, we may want to find two points, $c$ and $d$, such that we are 90 percent confident that $\mu$ lies in the interval between these points. In other words, on the average, 90 such intervals out of every 100 calculated on the basis of samples of size $n$ will include the population mean $\mu$.

The use of the future tense in explaining a confidence interval is very important because once such an interval based on a sample is determined, either the true parameter lies in the interval or it does not. The value of $\mu$ does not have a probability of 0.90 of being within the interval since it is not a random variable, but a constant. If it is in a given interval, then the probability that it is in the interval is 1.0; if not, the probability that it is within the interval is 0.0.

To better distinguish between the two types of intervals, consider a problem in process control. Suppose a manufacturer makes large tile pipes and the interior diameter of the pipe at its smallest rim is normally distributed with mean $\mu = 24$ and $\sigma = \frac{1}{4}$, or $N(24, 0.25^2)$. At random points in time, a sample of four pipe segments is selected from the production process, to check on the average diameter of all the pipe segments in the population being produced. Let $\theta$ represent the population mean and $\hat{\theta} = \bar{x}$ represent an unbiased estimator with a normal distribution.

Figure 7.2 represents the two types of intervals discussed above. Part (a) of the figure illustrates a probability interval with its center at the population mean, $\mu = 24$ inches, and values $a = 23.5$ and $b = 24.5$ chosen so that 95 percent of the values of $\bar{x}$ will lie between the endpoints $a = 23.5$ and $b = 24.5$. [These endpoints are found, as before, using Table III and the process described on page 226 (in Chapter 6). In terms of the cumulative distribution function $F(\hat{\theta})$,

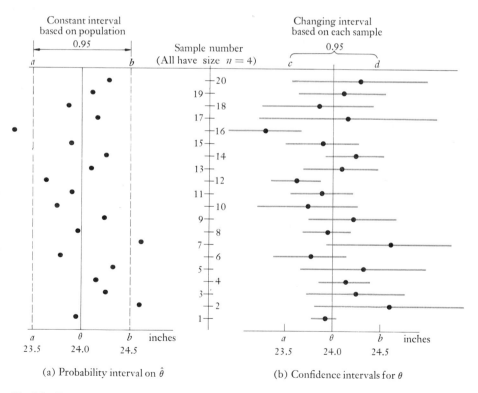

**Fig. 7.2.** Illustration of probability interval and of confidence interval for an unbiased estimator $\hat{\theta}$ with a normal distribution.

this means that $F(a) = 0.025$, $1.0 - F(b) = 0.025$, and $F(b) - F(a) = 0.95$.] In repeated samples of size 4, different values of $\hat{\theta}$ would be calculated. The 95-percent probability interval means that on the average, 95 out of 100 of these values of $\hat{\theta}$ should be between $a$ and $b$. Figure 7.2(a) illustrates twenty different values of $\hat{\theta}$ calculated from 20 different samples of size 4. Theoretically, 95 percent (or 19) of these should lie within the interval. Actually 17 of them do and three do not (sample numbers 2, 7, and 16). If a large number of the values of $\hat{\theta}$, or several consecutive ones, do not lie within the interval, it would be a sign that the production process was getting "out of control" and should be stopped and corrected.

Part (b) of Fig. 7.2 illustrates the 95-percent confidence intervals based on these same 20 samples of size 4 each. Since each sample of size 4 can have different values, each sample can have a different sample mean and variance. Each interval depends on these values, so even though the underlying sampling distribution for $\hat{\theta}$ is the same in each instance, the center and spread of the interval calculated can be different. In Fig. 7.2(b), the center of each interval is at $\hat{\theta} = \bar{x}$ and the endpoints representing $c$ and $d$ for each interval are shown. The values of $a$ and $b$ are not relevant to these intervals but are merely shown for scale comparison with Fig. 7.2(a). The meaning of a 95-percent confidence interval is that on the average, the true parameter $\theta$ will be included within such intervals in 95 out of 100 such calculations. Once a given interval is calculated, the probability that the parameter $\theta$ lies in the interval is 1.0 or 0.0. Among our 20 samples, eighteen do include the value $\theta$, two do not (sample numbers 12 and 16).

### The Probability of an Error, $\alpha$

Instead of expressing the probability that a population parameter $\theta$ will be contained in an interval to be calculated, it will be convenient to refer to the probability that this interval will *not* include the parameter. This probability is usually denoted by the letter $\alpha$ (the Greek letter for alpha). The value of alpha is often referred to as the probability of making an error, since it indicates the proportion of time that one will be incorrect, or in "error," in assuming that the specified interval contains the population parameter. Since the events, "the confidence interval will include $\theta$" and "the confidence interval will not include $\theta$" are complements, and since $\alpha$ is the probability of the latter event, $(1.0 - \alpha)$ equals the probability that the interval will include $\theta$. It is customary to refer to confidence intervals as being of size, "$100(1 - \alpha)\%$". Thus, if $\alpha$ is specified to be 0.05, the associated interval is a $100(1 - 0.05)\%$ or 95-percent confidence interval; if a 90-percent confidence interval is specified, then $\alpha = 1.0 - (90/100) = 0.10$. In our previous example of the process-control problem, $\alpha$ equals 0.05, which is the probability of error in saying that the interval contains the parameter (process being in control) when it really does not (the process needing adjustment). In only five percent of such intervals, on the average, would we be wrong in assuming that the interval includes the parameter $\theta$.

There is an obvious trade-off between the value of $\alpha$ and the size of the confidence interval: the lower the value of $\alpha$, the larger the interval must be, all other
things being equal. That is, if one need not be very confident that the population
parameter is within the interval, then a relatively small interval will suffice; in
order to be quite confident about $\theta$ being in the interval, a relatively large interval
will be necessary. The value of $\alpha$ is often set at 0.05 or 0.01, representing 95- and
99-percent confidence intervals, respectively. This procedure, although widely
used, does not necessarily lead to the optimal trade-off between the size of the
confidence interval and the risk of making an error.

In general, confidence intervals are constructed on the basis of sample information, so that not only do changes in $\alpha$ affect the size of the interval, but so do
changes in $n$, the sample size. The more observations collected, the more confident
one can be about the estimate of $\theta$, and as a result, the smaller the interval needed
to assure a given level of confidence. Although it is usually desirable to have as
small a confidence interval as possible, the size of the interval must be determined
by considering the costs of sampling and how much risk one wants to assume of
making an error. We shall return several times to this problem of determining the
optimal trade-off between the risks of making an error and the sample size. For
now we merely caution the reader to be aware that the task of determining an
"optimal" trade-off may not be an easy one.

### Determining a Confidence Interval

Usually, one of the first steps in constructing a confidence interval is to specify
how much confidence one wants to have that $\theta$ will fall in the resulting interval,
and the size of the sample. In other words, both $\alpha$ and $n$ are usually fixed in advance.
It is possible, however, under certain conditions to consider either $\alpha$ or $n$ as an
unknown and to solve for the value of this unknown.

In addition to specifying $\alpha$ in advance, one must also specify that proportion
of the probability of making an error attributable to the fact that $\theta$ will sometimes
be larger than the upper bound of the confidence interval, and that proportion
attributable to the fact that $\theta$ will sometimes be smaller than the lower bound of
the confidence interval. As we indicated previously, in determining a probability
interval, the *smallest* interval is obtained by dividing the probability to be excluded
outside the interval *equally* between the upper and the lower tails of the distribution. For a confidence interval, however, the decision on how to divide $\alpha$ into these
two parts should depend on how serious or costly it is to make errors on the high
side relative to errors on the low side. Since we normally want to avoid the more
costly errors, $\alpha$ should be divided so that these errors occur less frequently. Unfortunately, determining the costs of making an error may be quite difficult, so
that the common procedure in determining confidence intervals is the same as for
probability intervals—i.e., to exclude one-half of $\alpha$, or $\alpha/2$, on the high side and
one-half of $\alpha$ on the low side. Thus, if $\alpha = 0.05$, a value of $\alpha/2 = 0.025$ would be

the probability that $\theta$ will exceed the upper bound of the confidence interval, and $\alpha/2 = 0.025$ would be the probability that the lower bound will exceed $\theta$. Such a procedure assumes that errors on the high side are equally as serious as errors on the low side.

In the above discussion, we have noted that a given confidence interval will depend on $\alpha$ (and the way $\alpha$ is divided) and on the size of the sample. The final factor influencing the boundaries of a particular confidence interval is the sampling distribution of the statistic used to estimate the population parameter. Normally, the procedure for establishing a confidence interval for a population parameter $\theta$ is *first to find a point estimate* of this parameter. One's uncertainty about this point estimate is then determined by finding that interval of values about the point estimate which, according to the sampling distribution of the statistic used to estimate $\theta$, yields the desired degree of confidence. Since different sampling distributions are used for estimating a population mean, the binomial parameter $p$, and a population variance, we shall describe the process of constructing intervals for each of these cases in separate sections.

### 7.4 CONFIDENCE INTERVALS FOR μ (σ KNOWN)

Suppose we start by constructing a confidence interval for the population parameter $\mu$ based on a random sample drawn from a normal parent population with *known* standard deviation. The natural sample statistic for estimating $\mu$ is $\bar{x}$, the sample mean, for reasons which we discussed in Section 7.2. Recall the sampling distribution of $\bar{x}$ under the conditions listed above: the expected value of $\bar{x}$ equals $\mu$, and $\bar{x}$ has a standard deviation of $\sigma/\sqrt{n}$. Also recall from Formula (6.7) that the variable $z = (\bar{x} - \mu)/(\sigma/\sqrt{n})$ has a standardized normal distribution. Now, suppose we let the symbol $z_\alpha$ represent that value of the standardized normal variable $z$ such that the probability of observing values of $z$ greater than $z_\alpha$ is $\alpha$, or $P(z > z_\alpha) = \alpha$. Similarly, let $-z_\alpha$ equal the point such that $P(z < -z_\alpha) = \alpha$. Two such points are shown in Fig. 7.3.

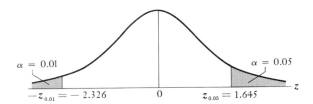

$\alpha = 0.01$ ............ $\alpha = 0.05$

$-z_{0.01} = -2.326$ ........ $0$ ........ $z_{0.05} = 1.645$ ........ $z$

**Fig. 7.3.** The value of $z_\alpha$, cutting off $\alpha$ percent from the standardized normal distribution.

Rather than finding those points cutting off an area of size $\alpha$ in each tail of the standardized normal distribution, it will be more convenient to find those points cutting off $\alpha/2$ probability in each tail (so that the total area cut off equals $\alpha$). To do this, let $z_{\alpha/2}$ represent the value such that the probability $P(z > z_{\alpha/2}) = \alpha/2$, and let $-z_{\alpha/2}$ equal the point such that $P(z < -z_{\alpha/2}) = \alpha/2$. If, for example, $\alpha = 0.05$, then from Table III the value of $z_{\alpha/2}$ satisfying $P(z > z_{\alpha/2}) = 0.025$ is seen to be $z_{\alpha/2} = 1.96$. The value of $-z_{\alpha/2}$ must be $-1.96$, since the normal distribution is symmetric. The probability that $z$ falls between the two limits, $-1.96$ and $+1.96$, is

$$P(-1.96 < z < 1.96) = 1.0 - 2(\alpha/2) = 1.0 - \alpha$$

$$= 1.00 - 0.05 = 0.95,$$

as shown in Fig. 7.4.

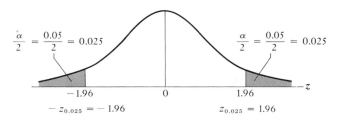

**Fig. 7.4.** The value of $z_{\alpha/2}$, cutting off a total area of $\alpha = 0.05$ from the standardized normal distribution, leaving an interval including $100(1 - \alpha)\% = 95$ percent of the probability.

In more general terms the probability that $z$ falls between two limits $-z_{\alpha/2}$ and $+z_{\alpha/2}$ can be written in the following form:

$$P(-z_{\alpha/2} < z < z_{\alpha/2}) = 1 - \alpha.$$

Note that the interval $-z_{\alpha/2} < z < z_{\alpha/2}$ is a $100(1 - \alpha)\%$ confidence interval for $z$. It is also a $100(1 - \alpha)\%$ confidence interval for $(\bar{x} - \mu)/(\sigma/\sqrt{n})$, since $z = (\bar{x} - \mu)/(\sigma/\sqrt{n})$. Unfortunately, this is not the confidence interval we originally set out to derive, for we wanted a $100(1 - \alpha)\%$ confidence interval for $\mu$, not for $(\bar{x} - \mu)/(\sigma/\sqrt{n})$. Fortunately the difference is not hard to resolve, for it is possible to find an equivalent expression which represents the confidence interval for $\mu$ by rearranging the terms in the confidence interval for $(\bar{x} - \mu)/(\sigma/\sqrt{n})$. That is, by rewriting the inequalities in the expression $-z_{\alpha/2} < z < z_{\alpha/2}$ we can show that this expression is equivalent to the following inequalities:*

$100(1 - \alpha)\%$ *confidence interval for $\mu$, where $\sigma$ is known and the parent population is normal (or $n > 30$):*

$$\bar{x} - z_{\alpha/2} \frac{\sigma}{\sqrt{n}} < \mu < \bar{x} + z_{\alpha/2} \frac{\sigma}{\sqrt{n}}. \tag{7.1}$$

To illustrate the use of Formula (7.1), let's return to the problem of estimating the mean I.Q. of that well-known eastern university. In this case we want to construct a 95-percent confidence interval for $\mu$ on the basis of a random sample of size $n = 25$. Furthermore, let's now assume we know that $\sigma = 5.4$, and (as before) the parent population is normal. The random sample of $n = 25$ yields $\bar{x} = 127$. Now, since we want a 95-percent confidence interval, the appropriate $z$-values are $z_{\alpha/2} = 1.96$ and $-z_{\alpha/2} = -1.96$. Substituting these two values, as well as $n = 25$, $\sigma = 5.4$, and $\bar{x} = 127$ into Formula (7.1), we get the desired 95-percent confidence interval:†

$$\bar{x} - z_{\alpha/2} \frac{\sigma}{\sqrt{n}} < \mu < \bar{x} + z_{\alpha/2} \frac{\sigma}{\sqrt{n}},$$
$$127 - (1.96)\frac{5.4}{\sqrt{25}} < \mu < 127 + (1.96)\frac{5.4}{\sqrt{25}},$$
$$127 - 2.12 < \mu < 127 + 2.12,$$
$$124.88 < \mu < 129.12.$$

---

* The equivalency is shown as follows:

$-z_{\alpha/2} < z < z_{\alpha/2}$

$-z_{\alpha/2} < \dfrac{\bar{x} - \mu}{\sigma/\sqrt{n}} < z_{\alpha/2}$      By substitution

$-z_{\alpha/2} \dfrac{\sigma}{\sqrt{n}} < (\bar{x} - \mu) < z_{\alpha/2} \dfrac{\sigma}{\sqrt{n}}$      By multiplying each term by $\sigma/\sqrt{n}$

$-\bar{x} - z_{\alpha/2} \dfrac{\sigma}{\sqrt{n}} < -\mu < -\bar{x} + z_{\alpha/2} \dfrac{\sigma}{\sqrt{n}}$      By adding $(-\bar{x})$ to each term

$\bar{x} + z_{\alpha/2} \dfrac{\sigma}{\sqrt{n}} > \mu > \bar{x} - z_{\alpha/2} \dfrac{\sigma}{\sqrt{n}}$      By multiplying each term by $(-1)$, thus changing the direction of both inequalities.

† For this problem, and the subsequent ones in this chapter, we assume that sampling is from an infinite population, or from a finite population with replacement. When this is not the case, the finite population correction factor must be used; that is, $\sigma/\sqrt{n}$ must be multiplied by $\sqrt{(N - n)/(N - 1)}$.

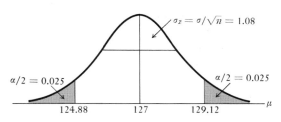

$$\sigma_{\bar{x}} = \sigma/\sqrt{n} = 1.08$$

$\alpha/2 = 0.025$          $\alpha/2 = 0.025$

124.88          127          129.12          $\mu$

**Figure 7.5**

This 95-percent confidence interval is illustrated in Fig. 7.5. We can infer from this analysis, with 95-percent confidence, that the value of $\mu$ lies between 125.88 and 129.12. That is, on the average, in 95 out of 100 such samples of size $n = 25$, an interval calculated in this manner will include the true population mean $\mu$. We do not know, of course, whether the above interval is one of the correct ones, or one of the incorrect ones, since $\mu$ is unknown.

For the confidence interval calculated above, the probability of making an error in assuming that $\mu$ is in the specified interval is $\alpha = 0.05$. If one desires a smaller risk of error $\alpha$, then a larger confidence interval must be used. For example, in order to have $\alpha = 0.01$ (i.e., a 99-percent confidence interval), the appropriate $z$-values are found in Table III to be $z_{\alpha/2} = 2.576$ and $-z_{\alpha/2} = -2.576$ (for $\alpha/2 = 0.005$). The new confidence interval is thus:

$$127 - (2.576)\frac{5.4}{\sqrt{25}} < \mu < 127 + (2.576)\frac{5.4}{\sqrt{25}},$$
$$127 - 2.78 < \mu < 127 + 2.78,$$
$$124.22 < \mu < 129.78.$$

Since this interval is wider than the previous one, we have greater confidence that it includes the population parameter $\mu$. We could further increase our confidence, and decrease the risk of error, $\alpha$, by extending the interval even more. Of course, there is a limit to the usefulness of the interval when it gets too large. We have almost 100-percent confidence, for example, that the mean I.Q. of college students lies in the interval $83 < \mu < 177$, but we could have made such a statement without doing any sampling or statistical inference. We sample to get useful information that is more precise. To obtain it, we must be willing to subject our conclusions or inferences to a small controlled level of risk $\alpha$.

**Relax the Assumption of Normality for the Population**

The relationship expressed in Formula (7.1) depends on the assumptions that $\sigma$ is known and that the parent population is normal. When these assumptions are met, Formula (7.1) holds for any sample size, no matter whether $n$ is large or small.

But suppose the parent population is *not normal*. In this case, if the sample size $n$ is small, then the distribution of $(\bar{x} - \mu)/(\sigma/\sqrt{n})$ is not normal, and there is no convenient way to determine a confidence interval. On the other hand, when $n$ is large (usually $n > 30$), we know by the central-limit theorem that $(\bar{x} - \mu)/(\sigma/\sqrt{n})$ is *approximately* normally distributed; hence the confidence interval specified by Formula (7.1) is still appropriate.

Suppose that, in our I.Q. example, we now drop the assumption of normality and assume the sample size was $n = 64$ rather than 25. Again, if $\sigma = 5.4$ and $\bar{x} = 127$, then a 95-percent confidence interval is:

$$127 - (1.96)\frac{5.4}{\sqrt{64}} < \mu < 127 + (1.96)\frac{5.4}{\sqrt{64}},$$
$$127 - 1.32 < \mu < 127 + 1.32,$$
$$125.68 < \mu < 128.32.$$

### 7.5 CONFIDENCE INTERVALS FOR μ (σ UNKNOWN)

The major assumption specified at the beginning of Section 7.4 was that the population standard deviation, $\sigma$ is known. This may not be very realistic for many applied problems. When the mean of the population is unknown and must be estimated, it is unlikely that the standard deviation about that unknown mean will be known. Instead, the population standard deviation often must be estimated from the sample standard deviation. Under these circumstances, $(\bar{x} - \mu)(s/\sqrt{n})$ has a *t*-distribution, with $(n - 1)$ degrees of freedom (assuming that the parent population is normal).

The procedure for determining a $100(1 - \alpha)\%$ confidence interval using the *t*-distribution follows the same pattern as when the normal distribution holds, except that different limits must be used. The limits are found using *t*-values from Table VII rather than *z*-values from Table III. Let $t_{\alpha/2, \, v}$ represent that value of the *t*-distribution with $v = n - 1$ degrees of freedom that excludes $\alpha/2$ of the probability in the upper tail. That is, $P(t > t_{\alpha/2, \, v}) = \alpha/2$; and by symmetry, $P(t < -t_{\alpha/2, \, v}) = \alpha/2$. Then, $P(-t_{\alpha/2, \, v} < t < t_{\alpha/2, \, v}) = 1 - \alpha$, or, by substitution, since $t = (\bar{x} - \mu)/(s/\sqrt{n})$,

$$P(-t_{\alpha/2, \, v} < (\bar{x} - \mu)/(s/\sqrt{n}) < t_{\alpha/2, \, v}) = 1 - \alpha.$$

Solving the inequalities as before for $\mu$, it can be shown that

---

$100(1 - \alpha)\%$ *confidence interval for* $\mu$, *population normal,* $\sigma$ *unknown:*

$$\bar{x} - t_{\alpha/2, \, v}\frac{s}{\sqrt{n}} < \mu < \bar{x} + t_{\alpha/2, \, v}\frac{s}{\sqrt{n}}.$$

(7.2)

---

To illustrate the use of the relationship described by Formula (7.2), suppose that a student has a discussion with a friend involving the average number of points scored by the high-school basketball teams in his home area. He really does not know the true population average nor does he know the population standard deviation. However, he sees no reason why the points scored per game might not be validly represented by a random variable $x$ with a normal distribution. The student then plans to use a random-number table and pick 10 scores at random from a weekly listing of results in the Sunday newspaper. To avoid some possible bias of always selecting scores of winning teams, or of selecting scores of both teams in the same game, he stratifies the sample so as to pick at random five scores each of winning and losing teams. He decides (correctly) that the $t$-distribution is appropriate and hence will use Formula (7.2). He also decides to obtain 99-percent confidence in his estimate of the mean score per game, $\mu$.

Column (1) of Table 7.2 presents the sample selected. Since the sample mean is an integer, $\bar{x} = 70$, he elects to compute the sample variance according to the definition, $s^2 = (1/(n - 1)) \sum(x_i - x)^2$, and finds $s^2 = 65.55$. The sample standard deviation is $\sqrt{65.55} = 8.10$, and $s/\sqrt{n} = 8.10/\sqrt{10} = 2.563$.

**Table 7.2**

| $x$ (score) | $(x - \bar{x})$ | $(x - \bar{x})^2$ | |
|---|---|---|---|
| 69 | $-1$ | 1 | |
| 81 | 11 | 121 | |
| 67 | $-3$ | 9 | |
| 80 | 10 | 100 | |
| 71 | 1 | 1 | |
| 70 | 0 | 0 | $\bar{x} = (1/10)700 = 70$ |
| 78 | 8 | 64 | |
| 68 | $-2$ | 4 | $s = \sqrt{(1/9)(590)} = \sqrt{65.56} = 8.10$ |
| 57 | $-13$ | 169 | |
| 59 | $-11$ | 121 | $s_{\bar{x}} = s/\sqrt{n} = 8.10/3.16 = 2.563$ |
| Sum    700 | 0 | 590 | |

From Table VII for the $t$-distribution, he next has to find the critical values for his 99-percent confidence interval, $\pm t_{\alpha/2, \, v}$. Using the row for $v = (n - 1) = 9$ degrees of freedom, and the columns for $F(t) = 1.0 - \alpha/2 = 1.0 - 0.005 = 0.995$, he obtains $\pm t_{0.005, \, 9} = \pm 3.25$; substituting the appropriate values of $x$, $s$, $t$, and $\sqrt{n}$ into Formula (7.2) gives

$$\bar{x} - t_{\alpha/2, \, v}\left(\frac{s}{\sqrt{n}}\right) < \mu < \bar{x} + t_{\alpha/2, \, v}\left(\frac{s}{\sqrt{n}}\right),$$

$$70 - 3.25\left(\frac{8.10}{3.16}\right) < \mu < 70 + 3.25\left(\frac{8.10}{3.16}\right),$$
$$70 - 3.25(2.563) < \mu < 70 + 3.25(2.563),$$
$$70 - 8.33 < \mu < 70 + 8.33, \quad \text{and} \quad 61.67 < \mu < 78.33.$$

The interval from 61.67 to 78.33 thus represents a 99-percent confidence interval for the mean points per game by high-school basketball teams. The student may be disappointed that the interval is so large. He feels that he could have guessed that the mean was between 60 and 80. This 99-percent confidence interval is not much different from his nonstatistical guess. However, the statistical work has given him a precise method of inference, a precise point estimate of $\mu$, and exact knowledge about the chance of error in assuming that the true mean does lie in this interval, namely $\alpha = 0.01$. He can decrease the size of his interval in two ways if he wishes—either by allowing a greater chance of error $\alpha$, or by putting more time and effort (cost) into the sampling to increase $n$.

Let us briefly examine the effects of each alternative. First, if he is willing to let $\alpha = 0.10$ and obtain only a 90-percent confidence interval, he finds in Table VII, $\pm t_{\alpha/2, \nu} = \pm t_{0.05, 9} = \pm 1.833$. Substituting his same sample values into Formula (7.2), he obtains the interval,

$$70 - 1.833(2.563) < \mu < 70 + 1.833(2.563),$$
$$65.30 < \mu < 74.70.$$

As a second alternative, suppose he took a sample three times larger than before, $n = 30$, and kept the confidence level at 99%. For comparison purposes, let's assume the sample of size 30 also had $\bar{x} = 70$ and $s = 8.10$, so that the only difference from the original problem is the change in sample size. The new standard deviation of the sampling distribution for $\bar{x}$ is $s/\sqrt{n} = 8.10/\sqrt{30} = 1.48$. The new critical values are $\pm t_{\alpha/2, \nu} = t_{0.005, 29} = \pm 2.756$. The new interval is thus

$$70 - 2.756(1.48) < \mu < 70 + 2.756(1.48),$$
$$65.92 < \mu < 74.08.$$

Clearly, allowing a greater risk of error or incurring a greater cost in sampling more items results in a smaller confidence interval. Note that each interval estimate is centered at the best point estimate, $\bar{x} = 70$. The reader may wish to determine the interval estimate for this situation if one is willing to have *both* the risk of $\alpha = 0.10$ and the larger sample size, $n = 30$.

### 7.6 CONFIDENCE INTERVALS FOR THE BINOMIAL PARAMETER $p$, USING THE NORMAL APPROXIMATION

The random variable $x/n$ was introduced earlier in this chapter as an estimator of $p$, the population parameter in a binomial distribution. This statistic, referred to as the *sample proportion*, can be used to determine confidence intervals for populations in applications involving the binomial distribution, such as the proportion of people in a given population who smoke cigarettes, the proportion of voters favoring a certain candidate, or the proportion of defective items produced in a production process.

Recall from Chapter 5 that we showed that when $n$ is large, the number of successes in $n$ independent Bernoulli trials is approximately normally distributed. We approximate the number of successes, $x$, by using the standardized normal variable $z = (x - np)/\sqrt{npq}$, where $E[x] = np$ and $V[x] = npq$. Now, if the *number* of successes, $x$, is normal, the *proportion* of success in $n$ trials, $x/n$, must also be normally distributed. Hence, at this point we want to show how the standardized variable $z$ can be used to approximate the *proportion* of successes in $n$ trials in precisely the same manner as it approximates the number of successes.

First, we denote our estimator of the population proportion parameter $p$ as $\hat{p}$, where $\hat{p} = x/n$. This estimator has already been shown earlier in this chapter to be unbiased; that is, $E[\hat{p}] = E[x/n] = p$. We also showed that the variance of $\hat{p}$ is $V[\hat{p}] = pq/n$. Unfortunately, we can't use this variance in constructing a confidence interval because it depends on the unknown parameter, $p$, which we are trying to estimate. The next best thing we can do is use our point estimate of $p$, $\hat{p} = x/n$, in place of $p$, and $(1 - \hat{p}) = 1 - x/n$ in place of $q$. Thus, our estimate of the variance is:

$$\text{Estimated variance:} \qquad \frac{\hat{p}(1 - \hat{p})}{n}. \tag{7.3}$$

Since a standardized normal variable is obtained by taking a normal variable ($x/n$ in this case), subtracting its mean ($E[x/n] = p$) and dividing by its (estimated) standard deviation ($\sqrt{\hat{p}(1 - \hat{p})/n}$), the appropriate $z$-variable is:

$$z = \frac{\hat{p} - p}{\sqrt{(\hat{p}(1 - \hat{p})/n)}}. \tag{7.4}$$

Formula (7.4) can now be used to construct a confidence interval in almost the same fashion as we did for $\mu$ in Section 7.4. First, form a $100(1 - \alpha)\%$ confidence interval for $p - \hat{p}$:

$$-z_{\alpha/2} \sqrt{\frac{\hat{p}(1 - \hat{p})}{n}} < p - \hat{p} < z_{\alpha/2} \sqrt{\frac{\hat{p}(1 - \hat{p})}{n}}.$$

Solving this expression for $p$ yields the desired $100(1 - \alpha)\%$ confidence interval.

---

$100(1 - \alpha)\%$ *confidence interval for* $p$:

$$\hat{p} - z_{\alpha/2} \sqrt{\frac{\hat{p}(1 - \hat{p})}{n}} < p < \hat{p} + z_{\alpha/2} \sqrt{\frac{\hat{p}(1 - \hat{p})}{n}}. \tag{7.5}$$

To illustrate Formula (7.5), suppose we would like to estimate the proportion of families, from some population, who own two or more cars. A random sample of $n = 144$ families shows that $x = 48$ families have two or more cars. Thus, the best point estimate for the population proportion is the sample proportion $\hat{p} = x/n = 48/144 = 1/3$. Our best estimate of the population variance is thus

$$\sqrt{\frac{\hat{p}(1 - \hat{p})}{n}} = \sqrt{\frac{(\frac{1}{3})(\frac{2}{3})}{144}} = 0.0393.$$

Suppose we now use these values and Formula (7.5) to construct a 95-percent confidence interval for $p$ (remember, $z_{\alpha/2} = 1.96$ for $\alpha/2 = 0.025$):

$$\tfrac{1}{3} - 1.96(0.0393) < p < \tfrac{1}{3} + 1.96(0.0393),$$
$$0.333 - 0.077 < p < 0.333 + 0.077,$$
$$0.256 < p < 0.410.$$

We can thus conclude, with 95-percent confidence, that the population proportion of families who own two or more cars is between 25.6 and 41.0 percent.

In some applications, especially in product control, a confidence interval for $p$ can be used to find the confidence interval for the *number* of defectives in a given shipping lot of very large size. Consider a problem of a manufacturer who produces ballpoint pen cartridges. Suppose each shipment includes 10,000 cartridges. The producer desires some control over these shipments so that no shipment will contain an excessive number of defective cartridges. He decides to take a random sample of 400 cartridges for inspection from a shipping lot of 10,000, and finds nine defectives ($x = 9$). On the basis of this result, he desires to obtain a 90% confidence interval for the number of defectives in the entire shipment. The random variable $x$ = number of defectives can be approximated by the normal distribution; hence a 90% confidence interval on the proportion defective is obtained by using Formula (7.5). In this case $n = 400$ and $x = 9$, so $\hat{p} = x/n = 9/400 = 0.0225$; since $\alpha = 0.10$, $\pm z_{\alpha/2} = \pm z_{0.05} = \pm 1.645$, and the appropriate confidence interval is:

$$0.0225 - 1.645 \sqrt{\frac{(0.0225)(0.9775)}{400}} < p < 0.0225 + 1.645 \sqrt{\frac{(0.0225)(0.9775)}{400}},$$
$$0.0225 - 1.645(0.0074) < p < 0.0225 + 1.645(0.0074),$$
$$0.0103 < p < 0.0347.$$

He may thus conclude, with 90-percent confidence, that the population contains between 1.03 and 3.47 percent defectives.

Several considerations and qualifications related to this result are worth mentioning. A rather large $\alpha = 0.10$ was used in this case because the producer is willing to put up with a fairly large risk of being in error in assuming the true proportion defective lies in the interval. His cost of producing a ballpoint pen cartridge is quite small, so a rough interval estimate is sufficient for him. If he were inspecting \$40 stereo cartridges rather than \$0.40 ballpoint-pen cartridges, he might want a much smaller risk. We know, in the present case, that the sampling was from a finite population without replacement, and the hypergeometric distribution would be technically accurate, not the binomial. However, $N = 10,000$ is so large that it makes a negligible difference. Finally, using our lowest extreme estimate for $p$, 0.0103, we check the reasonableness of our normal approximation by computing $npq = 400(0.0103)(0.9897) = 4.08 > 3$. The approximation is acceptable although a larger sample might be recommended for greater precision.

Now to complete our solution to the problem, it is easy to convert our result to a 90-percent confidence interval on the number of defectives in the shipment of size 10,000. If $p$ is the proportion defective, then the number defective is $Np = 10,000p$. Using the upper and lower limits for $p$, we conclude with 90-percent confidence, that the number of defectives in the shipment is between 103 and 347. If the producer thinks that this is a tolerable number, then he can release the lot for shipment. He has saved much in inspection costs and has maintained some quality control over his shipment. He is accepting a risk of $\alpha = 0.10$ that this interval is wrong. Realistically, he is satisfied if the true number of defectives is really less than 103, so he is only accepting a risk of $\alpha/2 = 0.05$ that the shipment contains more defectives than 347. That is, on the average, over many such shipments, samples, and statistical inferences, he would be releasing shipments with more defectives than he estimated only five times out of 100. In terms of a one-sided confidence interval, he has 95% confidence that the number of defectives is less than 347 in this example. Such one-sided statements are often used, rather than the two-sided confidence interval, in situations where the error in one direction is unimportant.

## 7.7 DETERMINING THE SIZE OF THE SAMPLE ($n$)

Thus far we have calculated the width of each confidence interval based on the assumption that the sample size, $n$, is known. In many practical situations, however, the decision-maker does not know what size sample is best for him. Instead, he may prefer to specify the width of the interval he wants, and use this information to solve for $n$. The conventional approach to this problem of solving for $n$ in these cases is to ask the decision-maker two questions:

1. What level of confidence does he want to have (i.e., the value of $100(1 - \alpha)$)?

2. What is the *maximum* difference ($D$) he wants to permit between the estimate of the population parameter, $\hat{\theta}$, and the true population parameter, $\theta$ (i.e., the value of $D$ is the amount of "error" he wants to allow in estimating $\theta$, where $|\hat{\theta} - \theta| \leqslant D$)?

We will consider confidence intervals for both the population mean $\mu$ and the population proportion $p$.

### For Statistical Inference on $\mu$

*Population normal, $\sigma$ known.* First, we consider the problem of determining $n$ when the decision-maker wants a $100(1 - \alpha)\%$ confidence interval for $\mu$, given that the parent population has a normal distribution with a known standard deviation. In this case we know the variable

$$z = \frac{\bar{x} - \mu}{\sigma/\sqrt{n}}$$

is $N(0, 1)$. Now if the required level of confidence is $1 - \alpha$, then the above equation results in the following $100(1 - \alpha)\%$ confidence interval for $\bar{x} - \mu$:

$$-z_{\alpha/2}\frac{\sigma}{\sqrt{n}} \leqslant \bar{x} - \mu \leqslant z_{\alpha/2}\frac{\sigma}{\sqrt{n}}. \tag{7.6}$$

Since the normal distribution is symmetric, we can concentrate on the righthand inequality, $\bar{x} - \mu \leqslant z_{\alpha/2}(\sigma/\sqrt{n})$. This inequality says that the largest value that $\bar{x} - \mu$ can assume is $z_{\alpha/2}(\sigma/\sqrt{n})$. But we also know that our decision-maker says that the largest value he wants $\bar{x} - \mu$ to assume is some amount $D$. Hence, we can write

$$D = z_{\alpha/2}\frac{\sigma}{\sqrt{n}}.$$

Solving this relationship for $n$ gives the value of $n$ which will assure him, with $100(1 - \alpha)\%$ confidence, that $\bar{x} - \mu$ will be no larger than $D$, on the average. Solving we get:

$$\boxed{\text{Required sample size:} \qquad n = \frac{z_{\alpha/2}^2 \sigma^2}{D^2}} \tag{7.7}$$

To illustrate (7.7), suppose that, in the I.Q. problem of Section 7.4, we wanted to find a 95-percent confidence interval for the mean I.Q. in such a manner that our sample result, $\bar{x}$, and the population mean differ by no more than $\frac{1}{2}$ a point; that is $|\bar{x} - \mu| < D = \frac{1}{2}$. Assuming, as before, that the parent population is normal, and $\sigma = 5.4$, how large should $n$ be to satisfy these conditions?

From Table III, $z_{\alpha/2} = 1.96$ when $\alpha/2 = 0.025$. Substituting this value and $D = \frac{1}{2}$, $\sigma = 5.4$, into Formula (7.7) yields the appropriate value for $n$:

$$n = \frac{(1.96)^2(5.4)^2}{(1/2)^2} = 448.08.$$

We always round *up* in this type of a problem to be assured that sample size is large enough; hence, a random sample of at least 449 students is needed to be assured that 95-percent of the time the value of $\bar{x}$ will be within $\frac{1}{2}$ of a point of the true population mean, $\mu$.

*Population not normal, $\sigma$ known.* If the population is not assumed to be normal but the standard deviation is known, the same method as above can be used to determine the minimal sample size necessary to satisfy the conditions of confidence and accuracy. By the central-limit theorem, we know that the distribution of sample means approaches the normal distribution. Thus, once the necessary sample size is obtained, we can check to see if that size $n$ exceeds 30 and if it does, then we are confident our method of solution was appropriate.

*Population normal, $\sigma$ unknown.* If the population is normal but the standard deviation is unknown, then the appropriate statistic to use is the *t* variable, $t = (\bar{x} - \mu)/(s\sqrt{n})$. Again, the maximum difference $|\bar{x} - \mu| = D$ is obtained from the decision-maker. In this case we are stuck for a value of $s$, since $s$ must be calculated from a sample, and we haven't taken a sample yet (the whole purpose is to decide what sample size to take). To make matters even worse, the appropriate *t*-value to use in calculating a $100(1 - \alpha)\%$ confidence interval is $t_{\alpha/2,\ \nu}$, which again depends on the unknown sample size. To make a long story short, we conclude this paragraph by saying that the solution for $n$ in this case is not a direct process, but can be achieved by a succession of iterative steps using sequential sampling, which we will not present here.

### For Statistical Inference on *p*

The size of the sample can be determined in a confidence interval for the binomial parameter $p$ if the maximum difference between $\hat{p}$ and $p$, $|\hat{p} - p| = D$, is specified in advance. In sampling the number of two-car owners, for instance, it may be desirable to restrict the maximum error to, say, three percentage points, $D = 0.03$. In this case we can *not* use the estimate of the standard deviation used previously in working with a population proportion, $\sqrt{\hat{p}(1 - \hat{p})/n}$, as $\hat{p}$ isn't known yet (we haven't taken the sample). If there is a good estimate of $p$ available from other sources, then these may be used. If not, then the estimate of standard deviation to be used is $\sqrt{1/4n}$. This estimate is used because it represents the largest (and hence most conservative) value the standard deviation can ever assume, since $\sqrt{pq/n}$ will always be at its maximum when $p = q = \frac{1}{2}$ (try several values to convince yourself). Now, if we substitute $\sqrt{1/4n}$ in place of the denominator in the variable $z = (\hat{p} - p)/\sqrt{pq/n}$, we obtain

$$z = \frac{(\hat{p} - p)}{\sqrt{1/4n}}.$$

This relationship can be used to derive the following $100(1 - \alpha)\%$ confidence interval for $\hat{p} - p$:

$$-z_{\alpha/2}\sqrt{1/4n} \leqslant \hat{p} - p \leqslant z_{\alpha/2}\sqrt{1/4n}.$$

Again, setting $\hat{p} - p$ at its maximum value $D$, we have

$$D = z_{\alpha/2}\sqrt{1/4n},$$

or

$$D^2 = z_{\alpha/2}^2 \frac{1}{4n}.$$

Therefore,

$$\textit{Required sample size:} \quad n = \frac{z_{\alpha/2}^2}{4D^2}. \tag{7.8}$$

Returning to the example problem of estimating "the proportion of families owning two or more cars" within three percentage points with 95-percent confidence, we know $z_{\alpha/2} = 1.96$ (since $\alpha = 0.05$), and $D = 0.03$. Thus,

$$n = \frac{1}{4}\left(\frac{1.96^2}{0.03^2}\right) = 1067.1$$

We thus conclude that a sample of 1068 people is needed to assure an error of less than three percentage points 95 percent of the time.

## *7.8  CONFIDENCE INTERVAL FOR $\sigma^2$

Under some circumstances it may be desirable to construct a confidence interval for an estimate of an unknown population variance. As we said before, the telephone company is often interested in the *variability* of the length of telephone conversations, and a contractor purchasing steel girders may be interested in the *variance* of their tensile strengths. Or a government economist may be just as concerned about the *variability* of taxes paid among individuals as he is about the average tax paid, because the income-redistribution effect of taxation is very important. In these cases it may be important to establish limits on just how large or small $\sigma^2$ might be; that is, to determine a confidence interval for the population variance.

To construct a $100(1 - \alpha)\%$ confidence interval for $\sigma^2$ when sampling from a *normal* population, recall, from Section 6.9, that the variable $(n - 1)s^2/\sigma^2$ has a chi-square distribution with $(n - 1)$ degrees of freedom. Denote that point cutting off $\alpha/2$ of the area of the righthand side of a chi-square distribution with $v = n - 1$

---

* This section, which may be omitted without loss in continuity, assumes the reader has covered Section 6.9.

degrees of freedom as $\chi^2_{\alpha/2,\,v}$ [that is, $P(\chi^2 > \chi^2_{\alpha/2,\,v}) = \alpha/2$]. Since the chi-square distribution is not symmetrical, $-\chi^2_{\alpha/2,\,v}$ does *not* give the appropriate value cutting off $\alpha/2$ of the lefthand side of this distribution. The point that does give the correct probability is that value of $\chi^2$ cutting off $1 - \alpha/2$ of the righthand tail, or $\chi^2_{1-\alpha/2,\,v}$. That is,

$$P(\chi^2 < \chi^2_{\alpha/2,\,v}) = 1.0 - F(\chi^2_{\alpha/2,\,v}) = \alpha/2$$

and

$$P(\chi^2_{1-\alpha/2,\,v} < \chi^2) = F(\chi^2_{1-\alpha/2,\,v}) = \alpha/2.$$

The interval between the points $\chi^2_{\alpha/2,\,v}$ and $\chi^2_{1-\alpha/2,\,v}$ thus contains $(1 - \alpha)$ probability. It is now possible to define a $100(1 - \alpha)\%$ confidence interval for the variable $(n - 1)s^2/\sigma^2$:

$$\chi^2_{1-\alpha/2,\,v} < \frac{(n-1)s^2}{\sigma^2} < \chi^2_{\alpha/2,\,v}. \tag{7.9}$$

Again, solving these inequalities for the unknown parameter $\sigma^2$ gives:

---

$100(1 - \alpha)$ *percent confidence interval for* $\sigma^2$, *parent population normal:*

$$\frac{(n-1)s^2}{\chi^2_{\alpha/2,\,v}} < \sigma^2 < \frac{(n-1)s^2}{\chi^2_{1-\alpha/2,\,v}}. \tag{7.10}$$

---

It is important to note that, in going from (7.9) to (7.10), the *sense* of the inequalities has changed. Thus, the larger value of $\chi^2$ now appears in the denominator of the *lower* endpoint for $\sigma^2$, while the smaller value of $\chi^2$ is in the denominator of the term giving the upper endpoint for $\sigma^2$.

Consider once again the population of basketball scores, and assume that this time we are interested in estimating the variance of the population, using the same sample of size $n = 10$ reported in Table 7.2. The best point estimate of $\sigma^2$ is $s^2$ which is computed from this sample to be $s^2 = \frac{1}{9}(590) = 65.55$. To achieve an interval for $\sigma^2$ with a known level of confidence and a known risk of error, let us compute a 95% confidence interval for $\sigma^2$.

Figure 7.6 illustrates the relevant chi-square distribution for $v = n - 1 = 9$ degrees of freedom and the cutoff values for $\alpha/2 = 0.025$ and $(1 - \alpha/2) = 0.975$. These values are not equidistant from the mean of the chi-square $(E[\chi^2] = v = 9)$ because the $\chi^2$ is a skewed distribution. The cutoff values are found from Table IV in terms of the cumulative distribution for $v = 9$ with $F(\chi^2) = 0.975$ and $F(\chi^2) = 0.025$, to be 19.02 and 2.70, respectively. Thus $P(2.70 < \chi^2_9 < 19.02) = 1 - \alpha = 0.95$.

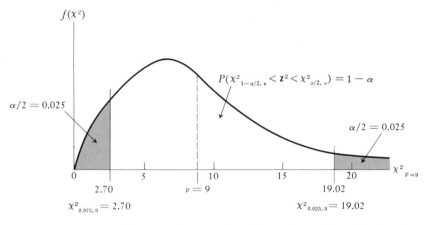

**Fig. 7.6.** The values of $\chi^2_{\alpha/2, v}$ and $\chi^2_{1-\alpha/2, v}$ , cutting off a total area of $\alpha = 0.05$ from the chi-square distribution with $n = 10$, leaving an interval including 95 percent of the probability in the middle.

Substituting the values for $\chi^2_{0.025,\, 9} = 19.02$, $\chi^2_{0.975,\, 9} = 2.70$, $n = 10$, and $s^2 = 65.55$ into Formula (7.5) gives,

$$\frac{9(65.55)}{19.02} < \sigma^2 < \frac{9(65.55)}{2.70},$$

$$31.02 < \sigma^2 < 218.50.$$

On the basis of this sample of size 10, one can infer with 95-percent confidence that the population variance lies between 31.02 and 218.50. There is a 2.5 percent chance of error that the true $\sigma^2$ may be greater than 218.50 and a 2.5 percent chance of error that it may be less than 31.02.

### 7.9 SUMMARY

Table 7.3 summarizes the test statistics and confidence intervals described in this chapter for various situations of statistical inference.

### REVIEW PROBLEMS

1. Describe briefly three properties of a "good estimator."
2. a) What factors must be established before a confidence interval can be constructed?
   b) What does the Greek letter $\alpha$ represent? How would you suggest the level of $\alpha$ be established?

**Table 7.3** Summary of test statistics and confidence intervals for statistical inference

| Unknown parameter | Population characteristics and other description | Reference sections | Test statistic involving the best sample estimator | Endpoints for $100(1 - \alpha)\%$ confidence interval | Reference formula |
|---|---|---|---|---|---|
| (1) Mean $\mu$ | Population $N(\mu, \sigma^2)$ or sample size $n \geqslant 30$, $\sigma$ *known* | 6.5 7.4 | $z = (\bar{x} - \mu)/(\sigma/\sqrt{n})$ | $\bar{x} \pm z_{\alpha/2}(\sigma/\sqrt{n})$ | (7.1) |
| (2) Mean $\mu$ | Population $N(\mu, \sigma^2)$ *σ unknown* | 6.8 7.5 | $t_{v.\mathrm{d.f.}} = (\bar{x} - \mu)/(s/\sqrt{n})$ where $v = n - 1$ | $\bar{x} \pm t_{\alpha/2,\, v}(s/\sqrt{n})$ | (7.2) |
| (3) Variance $\sigma^2$ | Population $N(\mu, \sigma^2)$ | 6.9 7.8 | $\chi^2_{v.\mathrm{d.f.}} = \dfrac{(n-1)s^2}{\sigma^2}$ where $v = n - 1$ | $\dfrac{(n-1)s^2}{\chi^2_{\alpha/2,\, v}} ; \dfrac{(n-1)s^2}{\chi^2_{1-\alpha/2,\, v}}$ | (7.10) |
| (4) Proportion $p$ | Repeated independent trials, $npq \geqslant 3$, $\hat{p} = x/n$ | 7.6 | $z = (\hat{p} - p)/\sqrt{\dfrac{pq}{n}}$ | $\hat{p} \pm z_{\alpha/2}\sqrt{\dfrac{\hat{p}(1 - \hat{p})}{n}}$ | (7.5) |

3. Differentiate between:

   a) a point estimate and an interval estimate;

   b) unbiasedness and consistency.

4. Find 90% confidence limits for the population mean based on a sample of size 36, if the sample mean is 950 and the population standard deviation is 70. Explain what this confidence interval means. Assume a normal parent population.

5. a) What is meant by a "$100(1 - \alpha)\%$ confidence interval for $\mu$"?

   b) Write down the expression for calculating a $100(1 - \alpha)\%$ confidence interval for $\mu$ on the basis of a random sample of size $n$ (large) from a normal population with mean $\mu$ and standard deviation $\sigma$.

   c) Write down the expression for calculating a $100(1 - \alpha)\%$ confidence interval for $\mu$ on the basis of a random sample of size $n$ (small) from a normal population with mean $\mu$ and unknown standard deviation.

6. Suppose the following nine values represent random observations from a normal parent population: 1, 5, 9, 8, 4, 0, 2, 4, 3. Construct a 99-percent confidence interval for the mean of the parent population.

7. A production assembly process is scheduled as a 20-minute operation. A time study, using 9 randomly selected observations unknown to the employees involved, shows a sample average of 24.3 minutes; the standard deviation is $\sigma = 6$ min. Find 90% confidence limits on the true average time used for this operation, and argue for or against revising the 20-minute standard in light of your results.

8. In one university the average height in a random sample of 256 male students was 70 in., with a standard deviation of 2 in.

   a) Construct 95- and 99-percent confidence intervals for the average height of all male students.

   b) Repeat part (a), assuming the sample size had been 100 rather than 256.

9. A survey of 16 doctors selected at random revealed that the average annual consumption of aspirin tablets per doctor was 84 with a standard deviation of 20. Establish 90% confidence limits for the average annual per capita consumption of aspirin tablets of all doctors.

10. Twenty-five loan applications in a bank were randomly selected for the purpose of determining the average dollar amount requested for each loan.

    a) Construct a 95-percent confidence interval for $\mu$, assuming that the sample mean was $\bar{x} = \$900$ and the sample standard deviation was $\$150$. Use the $t$-distribution.

    b) Repeat part (a), using a normal-distribution approximation rather than the $t$-distribution. Compare your answer to part (a) with the confidence interval in this part.

11. A survey result of 900 *Playboy* readers indicates that 40% finished at least two years of college. Set 95% confidence limits on the true proportion of all *Playboy* readers with this background.

12. In order to estimate the percent of all housewives who use "Wash Away" detergent, 196 housewives were randomly selected and interviewed. If 108 of these housewives use this

product, what would be a 99-percent confidence interval for the population percent of housewives who use "Wash Away"?

13. A television manufacturer would like to know what proportion of television set owners have color sets. In a sample of 100 randomly selected owners, 40 percent were found to own color sets. Construct a 95-percent confidence interval for the population proportion of television owners who have color sets.

14. The total particle removal (TPR) in a certain cigarette filter can be measured on a smoking machine, using a sample of these cigarettes. The standard deviation for TPR for the population of such cigarettes is 2 mg. What size sample is necessary to estimate the average TPR for all such cigarettes within 0.4 mg with 90% confidence?

15. A population of families have an unknown mean income $\mu$; the standard deviation of these incomes is known to be $1,000. How large a random sample would be needed to determine the mean income if it is desired that the probability of a sampling error of more than $50 be less than 5 percent?

16. Assume that a variable $x$ has a normal distribution with a mean of 30 and standard deviation of 3.
   a) If a random sample of size 36 is drawn from this population, what is the probability that the sample mean is less than 30.75?
   b) What size sample would I need to draw to have the same probability as you obtained in part (a), but this time applying to the statement, "$\bar{x} < 30.45$"?

17. A supermarket manager wishes to make a sample estimate of the average time a customer spends at the checkout register. He knows from past experience that the standard deviation is $2\frac{1}{2}$ minutes, and he would like to estimate the mean checkout time within plus-and-minus $\frac{1}{2}$ minute. He specifies the confidence level of 99%. What is the sample size required for him to obtain this estimate?

18. I wish to estimate the proportion of defectives in a large production lot within 0.005 of the true proportion, with 90% confidence. From past experience, it is believed that the proportion defective is about $p = 0.02$. How large a sample must be used?

19. You want to find some measure of dispersion for the distribution of a certain population. The best measure would be __(a)__. This is a better measure of dispersion than __(b)__ because __(c)__. However, you cannot take time to measure all the items in the entire population. Thus, you will use the logical process of __(d)__ to make conclusions about the population dispersion. In following this process, you will select a sample and measure __(e)__. This will be a good estimate of (a) because __(f)__.

20. A sample of size 15 has standard deviation 3. Find 90% confidence limits on the true population standard deviation.

21. A manufacturer of steel washers periodically samples the washers being produced as a check on the variability of the inside diameter of the washers. A sample of size 20 was checked and found to have a standard deviation of 0.002 in. On the basis of this sample, find a 95-percent confidence interval for the true variance.

## EXERCISES

22. Suppose that $x_1$ and $x_2$ are independent random variables having the Poisson probability distribution with parameter $\lambda$. Show that the mean of these variables is an unbiased estimator of $\lambda$.

23. If $\hat{\theta}$ is an unbiased estimator of $\theta$, under what conditions will $\hat{\theta}^2$ be an unbiased estimator of $\theta^2$?

24. Which of the properties of a good estimator does $\bar{x}$ have? Prove as many of these as you can.

25. At a certain university, there are 53 fraternities and 46 sororities. The average grade-point index of members of this group is 2.25, with a standard deviation of 0.30, and the distribution is quite positively skewed. Suppose nine of these members are selected at random for a certain project; what is the probability that their average grade-point index is better than 2.55? Can you tell me with 99% confidence, what values I should expect this average to fall between?

26. The quarterly sales of five different Softee Hamburger drive-in restaurants are 170, 160, 140, 180, and 140 (in thousands of dollars), respectively. You desire to make an estimate of the average sales for this quarter, for each such outlet in the total chain.

    a) Can you make an estimate on the basis of this sample information, which will have probability equal to 1.00 of being correct? Why or why not?

    b) Make an interval estimate by using the sample mode plus-or-minus $\frac{1}{3}$ its range.

    c) Make an interval estimate by using the mean plus-or-minus the standard deviation of the sample.

    d) In probability, which of these estimates is best and why, and how could the best one be made even better?

27. Suppose you are purchasing agent for the State, and you contract with Batpower Company to buy 10,000 batteries according to specifications in the contract. You check 300 batteries and find 42 defective. Considering sampling error, can you be 99% confident that the entire lot of 10,000 is more than 10% defective?

28. Suppose the annual earnings of college graduates has an unknown distribution with $\sigma = \$1200$.

    a) Based on a sample of $n = 16$ with $\bar{x} = 8000$, find a 98% confidence interval on $\mu$.

    b) How many graduates would I need to sample, so that I can have 98% confidence that the sample mean is within sixty dollars of the true mean?

29. Repeat part (a) of Exercise 28 if $\sigma$ is unknown, but the sample standard deviation is 1200.

30. Suppose that Senator Fogbound has engaged a team of public-opinion surveyors in an effort to determine the percent of the population who favor his stand on a current issue. The survey company will conduct a random survey of public opinion at a cost of 35 cents per interview. How much will it cost the senator if he insists that the sampling error be less than 5 percent 95% of the time, and if he has no idea of the percent of the population favoring his stand?

31. A random sample of the I.Q. of 5 percent of the students in a small university resulted in a mean I.Q. of 120 with a standard deviation of 3. How many students are enrolled at this university if a 99-percent confidence interval for the mean I.Q. of all students in the university extends three-tenths of a unit to either side of the sample mean? (Assume that sampling is done with replacement.)

32. A machine producing ball bearings is stopped periodically so that the diameter of the bearings produced can be checked for accuracy. In this particular case it is not the mean diameter which is of concern, but the variability of the diameters which must be checked. Suppose that a sample of size $n = 61$ is taken and the variance of the diameters of the bearings sampled is found to be 9.4 mm.

   a) Construct a 95-percent confidence interval for $\sigma^2$.
   b) Assume that, if this machine is working properly, the variance of the bearings produced will be 8.0 mm. Does this sample indicate that the machine is working improperly? Explain.

33. Redo Exercise 31, assuming that a random sample of 50 percent of all students was tested, and that sampling was without replacement (remember to correct for the finite population).

# Hypothesis Testing: One-sample Tests

## 8.1 INTRODUCTION AND BASIC CONCEPTS

The procedures presented in Chapter 7 describe the process of making both point and interval estimates of population parameters. As we saw then, the advantage of using an interval estimate is that its construction permits the expression of uncertainty about the true value of the population parameter. Another advantage of confidence intervals concerns testing the validity of *assumed* values of the population parameters. These assumptions are usually referred to as statistical hypotheses, and determining the validity of an assumption of this nature is called the *test of a statistical hypothesis*, or simply *hypothesis testing.*

The purpose of hypothesis testing is to choose between two conflicting hypotheses about the value of a population parameter. The choice cannot be made with certainty unless the entire population is examined. Usually both the value of an estimator (based on a sample drawn from the population) and the theoretical probability distribution of that estimator are used to make the judgment between the conflicting hypotheses. Thus, the application of hypothesis testing brings together many of our previous topics—calculation of sample statistics, random variables, probability distributions, and statistical inference.

For example, an engineer in charge of the quality of items produced on an assembly line may wish to choose between the two conflicting hypotheses, $p \leqslant 0.10$ or $p > 0.10$, where $p$ is the proportion defective in the population of items produced. Let's assume he cannot choose to check every item of the population because the cost and time delay would be too expensive. Instead, he wishes to statistically choose between the hypotheses on the basis of a sample in which he has observed the proportion of defective sample items, $\hat{p}$. From knowledge of the probability distribution of $\hat{p}$ (i.e., the binomial distribution) and knowledge of statistical inference (i.e., $\hat{p}$ is an unbiased estimator of $p$), he can construct a test procedure to analyze the sample evidence and judge which of the two hypotheses is more likely to be acceptable and which should be rejected (accepting some risk of error in his decision). In the following section the detailed general procedure for constructing such a test of hypothesis is presented. Subsequent sections give specific examples of the use of this procedure in some commonly occurring situations.

### Types of Hypotheses

In specifying the conflicting hypotheses about the values that a population parameter might assume, it is convenient to distinguish between *simple hypotheses* and *composite hypotheses.* In a simple hypothesis, only one value of the population parameter is specified. If the engineer hypothesizes in his test that the probability of a defective item is $p = 0.10$, this represents a simple hypothesis. A psychologist investigating the hypothesis that the mean I.Q. of a group is $\mu = 115$ is testing a simple hypothesis. If the exact difference between two population parameters is specified such as $\mu_1 - \mu_2 = 0$, this also represents a simple hypothesis. Composite hypotheses, on the other hand, specify not just one value but a *range* of values which the population parameter might assume. The hypotheses that $p \leqslant 0.10$, that $\mu \neq 100$, and that $\mu_1 - \mu_2 \neq 0$ all represent composite hypotheses because in each case more than one value is specified. As you might suspect, assumptions in the form of simple hypotheses are, in general, easier to test than are composite hypotheses. In the former case we need to determine only whether or not the population parameter equals the specified value, while in the latter case it is necessary to determine whether or not the population parameter takes on any one of what may be a very large (or even infinite) number of values.

The two conflicting (i.e., mutually exclusive) hypotheses in a statistical test are normally referred to as the *null hypothesis* and the *alternative hypothesis*. The term "null hypothesis" developed from early work in the theory of hypothesis testing, in which this hypothesis corresponded to a theory about a population parameter which the researcher was *hoping would be rejected*. That is, the null hypothesis specified those values which the researcher thought did *not* represent the true value of the parameter (hence the word "null," which means invalid, void, or amounting to nothing). The *alternative hypothesis* specifies those values of the parameter that the researcher believes *do* hold true, and, of course, he hopes that the sample data lead to acceptance of this hypothesis as true.

Nowadays it is usually accepted common practice *not* to associate any special meaning to the null or alternative hypotheses, but merely to let these terms represent two different assumptions about the population parameter. We shall see in a moment that for statistical convenience it may make a difference which hypothesis is called the null and which is called the alternative. The null and alternative hypotheses are distinguished by the use of two different symbols, $H_0$ representing the null hypothesis and $H_a$ the alternative hypothesis. Thus, a psychologist who wishes to test whether or not a certain class of people have a mean I.Q. higher than 100 might establish the following null and alternative hypotheses:

$$H_0: \mu \leqslant 100 \qquad \text{(Null hypothesis)},$$
$$H_a: \mu > 100 \qquad \text{(Alternative hypothesis)}.$$

On the other hand, if he wishes to test for differences between the mean I.Q. of two groups, this psychologist might want to establish the null hypothesis that the two groups have equal means and the alternative hypothesis that their means are not equal:

$$H_0: \mu_1 - \mu_2 = 0 \qquad \text{(Null hypothesis)},$$
$$H_a: \mu_1 - \mu_2 \neq 0 \qquad \text{(Alternative hypothesis)}.$$

The null hypothesis and the alternative hypothesis can both be either simple or composite. The simple null hypothesis $H_0$: $\mu = 100$, for example, may be tested against a simple alternative hypothesis such as $H_a$: $\mu = 120$ or $H_a$: $\mu = 75$, or against a composite hypothesis such as $H_a$: $\mu \neq 100$, $H_a$: $\mu > 100$, or $H_a$: $\mu < 75$. Similarly, the composite null hypothesis $H_0$: $\mu < 100$ may be tested against a simple alternative, such as $H_a$: $\mu = 100$ or $H_a$: $\mu = 120$, or against a composite hypothesis, such as $H_a$: $\mu \geq 100$ or $H_a$: $\mu \geq 120$.

Regardless of the form of the two hypotheses, it is extremely important to remember that the true value of the population parameter under consideration *must* be either in the set specified by $H_0$ or in the set specified by $H_a$. By testing $H_0$: $\mu = 100$ against $H_a$: $\mu = 120$, for example, one is asserting that the true value of $\mu$ equals either 100 or 120, and that *no other values are possible*. One means for assuring that either $H_0$ or $H_a$ contains the true value of $\theta$ is to let these two sets be *complementary*. That is, if the null hypothesis is $H_0$: $\mu = 100$, then the alternative hypothesis would be $H_a$: $\mu \neq 100$; or if $H_0$: $\mu \leq 100$, then $H_a$: $\mu > 100$. From a statistical point of view, the easiest form to handle is a simple null hypothesis versus a simple alternative hypothesis. Unfortunately, most real-world problems cannot be stated in this form, but instead involve a composite null or a composite alternative hypothesis, or both. If a particular problem cannot be stated as a test between two simple hypotheses, then the next best alternative is to test a simple null hypothesis against a composite alternative. In other words, one should always try to structure the problem so that the null hypothesis is a simple rather than a composite hypothesis.

**One- and Two-sided Tests**

If one is fortunate enough to be able to construct the test of hypotheses so that the null hypothesis is simple, then the composite alternative hypothesis may specify one or more values for the population parameter, and these values (or value) could lie entirely above, or entirely below, or on both sides of the value specified by the null hypothesis. A statistical test in which the alternative hypothesis specifies that the population parameter lies entirely above or entirely below the value specified in the null hypothesis is called a *one-sided test*; an alternative hypothesis which does not specify that the parameter lies on one particular side of the value indicated by $H_0$ is called a *two-sided test*. Thus, $H_0$: $\mu = 100$ tested against $H_a$: $\mu > 100$ is a one-sided test since $H_a$ specifies that $\mu$ lies on one particular side of 100. The same null hypothesis tested against $H_a$: $\mu \neq 100$ is a two-sided test since $\mu$ can lie on *either* side of 100.

**The Form of the Decision Problem**

The decision problem involved in hypothesis testing is a choice between two mutually exclusive propositions about a population parameter when we are faced with the uncertainty inherent in sampling from a population. The decision-maker has only the sample evidence on which to base his choice of accepting the null hypothesis (which is equivalent to rejecting the alternate hypothesis) or rejecting the null hypothesis (accepting $H_a$). The standard method of solving this decision problem is to first assume that the null hypothesis is true (just as we presume innocence until proven guilty in a court of law). Then, using the probability theory from Chapters 4, 5, and 6, we can establish the criteria which will be used to decide whether we think there is sufficient evidence to declare $H_0$ false. A sample is then taken, and the sample evidence is compared to the criteria, to decide whether to accept or reject $H_0$.

Contrary to practice in a court of law, where innocence is maintained as long as any reasonable doubt about guilt remains, in hypothesis testing we reject $H_0$ on the basis of only a reasonable doubt about its truth. With such a procedure, the probability value which we use to conclude that there is a reasonable doubt about the truth of $H_0$ is critical. Moreover, since the decision to accept or reject $H_0$ is based on probabilities and not on certainty, there are chances of error in the decision. Specifically, there are two types of error: (1) one might decide to accept the null hypothesis when this hypothesis is not true (called a *Type II error*); or (2) one might decide to reject the null hypothesis when this hypothesis is, in fact, true (called a *Type I error*).

|  | The true situation may be: | |
|---|---|---|
|  | $H_0$ is true | $H_0$ is false |
| Accept $H_0$ | Correct decision | Incorrect decision (Type II error) |
| Reject $H_0$ | Incorrect decision (Type I error) | Correct decision |

Fig. 8.1. The four possible decision outcomes in hypothesis testing.

There are four possible situations in hypothesis testing, as shown in Fig. 8.1. This figure presents the basic decision problem in hypothesis testing with reference only to the null hypothesis. A good exercise for the reader would be to construct a similar figure, but making reference only to $H_a$ (remembering that accepting $H_0$ implies rejecting $H_a$, and vice versa).

A Type I error is committed by the psychologist who rejects the null hypothesis that the mean I.Q. of a given group equals 100 (in favor of some alternative hypothesis) when the true population mean does equal 100. Or, if he establishes the null hypothesis $H_0$: $\mu < 100$, and this hypothesis is rejected (in favor of the alternative hypothesis $H_a$: $\mu \geqslant 100$), then a Type I error is committed if the true mean is less than 100. On the other hand, a Type II error would be committed if the null hypothesis $H_0$: $\mu < 100$ is accepted when the alternative hypothesis $H_a$: $\mu \geqslant 100$ is true. Another example of Type I and II errors can be given by again recalling the quality-control problem of Section 4.3, where defective items occur in a production process with either probability $p = 0.10$ or $p = 0.25$. Suppose we establish the null hypothesis $H_0$: $p = 0.10$ and the alternative hypothesis $H_a$: $p = 0.25$. A Type I error is committed if our decision rule leads to acceptance of the fact that a major adjustment is needed when only a minor adjustment ($p = 0.10$) is necessary, and a Type II error is committed if the decision rule leads to only minor adjustments when a major adjustment is needed.

Clearly, from Fig. 8.1, two correct decisions are possible and two incorrect decisions are possible. The structure and method of hypothesis testing has been designed so that the concepts of probability and statistics can be applied to this type of problem in such a way that the probability of making the correct decisions is large and the probability of making the incorrect decisions is small. Some of the considerations involved in measuring the probability of the errors, of ensuring that they are small while the probability of correct decisions is large, have already been presented in the context of *confidence intervals*.

It should be clear that the probabilities of Type I and Type II errors are *conditional probabilities*, in that the former depends on the condition that $H_0$ is true, and the latter on the condition that $H_a$ is true. The conditional probability

$$P(\text{Type I error}) = P(\text{Reject } H_0 \mid H_0 \text{ is true})$$

is commonly denoted by the lower case Greek letter alpha ($\alpha$), and is called *the level of significance*. The level of significance of a statistical test is directly comparable to the probability of an error that we used in Chapter 7 (also called $\alpha$) in describing a confidence interval. In fact, since the probability

$$P(\text{Reject } H_0 \mid H_0 \text{ is true})$$

must be the complement of $P(\text{Accept } H_0 \mid H_0 \text{ is true})$, we can thus write

$$P(\text{Accept } H_0 \mid H_0 \text{ is true}) = 1 - \alpha.$$

This probability corresponds to the concept of a $100(1 - \alpha)\%$ confidence interval. In constructing a statistical test, we would obviously like to have a small probability of making a Type I error; hence one objective is to construct the test to *minimize $\alpha$*.

The probability $P(\text{Type II error}) = P(\text{Accept } H_0 \mid H_0 \text{ is false})$ is usually denoted by the Greek letter beta ($\beta$). Thus,

$$P(\text{Reject } H_0 \mid H_0 \text{ is false}) = 1 - \beta.$$

| | The true situation may be: | |
|---|---|---|
| | $H_0$ is true | $H_0$ is false |
| Accept $H_0$ | $1 - \alpha$ (Confidence level) | $\beta$ (beta) |
| Reject $H_0$ | $\alpha$ (alpha) | $1 - \beta$ (Power of the test) |
| Sum | 1.00 | 1.00 |

Action {

**Fig. 8.2.** The probability of each decision outcome in a test of hypothesis.

This probability (that is, $1 - \beta$) is known as the *power* of a statistical test since it indicates the ability (or "power") of the test to *correctly* recognize that the null hypothesis is false (and hence $H_0$ should be rejected). Thus, one always desires to construct a test with a large power (close to one), or equivalently, with a low value of $\beta$. Figure 8.2 presents the same decision problem as that shown in Fig. 8.1, except that we now identify the *probability* associated with each of the four cells.

Note that the probability of each decision outcome is a *conditional* probability, and that the elements in the same column sum to 1.0, since the events with which they are associated are *complements*. By now it should be apparent that $\alpha$ and $\beta$ need not add to unity, as these two probabilities are not complementary. Thus, a one-unit change in $\alpha$ does not imply a corresponding one-unit change in $\beta$, or vice versa. However, $\alpha$ and $\beta$ are not independent of each other, nor are they independent of the sample size $n$. When $\alpha$ is lowered, $\beta$ normally rises, and vice versa (if $n$ remains unchanged). If $n$ is increased, it is possible for both $\alpha$ and $\beta$ to decrease, because sampling error is potentially decreased. Since increasing $n$ usually costs money, the researcher must decide just how much additional money he is willing to spend on increasing the sample size in order to reduce the size of $\alpha$ and $\beta$. Such an analysis, concerned with balancing the costs of increasing the sample size against the costs of Type I and Type II errors, is a fairly complex subject. We will cover this topic briefly in Section 8.7. Until then we will assume that the sample size is fixed at some predetermined value $n$.

## 8.2 THE STANDARD FORMAT OF HYPOTHESIS TESTING

There are many different population parameters, many different potential forms of hypotheses, and many different sample statistics, random variables, and probability distributions that might be involved in testing hypotheses. It is therefore not feasible to catalogue all such tests. However, they all follow a similar procedure, which can be learned and then applied to different situations as they arise. This procedure can be presented in the following five steps:

1. State the null and alternative hypotheses;
2. Determine a suitable test statistic;
3. Determine the critical region;
4. Compute the value of the test statistic; and
5. Make the statistical decision and interpretation.

We will illustrate this procedure by a simple example that will continue throughout all five steps.

Suppose it has been asserted that the mean I.Q. of all the students in a certain university is 130, although some people claim the mean is *not* 130. To begin with, we will assume that the I.Q.'s of the students in this population are known to be normally distributed with a standard deviation of $\sigma = 5.4$. A random sample of size $n = 25$ is proposed to determine whether or not $\mu = 130$. We now use the five steps outlined above to make a decision about the mean I.Q. of this population.

*Step 1. State the null and alternative hypotheses*

In every hypothesis-testing problem the two conflicting hypotheses need to be specified clearly. It is usually convenient to formulate the null hypothesis as a simple hypothesis and the alternative hypothesis as a composite hypothesis, although it is not necessary to do so. In any case the two conflicting hypotheses must be mutually exclusive, and they must be formulated so that the true value of the population parameter is included in either the null or the alternative hypothesis (i.e., it is not possible for *both* hypotheses to be false).

For our I.Q. example the parameter in the test is the population mean $\mu$. One hypothesis is that $\mu = 130$, and the other is that $\mu \neq 130$. If we let the simple hypothesis be $H_0$, then:

$$\text{Null hypothesis: } H_0: \mu = 130,$$
$$\text{Alternative hypothesis: } H_a: \mu \neq 130.$$

Notice that one of these hypotheses must be true since $H_a$ is the complement of $H_0$. In this case $H_a$ is a *two-sided* alternative, since it involves values on both sides of $H_0$.

*Step 2. Determine a suitable test statistic*

The second step in testing hypotheses is to determine which of the random variables we have studied can be used to decide whether to accept or reject $H_0$. This random variable, called the *test statistic*, is often the $z$ variable $z = (\bar{x} - \mu)/(\sigma/\sqrt{n})$, or the *t*-variable $t = (\bar{x} - \mu)/(s/\sqrt{n})$, although other variables are frequently used (such as the $\chi^2$ or $T$ variables). In general, the value of $\mu$ used in the formula for $z$ or $t$ is that value specified by the null hypothesis.* The remaining values in the formula will either be known in advance (such as $\sigma$ or $\sqrt{n}$), or calculated from the sample (such as $\bar{x}$ or $s$).

In our I.Q. example, the test statistic is the random variable $z = (\bar{x} - \mu)/(\sigma/\sqrt{n})$, where $\mu = 130$ (from the null hypothesis) and $\sigma = 5.4$. Since we have assumed the population to be normally distributed (i.e., the $x$'s are normal), we know that $z = (\bar{x} - 130)/(5.4/\sqrt{n})$ will also be normally distributed, and have a mean of 0 and a standard deviation of 1.0 [that is, $z$ is $N(0, 1)$]. If $\bar{x}$ is near 130, then $z$ will have a value near zero, and we will want to accept $H_0: \mu = 130$. On the other hand, if $\bar{x}$ is not near 130, then $z$ will have a large value (either positive or negative), and we will want to accept $H_a: \mu \neq 130$.

### Step 3. Determine the critical region

As we indicated above, certain values of the test statistic lead to acceptance of $H_0$ (e.g., values of $z$ close to zero), while other values lead to rejection of $H_0$ (e.g., values of $z$ not close to zero). In most statistical tests it is important to specify, *before the sample is taken*, exactly which values of the test statistic will lead to rejection of $H_0$ and which will lead to acceptance of $H_0$. The former set (leading to rejection of $H_0$) is called the *critical region. The boundary value* is that point which separates the critical region from the acceptance region. When the alternative hypothesis is two-sided (as in our I.Q. example), these regions would look like the accompanying diagram.

| Critical region<br>(reject $H_0$) | Boundary values<br>Acceptance region<br>(accept $H_0$) | Critical region<br>(reject $H_0$) |
|---|---|---|
| Large negative $z$ | $z \doteq 0$<br>$z$ values close to zero | Large positive $z$ |

The problem is to determine the *exact* location of the two boundary values shown in the diagram. Their location depends partially on the level of risk that one is willing to take that a Type I error might occur (i.e., on the size of $\alpha$). In general, the larger the risk one is willing to take, the closer together the boundaries will be. This is because $\alpha = P(\text{Reject } H_0 \mid H_0 \text{ is true})$ can be written as

$$\alpha = P(\text{Test statistic value falls in the critical region} \mid H_0 \text{ is true});$$

---

* This is why testing is easier when $H_0$ is a simple rather than a composite hypothesis. If many values of the population parameter could be true under $H_0$, then the test statistic is not so easily stated nor easily used.

hence, the larger $\alpha$ is, the larger the critical region will be. In other words, when we make $\alpha$ smaller, we increase $(1 - \alpha)$, and this means that we are increasing the size of our acceptance region.

Since the value of $\alpha$ is dependent on the value of $\beta$ and the sample size, the determination of a critical region depends on how one wants to balance off the associated costs. If the sample size is fixed in advance, then the optimal critical region is one which minimizes *both* $\alpha$ and $\beta$. But since $\alpha$ increases as $\beta$ decreases, and vice versa, some other criterion is necessary for finding the best critical region.

The traditional method for selecting a region corresponding to the rejection of $H_0$ is first to establish a value for $\alpha$ and then to choose that critical region which yields the smallest value of $\beta$. The rationale behind this procedure is that it is important to first establish the risk one wants to assume of incorrectly rejecting a true null hypothesis. In other words, the size of the Type I error in this approach is viewed as so much more important than the size of the Type II error that the size of $\beta$ is considered only after $\alpha$ has been fixed at some predetermined level. The practice of selecting a critical region in this manner stems from the early research on hypothesis testing, in which the null hypothesis usually represented "current opinion" on an issue and the alternative hypothesis represented a viewpoint of the researcher contrary to that commonly accepted. In testing a new drug, such as a cure for cancer, the drug must be assumed to be of no benefit, or even harmful, until it is proven otherwise. The alternative hypothesis is that the drug is indeed beneficial. A most serious Type I error would be made if a harmful drug ($H_0$ true) were certified as beneficial.

The value of $\alpha$, or the level of significance, indicates the importance (i.e., "significance") that a researcher attaches to the consequences associated with incorrectly rejecting $H_0$. Researchers in the social sciences often use a level of significance of $\alpha = 0.05$, indicating that they are willing to accept a five-percent chance of being wrong when they reject $H_0$. Although $\alpha$ is usually never set higher than $\alpha = 0.05$, and values such as $\alpha = 0.025$, $\alpha = 0.01$ and $\alpha = 0.001$ are frequently used, there is no reason that any other value could not be used. If $\alpha$ is set at some predetermined level, then the optimal critical region is the one which minimizes the probability of a Type II error ($\beta$). This optimal critical region is called the *uniformly most powerful* critical region, since minimizing $\beta$ is equivalent to maximizing the power of the test $(1 - \beta)$. It can be shown that when the level of significance is $\alpha$, the optimal critical region for a two-sided test consists of that $\alpha/2$ percent of the area in the righthand tail of the distribution, plus that $\alpha/2$ percent in the lefthand tail. Thus, establishing a critical region is quite similar to determining a $100(1 - \alpha)\%$ confidence interval.

In our I.Q. example, a Type I error occurs when it is concluded that $\mu \neq 130$ when $\mu$ is really 130. Suppose we assume that the researcher uses an $\alpha = 0.05$ level of significance; i.e., he wants to have no more than a 5-percent chance of rejecting

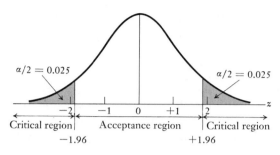

**Fig. 8.3.** Critical regions for testing $H_0$: $\mu$ = 130 against $H_a$: $\mu \neq$ 130 with $\alpha$ = 0.05, $\sigma$ = 5.4, $n$ = 25.

$H_0$: $\mu$ = 130 when this hypothesis is true (a Type I error). Since our test statistic, $z = (\bar{x} - 130)/(5.4/\sqrt{25})$, is $N(0, 1)$, we need to find the boundary values for $\alpha/2$ = 0.05/2 = 0.025 in each tail of the standardized normal distribution. From Table III, these boundary values are easily seen to be $-1.96$ and $+1.96$, as shown in Fig. 8.3.

The boundary values shown in Fig. 8.3 are in terms of the $z$-distribution. We can easily transform these boundary values into their comparable values for $\bar{x}$ by merely solving for $\bar{x}$ in the formula $z = (\bar{x} - 130)/(5.4/\sqrt{25})$. If we let $z = -1.96$ and solve $-1.96 = (\bar{x} - 130)/(5.4/\sqrt{25})$, we get $\bar{x}$ = 127.88. Similarly, when $z = +1.96$, $\bar{x}$ = 132.12. Thus, the appropriate critical regions are those shown in Fig. 8.4.

From Fig. 8.4 we know, for random samples of size $n$ = 25, that 2.5 percent of the time, $\bar{x}$ will be less than 127.88, and 2.5 percent of the time, $\bar{x}$ will be greater than 132.12.

We must emphasize that Figs. 8.3 and 8.4 are two *equivalent* ways of presenting the same critical regions. The only difference is that the former is in terms of the test statistic $z$, while the latter is in terms of the sample mean itself $\bar{x}$. Since it is

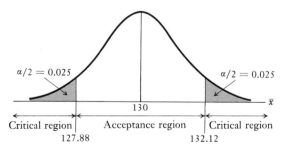

**Fig. 8.4.** Critical regions for testing $H_0$: $\mu$ = 130 against $H_a$: $\mu \neq$ 130 with $\alpha$ = 0.05, $\sigma$ = 5.4, $n$ = 25 (in terms of $\bar{x}$).

usually much more convenient to use standardized units in describing critical regions, we will hereafter use the format of Fig. 8.3.

*Step 4. Compute the value of the test statistic*

Now that we have specified the two conflicting hypotheses, determined the appropriate test statistic, and found the critical regions, we need to see whether the sample result falls in the critical region or in the acceptance region. To do this it is necessary to calculate the value of the test statistic (the $z$-variable) which corresponds to the observed sample results ($\bar{x}$).

For our I.Q. example, let's suppose that the mean of the random sample of $n = 25$ was $\bar{x} = 133$. When we substitute this value into our test statistic $z = (\bar{x} - 130)/(5.4/\sqrt{25})$, we obtain the following computed value:

$$z = \frac{133 - 130}{5.4/\sqrt{25}} = +2.78.$$

*Step 5. Make the statistical decision and interpretation*

If the calculated value of the test statistic lies in the critical region, then $H_0$ is rejected. When the calculated value falls in the acceptance region, then $H_0$ is accepted. In either case, it is important for the researcher to summarize or interpret his decision (to reject or accept $H_0$) *in terms of the original problem*, because the results of statistical tests in business and economics are often presented to and utilized by people who may not understand statistical terminology.

In the I.Q. example, the calculated value $z = +2.78$ is greater than the boundary value of $+1.96$ (shown in Fig. 8.3); hence the sample result falls in the critical region. This leads us to reject $H_0: \mu = 130$. In interpreting this result we can conclude that the sample mean ($\bar{x} = 133$) is too far above $\mu = 130$ for us to believe it occurred by chance in our sample of $n = 25$. We thus conclude that the true population mean is probably not 130, although we must admit there is some risk (5 percent) that this conclusion is not true.

**Further Comments**

The five steps outlined above occur in almost all tests of hypotheses, even though the details of the process may change from example to example. We will illustrate the basic process in a number of examples in this chapter and in the following chapters. Before doing so we should point out that a modification of the procedure is sometimes necessary when the level of significance $\alpha$ cannot be determined in advance of the test. This may happen when the decision-maker is someone different from the person carrying out the research. In such circumstances, the researcher may want to interchange steps 3 and 4 in the testing procedure. That is, since an $\alpha$-value cannot be specified, he would not be able to define a critical region, but he could still compute the value of the test statistic (recall that the test statistic does not depend on $\alpha$). It would then be up to the decision-maker to determine whether or not this computed value falls in the critical region or the acceptance region based on the level of risk *he* wants to assume (i.e., $\alpha$). We must add that this procedure might be referred to as "passing the buck," and it is not always looked upon favorably by executives who like their researchers to reach a conclusion.

At this point it may be worthwhile to mention the similarities and differences between the procedures presented in Chapter 7 for determining confidence intervals and the methods of this chapter for testing hypotheses. In Chapter 7 our objective was to express the uncertainty about an *observed* value of the population parameter (a point estimate) by constructing an interval about that value, and then to express our degree of belief that the population parameter was contained within that interval. The value of $\alpha$ in that context represented the probability of making an error in assuming that the confidence interval does, in fact, contain the true value of the population parameter. In this chapter we are, in effect, establishing a confidence interval (an acceptance region) about an *assumed* value of the population parameter in order to test hypotheses about this value. In this case the value of $\alpha$ represents the probability of making an error if, in fact, the null hypothesis is true. Thus, although the objectives are different, the process described in Chapter 7 for constructing a confidence interval is almost identical to the method in this chapter for determining an acceptance region.

The practice of establishing $\alpha$ at some (arbitrary) predetermined level and then finding the critical region minimizing $\beta$ is widely used in current research. Most modern statisticians, however, do not condone the indiscriminate use of this practice for a number of reasons. First, as we mentioned earlier, it is no longer considered necessary or even good statistical practice to always establish a null hypothesis representing a theory that the researcher is trying to disprove. Such a convention may give an artificiality to the research which distorts its true purpose or hinders the effectiveness of the testing procedure. If the null hypothesis no longer represents "current theory" on an issue, then there is little justification for focusing attention primarily on the probability of a Type I error. A much more important objection to the approach of selecting a critical region by first establishing a value of $\alpha$ is that many researchers set $\alpha$ at some arbitrary level, such as $\alpha = 0.05$, or $\alpha = 0.01$, without considering the associated level of $\beta$. This practice certainly does not represent an optimal approach to balancing the risks of an incorrect decision, for it ignores the risks associated with a Type II error. While it is not incorrect to set $\alpha$ at some predetermined level, this value should be established by considering the risks of *both* Type I and Type II errors.

We will discuss the problems mentioned above in the examples to follow.

These examples illustrate tests of hypotheses involving a number of different population parameters. The most common tests of hypotheses involve $\mu$, the population mean. Tests about $\mu$ are usually designed to indicate, on the basis of a *single* sample, which of two hypothesized values of $\mu$ should be rejected. In other circumstances, one may be interested in designing a test to indicate, on the basis of a sample from each of *two different* populations, whether or not these *two* samples were drawn from populations having equal means. This breakdown between one- and two-sample tests will be a convenient one for us to follow. Some common one-sample tests are discussed in this chapter; two-sample tests are presented in Chapter 10.

## 8.3  ONE-SAMPLE TESTS ABOUT  $\mu$ ($\sigma$ KNOWN)

Suppose a null and an alternative hypothesis have been formulated about the value of a population mean, and it is desirable to test these hypotheses by sampling from the population. Let $\mu$ be the *true* value of the population mean, and let $\mu_0$ be the value of the population mean specified by the null hypothesis. As we pointed out previously, the most convenient null hypothesis is a simple hypothesis, such as $H_0: \mu = \mu_0$ ($\mu_0$ was 130 in our I.Q. example). To test this hypothesis we need to know the distribution of the test statistic, and this distribution will depend on the sampling distribution of $\bar{x}$. First, we know that if the parent population ($x$) is normal, then $\bar{x}$ will also be normal; and even when $x$ is not normal, if $n$ is large, $\bar{x}$ can be assumed to be approximately normally distributed with mean $\mu_0$ and standard deviation $\sigma/\sqrt{n}$ because of the central limit theorem. Transforming $\bar{x}$ into standardized units yields the test statistic $z = (\bar{x} - \mu_0)/(\sigma/\sqrt{n})$, which we know is $N(0, 1)$.

To again illustrate the use of this test statistic, let's assume the alternative hypothesis in the I.Q. example had been one-sided rather than two-sided, as follows: $H_a: \mu > 130$. Perhaps the school has noticed that the admission test scores have been increasing and they want to see if this means that the average I.Q. has gone up. By using $H_a: \mu > 130$, they are saying that $\mu < 130$ is not possible (or at least of no interest); hence there is no need to include it in either $H_0$ or $H_a$. Whenever the alternative hypothesis is one-sided, there will be just one critical region, and the side of the test-statistic distribution on which the critical region lies will be the same as that on which $H_a$ lies, relative to $H_0$. For example, in our new I.Q. problem, we are assuming that the only values of $\mu$ specified by $H_a$ are those greater than 130; hence we reject $H_0$ only when $\bar{x}$ is significantly greater than 130.

When the level of significance is $\alpha$, the uniformly most powerful critical region is that $\alpha$ percent of the distribution lying in the tail of the distribution specified by $H_a$. For example, if we again let $\alpha = 0.05$ and $H_a: \mu > 130$, then the appropriate boundary point from the standardized normal distribution (Table III) is $+1.645$. The corresponding boundary value in terms of $\bar{x}$ is found by setting $z = +1.645$ and solving for $\bar{x}$ in our test-statistic formula (remember, $\mu_0 = 130$, $\sigma = 5.4$, and $n = 25$):

$$+1.645 = (\bar{x} - 130)/(5.4/\sqrt{25}) \qquad \text{yields } \bar{x} = 131.78.$$

These two equivalent boundary points (that is, $z = 1.645$ and $\bar{x} = 131.78$) are shown in Fig. 8.5.

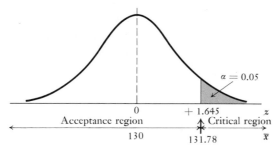

**Fig. 8.5.** The critical region for the one-sided test when $\alpha = 0.05$.

## Another Example

As another illustration of the testing procedure for $\mu$, let's reconsider the example of Sections 6.6 and 6.7 where a sample of 36 dormitory residents was taken to judge whether the average monthly phone charge, claimed to be \$15.30, seems to be true for the population of 300 dormitory phone subscribers. The standard deviation for this population was assumed to be $\sigma = \$4.10$. Let us assume that the null hypothesis is to be the simple hypothesis, and

$$H_0: \mu = 15.30, \qquad H_a: \mu \neq 15.30.$$

From the central-limit theorem we know that $\bar{x}$ will be approximately normal; hence we can again use the standardized normal test statistic $z = (\bar{x} - \mu_0)/(\sigma/\sqrt{n})$, where $\mu_0 = \$15.30$, $\sigma = \$4.10$, and $n = 36$. In this case it is not completely accurate to use a standard error of $\sigma/\sqrt{n}$, since we are sampling without replacement from a *finite* population. Rather, the finite population correction factor, $\sqrt{(N - n)/(N - 1)}$, should be used, so that the appropriate test statistic is now

$$z = \frac{(\bar{x} - \mu_0)}{(\sigma/\sqrt{n})\sqrt{(N - n)/(N - 1)}},$$

where $N = 300$. For this example, let's assume the researcher doesn't know the appropriate level of significance for the decision-maker. Hence, we now proceed directly to determination of the computed value of the test statistic. Assume that a random sample of $n = 36$ yields an average monthly phone bill of $\bar{x} = \$16.90$. The calculated $z$ value is thus:

$$z = \frac{\bar{x} - \mu_0}{(\sigma/\sqrt{n})\sqrt{(N - n)/(N - 1)}} = \frac{\$16.90 - 15.30}{(4.10/\sqrt{36})\sqrt{264/299}} = +2.49.$$

The researcher can now report that if his boss is willing to assume a risk of $\alpha = 0.02$ (or larger), then this result will lead to rejection of $H_0$ (because the appropriate boundary values for $\alpha/2 = 0.01$ is $z = \pm 2.326$). If his boss selects a value of $\alpha = 0.01$ (or lower), then this sample would lead to acceptance of $H_0$ (since the boundary points for $\alpha/2 = 0.005$ would then be $z = \pm 2.576$). In other words, if the decision-maker concludes that phone charges are not equal to \$15.30, then he runs a risk of being wrong of 0.02.

### 8.4  ONE-SAMPLE TESTS ABOUT μ (σ UNKNOWN)

In our previous examples of testing hypotheses about a population mean, we assumed that the standard deviation of the population was known. In most situations when $\mu$ is unknown, the population standard deviation is also unknown, since $\sigma$ depends on the size of the squared deviations about the mean.

In Section 6.8 we explained that the best estimate for $\sigma$ in the formula $z = (\bar{x} - \mu)/(\sigma/\sqrt{n})$ is the sample standard deviation $s$. We also explained that when the parent population is normal, the substitution of $s$ for $\sigma$ in the ratio $(\bar{x} - \mu)/(\sigma/\sqrt{n})$ yields a random variable known to have a $t$-distribution with $(n - 1)$ degrees of freedom which we write as $t_{(n-1)}$. Thus, the appropriate test statistic is now:

$$t_{(n-1)} = \frac{\bar{x} - \mu_0}{s/\sqrt{n}}. \tag{8.1}$$

As an example of a test about $\mu$ when $\alpha$ is unknown, suppose a promotions expert suggests to an automobile manufacturer (Drof Motor Co.) that they extend their service guarantee to 24,000 miles on transmissions, muffler systems, and brakes. He says this change would make good advertising copy and be relatively costless to Drof Motors because he claims that such parts seldom require service during this period anyway. He feels that the average car will run longer than this before the cost of such repairs exceeds \$50.

To test this claim, let's assume Drof Motors has asked you (as their "expert" on hypothesis testing) to sample a few car owners, check their service records, and give the company advice on extending its guarantee. Now, suppose you decide to check the service record of 15 randomly surveyed Drof car-owners; you will let the variable $x$ represent the number of miles driven (since purchase) when the cumulative service repair cost on the parts under study exceeds \$50. Drof is interested in determining whether the population mean value, $\mu$, equals 24,000 ($H_0$: $\mu$ = 24,000) or if it is less than 24,000 ($H_a$: $\mu$ < 24,000). They use a simple null hypothesis since, at this point, they are not concerned with the possibility that $\mu$ might exceed 24,000 (technically, however, $H_0$ is $H_0$: $\mu \geq$ 24,000).

A Type I error in this case would occur if it is concluded that $\mu$ < 24,000 when this hypothesis is not true. In this case, Drof would be afraid to extend the guarantee policy since it *could* do so. A Type II error would occur if $\mu$ = 24,000 is accepted when $\mu$ < 24,000; this error might result in excessive service costs if Drof decided to extend the guarantee. Suppose that, in conferring with the owners of Drof Motors, you find they are willing to assume a risk of 0.025 of incorrectly rejecting $H_0$: $\mu$ = 24,000; thus, $\alpha$ = 0.025.

Based on past experience, Drof expects the distribution of $x$ to be normal, but they don't know its standard deviation; hence, the appropriate test statistic is the $t$-distribution, $t = (\bar{x} - \mu)/(s/\sqrt{n})$. The critical region is that 2.5 percent of the $t$-distribution lying in the lefthand tail (since $H_a$ is to the "left" of $H_0$). The boundary point for the acceptance region is found in Table VII under the heading 0.975 and in the row $v = 14$ (since $n - 1 = 14$) to be $t = -2.145$. This critical region is shown in Fig. 8.6.

Now let's assume that analysis of the random sample of 15 Drof owners results in the values $\bar{x}$ = 22,500 and $s$ = 4000. The calculated value of $t$ for $(n - 1)$d.f. is:

$$t_{(14)} = \frac{(\bar{x} - \mu_0)}{s/\sqrt{n}} = \frac{22,500 - 24,000}{4000/\sqrt{15}} = \frac{-1,500}{1033} = -1.452.$$

Since the calculated value $t = -1.452$ lies in the acceptance region (it is not less than $-2.145$), we conclude that Drof Motors should accept the claim that $\mu$ = 24,000. Even though our sample result indicated a value less than 24,000, this value is not enough lower than 24,000 to reject $H_0$ at the 0.025 level of significance. This result, however, might be encouraging enough to the promotions manager for him to consider taking another sample with a larger $n$ and determining the value of the $\beta$ error in accepting $H_0$.

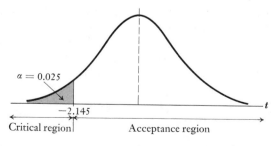

**Fig. 8.6.** Critical region for $H_0$: $\mu$ = 130 vs. $H_a$: $\mu$ < 130 when $\sigma$ is unknown and $n$ = 15.

Most of our examples thus far have assumed that the value of $\alpha$ is specified in advance; we have not discussed how this assumption affects the probability of a Type II error $\beta$. The following section covers this subject.

### *8.5 MEASURING $\beta$ AND THE POWER OF A TEST

One of the reasons we have been avoiding discussion of the calculation of $\beta$ is that the alternative hypothesis is generally a composite hypothesis. This means that we can't calculate one value for $\beta$ because there is no one value specified for $H_a$ to be true. To illustrate, recall in our test involving telephone charges, that $H_0: \mu = 15.30$ was tested against $H_a: \mu \neq 15.30$. A Type I error (incorrectly rejecting $H_0$) in this case is well defined, since we know it will occur only when $\mu = \$15.30$. But a Type II error (incorrectly accepting $H_0$) can occur for any value of $\mu$ not equal to 15.30. It should perhaps be obvious that the probability of incorrectly accepting $H_0: \mu = \$15.30$ is much higher when the true value of $\mu$ is $\$15.00$ than when the true value is $\mu = \$25.00$. In other words, the value of $\beta$ would be different for these two situations.

The different probability values for $\beta$ which occur when $H_a$ is composite can be presented in a table, or graphed, or described by a functional relationship. Often, however, it is more useful to present the values of $(1 - \beta)$. A function describing such probabilities is called a *power function*, since it indicates the ability (or "power") of the test to correctly reject a false null hypothesis. In general, test statistics and critical regions having the highest power are preferred. Although it is beyond the scope of this book to examine the concepts involved in finding a power function for most statistical tests, we need to emphasize that the tests presented thus far have made use of these concepts in that we always selected the *uniformly most powerful critical region*. The complexity involved in finding the power of a statistical test again emphasizes the rationale for making the null hypothesis a simple test. If $H_0$ were a composite hypothesis, then we would also have to use a function to describe the values of $\alpha$, instead of having just a single value.

To illustrate the calculation of $\beta$ (or $(1 - \beta)$) and the trade-offs between $\alpha$, $\beta$, and the sample size, consider a firm manufacturing rubber bands. Control over the number of rubber bands placed in each box is kept by sampling and testing hypotheses, rather than by a counting procedure. When the production process is working correctly, the number of good bands placed in each box $(x)$ has a mean of $\mu = 1000$, with a standard deviation of 37.5. This variable $x$ is presumed to have an approximately normal distribution; that is, $x$ is $N(1000, 37.5^2)$.

In this case the company wants to test $H_0: \mu = 1000$ against $H_a: \mu \neq 1000$, and the appropriate test statistic is $z = (\bar{x} - \mu_0)/(\sigma/\sqrt{n})$. A Type I error occurs

---

* This section may be omitted without loss in continuity.

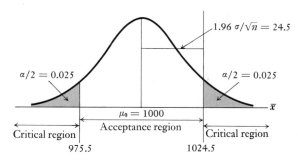

**Fig. 8.7.** Critical region for test on $H_0: \mu = 1000$ against $H_a$: $\mu \neq 1000$, $\sigma = 37.5$, $n = 9$, and $\alpha = 0.05$.

whenever the test results suggest that the process is out of control ($\mu \neq 1000$), when it actually *is* in control ($\mu = 1000$). A Type II error occurs whenever the process is judged to be in control ($\mu = 1000$) and it actually is out of control ($\mu \neq 1000$). In constructing a test of these hypotheses, let's assume that the company periodically selects a random sample of size $n = 9$, and the company policy is to let $\alpha = 0.05$. For these values the boundary points of the acceptance region are

$$\mu_0 \pm z_{\alpha/2}\sigma/\sqrt{n} = 1000 \pm 1.96(37.5/\sqrt{9}) = 1000 \pm 24.5.$$

This acceptance region is shown in Fig. 8.7.

The question we turn to now is how to calculate $\beta$ for this problem. Since $\beta$ is a conditional probability which depends on the value of $\mu$, suppose we assume that $\mu = 990$. We can now write $\beta = P(\text{Accept } H_0: \mu = 1000 \mid \mu = 990)$. From Fig. 8.7 we see that $H_0$ is accepted whenever $\bar{x}$ lies between 975.5 and 1024.5. Hence, $\beta = P(975.5 \leqslant \bar{x} \leqslant 1024.5 \mid \mu = 990)$. This probability can be determined by using the same procedure we learned in Chapter 6 in working with $z = (\bar{x} - \mu)/(\sigma/\sqrt{n})$. First, we transform the problem to standardized normal terms by letting $\mu = 990$, $\sigma = 37.5$ and $\sqrt{n} = \sqrt{9}$, and then use Table III to find the appropriate probabilities.

$$\begin{aligned} P(975.5 \leqslant \bar{x} \leqslant 1024.5) &= P\left(\frac{975.5 - 990}{37.5/\sqrt{9}} \leqslant \frac{\bar{x} - \mu}{\sigma/\sqrt{n}} \leqslant \frac{1024.5 - 990}{37.5/\sqrt{9}}\right) \\ &= P(-1.16 \leqslant z \leqslant 2.76) = F(2.76) - F(-1.16) \\ &= 0.9971 - 0.1230 = 0.8741. \end{aligned}$$

Thus, $P(\text{Type II error}) = 0.8741$, as shown in Fig. 8.8. This means our test procedure is such that when $\mu = 990$, we will *incorrectly* accept $H_0: \mu = 1000$ as being true 87.41 percent of the time. The *power* of this test is $1 - \beta = 0.1259$, which means that this test will *correctly* recognize a false null hypothesis 12.59 percent of the time.

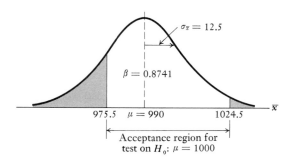

**Fig. 8.8.** Probability of $\beta$ for the critical region shown in Fig. 8.7 if the true mean is $\mu = 990$.

Instead of using the value $\mu = 990$ to calculate $\beta$, we might have used $\mu = 1010$. These two values of $\mu$ are both an equal distance (10 units) away from $H_0: \mu = 1000$; so it should not be surprising to learn that the value of $\beta$ is the same in both cases ($\beta = 0.8741$). Figure 8.9(a) shows the area corresponding to $\beta$ for $\mu = 1010$, while parts (b) and (c) of this figure show the area for $\beta$ corresponding to $\mu = 970$ and $\mu = 950$, respectively. The calculation of $\beta$ when $\mu = 970$ is shown below:

$$P(\text{Type II error} \mid \mu = 970) = P(975.5 \leqslant \bar{x} \leqslant 1024.5 \mid \mu = 970)$$
$$= P\left(\frac{975.5 - 970}{37.5/\sqrt{9}} \leqslant z \leqslant \frac{1024.5 - 970}{37.5/\sqrt{9}}\right)$$
$$= P(0.44 \leqslant z \leqslant 4.36)$$
$$= F(4.36) - F(0.44) = 1.000 - 0.670 = 0.3300.$$

**Table 8.1** Value of $\beta$ and power for given true values of $\mu$

| $\mu$ | $P(\text{Accept } H_0) = \beta$ | $P(\text{Reject } H_0) =$ Power $= 1 - \beta$ |
|---|---|---|
| 950 | 0.0207 | 0.9793 |
| 970 | 0.3300 | 0.6700 |
| 980 | 0.6406 | 0.3594 |
| 990 | 0.8741 | 0.1259 |
| 1000 | $1 - \alpha = 0.95$ | $\alpha = 0.05$ |
| 1010 | 0.8741 | 0.1259 |
| 1020 | 0.6406 | 0.3594 |
| 1030 | 0.3300 | 0.6700 |
| 1050 | 0.0207 | 0.9793 |

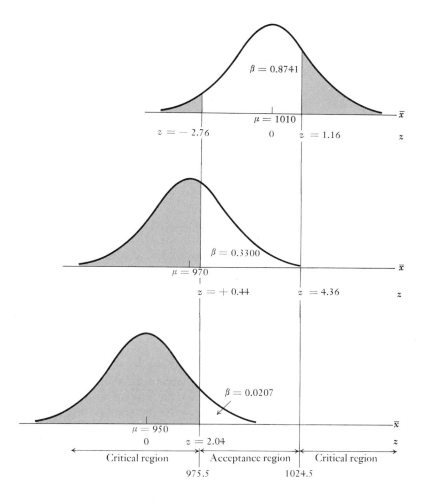

**Fig. 8.9.** Values of β for different μ's, α = 0.05, n = 9, σ = 37.5.

Note that in Fig. 8.9 the size of β decreases as the value of μ gets farther away from μ = 1000. That is, the more incorrect $H_0$ is, the lower will be the value of β [and the higher $(1 - β)$]. This fact is shown in Table 8.1, where we present the value of β and $(1 - β)$ for eight different values of μ. The row corresponding to μ = 1000 is placed in a box to emphasize that this is the one case when $H_0$ is

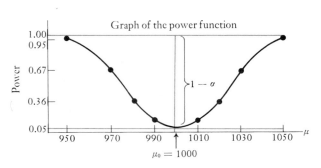

**Figure 8.10**

true; hence $\beta$ is not defined for $\mu = 1000$. Figure 8.10 is a graph of the power function $(1 - \beta)$.

In comparing the power functions of a number of different tests, we look for tests where the power function rises quickly as the value of $\mu$ differs by small amounts from $\mu_0$. Thus, the most powerful test would be the one with the steepest ascending power function.

### The Trade-offs between $\alpha$ and $\beta$.

We have emphasized that when the sample size is fixed, $\alpha$ and $\beta$ have an inverse relationship. To illustrate this trade-off, we will use the same production-process example described above, but we now will change $\alpha$ from 0.05 to 0.10. Since $\alpha$ has increased, the effect should be to reduce $\beta$. Figure 8.11 shows the new acceptance region, calculated by letting $z_{\alpha/2} = z_{0.05} = 1.645$. In this situation, the null hypothesis is accepted if $979.5 \leqslant \bar{x} \leqslant 1020.5$. The result is that the size of $\beta$ is reduced from 0.3300 (shown in Table 8.1) to the new value, $\beta = 0.2236$. The power of the test has increased correspondingly. These values are shown in Fig. 8.11.

If we calculated additional values of the power function, we would find all of them larger when $\alpha = 0.10$ than when $\alpha = 0.05$. As we increase $\alpha$, we narrow the acceptance region, and hence make our test more powerful. Similarly, if we decrease the value of $\alpha$, then $\beta$ will rise and the power of the test will fall. Thus, the size of $\alpha$ is *inversely* related to the size of $\beta$ and directly to the size of the power, but the trade-off is not one-to-one. In this example, $\alpha$ was increased by 0.05 (0.05 to 0.10), but $\beta$ decreased by more than 0.10 (0.3300 to 0.2236).

### Decreasing $\alpha$ and $\beta$ by Increasing $n$

Until now our discussion has assumed that the size of the sample is fixed in advance. If $n$ is changed, however, the size of both $\alpha$ and $\beta$ may be changed, because the size of $n$ affects the location of the acceptance and critical regions. To illustrate this effect, suppose we return $\alpha$ to its previous level of 0.05, and increase $n$ from

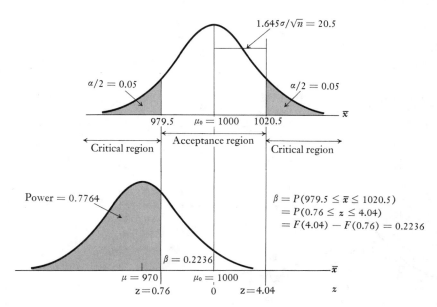

**Fig. 8.11.** Critical region for $H_0$: $\mu = 1000$ against $H_a$: $\mu \neq 1000$ given $\sigma = 37.5$, $n = 9$, $\alpha = 0.10$; and the representation of $\beta$ given that the true mean is $\mu = 970$.

9 to 36. The new critical region for our test, and the determination of $\beta$ when the true value of $\mu$ is 970, are shown in Fig. 8.12.

We see that the increased sample size makes our test more sensitive in distinguishing between $H_0$ and $H_a$ since the standard error of the mean, $\sigma/\sqrt{n}$, is now half its former value (from $37.5/\sqrt{9} = 12.5$ to $37.5/\sqrt{36} = 6.25$). The null hypothesis will be accepted in this test if $987.75 < \bar{x} < 1012.25$. The probability that $H_0$ will be accepted given that $\mu = 970$ is now only $\beta = 0.0023$. By comparing Figs. 8.11 and 8.12, we see that we have reduced $\alpha$ from 0.10 to 0.05 and reduced $\beta$ from 0.2236 to 0.0023 merely by increasing $n$ from 9 to 36. Unfortunately, larger sample sizes are more time-consuming and often quite costly, so the researcher is faced with the task of balancing off the costs of making incorrect decisions against the costs of sampling. We will return to this consideration in Section 8.7 and in Chapter 9.

## 8.6 TEST ON THE BINOMIAL PARAMETER

The population parameter being examined in a test of hypothesis need not always be the population mean $\mu$. In many situations, the parameter in question is the *proportion* of observations having a certain attribute. When the observations are

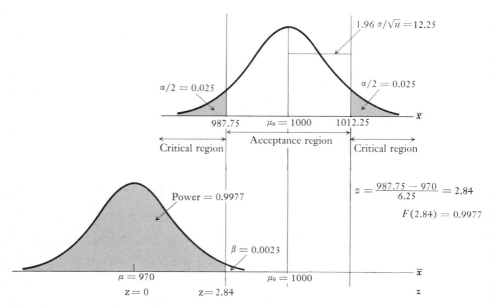

**Fig. 8.12.** Critical region for the test on $H_0$: $\mu = 1000$ against $H_a$: $\mu \neq 1000$, $\sigma = 37.5$, $n = 36$, $\alpha = 0.05$; and the representation of $\beta$ when the true mean is $\mu = 970$.

independent of each other (i.e., they are randomly selected) and the attribute of interest either occurs or does not occur in each observation, then the appropriate test statistic follows the *binomial distribution*. In this case, the best estimator of the population proportion $p$ is the sample proportion $\hat{p} = x/n$; the mean and variance in a binomial distribution are $\mu = np$ and $\sigma^2 = npq$, where $q = 1 - p$, $n$ is the total number of observations, and $x$ is the number of occurrences.

A test involving an unknown population proportion may be one-sided or two-sided, and the values hypothesized for $p$ may range from zero to one. The formal hypothesis-testing procedure given in Section 8.2 also applies to tests of the binomial parameter. To illustrate this procedure, we will again use the example in Section 4.3. Recall that in this example a quality engineer is going to sample 20 items from a malfunctioning production process to determine whether $H_0$: $p = 0.10$ is true (a minor adjustment) or $H_a$: $p = 0.25$ is true (a major adjustment). If the reader will refer to Table 4.3 (page 145), he will see that the decision rule was to reject the null hypothesis, $H_0$: $p = 0.10$ when four or more defectives appear in the sample of 20, and to reject the alternative hypothesis, $H_a$: $p = 0.25$, when three or less defectives are received. Should the sample yield four or more defectives (the critical region) when $p = 0.10$, then a Type I error will result. A Type II error will be committed if $p = 0.25$, but the sample value falls in the acceptance region (less than four defectives). The critical and acceptance regions for this problem are shown in Fig. 8.13, where, again, the probabilities come from Table 4.3.

The shaded areas represent the probabilities of Type I and Type II errors; $\alpha$

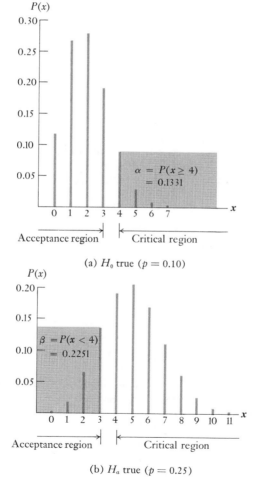

(a) $H_0$ true ($p = 0.10$)

(b) $H_a$ true ($p = 0.25$)

**Fig. 8.13.** Critical and acceptance regions to test $H_0$: $p = 0.10$ versus $H_a$: $p = 0.25$.

is the probability that $x \geqslant 4$ when $H_0$ is true, and $\beta$ is the probability that $x < 4$ when $H_a$ is true. The quality engineer only needs to determine the number of defectives in the sample of 20 and make his decision for a minor or major adjustment according to the critical region shown in Fig. 8.13. Also, he knows precisely his risks of making an error.

The only substantive difference between tests on the binomial parameter $p$ and the previous tests on the population mean $\mu$ is that here we are dealing with a discrete rather than a continuous distribution. It is important for us to be more specific about the process of choosing a critical region when the sampling distribution is discrete. Clearly, if $p_0$ represents some hypothesized value of $p$, then it is possible to test $H_0: p = p_0$ against $H_a: p \neq p_0$, or $H_a: p < p_0$, or $H_a: p > p_0$. Suppose we decide to test the hypothesis $H_0: p = p_0$ against $H_a: p > p_0$ at some level of significance $\alpha$. The problem may be that there is no critical region which cuts off *exactly* $\alpha$ percent of the righthand tail of the binomial distribution, since only integer values of $x$ can be used. For example, if we want $\alpha = 0.05$ it may be that one value of $x$ cuts off 6 percent of the distribution, and the next value of $x$ cuts off 4 percent. In these cases the usual procedure is to pick the critical region cutting off the *smaller* area in order to be assured that the probability of a Type I error is no larger than $\alpha$. This procedure is equivalent to finding the *smallest* value of $x$ which satisfies the following inequality:

$$P(x \text{ or more successes in } n \text{ trials}) \leqslant \alpha. \tag{8.2}$$

The smallest value of $x$ satisfying Formula (8.2) represents the appropriate boundary point.

To illustrate Formula (8.2), suppose we want to test $H_0: p = 0.25$ against $H_a: p > 0.25$; a sample of size $n = 20$ is taken and $\alpha$ is assumed to be 0.05. Table I can be used to determine the smallest value of $x$ cutting off probabilities equal to or less than 0.05. Reading under $n = 20$ and $p = 0.25$, we see that if $x = 13$, then 0.0002 is cut off; if $x = 12$, then $0.0008 + 0.0002 = 0.0010$ is cut off. Continuing in this fashion, if $x = 9$, the critical region is of size 0.0410, which is the smallest area less than 0.05. The next value of $x$, $x = 8$, determines a critical region of size $0.0410 + 0.0609 = 0.1019$, which is too large. Hence, to assure an $\alpha$ no larger than $0.05$, $x \geqslant 9$ is the appropriate critical region. Any sample that results in 9 or more successes in 20 trials leads to rejection of $H_0: p = 0.25$.

### Using the Normal Approximation when $n$ is Large

Recall that it is not always convenient to work with binomial tables directly, because the arithmetic may be tedious, or tables may not be available for certain values of $p$. Fortunately, when the value of $n$ is large and the value of $p$ is not close to either zero or one, then the test on the binomial parameter $p$ may be structured using the standard normal distribution. Whenever $npq \geqslant 3$, the normal approximation to the binomial is satisfactory for most purposes. The test statistic in this case is $z = (\hat{p} - p_0)/\sqrt{p_0(1 - p_0)/n}$.

To illustrate this test procedure, suppose a city mayor, during a press conference, rejects a suggestion that the city needs a strict antinoise and antipollution law because he says his own personal survey indicates that only 20% of the residents are concerned over this matter. Some concerned citizens, however, finance a survey of adult residents, asking whether they believe that such a law should be an important high-priority matter at this time. A test of hypotheses structured to either counter or support this claim would be $H_0: p = 0.20$ versus $H_a: p > 0.20$. A one-sided test is used because we are not interested in distinguishing between $p = 0.20$ and $p < 0.20$; that is, the mayor's claim can be refuted only if we can accept the alternative hypotheses $H_a: p > 0.20$. Now, let's assume that the survey included 81 respondents, which means that it is appropriate to use the standard normal approximation (since $npq = 81(0.20)(0.80) = 12.96 > 3$).

The critical region for this test is determined by establishing $\alpha$ and using the probability distribution for $z = (\hat{p} - p_0)/\sqrt{p_0(1 - p_0)/n}$. Assume $\alpha = 0.01$. From Table III the appropriate boundary value of $z$ for this one-sided test is 2.326, as shown in Fig. 8.14.

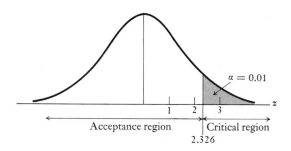

**Fig. 8.14.** Critical region for the test on $H_0: p = 0.20$ against $H_a: p > 0.20$, $npq \geqslant 3$, $\alpha = 0.01$.

Suppose the number of people favoring the new law is 33. This means the sample proportion is $\hat{p} = 33/81 = 0.407$, and the calculated value of the test statistic is:

$$z = \frac{(\hat{p} - p_0)}{\sqrt{(p_0(1 - p_0))/n}} = \frac{0.407 - 0.200}{\sqrt{(0.20)(0.80)/81}} = +4.66.$$

Since $z = 4.66$ falls in the critical region shown in Fig. 8.14, we can conclude that more than 20% of the residents would favor this type of law.

## 8.7 BALANCING THE RISKS AND COSTS OF MAKING A WRONG DECISION

We now return to the important problem of how to choose the "best" critical region to balance off the risks and costs associated with making a Type I error against those associated with making a Type II error. As we pointed out in Chapter 6, the objective of sampling can be stated as one of minimizing the cost of making an incorrect decision (including sampling costs, which will be considered later). Unfortunately, in many circumstances there may be no easy way even to determine what these costs are, much less to try to balance them. In medical research, for example, it may be difficult if not impossible to assess the costs associated with an incorrect decision involving a new drug or surgical technique. But even if these costs could somehow be assessed, how does one go about balancing costs which may include pain, suffering, and even loss of life? On the other hand, if it is possible to identify the relevant costs and to express them in terms of some comparable basis, such as dollars, then we may be able to balance these costs quite explicitly. The following example is given to describe the process of finding the critical region which minimizes the expected cost of making an incorrect decision. It uses the quality-control problem of Section 8.6. It should be noted that, for the first part of this example, we assume the sample size to be fixed at $n = 20$; later we relax this assumption and consider a sample of size $n = 50$. Suppose that all possible costs associated with an incorrect decision (i.e., loss in profit, goodwill, etc.) are those shown in Fig. 8.15. A correct decision is assumed to result in no loss in profit or goodwill.

For each possible critical region we can calculate the expected cost (per sample of 20) which will result if the process is (1) producing 10 percent defectives or (2) producing 25 percent defectives. These expected costs are calculated by multiplying the probability of making each type of error by the cost of making that error. We have calculated these values for two critical regions, the one used in

|  | Percent defectives | |
|  | $p = 0.10$<br>$H_0$ is true | $p = 0.25$<br>$H_0$ is false |
|---|---|---|
| Accept $H_0$<br>make Minor<br>adjustment | Correct<br>decision | \$200 |
| Reject $H_0$<br>make Major<br>adjustment | \$500 | Correct<br>decision |

**Fig. 8.15.** Costs of making an incorrect decision.

Fig. 8.13 of $x \geqslant 4$, and an alternate one, $x \geqslant 5$, which gives a smaller $\alpha$ at the expense of a larger $\beta$.

A.  *Critical region* $x \geqslant 4$ ($\alpha = 0.1331$ *and* $\beta = 0.2251$)

1.  For $p = 0.10$:

Expected cost $= P(\text{Type I error}) \times$ Cost of a Type I error
$$= 0.1331(\$500) = \$66.55.$$

2.  For $p = 0.25$:

Expected cost $= P(\text{Type II error}) \times$ Cost of a Type II error
$$= 0.2251(\$200) = \$45.02.$$

B.  *Critical region* $x \geqslant 5$ ($\alpha = 0.0433$ *and* $\beta = 0.4148$)*

1.  For $p = 0.10$:

Expected cost $= P(\text{Type I error}) \times$ Cost of a Type I error
$$= 0.0433(\$500) = \$21.65.$$

2.  For $p = 0.25$:

Expected cost $= P(\text{Type II error}) \times$ Cost of a Type II error
$$= 0.4148(\$200) = \$82.96.$$

The expected costs given above are conditional values, in that each one was calculated by assuming that either $p = 0.10$ or $p = 0.25$. In order to be able to determine which of these critical regions is better, we need to know how often the process is expected to be producing 10 percent defectives, relative to the number of times it will be producing 25 percent defectives. Suppose that, if an adjustment is required, the probability that it will be a minor adjustment is 0.70, while the probability that a major adjustment is necessary is 0.30. The *total* expected costs associated with each of the two critical regions can now be calculated by taking the product of the expected costs determined above and the probability that each of these costs will be incurred.

A.  *Critical region* $x \geqslant 4$:

Total expected cost
$$= P(\text{Major adjustment needed}) \times \text{Expected cost of Type II error}$$
$$+ P(\text{Minor adjustment needed}) \times \text{Expected cost of Type I error}$$
$$= 0.30(\$45.02) + 0.70(\$66.55)$$
$$= \$60.09.$$

---

* The reader can quickly calculate these probabilities of errors from Table 4.3 by moving the decision line down one row and recalculating the sum of terms then included in the brackets. Table I for the binomial, with $n = 20$, $p = 0.25$ and $p = 0.10$ may, of course, be used directly to get the same values.

*B. Critical region $x \geqslant 5$:*

Total expected cost
$$= P(\text{Major adjustment needed}) \times \text{Expected cost of Type II error}$$
$$+ P(\text{Minor adjustment needed}) \times \text{Expected cost of Type I error}$$
$$= 0.30(\$82.96) + 0.70(\$21.65)$$
$$= \$40.44.$$

Thus, when the process is malfunctioning, the total expected cost for each sample of 20 equals \$60.09 if the critical region is $x \geqslant 4$, and \$40.04 if the critical region is $x \geqslant 5$. We leave it as an exercise for the reader to determine that $x \geqslant 5$ is, in fact, the *optimal* critical region for this problem, with total expected cost smaller than any other critical region.

**Changing the Sample Size**

Suppose, in the above sample, that the sample size could have been increased to $n = 50$ at a cost of \$10. Our discussion of trade-offs between $\alpha$, $\beta$, and $n$ in Section 8.5 suggests that this increase in sample size can lead to a decrease in both $\alpha$ and $\beta$ if an appropriate critical region is used. The question is whether or not the decreased probability of making an error is worth the increased sampling costs of \$10. In order to answer this question, let us select a critical region for the new situation (with $n = 50$) and then determine $\alpha$ and $\beta$ and the total expected cost for this critical region. This total expected cost can then be compared with the preceding optimal of \$40.44.

Since $n$ has increased 2.5 times from 20 to 50, let us arbitrarily try a new boundary value which is 2.5 times the initial one ($x = 4$); that is, $x = 10$. We leave it as an exercise for the reader to determine if a better critical region than $x \geqslant 10$ could be found, perhaps by using $x \geqslant 12$ or $x \geqslant 13$ (which are approximately 2.5 times the previous optimal value of $x = 5$). The probability of observing a specific number of defectives when $n = 50$ under the two hypotheses, $H_0: p = 0.10$ and $H_a: p = 0.25$, is shown in Table 8.2. From these values the probabilities of Type I and Type II errors are seen to be $\alpha = 0.0245$ and $\beta = 0.1636$. We thus see that increasing $n$ from 20 to 50 has reduced both $\alpha$ and $\beta$ (see Table 8.2).

*Critical region $x \geqslant 10$*

1. For $p = 0.10$:
   Expected cost $= P(\text{Type I error}) \times \text{Cost of a Type I error}$
   $$= 0.0245(\$500) = \$12.25.$$

2. For $p = 0.25$:
   Expected cost $= P(\text{Type II error}) \times \text{Cost of a Type II error}$
   $$= 0.1636(\$200) = \$32.72.$$

Total expected cost

$$= P(\text{Major adjustment needed}) \times \text{Expected cost of an error}$$
$$+ P(\text{Minor adjustment needed}) \times \text{Expected cost of an error}$$
$$= 0.30(32.72) + 0.70(\$12.25) = \$26.58.$$

Thus, if we have to choose between a sample of 20 and critical region $x \geqslant 5$ (with $\alpha = 0.0433$ and $\beta = 0.4148$) in which the costs will average $40.04, and a sample of 50 and critical region $x \geqslant 10$ (with $\alpha = 0.0245$ and $\beta = 0.1636$) in which the costs will average $26.58 for the incorrect decisions and $10.00 for the additional observations, it would be better to take the larger sample. It may be, of course, that some other critical region will be even better than $x \geqslant 10$, or that some other sample size gives a lower expected cost. Given information on the cost of all possible sample sizes, the "optimal" sample size and its associated critical region could be determined for this problem. Again, parts of this task are left as an exercise for the reader. In Chapter 9 we shall return to an extended version of this type of problem and study in more detail the question of sample size.

**Table 8.2** Determining $\alpha$ and $\beta$ when $n = 50$ and the critical region is $x \geqslant 10$

| $x$ | Decision | If $p = 0.10$, $H_0$ is true $P(x) = \binom{50}{x}(0.10)^x(0.90)^{50-x}$ | | If $p = 0.25$, $H_a$ is true $P(x) = \binom{50}{x}(0.25)^x(0.75)^{50-x}$ |
|---|---|---|---|---|
| 0 | | 0.0052 | | 0.0000 |
| 1 | | 0.0286 | | 0.0000 |
| 2 | Accept $H_0$, | 0.0779 | | 0.0001 |
| . | make | . | Acceptance | . |
| . | minor | . | region    $\beta = 0.1636$ | . |
| . | adjustment | . | | . |
| 8 | | 0.0643 | | 0.0463 |
| 9 | | 0.0333 | | 0.0721 |
| 10 | | 0.0152 | | 0.0985 |
| 11 | | 0.0061 | | 0.1194 |
| 12 | Reject $H_0$, | 0.0022 | | 0.1294 |
| . | make | . | Critical | . |
| . | major | . | $\alpha = 0.0245$   region | . |
| . | adjustment | . | | . |
| 24 | | 0.0000 | | 0.0002 |
| 25 | | 0.0000 | | 0.0001 |
| 26–50 | | 0.0000 | | 0.0000 |
| Sum | | 1.0000 | | 1.0000 |

## *8.8 TESTS USING THE CHI-SQUARE DISTRIBUTION

Once the procedure of testing hypotheses has been mastered, it can be applied to many other test situations if an appropriate test statistic can be developed. In this section, two further examples of tests are given, which utilize test statistics with a chi-square distribution. Table IV, which defines values of the cumulative chi-square distribution, is used to determine the critical regions for these tests.

### Test on a Population Variance, $\sigma^2$

In Section 6.9 we formulated a chi-square random variable,

$$\chi^2_{(n-1)\,\text{d.f.}} = (n-1)s^2/\sigma^2,$$

for a situation where a random sample of size $n$ is taken from a normal population with standard deviation $\sigma^2$. In Section 7.8, a confidence interval for $\sigma^2$ was developed using this variable. We now illustrate the use of the $\chi^2$ variable in testing hypotheses about the unknown population variance, $\sigma^2$ based on a sample estimator $s^2$.

Consider a distribution of monthly profits from franchises for Corporal Smoothers Missouri Fried Catfish. We are interested in the risk associated with operating such a franchise, and believe that the variance of monthly profits would be a good measure of this risk. Suppose the distribution is normal, with a mean near \$2200. We believe that a level of risk represented by $\sigma = \$800$, or $\sigma^2 = 640,000$, is too great for investment in such a franchise; and we wish to test whether the true variance might be significantly smaller than 640,000. The appropriate null and alternate hypotheses are:

$$H_0: \sigma^2 = 640,000, \qquad H_a: \sigma^2 < 640,000.$$

For this type of problem, the test statistic is

$$\chi^2 = (n-1)s^2/\sigma_0^2,$$

where $\sigma_0^2$ is the hypothesized value of $\sigma^2$ when $H_0$ is true, and $s^2$ is the variance calculated from the random sample (of size $n$). The critical region for such a one-sided test is that $\alpha$ percent of the values in the left tail of a chi-square distribution. For example, suppose we choose $\alpha = 0.05$ and a sample of size 12 (perhaps one observation for each month of the year, chosen randomly among the set of franchises). From Table IV the boundary value is $\chi^2 = 4.57$. The critical region for this example is thus all values of $\chi^2$ less than 4.57, as shown in Fig. 8.16.

Suppose the sample of 12 values of monthly profits results in a sample variance

---

* This section, which may be omitted without loss of continuity, assumes that Sections 6.9 and 7.8 have been covered.

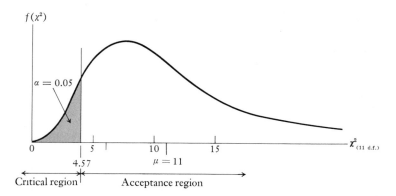

**Fig. 8.16.** Critical region for a one-sided chi-square test with $v = n - 1 = 11$ and $\alpha = 0.05$.

equal to $s^2 = 360{,}000$ ($s = 600$). The calculated value of the test statistic is,

$$\chi^2_{11} = \frac{(n-1)s^2}{\sigma^2} = \frac{11(360{,}000)}{640{,}000} = 6.18.$$

The sample result thus falls in the acceptance region. This means that, even though the sample variance is much less than the hypothesized value for the population variance, it is not sufficiently smaller for us to reject $H_0$. Hence, we are wary of an investment in Corporal Smoothers Missouri Fried Catfish because it is sufficiently likely that the variance of monthly profits may be $640{,}000$ ($\sigma = \$800$).

The reader should note that, while the standard deviation may be the more easily interpreted measure of risk or variability, the $\chi^2$ test deals with the square of $\sigma$, the variance. Similar tests on $\sigma^2$ in other situations may be one-sided high tests or two-sided tests, all of which can be done following the same general procedure.

### Goodness-of-Fit Test

Recall that when the set of outcomes in an experiment can be divided into two categories (such as a success or a failure, make a sale or do not, watch pro football or professional baseball on television, etc.), then the appropriate test statistic is the binomial variable. When more than two categories or classes of outcomes are involved, then the appropriate statistic is the chi-square variable. Although the chi-square test can also be used when there are only two categories, the binomial is preferred in this case because it is more powerful.

The chi-square variable is used in this situation to test how closely a set of observed frequencies corresponds to a given set of expected frequencies. The expected frequencies can be thought of as the average number of values expected to fall in each category, based on some theoretical probability distribution. For example, one probability distribution which is often useful is one such that the expected frequencies in the various categories will all be equal. The observed frequencies can be thought of as a sample of values from some probability distribution. The chi-square variable can be used to test whether the observed and expected frequencies are close enough for us to conclude that they came from the same probability distribution. For this reason the test is called a "goodness-of-fit" test.

Suppose we assume that there are $c$ categories ($c > 1$) and the *expected* frequency in each of these categories is denoted as $E_1, E_2, \ldots, E_c$, or equivalently, $E_i$ ($i = 1, 2, \ldots, c$). Similarly, the $c$ *observed* frequencies will be denoted as $O_1$, $O_2, \ldots, O_c$, or $O_i$ ($i = 1, 2, \ldots, c$). To test the goodness of fit of the observed frequencies ($O_i$) to the expected frequencies ($E_i$), we use the following chi-square variable with $c - 1$ degrees of freedom.

$$\chi^2_{(c-1)\text{ d.f.}} = \sum_{i=1}^{c} \frac{(O_i - E_i)^2}{E_i}. \tag{8.3}$$

Formula (8.3) measures the goodness of fit between the values of $O_i$ and $E_i$ as follows: when the fit is good (that is, $O_i$ and $E_i$ are generally close), then the numerator of (8.3) will be relatively small, and hence the value of $\chi^2$ will be low. Conversely, if $O_i$ and $E_i$ are not close, then the numerator of (8.3) will be relatively large, and the value of $\chi^2$ will also be large. Thus, the critical region for the test statistic given by (8.3) will always be in the *upper* tail because we want to reject the null hypothesis whenever the difference between $E_i$ and $O_i$ is relatively large. For example, suppose, in a particular problem involving 16 categories, that the fit between the 16 values of $O_i$ and $F_i$ from Formula (8.3) yields $\chi^2 = 25.0$. From Table IV in the row corresponding to $c - 1 = 15$ d.f., we find that $P(\chi^2 \geqslant 25) = 0.05$. Thus, at the $\alpha < 0.05$ level of significance, we can reject the null hypothesis that the observed values came from the same distribution as the expected values.

To illustrate the use of the chi-square test, suppose that an automobile dealer, in trying to arrange vacations for his salesmen, decides to test the (null) hypothesis that his sales of new cars were equally distributed over the first six months of last year. His expected frequency distribution thus specifies that $E_1 = E_2 = \cdots = E_6$. The alternative hypothesis is that sales were not equally distributed over the six months. Since he sold 150 new cars in this period, the expected frequency under the null hypothesis would be 25 cars sold in each month. The observed sales are given in Table 8.3.

The null and alternative hypotheses are:

$$H_0: E_1 = E_2 = \cdots = E_5 = 25;$$
$$H_a: \text{The frequencies are not all equal.}$$

**Table 8.3** Monthly new car sales

| | | | Months | | | | |
|---|---|---|---|---|---|---|---|
| | Jan. | Feb. | Mar. | Apr. | May | June | Total |
| Expected sales ($E_i$) | 25 | 25 | 25 | 25 | 25 | 25 | 150 |
| Observed sales ($O_i$) | 27 | 18 | 15 | 24 | 36 | 30 | 150 |

The chi-square statistic for this example has $c - 1 = 5$ degrees of freedom. If we let $\alpha = 0.025$, then the appropriate critical region, shown in Fig. 8.17, is derived from Table IV in the row $v = 5$. From this figure we see that the null hypothesis will be rejected in favor of the alternate hypothesis of unequal frequencies in monthly sales if the calculated value of $\chi^2$ exceeds 12.83.

The value of $\chi^2$ can now be calculated as follows:

$$\chi^2 = \frac{(27 - 25)^2}{25} + \frac{(18 - 25)^2}{25} + \frac{(15 - 25)^2}{25}$$
$$+ \frac{(24 - 25)^2}{25} + \frac{(36 - 25)^2}{25} + \frac{(30 - 25)^2}{25}$$
$$= 12.0.$$

Thus, at the significance level of $\alpha = 0.025$, we fail to reject the null hypothesis that all monthly sales are equal (i.e., deviations from sales equal to 25 are due to

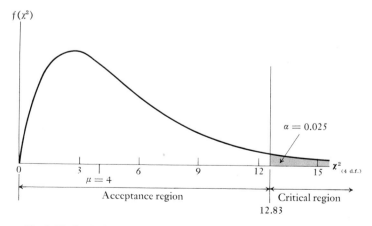

**Fig. 8.17.** Critical region for a goodness-of-fit test, $v = 5$, $\alpha = 0.025$.

random occurrences). However, from Table IV, note that for $c - 1 = 5$ degrees of freedom and $\alpha = 0.05$, the critical value for the test would be 11.07. Therefore, if $\alpha = 0.05$ had been selected as the level of significance for rejecting $H_0$, then the null hypothesis would have been rejected. In this case the choice of the $\alpha$-level is crucial to the decision. The costs of having too many or too few salesmen on hand in a given month should be given more consideration before setting $\alpha$ at an arbitrary level. Also, a larger number of past months may be sampled, if the sales pattern is presumed not to have changed. For example, the sum of sales for the past three years in the month of January, and February, and March, etc., might be used to test for equal monthly sales.

## 8.9 SUMMARY

The test statistics and test situations described in this chapter are summarized in Table 8.4, which repeats much of the information from Table 7.3.

**Table 8.4** Summary of one-sample test statistics

| Unknown parameter | Population characteristics and other description | Reference section | Test statistic |
|---|---|---|---|
| Mean $\mu$ | Population $N(\mu, \sigma^2)$ or sample size $n \geqslant 30$, $\sigma$ known | 8.3 7.4 6.5 | $z = (\bar{x} - \mu)/(\sigma/\sqrt{n})$ |
| Mean $\mu$ | Population $N(\mu, \sigma^2)$ $\sigma$ unknown | 8.4 7.5 6.8 | $t_{v\,\text{d.f.}} = (\bar{x} - \mu)/(s/\sqrt{n})$ where $v = n - 1$ |
| Proportion $p$ | Repeated independent trials $npq > 3$, $\hat{p} = x/n$ | 8.6 7.6 4.3 | $z = (\hat{p} - p_0)\Big/\sqrt{\dfrac{p_0(1 - p_0)}{n}}$ |
| Variance $\sigma^2$ | Population $N(\mu, \sigma^2)$ | 8.8 7.8 6.9 | $\chi^2_{v\,\text{d.f.}} = \dfrac{(n - 1)s^2}{\sigma^2}$ where $v = n - 1$ |
| Expected frequencies $E_i$ | Observed classification frequencies $O_i$ for $c$ classes | 8.8 | $\chi^2_{v\,\text{d.f.}} = \sum_{i=1}^{c} \dfrac{(O_i - E_i)^2}{E_i}$ where $v = c - 1$ |

## REVIEW PROBLEMS

1. Distinguish between the following terms:
   a) simple hypothesis and composite hypothesis,
   b) null hypothesis and alternative hypothesis,
   c) one-sided test and two-sided test,
   d) Type I and Type II errors,
   e) acceptance and rejection regions.

2. a) Given a population standard deviation of 2, test $H_0: \mu = 30$ vs. $H_a: \mu \neq 30$ at the 0.01 significance level. A sample of size 25 gives a sample mean of 34.
   b) In making your conclusion in the above test, what types of error might you be making?
   c) How can these errors be reduced?

3. A manufacturer of steel rods considers his manufacturing process to be working properly if the mean length of the rods is 8.6 in. The standard deviation of the rods always runs about 0.3 in. Suppose that a sample of 36 rods is taken, and the average length of these rods is found to be 8.7 in.
   a) What are the null and alternative hypotheses for this problem?
   b) Given the sample results, can the process be considered to be working properly? Let $\alpha = 0.05$.

4. A manufacturer of rope produces 12,500·ft of rope a day, in lengths of 50 ft each. Each day 16 of the 50-ft lengths are randomly selected and tested for tensile strength. The mean tensile strength of one sample of 16 observations was 340 pounds, with a standard deviation of $s = 16$. At what level of significance can the null hypothesis $H_0: \mu = 350$ be rejected in favor of $H_a: \mu \neq 350$? At what level can $H_0: \mu = 350$ be rejected if $H_a: \mu < 350$? What assumptions about the parent population are necessary to determine these probabilities?

5. In a study of family income levels, nine households in a certain rural county are selected at random. An average income of $8000 with a standard deviation of $2400 is reported from this sample survey. Use a statistical test to decide if the null hypothesis that $\mu = 10,000$ can be rejected in favor of $H_a$ that the true average income is less than $10,000 per household. Use the 0.05 significance level.

6. A survey of 36 homes in New York revealed that television is watched 27 hours per week, on the average. The sample standard deviation was 4 hours. Test whether this sample indicates that the number of hours New York families watch television equals the nation-wide average of 25 hours per week ($H_0$), or is more than the national average ($H_a$). You should be 99% confident of your answer.

7. You are interested in a site for a new restaurant, and a real estate developer claims that, on the average, male resident students at the University eat at least eight meals per week at establishments located more than one mile (beyond walking distance) from their residences. To test this claim, you set $H_a: \mu < 8$, randomly select 6 students, and record the number of times they ate beyond the "one-mile boundary" in one designated week; the results are 6, 5, 7, 4, 8, and 6. Test, at the 0.01 significance level, whether these results indicate rejection of $H_0: \mu \geqslant 8$.

8. A power shovel was designed to remove 31.5 cubic feet of earth per scoop. On a test run, some 25 sample scoops were made; the mean of the samples was 29.3 cubic feet. The standard deviation, as derived from the sample information, was three cubic feet. Test at the 99% level of confidence whether the design specifications for this equipment should be revised on the basis of the sample information. Describe the two types of error possible in this test.

9. The following nine observations were drawn from a normal population: 25, 17, 18, 22, 21, 27, 19, 15, 25.

   a) Test the null hypothesis $H_0: \mu = 24$ against the alternative hypothesis $H_a: \mu \neq 24$. At what level of significance can $H_0$ be rejected?

   b) At what level of significance can $H_0: \mu = 24$ be rejected when tested against $H_a: \mu < 24$?

10. a) The average daily output of a certain department within an industrial plant is scheduled to be 85 units. Twenty-five days are selected at random, and the output is observed for each day. From this sample, the average output is calculated to be 81 units, and the standard deviation is 8 units. Test with 99% confidence whether or not the average output is different from that scheduled.

    b) Explain the meaning of the beta-risk for this test.

11. A firm packaging a deluxe type of ornamental matches for fireplace use designed a process to place 18 matches per box. The process was started and allowed to operate for 30 minutes. A sample of 16 boxes was then drawn. On the basis of this sample, the number of matches per box averaged 17 while the standard deviation was calculated to be 2. Would a one-sided test accept the null hypothesis of a mean of 18 if alpha were set at 0.05?

12. Suppose that you decide to test the null hypothesis that the probability of a six on each toss of a single die is 0.17. You intend to toss this die 100 different times, and will reject the null hypothesis if fewer than 10 sixes or more than 24 sixes appear.

    a) For the critical region described above, what is the probability of making a Type I error using Table I?

    b) Use Formula (5.7) and Table III to obtain an approximation to your answer to part (a).

    c) What is the probability of a Type I error if the critical region is defined to be the occurrence of 22 or more sixes (a one-tailed test)?

    d) If the alternative hypothesis is $H_a: p = 0.25$, and the critical region in part (c) is used, what is the probability of a Type II error?

13. In one state in the U.S. a law was recently passed requiring all motorcyclists to wear protective helmets. Before this law, motorcyclists were fatally injured in 5 percent of all motorcycle accidents reported to the state police. During the month after the law was passed, there was only one fatal injury in the 100 accidents reported to the police. Is this sample sufficient to reject the null hypothesis $H_0: p = 0.05$ in favor of $H_a: p < 0.05$, using $\alpha = 0.01$?

14. One student spokesman claims there is more student support for the campus food services this year than last year, when 80% of the students were dissatisfied with the food service. Suppose that, in a current sample of 64 randomly selected students, there are 20 who

think the food services are satisfactory, and the others indicate some dissatisfaction. Test the claim at the 0.01 level of significance.

15. In Lake Opeongo, Ontario, two-fifths of the trout are supposed to be over 15 inches long. During two weeks of fishing, a group catches 100 trout, and finds 60 of them to be over 15 inches long. Can this group argue that the lake must contain a proportion larger than 2/5 of fish over 15 inches long? Test at the 0.005 significance level.

16. A reasonable proportion of excessive drinkers at home football games would be one-fifth, according to a local editorial. Sampling from a student section of 36 fans revealed nine in a condition indicating excessive alcohol intake. Would you conclude that too many fans are drinking excessively? Test at the 0.01 significance level.

17. The proportion of young egrets which survive and become full-grown under natural conditions is $\frac{1}{5}$. Because the birds are becoming scarce, a zoo curator traps some of them and attempts to hatch and raise the young in a controlled environment. Out of 64 eggs, he is able to raise 16 birds to full growth. Should the procedures of this curator be copied if we desire to increase the population of egrets? Allow for a 5% chance that you would decide on the expensive controlled procedure when, really, it is not significantly better.

18. Suppose that a supermarket has agreed to advertise through a local newspaper if it can be established that the newspaper's circulation reaches more than 50 percent of the supermarket's customers.
    a) What null and alternative hypotheses should be established in this problem in trying to decide, on the basis of a sample of customers, whether or not the supermarket should advertise in the newspaper?
    b) If a sample of size $n = 64$ is collected, and $\alpha = 0.01$, what is the critical value for making a decision whether or not to advertise?

19. a) Which is more serious, a Type I or a Type II error?
    b) Why is it necessary to be concerned with the *probability* of Type I and Type II errors?
    c) What is meant by the phrase "the costs of making an incorrect decision"? How are these costs related to the problem of balancing the risks of making an incorrect decision?
    d) How would you go about the process of testing hypotheses in circumstances in which it is impossible to associate, at least directly, any dollar values to the costs of making an incorrect decision? Assume that you are responsible for making an important decision (such as a doctor who must decide whether or not to operate on a patient).

20. What is meant by the phrase "goodness-of-fit test"? How is the $\chi^2$-distribution used to make this test?

21. Test the null hypothesis $H_0: \sigma^2 = 9$ vs. the alternative hypothesis that the population variance is less than 9, based on a sample of size 19, with variance 4. Use $\alpha = 0.025$.

22. Given a sample of size 12 with standard deviation 6, test if this result implies rejection of $H_0: \sigma^2 = 25$ or rejection of the alternative hypothesis that the population variance is greater than twenty-five. Use $\alpha = 0.05$.

23. In one of his classical experiments on heredity, Gregor Mendel observed the color of the plants bred from a purple-flowered and a white-flowered hybrid. Out of 929 plants, 705

were observed to have a purple flower, and 224 were observed to have a white flower. Test the hypothesis that the probability of a purple-flowered plant is $\frac{3}{4}$, by means of the $\chi^2$-test, using $\alpha = 0.05$.

24. A store wishes to determine whether there is any difference in frequency of customers among selected groups of people during a typical day. The observed frequencies are:

| Group | Observed frequency |
|---|---|
| Teenage | 12 |
| Young adult | 16 |
| Middle age | 24 |
| Senior citizen | 8 |

Use $\alpha = 0.05$, and test whether this evidence indicates a difference in frequencies among groups.

## EXERCISES

25. A sample survey firm is contracted by an advertising agency to determine whether or not the average income in a certain large metropolitan area exceeds $7500. The agency wants the results of this survey to reject the null hypothesis $H_0: \mu = \$7500$ in favor of $H_a$: $\mu > \$7500$ at the $\alpha = 0.05$ level of significance when the true mean is as small as $7600. If the population standard deviation of incomes in this area is assumed to be $1000, how large a sample will the survey firm have to take in order to meet the requirements of the advertising agency?

26. Suppose that, in Exercise 25, the survey firm charges $5 for each observation it collects for the advertising agency. How much will it cost the agency to be able to reject $H_0$: $\mu = \$7500$ at the $\alpha = 0.01$ level rather than at the $\alpha = 0.05$ level?

27. Suppose a test on the hypotheses $H_0: \mu = 300$ against $H_a: \mu > 300$ is done with $\alpha = 0.01$, $\sigma = 40$, and $n = 16$.

   a) What is the probability that the null hypothesis might be accepted when the true mean is really 310?
   b) Repeat part (a) if $\mu$ is really 330.
   c) How do these values of $\beta$ change if the test had used $\alpha = 0.05$?

28. Suppose that you are presented with an urn containing six balls, some of which are red, the rest of which are white. Let $\theta$ represent the number of red balls. You decide to test the null hypothesis $H_0: \theta = 3$ against $H_a: \theta \neq 3$, by drawing two balls from the urn, accepting $H_a$ if the two balls are the same color, and accepting $H_0$ if the two balls are different colors.

   a) If sampling occurs with replacement, find the probability of a Type I error. Find the probability of a Type II error, assuming that there are $\theta = 0, 1, 2, \ldots, 6$ red balls.
   b) Repeat part (a), assuming that sampling occurs without replacement.

29. Suppose that you are given the probability density function $f(x) = 1/\theta$,

$$f(x) = \begin{cases} 1/\theta & \text{for } 0 \leqslant x \leqslant \theta, \\ 0 & \text{otherwise.} \end{cases}$$

a) You decide to test the null hypothesis $H_0: \theta = 1$ against the alternative hypothesis $H_a: \theta = 2$ by means of a single observation. What is the value of $\alpha$ and $\beta$ if you select the interval $x \leqslant 0.5$ as the critical region? Sketch this density function under both the null and alternative hypotheses, and indicate the critical region on this graph.

b) What is the value of $\alpha$ and $\beta$ if you select $x \leqslant 0.75$ as the critical region?

c) Which of these two critical regions would be more appropriate if a Type II error is more serious than a Type I error?

30. It has been estimated that most United States families spend approximately 90 percent of their yearly income and save only 10 percent. Suppose that a random sample of 100 families with incomes exceeding $25,000 shows that these people spend only 80 percent of their yearly income, saving 20 percent.

a) Does this sample support the hypothesis that people with incomes exceeding $25,000 save more than 10 percent of their income? What is the null hypothesis in this case? Given these sample results, what is the probability that the null hypothesis is true?

b) Would you conclude from the sample in this problem that families with high incomes will tend to save more than families with more average incomes? Why or why not?

31. Senator Fogbound has taken a random sample of 20 of his constituents in order to decide whether a majority of the people favor the legislation he is proposing.

a) At what level of significance can he reject the null hypothesis $H_0: p = \frac{1}{2}$ in favor of $H_a: p \neq \frac{1}{2}$ if 15 of the 20 respondents indicate that they favor his legislation? At what level of significance can the null hypothesis be rejected if $H_a: p > \frac{1}{2}$?

b) What critical region would you establish to test the null hypothesis $H_0: p = 0.65$ against the alternative $H_a: p > 0.65$ at the 0.01 level of significance? What would the critical region be for testing this same null hypothesis against $H_a: p \neq 0.65$ at the 0.05 level of significance?

32. An industrial firm making small battery-powered toys periodically purchases a large number of flashlight batteries for use in the toys. The policy of this company has been never to accept a shipment of batteries unless it is possible to reject, at the 0.05 level of significance, the hypothesis that the batteries have a mean life of less than 50 hours. The standard deviation of the life of all shipments has been running about 3 hours.

a) What null and alternative hypotheses should be established to implement the company policy?

b) Should the company accept shipment if a sample of 64 batteries results in a mean life of 50.5 hours?

c) What is the minimum mean life this company should accept in a sample of 64 batteries?

33. In the discussion in Section 8.7, we calculated the probability of making an incorrect decision concerning adjustments to a production process. The values of $\alpha$ and $\beta$ were determined for the critical region $x \geqslant 4$ and for the critical region $x \geqslant 5$, assuming a sample size of 20.

a) Calculate $\alpha$ and $\beta$ for this problem for the critical region $x \geqslant 6$.
b) Calculate the expected cost associated with this critical region when the probability of a defective is 0.10. Do the same thing for the probability of a defective equal to 0.25.
c) Calculate the total expected cost associated with this critical region, assuming that the probability of a minor adjustment is 0.70 and the probability of a major adjustment is 0.30.
d) Is the cost you determined in part (c) better or worse than the cost calculated in Section 8.7 for the critical region $x \geqslant 5$? Do you think $x \geqslant 5$ is the best critical region for this sample size? Explain why, and then try to draw a graph relating the location of the critical region and the total expected costs (let the horizontal axis be the lower bound of the critical regions).

34. In Section 8.7 it was shown that, for the production process under investigation, a sample size of 50 results in a lower expected cost than a sample size of 20 if the additional observations cost only an extra $10. Suppose that we now have the opportunity to buy 50 more observations (for a total $n$ of 100) for $15 more (i.e., sampling cost of $25). Estimate, as best you can, the optimal critical region for this size of sample. Is the total expected cost in this case lower or higher than the cost when $n = 50$?

35. Return to Exercise 31, Chapter 4, and once again determine the Poisson approximation to the observed frequencies for deaths from the kick of a horse in the Prussian Army. Use the $\chi^2$-test to determine whether the observed frequencies and the expected frequencies under the Poisson distribution differ significantly. At what level of significance can $H_0$ be rejected?

36. A cafeteria proposes four main entrees. For planning purposes, the manager expects the proportions of each that will be selected by his customers to be:

| Selection | Hot dogs and chile | Roast beef | Steak | Fish |
|---|---|---|---|---|
| Proportion | 0.20 | 0.50 | 0.20 | 0.10 |

The first 50 customers select hot dogs and chile 15 times, roast beef 20, steak 5, and fish 10. The manager wonders whether he should revise his preparation schedule, or whether this deviation from his expectations is merely chance variation that should balance out overall. Make an appropriate test, at the 0.01 level of significance, to give advice to this manager.

# Statistical
# Decision Theory

## 9.1  INTRODUCTION TO A DECISION PROBLEM

The focus of this chapter, as the title implies, is on the process of making decisions from a statistical point of view. Decision-making in this context is often referred to as *decision-making under uncertainty*, because the consequences or payoffs resulting from each decision are not assumed to be known (in advance) with certainty. As you might suspect, there is also a set of techniques concerned with *decision-making under certainty*, where consequences of a decision are assumed to be known (in advance) with certainty. Most real-world decisions generally involve some element of uncertainty; hence it is convenient to have a formal procedure for analyzing each possible action a decision-maker might take in a given situation, and for selecting the best action.

The origins of statistical decision theory are relatively recent, dating back to only about the early 1950's. Since that time, this branch of statistics has grown rapidly in popularity, with a corresponding development of the theory and its applications. Because much of the analysis of statistical decision theory centers around a formula first published in 1763 by the Reverend Thomas Bayes (Bayes' rule), this approach is often referred to as the "Bayesian" approach. In fact there is considerable debate among statisticians about the appropriateness of the Bayesian approach compared to the "classical" approach to statistics (i.e., traditional procedures in sampling, estimation, and hypothesis-testing). We will not enter into this debate here, but instead will present both the advantages and disadvantages of statistical decision theory.

In analyzing a decision-making problem, it is necessary to be able to specify exactly what actions (or alternatives) are available. We will label actions as $a_1$, $a_2, a_3, \ldots$ In addition, we assume that each action yields a *payoff* (or some type of consequence), which depends on the value of a random variable called the *state of nature*. States of nature will be labeled as $\theta_1, \theta_2, \theta_3, \ldots$ For example, suppose you are considering buying 100 shares of one of four common stocks (actions $a_1, a_2, a_3$, or $a_4$), each of which costs $10 now. You intend to sell your stock at the end of one year. If you could somehow foresee the future and *knew* that one year from now the prices of these stocks would be $15, $11, $8, and $10, respectively, then your decision would be an easy one—buy the first stock (action $a_1$). Action $a_1$ yields a profit of $500 ($5 profit on 100 shares); $a_2$ gives a payoff of $100; $a_3$ results in a profit of $-$200 (a loss); and $a_4$ produces no profit at all. This situation represents decision-making under certainty, since there is no uncertainty about what state of nature (i.e., what set of prices) will occur a year from now.

Instead of knowing what prices will be in one year, you probably can only guess what the prices might be, based on your impression of the economy in general, your knowledge of various industries and firms within these industries, or perhaps merely based on a hot tip from a friend. There may be many different states of nature, each yielding a different set of payoffs for the various actions you could take. Thus, your problem is really one of decision-making under uncertainty.

**A Decision Problem Under Uncertainty**

To extend the stock problem to include uncertainty, suppose you decide there are three possible states of nature ($\theta_1$, $\theta_2$, and $\theta_3$). For instance, you might decide that stock prices one year hence are directly related to the stability of the economy during the year. In this case, $\theta_1$ might correspond to a mild recession, $\theta_2$ to a stable economy, and $\theta_3$ to a mild inflation. (Other states are also possible (e.g., a depression), but for convenience we will assume that there are only three.) Suppose that, for $\theta_1$ the prices a year from now will be those given previously ($15, $11, $8, and $10); in $\theta_2$ the prices will be $5, $12, $12, and $13; while in $\theta_3$ they will be $17, $11, $15, and $15. The payoffs that these prices reflect can be expressed in what is called a *payoff table*, as shown in Table 9.1.

**Table 9.1** Payoff table for stock example

|         |       | States of nature | | |
|---------|-------|------------|--------|--------|
|         |       | $\theta_1$ | $\theta_2$ | $\theta_3$ |
|         | $a_1$ | $500   | -$500  | $700   |
|         | $a_2$ | 100    | 200    | 100    |
| Actions | $a_3$ | -200   | 200    | 500    |
|         | $a_4$ | 0      | 300    | 500    |

*Dominant actions.* There is no one action for this payoff table which is obviously the "best" one. Action $a_1$ yields the largest payoff *if $\theta_1$ or $\theta_3$* occurs, while $a_4$ is optimal if $\theta_2$ occurs. Note that $a_3$ can never be the optimal action because $a_4$ always results in a payoff at least as large as, or larger than, $a_3$ no matter *what* state of nature occurs. When $\theta_1$ occurs, $a_4$ yields $0 and $a_3$ yields only -$200; for $\theta_2$, $a_4$ results in a payoff of $300 compared to only $200 for $a_3$; in $\theta_3$, both actions yield the same payoff, $500. An action such as $a_3$ in this case, which is no better than some other action no matter what state of nature occurs, is said to be a *dominated* action.

*Maximin solution.* It might also appear that $a_2$ can never be optimal for this payoff table. This action is optimal, however, under what is called the *maximin* criterion. This criterion says the decision-maker should focus only on the *worst* possible payoff that could happen for each action (that is, $-\$500$ for $a_1$, $\$100$ for $a_2$, $-\$200$ for $a_3$, and $\$0$ for $a_4$). He should then pick the action which gives him the *maximum* payoff among all these *minimum* values (hence, *maximin*). In the stock example, the maximin criterion leads to action $a_2$, since this stock's lowest payoff, $\$100$, is the largest among the *minimum* values.

*Minimax regret solution.* Closely related to the maximin approach is another criterion called the *minimax regret* criterion. Under this approach, the best action for the decision-maker is to select that alternative which minimizes his *maximum* (i.e., minimax) *regret*. We can illustrate how regret is calculated by referring to our stock example. Let's suppose $\theta_1$ occurs. If the decision-maker had selected $a_1$, then he would have *zero* regret, since $\$500$ is the best payoff he could receive under $\theta_1$. If, however, he had selected $a_2$, then he would have had a regret of $\$400$, since this is the amount of money he "lost" by not selecting $a_1$ (that is,

$$\$500 - \$100 = \$400).$$

Similarly, $a_3$ has a regret of $\$700$ and $a_4$ has a regret of $\$500$. For $\theta_2$ we see that $a_4$ is the action yielding *zero regret*, while $a_1$ gives $\$800$ regret, and $a_2$ and $a_3$ each yield $\$100$ regret. The regret values for all actions and states can be expressed in what is called a *regret table* (or *opportunity loss* table), like Table 9.2. The action *minimizing* the maximum regret is seen to be action $a_4$.

**Table 9.2** Regret table for stock example (based on Table 9.1)

|            |       | States of nature |            |            |            |          |
|------------|-------|------------------|------------|------------|------------|----------|
|            |       | $\theta_1$       | $\theta_2$ | $\theta_3$ | Max regret | Minimax  |
|            | $a_1$ | $\$\ \ 0$        | $\$800$    | $\$\ \ 0$  | $\$800$    |          |
| Actions    | $a_2$ | 400              | 100        | 600        | 600        |          |
|            | $a_3$ | 700              | 100        | 200        | 700        |          |
|            | $a_4$ | 500              | 0          | 200        | 500        | ←——————  |

## 9.2 EXPECTED MONETARY VALUE CRITERION

It is not difficult to see that the maximin and the minimax regret criteria are both essentially pessimistic approaches to decision-making, in that they focus only on the worst events that might take place. Since most of us do not believe that Lady Nature is "out to get us," we would like a procedure which not only takes into account all of the values in the payoff table, but also considers their *relative likelihood*. Fortunately, there is a procedure for accomplishing this, called the *Expected Monetary Value criterion*.

In order to be able to use the Expected Monetary Value (EMV) criterion, it is necessary to know (or be able to determine) the probability of each state of nature. If there is considerable "objective" evidence (e.g., historical data) or a theoretical basis for assigning probabilities, then this task may be a fairly easy one. The difficulty in many real-world problems is that there may be little or no historical data and no theoretical basis to use in making probability assessments. The answer to the question of how to assign probabilities in these circumstances is not an easy one. Bayesian statisticians usually suggest that, in assessing the *subjective probability* of an event, the decision-maker should ask himself *at what odds* he would be exactly indifferent between the two sides of an even bet. For instance, at what odds would you consider it to be a "fair" bet (i.e., you would be indifferent between the two sides of the bet) if you are asked to bet either that the Baltimore Colts win the American Football League Conference or that they do *not* win this conference? If you say, for example, that 4:1 odds against the Colts winning represents "fair" odds (that is, you would be willing to take either side of the bet at these odds), then, by definition, your subjective probability that the Colts will *win* the conference is 1/5, or 0.20. In a decision-making context, one could use this approach to assess the probability of each state of nature, being careful, of course, to see that the sum of these probabilities equals one.

You should recall from the material on expectations in Section 3.3 that an expected value is merely the mean (or arithmetic average) of a random variable. Since an expectation calculated in decision theory usually involves monetary values, such an expectation is generally referred to as an *expected monetary value* (or *EMV*). To illustrate the calculation of EMV's, let's suppose that, in the stock example, the probabilities of our three states of nature are

$$P(\theta_1) = 0.30, \qquad P(\theta_2) = 0.60, \qquad \text{and} \qquad P(\theta_3) = 0.10.$$

Each EMV of actions $a_1$, $a_2$, $a_3$, and $a_4$ is given by multiplying each payoff by its probability of occurrence, and then summing these values:

$$\text{EMV}(a_1) = \quad \$500(0.30) - \$500(0.60) + \$700(0.10) = -\$ 80;$$
$$\text{EMV}(a_2) = \quad \$100(0.30) + \$200(0.60) + \$100(0.10) = \quad \$160;$$
$$\text{EMV}(a_3) = -\$200(0.30) + \$200(0.60) + \$500(0.10) = \quad \$110;$$
$$\text{EMV}(a_4) = \quad \$ \ 0(0.30) + \$300(0.60) + \$500(0.10) = \quad \$230.$$

Suppose we let $\pi_{ij}$ represent the profit to the decision-maker if he selects strategy $a_i$ and the $j$th state of nature occurs $(j = 1, 2, \ldots, m)$. Using this notation, we can write the EMV of action $a_i$ as follows:

> *Expected monetary value of action $a_i$:*
>
> $$\text{EMV}(a_i) = \sum_{j=1}^{m} \pi_{ij} P(\theta_j). \tag{9.1}$$

Under the EMV criterion, the decision-maker selects the alternative that will yield the highest expected monetary value. For this example, action $a_4$ results in the highest EMV, with an average payoff of $230. (If the three states of nature had been assigned probabilities other than 0.30, 0.60, and 0.10, then some other action might be the optimal decision. For example, you should verify that 0.40, 0.20, and 0.40 lead to action $a_1$ as the optimal EMV.) No matter what values $P(\theta_1)$, $P(\theta_2)$, and $P(\theta_3)$ take on, however, $a_3$ can never yield the largest EMV because it is a *dominated action.*

### The Effect of Risk on the Optimal Decision

The EMV criterion itself suffers from one major weakness, namely, that it considers only the expected or mean profit, and does not take into account the variance in the payoffs. If the variance is fairly constant across the relevant alternatives, then this weakness will probably not cause any problems; but when the variability is large, the EMV criterion might indicate an action which won't be the most preferred for some people. For instance, suppose you must choose between two stocks ($a_1$ and $a_2$), and there are only two possible states of nature ($\theta_1$ and $\theta_2$), each having the *same* probability,

$$P(\theta_1) = 0.50 = P(\theta_2).$$

Let the payoff table be given by Table 9.3.

**Table 9.3** Example of decision affected by risk

|  |  | States of nature | | |
|---|---|---|---|---|
|  |  | $\theta_1$ | $\theta_2$ | EMV |
| Actions | $a_1$ | −$50 | $100 | 25 |
|  | $a_2$ | 10,500 | −10,000 | 250 |
| Probability |  | 0.50 | 0.50 |  |

Even though the EMV of $a_2$ is 10 times as large as that of $a_1$, most people (including the authors) would select $a_1$ if forced to pick between the two stocks, because they cannot afford to risk losing $10,000. Some people might prefer $a_2$ over $a_1$, which merely illustrates the fact that the value of a dollar to one person is not necessarily the same as the value of a dollar to some other person; neither does the value of a dollar necessarily *remain the same* to one person over time. We will see later in this chapter (Section 9.8) that it is possible to take the *value* of money into account in decision-making situations by using the expected utility criterion.

Our decision situations thus far have assumed that a single decision is to be made for the payoffs involved. This might not be the case for a business man who regularly makes decisions involving relatively large amounts of money, or a stock-broker who is continually investing in the stock market. In such cases, a decision-maker may be less concerned about the variance in payoffs for any one decision because he knows that, over a large number of decisions, his gains should offset his losses.

## 9.3 PERKINS PLASTICS—AN EXAMPLE

To illustrate the process involved in calculating EMV's in a more complicated setting, consider the following example about a small plastics firm:

Perkins Plastics, in its experiments with chrome plating on a plastic butterfly valve, has had adhesion problems because of irregularities in the electric flow during the plating process. Approximately 70% of the time the current is fairly uniform, in which case each batch of valves they produce (1000 at a time) will contain about 90% good valves and only 10% defective valves. The other 30% of the time, when the current is somewhat irregular, only 60% of the valves are good and 40% are defective. Unfortunately, there is no way the engineers at Perkins can determine how good the current flow is without testing each item in the batch. All they know is that each batch seems to have either 90% good or 60% good.

Perkins has several alternative ways to handle each batch. They can send the batch directly to the next operation (assembly), and hope for the best. Their records show that when they do this they incur costs of delay and adjustment of about $1000 for each batch with 90% good valves, and $4000 cost for the 60% good batches. Another alternative for Perkins is to rework the entire batch. This process ensures that the batch will be sufficiently free from defects so that no costs of delay and adjustments occur. However, this reworking costs $2000.

The decision facing Perkins at this point is whether they should send each batch directly to assembly, or rework it. The relevant data is shown in the payoff table on page 328.

In addition to constructing a payoff table, in many problems it is also helpful

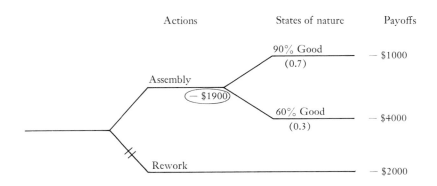

**Fig. 9.1.** Tree diagram for Perkins Plastics.

to diagram the situation in the form of a "decision tree." For Perkins Plastics the decision tree in Fig. 9.1 is appropriate.

Note in this diagram how the first set of "branches" of the tree represents the decision-maker's possible *actions,* and the second set the various *states of nature.* The circled value represents the EMV resulting from the decision to send the batch directly to assembly. Nonoptimal branches on a decision tree are usually marked with the symbol ✕, as shown on the rework branch.

**Table 9.4** Payoff table for Perkins Plastics

|  |  | States of nature | | |
|---|---|---|---|---|
|  |  | 90% Good | 60% Good | EMV |
| Actions | Assembly | −$1000 | −$4000 | −$1900 |
|  | Rework | −$2000 | −$2000 | −$2000 |
| Probability |  | 0.70 | 0.30 |  |

Thus far we have assumed that a decision-maker arrives at a probability value for each state of nature on the basis of his personal assessment of the situation; this assessment can be based on a large variety of factors, such as any historical data he has, his experience in similar situations, or just his "gut" feelings. In many circumstances, however, it often seems desirable to have more information about the probability of each state of nature before attempting to make a decision. Most people gather information of this type almost daily: we ask our friends about places to eat, or live, or courses to take; we read consumer reports before making major purchases; and we study the stock market before investing. In a very real sense, you could say we are *gathering sample evidence* about the relevant states of nature.

In our analysis of decisions we would like to be able to formally incorporate additional (sample) information into the process. Additional information is usually not free, however; if nothing else, it takes time and effort merely to gather and interpret this information. Thus, our decision process should not only be able to incorporate new information, but it should also be able to indicate how *much* additional information it is worthwhile to collect.

Our first step in evaluating information will be to consider the effect of sample information on the decision-maker's evaluation of the relative likelihood of each state of nature. We will then turn to the more general question of *how much* information should be collected.

## 9.4 THE REVISION OF PROBABILITIES

A decision-maker's assessment of the probability of each state of nature before seeing any sample information is called his *prior* (or *a priori*) probability for that state. Let's assume there are $m$ mutually exclusive and exhaustive states of nature labeled $\theta_1, \theta_2, \ldots, \theta_m$ (or $\theta_i$, $i = 1, 2, \ldots, m$). Furthermore, let $P(\theta_i)$ be the prior probability of the $i$th state, where

$$\sum_{i=1}^{m} P(\theta_i) = 1.$$

Finally, we assume that the different sample results which might occur can be represented by the random variable $x$, and $P(x = x)$ is the probability that the random variable $x$ assumes the specific value $x$.

Now, consider a decision-maker who has observed some sample result $x$, and wants to know how this information changes his prior probability of the state $\theta_k$ (where $1 \leqslant k \leqslant m$). That is, he would like to determine the probability of $\theta_k$, *given $x$*, which is written as $P(\theta_k \mid x)$ and is called the *posterior probability* of $\theta_k$ given $x$. In order to revise the prior probability $P(\theta_k)$, we need to know how likely it is that the sample result $x$ would take place under each one of the $m$ different states of nature. These probabilities, of the form

$$P(x \mid \theta_1), \qquad P(x \mid \theta_2), \qquad \ldots, \qquad P(x \mid \theta_m),$$

are called likelihoods. The relationship between the $m$ prior probabilities $[P(\theta_i)]$, the $m$ likelihoods $[P(x \mid \theta_i)]$, and the one posterior probability of interest, is given by Bayes' rule. In terms of these symbols, we rewrite Formula (2.11) as follows:

$$
\textit{Bayes' rule:} \quad P(\theta_k \mid x) = \frac{P(\theta_k)P(x \mid \theta_k)}{\displaystyle\sum_{i=1}^{m} P(\theta_i)P(x \mid \theta_i)}. \tag{9.2}
$$

Thus, a posterior probability is calculated as follows:

$$
\textit{Posterior probability} = \frac{(\text{Prior probability})(\text{Likelihood})}{\sum(\text{Prior probability})(\text{Likelihood})}.
$$

### Example Using Bayes' Rule

We can illustrate Bayes' rule by assuming that Perkins Plastics has recently learned they can use a device in their plating operation which is capable of testing a sample of the valves from each batch. The cost for this device, including its rent and all labor involved, is $5 for each valve tested. Perkins is currently considering testing a single valve from the current batch (larger samples will be considered later).

The prior probability of the 90% good and the 60% good batches for Perkins were given previously:

$$
P(90\%) = 0.70, \qquad P(60\%) = 0.30.
$$

The probability of drawing one good valve from the 90% batch is

$$
P(G \mid 90\%) = 0.90,
$$

while the probability of drawing a defective from this batch is

$$
P(D \mid 90\%) = 0.10.
$$

Similarly,

$$
P(G \mid 60\%) = 0.60 \qquad \text{and} \qquad P(D \mid 60\%) = 0.40.
$$

These four conditional probabilities are the likelihoods.

The problem for Perkins at this point is how to revise the prior probabilities 0.70 and 0.30 after seeing a sample of one valve. For example, if the sample is a good valve, then they want to calculate the following two posterior probabilities:

$$
P(90\% \mid G) \qquad \text{and} \qquad P(60\% \mid G).
$$

Using Bayes' rule,

$$
\begin{aligned}
P(90\% \mid G) &= \frac{P(90\%)P(G \mid 90\%)}{P(90\%)P(G \mid 90\%) + P(60\%)P(G \mid 60\%)} \\
&= \frac{(0.70)(0.90)}{(0.70)(0.90) + (0.30)(0.60)} = \frac{0.63}{0.81} = 0.7778.
\end{aligned}
$$

If $P(90\% \mid G) = 0.7778$, then, by the probability law for complements, $P(60\% \mid G) = 0.2222$. Thus, observing one good valve has increased the probability that the batch is 90% good from 0.70 (the *prior*) to 0.7778 (the *posterior*). The reader may wish to practice using Bayes' rule to verify that

$$P(90\% \mid D) = 0.3684 \qquad \text{and} \qquad P(60\% \mid D) = 0.6316.$$

*Marginal probabilities.* In the process of making decisions on the basis of posterior probabilities, we will need to determine how likely it is that each sample result will occur, that is, $P(G)$ and $P(D)$ in Perkins Plastics. These values are *marginal probabilities.* To calculate $P(G)$, we need to multiply the (prior) probability of having a 90% batch ($P(90\%) = 0.70$) times the probability of observing a good valve, given the batch is 90% good ($P(G \mid 90\%) = 0.90$), and add this to $P(60\%) \times P(G \mid 60\%)$. Referring to Formula (2.9),

$$P(G) = P(90\%)P(G \mid 90\%) + P(60\%)P(G \mid 60\%)$$
$$= (0.70)(0.90) + (0.30)(0.60) = 0.81.$$

Similarly,

$$P(D) = P(90\%)P(D \mid 90\%) + P(60\%)P(D \mid 60\%)$$
$$= (0.70)(0.1) + (0.30)(0.40) = 0.19.$$

Perhaps you noticed that we had already calculated $P(G) = 0.81$ in determining the denominator of $P(90\% \mid G)$ by Bayes' rule. The denominator of Bayes' rule will always be one of the marginal probabilities needed for decision-theory analysis. The marginal values 0.81 and 0.19 must sum to one, since no other sample outcomes are possible.

*Scheme for revising probabilities.* Bayes' rule thus provides us with a means for revising probabilities in the light of sample information. The prior probabilities represent a state of uncertainty before seeing any sample evidence, while the posterior probabilities represent the state of uncertainty *after* seeing a particular sample. It is important to emphasize that the terms prior and posterior probabilities relate only to a particular sample. A decision-maker, for example, may want to consider taking a second sample after observing the result of the first sample. The result of his first sample at that point represents "historical data;" hence, the posterior probabilities from the first sample become prior probabilities for the second sample. A diagram of this relationship appears in Fig. 9.2.

### 9.5 BAYES' RULE FOR LARGER SAMPLE SIZES

To illustrate the process of revising probabilities when the sample size is greater than one, let's assume that Perkins Plastics' first sample was a good valve, and

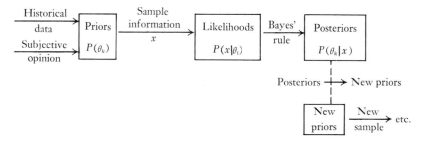

**Fig. 9.2.** The revision of probabilities.

they have now decided to sample one more valve. The prior probabilities for this second sample (the old posteriors) are

$$P(90\%) = 0.7778 \quad \text{and} \quad P(60\%) = 0.2222.$$

It is convenient to assume that the likelihoods for the second sample are the same as those for the first sample, since removing one valve (the first sample) from a batch of 1000 won't change the likelihoods very much. Now, suppose the second sample results in a defective valve (which we designate as $D_2$). The posterior probability $P(90\% \mid D_2)$ is:

$$P(90\% \mid D_2) = \frac{P(90\%)P(D_2 \mid 90\%)}{P(90\%)P(D_2 \mid 90\%) + P(60\%)P(D_2 \mid 60\%)}$$

$$= \frac{(0.7778)(0.10)}{(0.7778)(0.10) + (0.2222)(0.40)} = 0.467.$$

Thus we see that if the second valve is defective, it decreases the probability that the batch is 90% good from 0.7778 to 0.467.

An interesting property of Bayes' rule is that it makes no difference whether we revise probabilities after each observation, or whether two (or more) observations are grouped together and considered as a single sample of size two containing one *good*. For example, two separate samples of size one, the first containing a good valve and the second containing a defective valve, is exactly equivalent to a single sample of size two containing one defective and one good valve. To demonstrate this fact, let's calculate the posterior probability for the 90% batch, assuming a single sample of size two containing one good and one defective valve. That is, we want to calculate $P(90\% \mid 1G, 1D)$. If a single sample of two valves is equivalent to two separate samples of one valve, then the posterior probability $P(90\% \mid 1G, 1D)$ must equal the probability we just calculated,

$$P(90\% \mid D_2) = 0.467.$$

In calculating $P(90\% \mid 1G, 1D)$, the prior probabilities will be the same as in the original problem,

$$P(90\%) = 0.70 \quad \text{and} \quad P(60\%) = 0.30.$$

The likelihoods can be found in a table of binomial probabilities (see Table I) under the heading $n = 2$ (the number of valves in the sample), and in the columns $p = 0.90$ and $p = 0.60$: they are

$$P(1G, 1D \mid 90\%) = 0.18 \quad \text{and} \quad P(1G, 1D \mid 60\%) = 0.48.$$

These values can be used to calculate the posterior probability $P(90\% \mid 1G, 1D)$, as follows:

$$P(90\% \mid 1G, 1D) = \frac{P(90\%)P(1G, 1D \mid 90\%)}{P(90\%)P(1G, 1D \mid 90\%) + P(60\%)P(1G, 1D \mid 60\%)}$$

$$= \frac{(0.70)(0.18)}{(0.70)(0.18) + (0.30)(0.48)} = \frac{0.126}{0.270} = 0.467.$$

We see that the value 0.467 agrees with the value $P(90\% \mid D_2)$. Note that the posterior probability 0.467 is lower than the prior probability 0.70 because the sample $(1G, 1D)$ is more likely to have come from a 60% batch than a 90% batch.

As before, the probabilities given in the denominator of Bayes' formula are marginal probabilities. That is,

$$P(1G, 1D) = P(90\%)P(1G, 1D \mid 90\%) + P(60\%)P(1G, 1D \mid 60\%)$$
$$= (0.70)(0.18) + (0.30)(0.48) = 0.270.$$

Thus we have shown that, in binomial sampling, one can group the results of *several smaller samples* into one large sample. This means that the *order* in which the observations are received is not important; e.g., in a sample containing one defective and one good item, we don't care whether the good item or the defective item came first.

A good exercise for the reader at this point would be to calculate the other possible posterior probabilities for a single sample of $n = 2$, that is,

$$P(90\% \mid 2 \text{ Good}) \quad \text{and} \quad P(90\% \mid 2 \text{ Defective}).$$

The comparable posterior probabilities for the 60% batch are merely the *complements* of these values. So that you can verify your answers, the posterior and marginal probabilities are given in Table 9.5.

Thus we have seen how information, if it is available, can be used to revise the probability of each state of nature in a decision-making context. Our next task is to incorporate this information into our decision-making process, to see how much the information has helped us. Then we will be able to determine whether or not it is worthwhile to buy sample information, and if so, how much to buy.

Table 9.5 Probabilities for Perkins Plastics ($n = 2$)

|  |  | 2 Good | 1 Good, 1 Defective | 2 Defective |
|---|---|---|---|---|
| Posteriors | 90% | 0.840 | 0.467 | 0.127 |
|  | 60% | 0.160 | 0.533 | 0.873 |
| Marginals |  | 0.675 | 0.270 | 0.055 |

## 9.6 THE VALUE OF INFORMATION ($n = 1$)

The advantage of gathering information is that we can wait until we see the results of the sample, and *then* make the best decision in the light of the particular sample received. If a decision-maker always makes the same decision after gathering sample information as he would have before seeing the sample, then this information is of *no value* to him. In the Perkins Plastics problem, for example, the optimal decision, before seeing any sample information, is to send the batch to assembly. Unless one or more sample results lead to the decision to rework rather than sending the batch directly to assembly, the sample is worthless. The optimal decision after each possible sample result can be determined by using the posterior probability to determine the EMV of each action. To illustrate this process, let's return to the case where Perkins Plastics is considering one sample, of a single valve. We can now add a third alternative to the tree diagram in Fig. 9.1, a "sample" branch.

### The Decision Tree

The sequence of actions and states of nature in constructing a decision tree must always be arranged in exactly the same order as they occur to the decision-maker. In this case, his first decision is to take a sample of *one*. Next comes the sample results, either one good or one defective valve. After seeing this sample, the decision-maker must then decide either to rework, or to send directly to assembly. If he sends to assembly, the last thing he finds out is whether or not the batch is 90% good or 60% good. Note,·in the sample branch shown in Fig. 9.3, that the two marginal probabilities

$$P(G) = 0.81 \quad \text{and} \quad P(D) = 0.19,$$

as determined in Section 9.4, are shown below the lines representing the two sample results; and the four posterior probabilities (0.7778, 0.2222, 0.3684, 0.6316) are given below the events "90% good" and "60% good". The circled values are EMV's (rounded to the nearest dollar).

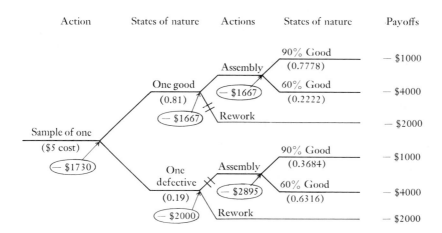

**Fig. 9.3.** Tree diagram for sample of one.

Calculation of the EMV's in a decision tree usually starts with the branches at the righthand margin, working backwards to the left. For example, we see that the expected cost of $-\$1667$ is derived from multiplying the posterior probability $P(90\% \mid G) = 0.7778$ times the cost of a 90% batch, $-\$1000$, and adding this to $P(60\% \mid G) = 0.2222$ times the cost of a 60% batch, $-\$4000$; that is,

$$-\$1667 = (0.7778)(-\$1000) + (0.2222)(-\$4000).$$

After a sample of one good valve, the optimal action is seen to be to send directly to assembly (at an expected cost of $-\$1667$), rather than to rework (at a cost of $\$2000$). If the sample valve is *defective*, however, the best strategy is to rework (at a cost of $\$2000$) rather than send to assembly at an expected value of $-\$2895$; that is,

$$(0.3684)(-\$1000) + (0.6316)(-\$4000) = -\$2895.$$

From the decision tree we see that the cost $-\$1667$ will occur with probability $P(G) = 0.81$, and the cost $-\$2000$ will occur with probability $P(D) = 0.19$. The expected cost of the sample branch equals the sum of these two costs times the probability that they will occur, or

$$(0.81)(-\$1667) + (0.19)(-\$2000) = -\$1730.$$

Add to this value the $5 cost of sampling and the total EMV is $-\$1735$.

**Evaluating the Results**

Recall that before any sample information was collected the optimal action was to send to assembly at a payoff of $-\$1900$. Since the optimal action now involves a payoff of $-\$1735$, the testing device will result in an average saving (per batch) of the difference between these two costs, or $165. This savings is called the *expected net gain from sampling* (ENGS). ENGS can be defined as follows:

$$\text{ENGS} = \text{EMV(Optimal after sample)} - \text{EMV(Optimal before sample).} \qquad (9.3)$$

In the case of Perkins Plastics,

$$\text{ENGS} = -\$1735 - (-\$1900) = \$165.$$

In decision-theory analysis, it is often useful to calculate an EMV without including sampling costs. This EMV is called the *expected value of sample information* (EVSI). EVSI can be calculated either directly from the decision tree, by not including the cost of sampling in the total EMV value, or by merely adding the cost of sampling to ENGS,

$$\text{EVSI} = \text{ENGS} + \text{Cost of sampling.} \qquad (9.4)$$

The value of EVSI represents the maximum amount the decision-maker would pay for the sample information. Perkins Plastics would thus pay no more than $170 for a sample of size one, since

$$\text{EVSI} = \text{ENGS} + \text{Cost of sampling}$$
$$= \$165 + \$5 = \$170.$$

### 9.7 ENGS FOR LARGER SAMPLE SIZES

If a sample of size one saves $165, a logical question is whether or not we can do even better with a sample of size two. Fortunately, we have already calculated (in Section 9.5) all the posterior and marginal probabilities necessary for this analysis. Because it is often easier to work with these probabilities in tabular form, we have reproduced them in Table 9.6 in a format that is convenient for calculating the posteriors. The decision tree for $n = 2$ is shown in Fig. 9.4.

*Results for sample size $n = 2$.* The total EMV for a sample of size two is seen in Fig. 9.4 to be $-\$1649$. Adding the $10 sampling cost to this value results in a total expected cost of $-\$1659$. As before, we can calculate ENGS by subtracting the

**Table 9.6** Probabilities for Perkins Plastics example ($n = 2$)

| Valves | State of nature | Prior | Likelihood | Joint | Posterior | |
|--------|-----------------|-------|------------|-------|-----------|---|
| 2 good | 90% batch | 0.70 | 0.81 | .567 | .567/.675 = | .840 |
|        | 60% batch | 0.30 | 0.36 | .108 | .108/.675 = | .160 |
|        |           | 1.00 | Marginal = | .675 |           | 1.000 |
| 1 good, | 90% batch | 0.70 | 0.18 | .126 | .126/.270 = | .467 |
| 1 defective | 60% batch | 0.30 | 0.48 | .144 | .144/.270 = | .533 |
|        |           | 1.00 | Marginal = | .270 |           | 1.000 |
| 2 defective | 90% batch | 0.70 | 0.01 | .007 | .007/.055 = | .127 |
|        | 60% batch | 0.30 | 0.16 | .048 | .048/.055 = | .873 |
|        |           | 1.00 | Marginal = | .055 |           | 1.000 |

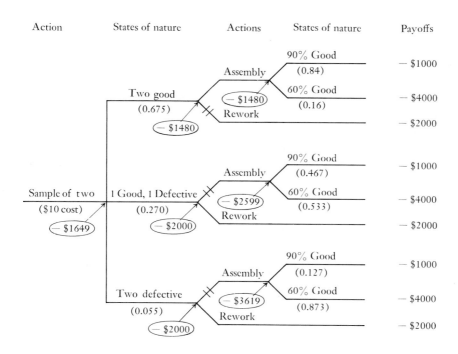

**Fig. 9.4.** Tree diagram for $n = 2$ based on Table 9.6.

cost of the *optimal policy before sampling* from this new total expected cost:

$$\text{ENGS}(n = 2) = \text{EMV(Optimal for } n = 2) - \text{EMV(Optimal before sample)}$$
$$= -\$1659 - (-\$1900) = \$241.$$

Since ENGS $(n = 1)$ was only \$165 and ENGS $(n = 2) = \$241$, a sample of two is better than a sample of one. EVSI can be calculated as follows:

$$\text{EVSI}(n = 2) = \text{ENGS}(n = 2) + \text{Cost of sampling}$$
$$= \$241 + \$10 = \$251.$$

The value \$251 represents the maximum amount Perkins should pay for a sample of two. The evaluations in Fig. 9.4 are summarized in Table 9.7.

**Table 9.7** Summary of decision analysis for $n = 2$ from Fig. 9.4

| Sample results | Marginal prob. | Optimal action | EMV | (EMV) × (Marginal prob.) |
|---|---|---|---|---|
| 2 good valves | 0.675 | Assembly | −\$1480 | −\$ 999 |
| 1 good, 1 defective | 0.270 | Rework | −\$2000 | −\$ 540 |
| 2 defective | 0.055 | Rework | −\$2000 | −\$ 110 |
| | | | Total EMV = | −\$1649 |

*Samples of size n > 2.* It should be clear by now that we can continue this process of analyzing larger and larger samples, and that the best sample size is the one giving the largest ENGS. Although the calculations become increasingly tedious as the sample size gets larger, it is often relatively easy to program these calculations on a computer. If this is done, then computer costs must be added to the analysis. At this point, it would be good practice for the reader to calculate ENGS for the Perkins Plastics problem for $n = 3$. You can check your calculation of marginal and posterior probabilities in Table 9.8. The ENGS for $n = 3$ is \$265.70. The values of ENGS and EVSI are given in Table 9.9 for $n = 1, 2, \ldots, 26$.

**Table 9.8** Probabilities for Perkins Plastics ($n = 3$)

| | | 3G | 2G, 1D | 1G, 2D | 3D |
|---|---|---|---|---|---|
| Posteriors | 90% | 0.8873 | 0.5676 | 0.1795 | 0.0352 |
| | 60% | 0.1127 | 0.4324 | 0.8205 | 0.9648 |
| Marginals | | 0.5751 | 0.2997 | 0.1053 | 0.0199 |

**Table 9.9** Values of EVSI and ENGS for selected sample sizes

| Sample size | EVSI | Sample cost | ENGS | Sample size | EVSI | Sample cost | ENGS |
|---|---|---|---|---|---|---|---|
| 1 | $170.00 | $ 5 | $165.00 | 14 | $494.53 | $ 70 | $424.53 |
| 2 | 251.00 | 10 | 241.00 | 15 | 506.82 | 75 | 431.82 |
| 3 | 280.70 | 15 | 265.70 | 16 | 513.03 | 80 | 433.03 |
| 4 | 281.51 | 20 | 261.51 | 17 | 514.30 | 85 | 429.30 |
| 5 | 340.80 | 25 | 315.80 | 18 | 523.77 | 90 | 433.77 |
| 6 | 380.05 | 30 | 350.05 | 19 | 533.60 | 95 | 438.60 |
| 7 | 400.04 | 35 | 365.04 | 20 | 539.21 | 100 | 439.21 |
| 8 | 405.35 | 40 | 365.35 | 21 | 541.33 | 105 | 436.33 |
| 9 | 423.85 | 45 | 378.85 | 22 | 543.92 | 110 | 433.92 |
| 10 | 450.49 | 50 | 400.49 | 23 | 551.80 | 115 | 436.80 |
| 11 | 451.96 | 55 | 406.96 | 24 | 556.66 | 120 | 436.66 |
| 12 | 472.33 | 60 | 412.33 | 25 | 559.01 | 125 | 434.01 |
| 13 | 479.94 | 65 | 414.94 | 26 | 559.26 | 130 | 429.26 |

From Table 9.9 we see that $n = 20$ is the optimal sample for Perkins Plastics.†
For other problems, the optimal size could be very large. On the other hand, it
may be that all values of ENGS are negative; this implies that the decision-maker
should not sample at all. In other instances, it may be better to take a sample,
observe the results of this sample, and *then* decide whether to stop sampling or to
continue with further observations. A discussion of such *stopping rules*, as they
are called, is beyond the scope of this book.

## *9.8  UTILITY ANALYSIS

As we pointed out earlier, the EMV criterion suffers from the weakness that it
fails to take into account the variability in profits of a decision. As early as the
18th century, Daniel Bernoulli investigated the fact that, for most people, the value
of a payoff does not always vary proportionally with its dollar amount. Bernoulli's
work can perhaps be considered as the first stage in the development of a method
permitting measurement of *relative* values in a decision-making context. This
method was developed by the late John von Neumann, a mathematician, and
Oskar Morgenstern, an economist, and was first published in 1944 in their now
classic book, *The Theory of Games and Economic Behavior*. Their method, in which

---

† The decision rule for $n = 20$ (which is not shown here) is to rework if the number of defectives
is 5 or larger.
* This section may be omitted without loss in continuity.

an index or scale of "utility," called a *utility function*, is constructed to measure relative values, is based on what is called the *theory of utility*.

### A Utility Scale

It is important to point out at this time that a utility scale is *unique to the individual* for whom it is constructed, and it is not meaningful to compare the values on one person's scale with the values on any other person's scale. We must also mention that, although the examples given below involve only monetary payoffs or consequences, nonmonetary factors can be taken into account as well.

A utility index, as determined by the von Neumann–Morgenstern approach, is measured on an *interval scale*. This type of scale is characterized by its lack of a predetermined zero-point (i.e., no specified origin) and the fact that the units of measurement can be arbitrarily selected (for example, Centigrade and Fahrenheit temperature scales represent interval measurement). It is because of this arbitrary choice of origin and unit that one cannot make interpersonal comparisons of utility.

The von Neumann–Morgenstern approach to determining a utility function is to ascertain the utility for a number of points between two values, and then use these points as the basis for sketching a continuous function over the entire range. Finding the utility for given dollar values is accomplished by asking the decision-maker to indicate when he is indifferent between the following two alternatives (where $A$, $B$, and $C$ are dollar values such that $\$A > \$B > \$C$):

> *Alternative I:* Receive $\$B$ for certain;
>
> *Alternative II:* Receive $\$A$ with probability $p$,
> Receive $\$C$ with probability $1 - p$.

Alternative II is sometimes referred to as the "standard lottery," while Alternative I is called the certainty equivalent. Suppose we denote the utility of Alternative I to the decision-maker as $U(\$B)$, and denote his expected utility for Alternative II as $pU(\$A) + (1 - p)U(\$C)$. If he is indifferent between these two alternatives, then

$$U(\$B) = pU(\$A) + (1 - p)U(\$C).$$

This equation is the basis from which a von Neumann–Morgenstern utility function is constructed.

### Constructing a Utility Function

Suppose we want to construct a utility function for the decision-maker in Perkins Plastics, that will associate his level of utility for profit values from $-\$5000$ to $\$5000$. In constructing a utility scale, two points must be arbitrarily assigned. Any dollar values can be assigned any utility values as long as the higher dollar value is assigned the higher utility (we assume everyone prefers more money to less). Let's rather arbitrarily set

$$U(\$0) = 0 \quad \text{and} \quad U(\$5000) = 100.$$

The unit of measurement in utility is called "utiles," so that $0 has a value of 0 utiles and $5000 has a value of 100 utiles.

There are a number of ways we can use these two values to determine additional points on the utility function. In one method, the value of $B$ is the decision variable. For example, we might let $A = \$5000$ and $C = \$0$, and $p = 1/2$, and ask the decision-maker what value of $B$ makes him exactly indifferent between the following two alternatives:

> *Alternative I:*  Receive $B for certain;
>
> *Alternative II:* Receive $5000 with $p = \frac{1}{2}$,
>
> Receive $0 with $1 - p = \frac{1}{2}$.

Assume our decision-maker says his certainty equivalent is reached when $B = \$1000$. The utility associated with $1000 can now be calculated as follows:

$$\begin{aligned} U(\$1000) &= \tfrac{1}{2}U(\$5000) + \tfrac{1}{2}U(\$0) \\ &= \tfrac{1}{2}(100) + \tfrac{1}{2}(0) \\ &= 50. \end{aligned}$$

Now we have three points on his utility curve (counting the original two points).

To determine a fourth point, we might use a second method in which the value of $C$ is the unknown value. Let $A = \$2000$, $B = \$0$, and $p = \frac{1}{2}$, and ask the decision-maker to choose between the following alternatives:

> *Alternative I:*  Receive $0 for certain;
>
> *Alternative II:* Receive $1000 with $p = \frac{1}{2}$,
>
> Receive $C with $1 - p = \frac{1}{2}$.

Assume he is indifferent when $C = -\$1500$. Then

$$\begin{aligned} U(\$0) &= \tfrac{1}{2}U(\$1000) + \tfrac{1}{2}U(-\$1500), \\ 0 &= \tfrac{1}{2}(50) + \tfrac{1}{2}U(-\$1500), \\ U(-\$1500) &= -50. \end{aligned}$$

This result gives a fourth point on his utility curve.

Although it is usually convenient to let either $A$, $B$, or $C$ be the value the decision-maker must adjust to the point of indifference, and let $p = \frac{1}{2}$ (it is easier to work with 50–50 odds), we can formulate a third method in which $p$ is the

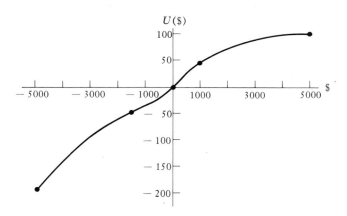

**Fig. 9.5.** Utility function for the decision-maker of Perkins Plastics.

decision variable. Let $A = \$0$, $B = -\$1500$, and $C = -\$5000$. The two alternatives are:

Alternative I:  Receive $-\$1500$ for certain;
Alternative II: Receive \$0 with probability $p$,
                Receive $-\$5000$ with probability $1 - p$.

If our decision-maker is indifferent when $p = 0.75$, then

$$U(-\$1500) = (0.75)U(\$0) + (0.25)U(-\$5000),$$
$$-50 = (0.75)(0) + (0.25)U(-\$5000),$$
$$U(-\$5000) = -200.$$

The examples above illustrate three variations on the process of determining points on a utility function. Once a sufficient number of such points have been calculated to assure that the function is accurately represented, then the utility function can be drawn by sketching a line between the points. We show, in Fig. 9.5, what the function for Perkins Plastics might look like when it is completed (the five points already calculated are marked with an ●).

**The Effect of Risk on the Utility Function**

Note several interesting characteristics about this function. First, the slope of the curve for most positive values is concave in shape, meaning that the decision-maker is a *risk avoider* in this region. That is, if two alternatives have equal EMV's, then he will prefer the alternative with the lower variance in payoff. This fact can be seen by our decision-maker's answer to the first set of alternatives presented him. At that time, he said that he was indifferent between receiving \$1000 for sure and

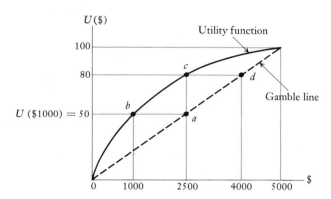

**Fig. 9.6.** A utility function for a risk-avoider.

a gamble between $5000 and $0, at 50–50 odds. Note that this gamble has an EMV of $2500. He is saying, in effect, that any amount, above $1000, received with certainty is preferred to the gamble. Even though the gamble has a higher EMV, he prefers the certainty equivalent because in the gamble he is taking a "risk" that his payoff will be only $0. Hence, he is called a "risk avoider." Perhaps Fig. 9.6 of our decision-maker's utility function between $0 and $5000 will help explain why his choices in this range represent risk avoidance.

The dashed line in this figure (the "gamble line") represents a standard lottery where $A = \$5000$ and $C = \$0$. When $p = \frac{1}{2}$, then the point $a$, which is halfway between the endpoints of the gamble line, represents the expected value of the gamble. We see that the EMV at point $a$ is $2500, and the expected utility is 50 utiles. Point $b$ represents the utility received by the decision-maker for $1000 received with certainty. Thus,

$$U(\$1000) = U(\text{Gamble}) = 50 \text{ utiles},$$

which is the answer this person indicated in response to our first set of alternatives.

Note from the diagram that our decision-maker's utility for $2500 received with certainty is 80 utiles (point $c$). One might ask at this time what value of $p$ would make this decision-maker indifferent between the standard lottery and $2500 for sure. As shown by point $d$ on his utility function, it takes a gamble with an EMV of $4000 to give him the 80 utiles that $2500 for sure yields him. Thus, the value of $p$ has to be 0.80 since

$$0.80(\$5000) + 0.20(\$0) = \$4000.$$

For payoffs from about $-\$500$ to $500, we see, in Fig. 9.5, that this decision-maker is a *risk-taker*, since his function is concave in shape in this region. His answer to our second set of alternatives illustrates this fact, for in it he indicated a preference for a gamble with an EMV of $-\$250$ unless the certainty equivalent was $0 or higher. Finally, we see that for large losses our decision-maker is again a *risk-avoider*, as his utility function is concave in shape in negative regions of the function. This fact is illustrated by his answer to our third set of alternatives. Had our decision-maker's function been linear over part of its range, then he would have been classified as *risk-neutral* in these portions of his curve. That is, his gamble line and his utility function would be the same line; hence, he would select the alternative giving him the highest EMV.

A utility function in which the decision-maker is willing to take risks for small amounts, but avoids risks for large losses or gains, is perhaps typical of many people. Most of us are willing to risk losing small amounts in poker games or by buying lottery tickets, but we avoid large losses by insuring our cars, homes, and businesses.

### Maximizing Expected Utility

Constructing a utility function is not an easy task, as we have seen. For important decisions, however, it may be a very worthwhile task, in that the substitution of utility values for monetary values may well change the optimal decision from one action to another. To illustrate the process of maximizing utilities, we have reproduced, in Fig. 9.7, the Perkins Plastics problem for $n = 1$ (see Figs. 9.1 and 9.3). The final number at the end of each branch is the appropriate utility value taken from Fig. 9.5 (that is,

$$-\$4000 = -150 \text{ utiles}, \quad -\$2000 = -60 \text{ utiles}, \quad \text{and} \quad -\$1000 = -40 \text{ utiles}).$$

The analysis is carried out in terms of expected utilities (instead of expected monetary value, as in Figs. 9.1 and 9.3).

The tree diagram in Fig. 9.7 shows that our decision-maker is enough of a risk-avoider so that he always decides to rework, even after a sample of one good valve. The information in a sample of $n = 1$ is thus worthless to him (EVSI $= 0$), and he should not pay the $5 this sample would cost. Of course, just because a sample of size $n = 1$ is not to his advantage doesn't mean that he should ignore the possibility of larger samples.

### 9.9  DECISION ANALYSIS FOR CONTINUOUS FUNCTIONS

Thus far only discrete functions have been used in our decision-theory analysis. That is, the number of actions have been discrete sets, as have the states of nature, the prior probabilities, the sampling distribution, the likelihoods, and the posterior

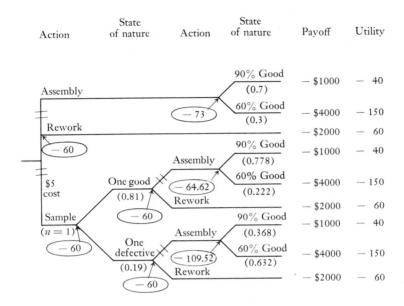

| Action | State of nature | Action | State of nature | Payoff | Utility |
|---|---|---|---|---|---|

Fig. 9.7. Utility analysis for Perkins Plastics ($n = 1$).

probabilities. Although most real-world problems involve only discrete functions, in many cases the number of alternatives is so large that a continuous function is much easier to handle. We will see in this section that decision analysis for continuous functions is a direct extension of much of the discrete theory.

Suppose a decision-maker is faced with a situation where one of the variables is too large to handle in discrete form. Perhaps the president of a company is trying to estimate what sales might be for a new product under a number of different assumptions about its price. Or a book club might be trying to determine whether or not to acquire the rights to a new book by estimating the percent of its members who will purchase this book. In both of these examples, it may be possible for sample information to be collected, and the result of the sample could be one of a large number of outcomes. The tree diagram on page 346 represents what the book club's decision problem might look like. The fans (≪) in this diagram represent variables involving a large number of outcomes or alternatives.

To begin an analysis of a decision situation such as that shown in Fig. 9.8, we must assume that the decision-maker can express his prior probabilities in the form of a probability density function. Although his density function can theoretically assume any shape he specifies, we will see that the process of revising these probabilities in the light of sample information becomes quite complicated unless

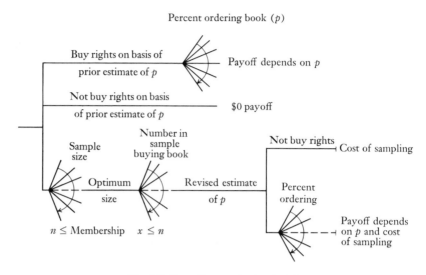

**Fig. 9.8.** Tree diagram for book club.

his prior distribution either is normally distributed or follows the beta distribution.* Limiting the prior distribution in this way is not as restrictive as it may seem, for these two distributions have been shown to be appropriate in many different situations. It is interesting to note that if a decision-maker's prior distribution is relatively flat (such as $x = 1, n = 2$ in the footnote), this means the decision-

---

* The beta distribution is a continuous function which is closely related to the binomial distribution. The shape of this distribution depends on its two parameters, $x$ and $n$. We will not discuss the beta distribution in this book, but rather present the following graph to indicate the variety of shapes the beta distribution can assume.

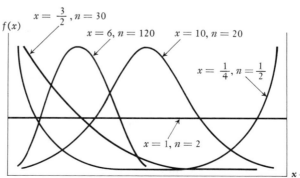

The beta distribution

maker has little or no knowledge about the relative likelihood of the various states of nature. Such a prior distribution is called a "diffuse" prior.

One of the most difficult problems involving continuous functions may occur in translating a decision-maker's prior knowledge and subjective opinion into either a normal or a beta distribution. The usual approach to this problem is similar to the approach for the discrete case—i.e., to have the decision-maker choose between a series of alternatives, until enough information is gathered to construct the density function. For example, suppose a sales forecast is being made, and a normal prior is assumed appropriate. If the decision-maker says he is indifferent at 50–50 odds that (1) sales will be greater than 2000, or (2) sales $\leqslant 2000$, then 2000 represents the mean of his distribution. Similarly, he might say that there is only one chance in twenty that sales will be larger than 2330. These values are consistent with a normal distribution with a mean of 2000 and a standard deviation of 200. This standard deviation was calculated by noting that his estimate of one chance in twenty corresponds to the 95th percentile, and the standardized normal equivalent for $F(z) = 0.95$ is $z = 1.65$. Solving

$$ z = \frac{x - \mu}{\sigma}, $$

where $x = 2330$, $\mu = 2000$, and $z = 1.65$, yields $\sigma = 200$.

**Bayes' Rule for Continuous Functions**

If a decision-maker wants to gather additional (sample) information, then Bayes' rule is again the means for transforming prior probabilities into posterior probabilities. Bayes' rule for continuous functions follows the same form as for discrete functions except that the continuous case involves integrating a continuous density function rather than summing a discrete probability function (refer to Section 9.4). If we let $f(\theta)$ be the prior density function, $f(x \mid \theta)$ represent the likelihood function, and $f(\theta \mid x)$ be the posterior density function, then Bayes' rule is:

*Bayes' rule for continuous random variables:*

$$ f(\theta \mid x) = \frac{f(\theta)f(x \mid \theta)}{\int f(\theta)f(x \mid \theta)\, d\theta}. \tag{9.5} $$

Although the random variables $f(\theta)$ and $f(x \mid \theta)$ can, at least theoretically, be any proper probability distributions, integration of the denominator of Bayes' rule is quite difficult unless $f(\theta)$ is a beta distribution and sampling is from a binomial distribution, or $f(\theta)$ is normally distributed and sampling is from a normal distribution. When the prior is a beta distribution, and the sampling binomial, the posterior distribution will also be a beta distribution. Since the beta distribution has not been discussed in this book, we will not attempt to illustrate how priors can be revised in this fashion (although the process is not a difficult one). Rather we turn to a more thorough analysis of the case where $f(\theta)$ and the sampling distribution are both normal.

*When the prior is normal and sampling is from a normal distribution, then the posterior distribution will also be normal.*

Furthermore, if the prior distribution has a mean $\mu_0$ and a variance $\sigma_0^2$, and a sample of size $n$ with sample mean $\bar{x}$ is taken from a normal distribution with variance $\sigma^2$, then the posterior distribution will have a mean $\mu_1$ and a variance $\sigma_1^2$, as follows:

*For normal distributions:*

$$\text{Posterior mean:} \quad \mu_1 = \frac{\mu_0\sigma^2 + n\bar{x}\sigma_0^2}{\sigma^2 + n\sigma_0^2}; \tag{9.6}$$

$$\text{Posterior variance:} \quad \sigma_1^2 = \frac{\sigma^2\sigma_0^2}{\sigma^2 + n\sigma_0^2}. \tag{9.7}$$

These formulas merely represent the process of taking a weighted average of the prior and sample evidence. The posterior mean $\mu_1$ will always lie between the prior mean $\mu_0$ and the sample mean $\bar{x}$. In addition, the posterior variance will always be smaller than the prior variance.* The more diffuse (i.e., flat) the decision-maker's prior distribution, the more his posterior distribution will depend on the sample data.

To illustrate the formulas for revising a normal prior, let's go back to the sales-forecasting problem in which the decision-maker's prior was

$$\mu_0 = 2000, \qquad \sigma_0^2 = 200^2.$$

Assume that he has taken a sample of size $n = 10$ from a population with

$$\sigma^2 = 400^2 \qquad \text{and obtains} \qquad \bar{x} = 1800.$$

The posterior values are computed from Formulas (9.6) and (9.7), to be:

$$\text{Posterior mean:} \quad \mu_1 = \frac{\mu_0\sigma^2 + n\bar{x}\sigma_0^2}{\sigma^2 + n\sigma_0^2}$$

$$= \frac{2000(400)^2 + 10(1800)(200)^2}{(400)^2 + 10(200)^2} = 1857;$$

---

* This result holds only for the normal process being discussed, and not for revising a beta prior by sampling from a binomial distribution.

$$\textit{Posterior variance:} \quad \sigma_1^2 = \frac{\sigma^2 \sigma_0^2}{\sigma^2 + n\sigma_0^2}$$

$$= \frac{(400)^2 (200)^2}{(400)^2 + 10(200)^2} = (106.9)^2.$$

Notice that the posterior mean (1857) lies between the prior mean (2000) and the sample mean (1800). The posterior variance $(106.9)^2$ is smaller than the prior variance $(200)^2$ or the sample variance $(400)^2$.

### Loss Functions and Decision-making

In order to be able to analyze decisions involving continuous variables, the decision-maker needs to specify a function comparable to the payoff table that we used previously. Since many problems in statistical decision theory are formulated in terms of losses (as was the example of Perkins Plastics), a *loss function* is commonly used, rather than a payoff table. The loss function can be of any form, although *linear functions are often used because they are fairly easy to manipulate and are appropriate in many different situations.* In sales forecasting, for example, losses will probably occur whenever the true sales figure is overestimated or underestimated. The linear loss function in this situation might look like that shown in Fig. 9.9, where the steeper sloping function represents the loss for underestimating sales, and the other function the loss for overestimation.

*Linear loss functions.* When the decision-maker's loss function is linear, the point estimation which minimizes his expected loss depends on the relative slopes of the two functions. If we let $C_u$ be the cost per unit of underestimation, and $C_0$ the cost per unit of overestimation, then *the optimal estimate is that value $\theta^*$ of*

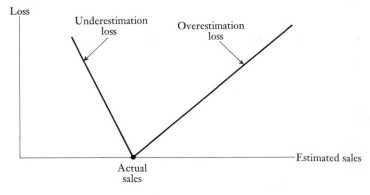

**Fig. 9.9.** Linear loss function.

*the decision maker's cumulative distribution of θ given by the following ratio:*

*Optimal value of F(θ) for minimizing expected loss:*

$$F(\theta^*) = \frac{C_\mathrm{u}}{C_\mathrm{u} + C_0}. \tag{9.8}$$

Let's assume in our sales forecasting problem that $C_\mathrm{u} = \$4$ per unit and $C_0 = \$2$ per unit. The optimal value of the decision-maker's cumulative distribution of θ is

$$\frac{\$4}{(\$4 + \$2)} = \frac{2}{3}.$$

That is, he should estimate that value $\theta^*$ for which

$$F(\theta^*) = P(\theta \leqslant \theta^*) = \tfrac{2}{3}.$$

Recall that his posterior distribution was normally distributed, with a mean of 1857 and a standard deviation of 106.9. Using the fact that for the standardized normal distribution

$$P(z \leqslant 0.43) = \tfrac{2}{3},$$

from Table III, we can solve for the optimal estimate $\theta^*$ by letting $z = 0.43$, as follows:

$$z = \frac{\theta^* - \mu}{\sigma}, \qquad 0.43 = \frac{\theta^* - 1857}{106.9};$$

therefore,

$$\theta^* = 1903.$$

Thus, the best sales estimate for this decision-maker is 1903 units.

*Quadratic loss function.* Another loss function often used in this type of problem is the "*quadratic loss function.*" For this function the decision-maker's loss is given by the *square* of the difference between the value he estimates ($a$) and the true value ($\theta$). If we let $l(a, \theta)$ represent the loss incurred when $a \neq \theta$, then the quadratic loss function is

$$l(a, \theta) = (a - \theta)^2.$$

This function can be diagrammed as shown in Fig. 9.10.

A quadratic loss function is popular not only because of its intuitive appeal, but because of its convenience mathematically in further analysis. *For a quadratic loss function, expected loss is minimized by using the mean of the decision-maker's distribution to estimate θ.* Thus, for our sales-forecasting problem, the optimal forecast under a quadratic loss function would be $1857.

By now you may have recognized that we have introduced enough continuous theory for a Bayesian analysis of a decision problem such as the book-club example

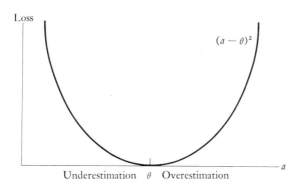

Fig. 9.10. Quadratic loss function.

shown in the tree diagram in Fig. 9.8. That is, we can assess prior probabilities, revise these probabilities using sample information, and evaluate various decision possibilities by using a loss function. If our simplifying assumptions of a beta or normal prior, binomial or normal sampling, and linear or quadratic loss functions are not met, then the analysis becomes more difficult, but not necessarily impossible. To study these topics in greater detail, the reader should consult the references on Bayesian analysis given at the end of this book.

## 9.10 BAYESIAN ANALYSIS: ADVANTAGES AND DISADVANTAGES

Most procedures normally associated with classical statistics, such as point and interval estimation, hypothesis testing and regression analysis can also be accomplished from a Bayesian point of view, although the two approaches differ considerably. The Bayesian approach is designed to result in optimal decisions for a given prior distribution and loss function. The major criticism of the Bayesian approach is that it requires a number of subjective evaluations whose validity is questioned by the classical statistician. How reasonable, for instance, is it to ask a decision-maker to form a prior distribution for a set of events for which he has little or no prior knowledge? And even if he *has* prior knowledge, is it really valid to ascertain his prior probability distribution by asking him to answer questions about alternatives (gambles) which he would never have to face in real life? Also, how is it possible for a decision-maker to formulate a loss function for events when he may have no idea of the consequences of a wrong decision, or for which he may not be able to express the losses in monetary terms (such as a surgeon trying to decide on an operation which may cost his patient's life).

The Bayesian answer to these questions is that a decision-maker is often making a number of similar evaluations in classical statistics and may not even be aware of it. For example, in hypothesis-testing, the classical statistician is assuming, in effect, a flat (diffuse) prior and basing his decisions on the sample evidence and a *quite subjective* value of the risk of making a Type I error ($\alpha$).

Despite the obvious advantages of statistical decision theory as an aid to the decision-making process, the number of businesses that have even experimented with its use remains quite small. Many companies simply have not been exposed to this "new" technique, while others have tried it and found it too time-consuming and/or costly to add to their decision-making process. On the other hand, a number of major companies such as DuPont, Pillsbury, General Electric, and the Ford Motor Company, have tried decision-theory analysis, and most often the users are pleased with it. Just how great an impact this approach will have on decision-making in the future is a question which cannot be answered at this time.

**REVIEW PROBLEMS**

1. Distinguish between decision-making under certainty and decision-making under uncertainty.
2. Describe, briefly, each of the following terms:
    a) maximin criterion
    b) minimax regret criterion
    c) dominated action
    d) EVSI
    e) ENGS
3. What are the disadvantages of the maximin and minimax regret criteria? How does the EMV criterion avoid these disadvantages? What is the disadvantage of the EMV criterion?
4. The Koseman Company is considering adding a new boiler to their factory in an attempt to avoid costly delays in case one of the present boilers must go down for repairs. They have determined that the following payoff table is appropriate for their present time horizon (5 years).

|  |  | States of nature | | |
|---|---|---|---|---|
|  |  | No repairs | Minor repairs | Major repairs |
| Actions | Do not add boiler | $0 | $-\$\,4{,}000$ | $-\$15{,}000$ |
|  | Add new boiler | $-\$10{,}000$ | $-\$10{,}000$ | $-\$10{,}000$ |
|  | Probability | 0.20 | 0.30 | 0.50 |

a) Find the optimal action using both the maximin and the minimax regret criteria.

b) Find the optimal action using the EMV criterion.

5. Draw the decision tree for Problem 4, indicating the optimal action by EMV.

6. A toy manufacturer is considering introducing a novelty item in time for the Christmas season. Because of a distribution agreement, the number of items produced must be either 1,000, 5,000, or 10,000 units. These units can be produced either by a labor-intensive process which involves a fixed cost of $2,000 plus a variable cost of $1.50 per unit, or by a capital-intensive process involving fixed costs of $5,000 and a variable cost of $1.00 per unit. The company sells this novelty item for $2.00 per unit. Their current estimates are that the probability of selling only 1,000 units is $\frac{1}{6}$, the probability of selling 5,000 units is $\frac{1}{2}$, and the probability of selling 10,000 units is $\frac{1}{3}$. Unsold units have no salvage value.

a) Find the optimal action using the maximin and the minimax regret criteria.

b) Find the optimal action using the EMV criterion.

c) Draw the decision tree, indicating the optimal action under EMV.

7. A lab test for diabetes has been shown to correctly identify the presence of this disease in 80% of all people who are diabetics. Thirty percent of the time, however, the test will indicate diabetes in someone who does not have this disease. If 10% of a given population has diabetes, what is the probability that a person whose lab test indicates diabetes actually has the disease?

8. Professor Ward Edwards of the University of Michigan conducted an experiment in which he asked college students to estimate the probability that a given sample of green and red balls came from one of two urns. Urn 1 contained 70 red balls and 30 green, while Urn 2 contained 70 green and 30 red balls. At the beginning of the experiment, one of these urns was selected at random, and then samples were drawn from this urn by randomly selecting a ball, noting its color, and then replacing the ball. The subjects did not know which urn had been selected.

a) Suppose one red ball is drawn. Without working it out, what is your *guess* as to posterior values $P(\text{Urn } 1 \mid \text{One red})$?

b) Suppose a sample of twelve balls is drawn, with 8 red and 4 green. What is your *guess* of the value of $P(\text{Urn } 1 \mid 8R, 4G)$?

c) In Edwards' experiments, a typical subject would guess the probability in part (a) to be about 0.60, and most estimated the probability in part (b) to be less than 0.80. Use Bayes' rule to calculate the actual posterior values for parts (a) and (b), and then assess how well you and Dr. Edwards' subjects did in estimating these probabilities.

9. A company can ship some equipment either by sea or by air. There is a possibility of a strike affecting either type of shipment. The cost matrix including shipment and delay costs is given below.

|  | Strike ($B_1$) | No strike ($B_2$) |
|---|---|---|
| Ship by air | 4000 | 3000 |
| Ship by sea | 6000 | 1000 |

a) If the probability of a strike is 0.4, what are the expected costs of each method of shipment?
b) Suppose some inside informer suggests that a strike will occur and the accuracy of this rumor (R) is given by

$$P(R \mid B_1) = 0.8, \quad \text{and} \quad P(R \mid B_2) = 0.3.$$

Find the revised probability of a strike, given this extra information.
c) Using the revised probability, find the best choice of shipment according to the expected monetary value criterion.

10. Suppose we receive a concession to sell popcorn at football games. We must decide whether to build one booth on the "Home" side or to build two booths, one each on the "Home" and "Visitors" sides of the stadium. If the games attract large crowds of visitors, it would be better to have two booths, but their cost and the cost of equipment within them would be a considerable expense. The payoff matrix is determined to be:

| Action | | Capacity crowds | Regular "home" crowd |
|---|---|---|---|
| $A_1$ | Build one booth | 350 | 300 |
| $A_2$ | Build two booths | 500 | 200 |

a) Suppose we consider the probability of capacity crowds to be 0.4 and the probability of regular sized "home" crowds to be 0.6. Which action is "best"?
b) Suppose a preseason forecast predicts a much improved team and preseason ticket sales are up by 30%. From previous experience, we determine that such situations have preceded games with capacity crowds 4 out of 10 times and have preceded games with regular crowds in 2 out of 10 cases. Find the revised probability of the two different-size crowds and determine which action is best in the light of the new information.

11. A corporation considers two levels of investment in a real-estate development, a low participation ($A_1$) or a high participation ($A_2$). Two states of nature are deemed possible, a partial success ($B_1$) or a complete success ($B_2$). The payoff matrix is estimated to be:

| | $B_1$ | $B_2$ |
|---|---|---|
| $A_1$ | -200 | 400 |
| $A_2$ | -500 | 1000 |

a) How large does the *a priori* probability of $B_1$ have to be in order to make action $A_1$ the "best" choice?
b) Suppose the states of nature are initially presumed to occur with probabilities $P(B_1) = 0.4$, $P(B_2) = 0.6$. Then a more careful study is made with the conclusion, x, that the project will be only a partial success. In previous relevant studies, this same conclusion was obtained in 8 out of 10 cases when similar projects were partial successes. Also,

this conclusion was obtained in four out of 12 cases when similar projects were *complete* successes. Find the revised probabilities of the states of nature, and determine which investment level is appropriate.

12. Suppose you are trying to choose between these investments ($a_1$, $a_2$, $a_3$), where the payoff table (in dollars) is as follows.

|  |  | States of nature | | |
|---|---|---|---|---|
|  |  | $\theta_1$ | $\theta_2$ | $\theta_3$ |
|  | $a_1$ | 0 | 0 | 1400 |
| Action | $a_2$ | 500 | 500 | 500 |
|  | $a_3$ | 400 | 100 | 900 |
| Probability |  | 0.30 | 0.30 | 0.40 |

What is the optimal action under EMV?

13. Joe Doakes is considering flying from New York City to Boston in the hopes of making an important sale to P. J. Bety, president of NOCO, Inc. If Joe makes this sale he will earn a commission of $1,100. Unfortunately, Joe figures there is a 50–50 chance that Bety will be called out of town at the last moment and he will have no chance at a sale. Even if he goes to see Bety, Joe estimates he has only one chance in five of making the sale. The trip will cost Joe $100 whether or not he gets to see Bety.

a) Draw the tree diagram for Joe, and determine whether Joe should fly to Boston or not.
b) Joe was heard to remark "I'd give my right arm to know if Bety will be in town." How much does he value his right arm?
c) Suppose an information service offers to tell Joe, before he decides to fly, whether or not they think Bety will be in town. The record of this company is such that if they say Bety will be in, the probability he will, in fact, be in is 0.70. If they say he will be out, they will be correct 90% of the time. If this service costs $10, and Joe figures the probability they will say Bety is in is 0.50, should he buy the service? Draw the tree diagram. Find ENGS.

14. a) Can EVSI ever assume a negative value? Explain.
b) What will be the value of EVSI if, no matter what sample result is observed, the same decision is optimal after sampling as was optimal before sampling?

15. The Dixon Corporation makes picture tubes for a large television manufacturer. Dixon is concerned because approximately 30% of their tubes have been defective. When the television manufacturer encounters a defective tube, Dixon is charged a $20 penalty cost (to pay for repairs and lost time). One way Dixon can avoid this penalty cost is to re-examine and fix each defective tube before shipping. This would cost an extra $7 per tube. Or, they can rent a testing device which costs $1 for each tube tested. Since this device is not infallible, its effectiveness was tested by running through it a large number of tubes,

some known to be good, and others known to be defective. The results of this study are shown below.

|  |  | State of tube | |
|---|---|---|---|
|  |  | Good | Defective |
| Test | Good | 0.75 | 0.20 |
| results | Def | 0.25 | 0.80 |
|  |  | 1.00 | 1.00 |

Draw the decision tree for Dixon, assuming they must decide between shipping directly, reexamining each tube, or testing each tube. Calculate ENGS and EVSI.

16. The Techno Corporation is considering making either minor or major repairs to a malfunctioning production process. When the process is malfunctioning, the percentage of defective items produced seems to be a constant, with either $p = 0.10$ or $p = 0.25$. Defective items are produced randomly, and there is no way Techno can tell for sure whether the machine needs minor or major repairs. If minor repairs are made when $p = 0.25$, the probability of a defective is reduced to 0.05. If minor repairs are made when $p = 0.10$, or major repairs made when $p = 0.10$ or $p = 0.25$, then the proportion of defectives is reduced to zero. Techno has recently received an order for 1,000 items. This item yields them a profit of $0.50 per unit, except that they have to pay a $2.00 penalty cost for each item found defective. Major repairs to the process cost $100 while minor repairs cost $60. No adjustment can be made to the production process once a run has started. Prior to starting the run, however, Techno can sample items from a "trial" run, at a cost of $1.00 per item.

   a) Find the optimal action for Techno if they are trying to decide between not sampling at all, and sampling one item. Draw the decision tree.
   b) Find the optimal action for Techno if they are willing to consider a sample of either one or two items. Draw the decision tree.

17. Suppose that a decision-maker has expressed an indifference between receiving $B$ for sure and a 50–50 gamble between $A$ and $C$ for each of the following values:

| A | C | B |
|---|---|---|
| $ 10 | −$ 5 | $ 0 |
| 10 | − 10 | − 5 |
| 20 | − 5 | 10 |
| 50 | − 10 | 10 |
| 100 | 0 | 20 |

   a) Let $U(\$10) = 10$ and $U(0) = 0$ and then find the utility for $-\$5$, $-\$10$, $\$20$, $\$50$, and $\$100$.
   b) Sketch this person's utility function. Can you classify him as a risk-taker, risk-neutral, or a risk-avoider?

18. Assume a friend of yours asks your advice on the following decision. He can take either $50 for certain, or participate in a gamble giving him an equal chance of winning $100, $40, or $0.
   a) What decision should you recommend (i.e., to gamble or not) if your friend wants to maximize his expected winnings?
   b) What decision should you recommend if he wants to maximize expected utility, and he tells you that he is indifferent between equal chances at $A$ and $C$, or $B$ for certain, as follows:

| $A$ | $B$ | $C$ |
| --- | --- | --- |
| 100 | 40 | 0 |
| 100 | 50 | 40 |

   c) Sketch your friend's utility function. Does he appear to be a risk-taker or a risk-avoider?

19. Suppose the decision-maker for the problem represented by Fig. 9.8 is attempting to assess his prior distribution of $p$, the percent of customers who will buy the book. He decides that the chances are only 1 out of 4 that $p$ will be less than 0.04, and the chances are 4 out of 5 that $p$ will not be greater than 0.08. Assuming his prior distribution is normal, find its mean and variance.

20. A food manufacturer is trying to determine the mean weight of his boxes of breakfast cereal. He feels that the weight is normally distributed with a standard deviation of 0.4, and his prior distribution is normal with a mean of 16.2 ounces and a standard deviation of 0.2. Find his posterior distribution.

21. The owner of a local fabric store is trying to prepare an order for a certain type of material, based on the mean number of yards purchased each month in his store. His prior distribution of the number of yards sold is normally distributed, with a mean of 120 yards and a variance of 25 yards. In making his order, the owner feels that the cost of underestimating sales (in yards) for a month is four times as costly as overestimating sales, and both are linear functions.
   a) What should his order be, based on his prior distribution?
   b) Assume the owner takes a sample of 4 months, and finds that sales are 120, 112, 122, and 118 yards. He feels the sampling distribution is normal with a variance of 20. Find his posterior distribution.
   c) What estimate should he make on the basis of his posterior distribution?
   d) Repeat parts (a) and (c) assuming a quadratic loss function.

**EXERCISES**

22. a) Draw the decision tree and find EVSI and ENGS for $n = 3$ for the Perkins Plastics problem (see Section 9.7).
   b) Prepare a flow chart for a computer program designed to solve the Perkins Plastics problem for $0 \leqslant n \leqslant 30$.
   c) Write and run the computer program for part (b).

23. The values of ENGS in Table 9.8 were calculated on a computer by first calculating the value for $n = 1$, then for $n = 2$, and so forth, up to $n = 26$.

   a) In writing a computer program to find the optimal sample size, why is it not possible to program the computer to stop as soon as ENGS begins to decrease? (*Hint:* See Table 9.8 for $n = 3$; $n = 4$.)

   b) Why, in Table 9.8, does EVSI increase only a small amount for some values of $n$ (for example, $n = 3$ to $n = 4$), while it increases a large amount for other values (for example, $n = 4$ to $n = 5$)?

   c) Design a decision rule for stopping a computer program which is calculating ENGS for successive sample sizes (keeping in mind your answer to part (a)).

24. A procedure for calculating EVSI which is equivalent to that presented in Section 9.6 is to focus *only* on those sample results which lead to a decision which is *different* from the optimal decision without sample information. The expected value of the sample information depends only on how much additional profit (on the average) is earned because of this change in decision. Thus, one needs to first calculate how much more expected profit the new decision yields over the old (using the posterior probabilities), and then multiply this value by the (marginal) probability of observing the sample result(s) which lead to a new decision.

   a) Note for the data in Fig. 9.3 that the previously optimal decision (shown in Fig. 9.1 to be to *send to assembly*) is changed to *rework* only after observing a defective valve. Since the expected value of sending to assembly after such a sample is $-\$2895$, and reworking costs only $-\$2000$, the decision to rework saves $\$895$. Verify that this savings, multiplied by the probability of observing one defective, equals the EVSI for $n = 1$.

   b) Verify by the above method that the EVSI for Fig. 9.4 is $\$251$.

25. Attempt to construct your own utility function for money for dollar values between $-\$500$ and $+\$1500$.

26. a) Sketch the utility function $U(m) = m^{1/2}$, where $m = $ money and $0 \leqslant m \leqslant 1600$.

   b) Does this function represent a risk-taker, risk-neutral, or a risk-avoider?

   c) Use this utility function to find the action maximizing expected utility in Exercise 12. Can you guess, before calculating EU, which action will be optimal? Explain.

   d) Will your answer to part (c) change if the utility function $U(m) = 100 + 10m^{1/2}$ is used instead of $U(m) = m^{1/2}$? Explain.

# 10

# Hypothesis Testing: Multi Sample Tests

## 10.1  INTRODUCTION

This chapter is a continuation of the presentation begun in Chapter 8 on hypothesis testing. That previous discussion outlined the concepts and general methods of hypothesis testing and gave examples of its use in situations involving a single sample. In many cases, inferences involving more than one population are needed in decision-making; these types of inference require the use of tests based on two or more samples, one taken from each population. The reader may recall that the procedure of hypothesis testing is quite standard over all test situations; however, as the circumstances of a problem differ, the appropriate test statistic and its underlying probability distribution may differ. In this chapter, several commonly used test statistics are presented which apply to frequently occurring classes of problems involving two or more samples. Neither this listing of multisample tests nor the earlier listing of single-sample tests in Chapter 8 is meant to be exhaustive. They merely intend to illustrate the powerful statistical tool of hypothesis testing in practical problems. Presumably, if the reader can understand and use these particular test statistics, he could also perform any other similar test for special situations as he needs them. Probably more than 100 such test statistics are easily found in any collection of statistical methods of hypothesis testing.

## 10.2  TEST ON DIFFERENCES BETWEEN MEANS OF TWO POPULATIONS ($\sigma_1$ AND $\sigma_2$ KNOWN, OR SAMPLE SIZES LARGE)

In Sections 8.3 and 8.4, tests about a hypothesized population mean $\mu$ based on one sample taken from that population were presented. The same approach can also be used to test whether two samples came from two populations with equal means or with a specified difference between their means. It is quite common to test for the effect of different treatments on two groups where one group is often a "control" group and the other is given special treatment, such as a new drug, a new approach to learning, or perhaps, just a different type of paint. In any case, the objective is to compare the two groups, each representing a sample from some population, to see if their means, on the attribute under investigation, differ significantly.

If we designate $\mu_1$ as the mean of one population and $\mu_2$ the mean of a second, a number of different null and alternative hypotheses are possible. If the null hypothesis is restricted to a simple hypothesis, then the most common form is $H_0: \mu_1 - \mu_2 = 0$, which asserts that the two means are equal. The alternative to this hypothesis might be that $\mu_1$ exceeds $\mu_2$ ($H_a: \mu_1 - \mu_2 > 0$), or that $\mu_2$ exceeds $\mu_1$ ($H_a: \mu_1 - \mu_2 < 0$), or perhaps merely that $\mu_1$ and $\mu_2$ are not equal ($H_a: \mu_1 - \mu_2 \neq 0$). It is also possible to hypothesize that the two means differ by some constant amount $k$, for which the appropriate null hypothesis would be $H_0: \mu_1 - \mu_2 = k$.

**The Standardized Normal Test Statistic**

In order to determine a critical region for testing for the difference between two population means, it is necessary to know the sampling distribution of the difference between two sample means, $\bar{x}_1 - \bar{x}_2$. From the rules on expectations in Section 3.3, we know that

$$\mu_{\bar{x}_1 - \bar{x}_2} = E[\bar{x}_1 - \bar{x}_2]$$
$$= E[\bar{x}_1] - E[\bar{x}_2] = \mu_1 - \mu_2.$$

The variance of the difference, $V[\bar{x}_1 - \bar{x}_2]$ (or $\sigma^2_{\bar{x}_1 - \bar{x}_2}$), can be derived as follows, assuming that the samples are drawn independently:

$$V[\bar{x}_1 - \bar{x}_2] = V[\bar{x}_1] + V[-\bar{x}_2]$$
$$= V[\bar{x}_1] + (-1)^2 \, V[+\bar{x}_2] \quad \text{(By Rule 7, Section 3.3)}$$
$$= \frac{\sigma_1^2}{n_1} + \frac{\sigma_2^2}{n_2}, \quad \text{(Since } V[\bar{x}] = \sigma^2/n).$$

where $\sigma_1^2$ and $n_1$ are the variance and sample size of the first population, and $\sigma_2^2$ and $n_2$ are the variance and sample size of the second population. If $\sigma^2_{\bar{x}_1 - \bar{x}_2} = \sigma_1^2/n_1 + (\sigma_2^2/n_2)$, then the standard deviation of the difference $\bar{x}_1 - \bar{x}_2$ must be $\sigma_{\bar{x}_1 - \bar{x}_2} = \sqrt{(\sigma_1^2/n_1) + (\sigma_2^2/n_2)}$. Finally, it can be shown that if the distributions of $\bar{x}_1$ and $\bar{x}_2$ are normal, then the distribution of the random variable representing their difference $(\bar{x}_1 - \bar{x}_2)$ will also be normally distributed. A normal variable $(\bar{x}_1 - \bar{x}_2)$ with mean $\mu_1 - \mu_2$ and standard deviation $\sqrt{(\sigma_1^2/n_1) + (\sigma_2^2/n_2)}$ can be written in standardized normal form by using the familiar transformation of subtracting from the random variable its mean and then dividing by its standard deviation:

$$z = \frac{(\bar{x}_1 - \bar{x}_2) - \mu_{\bar{x}_1 - \bar{x}_2}}{\sigma_{\bar{x}_1 - \bar{x}_2}} = \frac{(\bar{x}_1 - \bar{x}_2) - (\mu_1 - \mu_2)}{\sqrt{(\sigma_1^2/n_1) + (\sigma_2^2/n_2)}}. \tag{10.1}$$

For large samples, the substitution of the sample variances $s_1^2$ and $s_2^2$ as estimators of $\sigma_1^2$ and $\sigma_2^2$ will not change the fact that $\bar{x}_1 - \bar{x}_2$ is normally distributed (from the central-limit theorem) even if $\bar{x}_1$ and $\bar{x}_2$ are not themselves normally distributed. Thus, if the samples are independent and large, say $n_1$ and $n_2$ each exceeding 25, and if the variances of the population are unknown, the following test statistic may be approximated by the $z$-distribution when $n_1$ and $n_2$ are large:

$$z \doteq \frac{(\bar{x}_1 - \bar{x}_2) - (\mu_1 - \mu_2)}{\sqrt{(s_1^2/n_1) + (s_2^2/n_2)}}. \tag{10.2}$$

Frequently, this statistic rather than the one shown in Formula (10.1) will be used, since the true variances of the populations are generally unknown when the true means (about which the variance is determined) are unknown.

### A Test Example

To illustrate the use of Formula (10.2), suppose the hypothesis that starting salaries for college graduates working in the New York City area are equal to the starting salaries for college graduates working in the Chicago area is tested against the alternative hypothesis that these salaries are not equal. If $\mu_1$ represents the mean starting salary in the New York area and $\mu_2$ the mean starting salary in the Chicago area, then the above test is equivalent to testing

$$H_0: \mu_1 - \mu_2 = 0$$

against

$$H_a: \mu_1 - \mu_2 \neq 0.$$

This is a two-sided test whose critical value for a large sample at an $\alpha = 0.05$ level of significance is $z_{0.025} = 1.96$. Figure 10.1 illustrates the rejection region for this test on the $z$-scale.

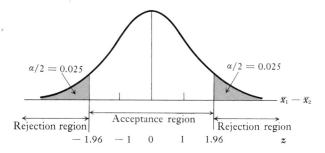

**Fig. 10.1.** Two-sided test on $\mu_1 - \mu_2$.

Suppose a random sample of 100 is taken from the New York area, with the result that $\bar{x}_1 = \$10,250$, with $s_1 = 200$. For the Chicago area a random sample of 60 yields a mean $\bar{x}_2 = \$10,150$, and $s_2 = 180$. The computed $z$-value is thus (from Formula (10.2)):

$$z = \frac{(10{,}250 - 10{,}150) - 0}{\sqrt{(200^2/100) + (180^2/60)}} = \frac{100}{\sqrt{940}} = 3.26.$$

Since this value of $z$ exceeds the critical value of 1.96, the null hypothesis must be rejected in favor of the alternative hypothesis. In other words, the difference in these two means may not be attributed entirely to chance, but rather there appears to be a systematic difference in starting salaries.

## 10.3  SMALL SAMPLE TEST ON DIFFERENCES BETWEEN MEANS OF TWO POPULATIONS ($\sigma_1^2$ AND $\sigma_2^2$ UNKNOWN BUT ASSUMED EQUAL)

The two-sample $z$-test described in Section 10.2 applies only for *large* samples, or when $\bar{x}_1$ and $\bar{x}_2$ are normally distributed and $\sigma_1^2$ and $\sigma_2^2$ are *known*. If the sample size is *small*, the $t$-distribution can be applied under certain circumstances. First, as was the case for one-sample tests, the distribution of the parent populations must be normal. In addition, this two-sample $t$-test assumes that the variances of the two variables are equal (i.e., $\sigma_1^2 = \sigma_2^2$), although there is a $t$-test we shall describe shortly which does not require this assumption. Fortunately, the $t$-test is a fairly "robust" test (as was pointed out in Section 6.8), so that deviations from the above assumptions may not destroy the usefulness of this approach.

**Using the $t$-distribution**

A $t$-test for the difference between two means based on small samples would seem to have the same form as Formula (10.2). However, there is a difficulty here because the $t$-test requires that the two population variances be equal, $\sigma_1^2 = \sigma_2^2$, whereas we usually have *two different* estimates of this value, $s_1^2$ and $s_2^2$, since two different samples were collected. In other words, even if $\sigma_1^2 = \sigma_2^2$, as the test requires, the two samples collected will not have exactly the same value of $s^2$ because of sampling error. But if $s_1^2$ and $s_2^2$ differ, which of these two values should be used to estimate the unknown population variance (let's call this variance $\sigma^2$, where $\sigma_1^2 = \sigma_2^2 = \sigma^2$)? The answer is that a weighted average of $s_1^2$ and $s_2^2$ is the best estimate, where the weights applied are their respective degrees of freedom relative to the total number of degrees of freedom. This type of weighting is used because in general the larger the sample, the better the estimate $s^2$ is of $\sigma^2$. Since there are $(n_1 - 1)$ degrees of freedom associated with $s_1^2$ and $(n_2 - 1)$ degrees of freedom associated with $s_2^2$, there are $(n_1 + n_2 - 2)$ total degrees of freedom. The *relative* weights assigned to each of the sample variances must therefore be $(n_1 - 1)/(n_1 + n_2 - 2)$ for $s_1^2$, and $(n_2 - 1)/(n_1 + n_2 - 2)$ for $s_2^2$. If we denote our weighted-average estimate of $\sigma^2$ as $s_{\bar{x}_1 - \bar{x}_2}^2$, then

$$s_{\bar{x}_1 - \bar{x}_2}^2 = \frac{n_1 - 1}{n_1 + n_2 - 2} s_1^2 + \frac{n_2 - 1}{n_1 + n_2 - 2} s_2^2.$$

Let's now return to Formula (10.1) and rearrange the denominator by letting $\sigma_1^2 = \sigma_2^2 = \sigma^2$, and factoring:

$$\sqrt{(\sigma_1^2/n_1) + (\sigma_2^2/n_2)} = \sqrt{\sigma^2(1/n_1 + 1/n_2)} = \sqrt{\sigma^2 \frac{n_1 + n_2}{n_1 n_2}}.$$

The righthand side of Formula (10.1) can now be rewritten in the following form:

$$\frac{(\bar{x}_1 - \bar{x}_2) - (\mu_1 - \mu_2)}{\sqrt{\sigma^2 \left(\dfrac{n_1 + n_2}{n_1 n_2}\right)}}.$$

Finally, if the value of the pooled estimate $s^2_{\bar{x}_1 - \bar{x}_2}$, is substituted for $\sigma^2$ in the denominator of the above expression, the resulting random variable can be shown to have a $t$-distribution with $(n_1 + n_2 - 2)$ degrees of freedom:

$$t_{n_1 + n_2 - 2} = \frac{(\bar{x}_1 - \bar{x}_2) - (\mu_1 - \mu_2)}{\sqrt{\left(\dfrac{(n_1 - 1)s_1^2 + (n_2 - 1)s_2^2}{n_1 + n_2 - 2}\right)\left(\dfrac{n_1 + n_2}{n_1 n_2}\right)}}. \tag{10.3}$$

The $t$-test using Formula (10.3) can be illustrated by using the previous example on starting salaries of college graduates, but now supposing that the two sample sizes are $n_1 = 11$ and $n_2 = 9$ and that it is not unreasonable that both $x_1$ and $x_2$ are normally distributed and have equal variances. For $n_1 + n_2 - 2 = 18$ degrees of freedom, and $\alpha = 0.05$, the critical ratio in a two-tailed test is $t_{(0.025, 18)} = 2.101$ from Table VII. The calculated value of $t$ is shown as follows:

$$t_{18} = \frac{(10,250 - 10,150) - 0}{\sqrt{\dfrac{10(200^2) + 8(180^2)}{18}\left(\dfrac{11 + 9}{11(9)}\right)}} = \frac{100}{\sqrt{(36622.2)(0.202)}} = 1.163.$$

Since the calculated value $t = 1.163$ is less than the critical value of 2.101, the alternative hypothesis that starting salaries in New York and Chicago are different must be rejected. The evidence of a \$100 difference based on these small samples is not as significant evidence as when based on large samples.

### The Case of Matched-pairs Samples

There is another way to test for significant differences between two samples involving small values of $n$ that does not require that the variances of the two populations be equal but still requires these populations to be normally distributed. In this test it is necessary that the observations in the two samples be collected in the form of what is called *matched pairs*. That is, each observation in the one sample must be paired with an observation in the other sample in such a manner that these observations are somehow "matched" or related, in an attempt to eliminate extraneous factors which are not of interest in the test. In our test for differences in starting salaries, for example, the graduates sampled in the New York area may be considerably older than the graduates sampled in the Chicago area, or they may represent a substantially different mix of undergraduate majors. If such differences are not of interest, then they can be systematically eliminated by selecting a sample in which each person in the New York area is carefully matched—in terms of age, sex, undergraduate major, or any other criterion—with a person in the Chicago area. One of the most widely used forms of matched pairs is to let a subject "serve as his own control," in which case the person is matched with himself at different points in time, or in a "before-and-after" treatment study.

If the observations can be collected in the form of matched pairs, then a $t$-test for differences between the two samples can be constructed on the basis of the *difference score* for each matched pair. This score is calculated by subtracting the score or value associated with the one person or object in each pair from the score of the other person or object in that pair. The $t$-test assumes that these difference scores are normally distributed and independent. If we denote the average difference in scores between the two populations by the capital Greek letter delta $\Delta$, then the hypothesis being tested is $H_0: \Delta = k$ where $k$ is the hypothesized average difference ($k = 0$ in a test of significance). If the values from the two matched samples are denoted by $x_i$ and $y_i$ and the difference score between matched pairs by $D_i = x_i - y_i$, then the average of $D_i$ is our best estimate of $\Delta$. The sample values of $D_i$ can be used in a test similar to a one-sample test on a mean with $\sigma$ unknown and the population assumed to be normal. The sample variance of the difference scores, $s_D$, and the sample mean of the difference scores, $\bar{D}$, are used to form:

$$t_{n-1} = \frac{\bar{D} - \Delta}{s_D/\sqrt{n}}, \tag{10.4}$$

where $n$ is the number of matched pairs in the two samples. Suppose, for example, that the observations in Table 10.1 represent the starting salaries for ten matched

**Table 10.1** Data for matched-pairs test

| Pair | $x_i$ New York City | $y_i$ Chicago | $D_i = x_i - y_i$ difference |
|------|------|------|------|
| 1 | $10,400 | $10,000 | $400 |
| 2 | 9,800 | 9,900 | −100 |
| 3 | 9,700 | 10,000 | −300 |
| 4 | 10,500 | 10,400 | 100 |
| 5 | 10,600 | 10,600 | 0 |
| 6 | 10,100 | 9,900 | 200 |
| 7 | 10,300 | 10,400 | −100 |
| 8 | 9,900 | 9,700 | 200 |
| 9 | 10,400 | 10,300 | 100 |
| 10 | 10,700 | 10,200 | 500 |
| Sum | 102,400 | 101,400 | 1,000 |
| Mean | 10,240 | 10,140 | 100 |

pairs from the New York and Chicago areas. The null hypothesis is that the average starting salaries are equal, $H_0: \Delta = 0$, against the two-sided alternate, $H_a: \Delta \neq 0$. The sample mean of the difference scores is $\bar{D} = 100$; the standard deviation of these ten difference scores can be shown to be $s_D = 240.0$. The calculated $t$-value is thus

$$t_9 = \frac{\bar{D} - \Delta}{s_D/\sqrt{n}} = \frac{100 - 0}{240/\sqrt{10}} = 1.32.$$

For $(n - 1) = 9$ degrees of freedom and an $\alpha$-level of 0.05, the critical ratio of a two-sided test is $t_{(0.025, 9)} = 2.262$. The calculated value of $t$ is smaller than this ratio; hence the null hypothesis that the difference scores were drawn from a population whose mean is equal to zero fails to be rejected. In other words, the difference between starting salaries is not sufficient to reject the null hypothesis on the basis of this sample.

### *10.4  A NONPARAMETRIC TEST ON DIFFERENCES BETWEEN TWO POPULATIONS

The statistical tests considered thus far have specified certain properties of the parent population which must hold before these tests can be used. A $t$-test, for example, requires that the observations come from a normal population, and if this test is used in testing for differences between means, the two populations must have equal variances. Although these tests are quite "robust" in the sense that the tests are still useful when the assumptions about the parent population are not exactly fulfilled, there are still many circumstances when the researcher cannot or does not want to make such assumptions. The statistical methods appropriate in these circumstances are called *nonparametric tests* because they do not depend on any assumptions about the parameters of the parent population.

**Measurement**

In addition to not requiring assumptions about the parameters of the parent population, most nonparametric tests do not require a level of measurement as strong as that necessary for parametric tests. By "measurement" we mean the process of assigning numbers to objects or observations, the level of measurement being a function of the rules under which the numbers are assigned. The measurement of quantifiable information usually takes place on one of four levels, depending on the strength of the underlying scaling procedure used. The four major levels of measurement are represented by nominal, ordinal, interval, and ratio scales.

The weakest type of measurement is given by a *nominal scale*, which merely sorts objects into categories according to some distinguishing characteristic and gives each category a "name" (hence nominal). Since classification on a nominal scale does not depend on the label or symbol assigned to each category, these symbols may be interchanged without affecting the information given by the scale.

---

\* This section can be omitted without loss of continuity.

Classifying automobiles by makes constitutes a nominal scale, as does distinguishing Republican from Democratic voters, or apples from oranges. In most nominal measurement, one is concerned with the number (or frequency) of observations falling in each of the categories.

An *ordinal scale* offers the next highest level of measurement, one expressing the relationship of order. Objects in an ordinal scale are characterized by relative rank, so that a typical relationship may be "higher," "greater," or "preferred to." Only the relations "greater than," "less than," or "equal to" have meaning in ordinal measurement. When a football team is "ranked" nationally, for example, such a measurement implies an ordinal scale if it is impossible (or meaningless) to say how *much* better or worse this team is compared to others. Most subjective attributes of objects or persons (e.g., flavor, beauty, honesty) are difficult if not impossible to consider on a scale higher than the ordinal. Distinguishing service personnel by rank (e.g., captain, major) is another example of ordinal measurement.

A third type of scale is given by *interval measurement*, sometimes called *cardinal measurement*. Measurement on an interval scale assumes an exact knowledge of the quantitative difference between objects being scaled. That is, it must be possible to assign a number to each object in such a manner that the difference between them is reflected by the difference in the numbers. Any size unit may be used in this type of measurement, and the choice of a zero point (origin) for the data can be made arbitrarily as long as a one-unit change on the scale always reflects the same change in the object being scaled. Temperature measured on either a Centigrade or a Fahrenheit scale represents interval measurement, as the choice of origin and unit for these scales is arbitrary. Temperature measured on an absolute scale, however, does not represent interval measurement, as this scale has a natural origin (the zero point is the point at which all molecular motion ceases). As another example, most I.Q. measures represent interval scales since there is no natural origin (zero intelligence?), and the choice of a unit can be made arbitrarily. The name "interval measurement" is used because this type of scale is concerned primarily with the distance *between* objects, that is, the "interval" between them.

The strongest type of measurement is represented by *ratio scales*, or scales which have all the properties of an interval scale *plus* a natural origin; only the *unit of measurement* is arbitrary. Fixing the origin (the zero point) permits comparisons not only of the intervals between objects, but of the absolute value of the number assigned to these objects. Hence, in this type of scale, "ratios" have meaning, and statements can be made to the effect that "$x$ is twice the value of $y$." Weight, length, and mass are all measured using a ratio scale. Distance, whether in terms of kilometers, miles, or feet, is an example of a ratio measurement, since all of these scales have a common origin (the zero point, representing no distance). Value measures of goods or income are also ratio measurement, whether the units are dollars, francs, D-marks, or yen. Zero earnings is the same in all currencies.

### Parametric vs. Nonparametric Tests

In addition to assuming some knowledge about the characteristics of the parent population (e.g., normality), parametric statistical methods require measurement equivalent to at least an interval scale. That is, in order to find the means and variances necessary for these tests, one must be able to assume that it is meaningful to compare intervals. It makes no sense to add, subtract, divide, or multiply ordinal scale values because the numbers on an ordinal scale have no meaning except to indicate rank order. There is no way to find, for example, the average between a captain and a major in terms of military rank.

The distinguishing characteristic of nonparametric tests is that usually no assumptions about the parameters of the parent population are necessary. To avoid the parametric assumptions normally required for tests based on interval or ratio scales, most nonparametric tests assume only nominal or ordinal data. That is, such tests ignore any properties of a given scale except ordinality. This means that if the data are, in fact, measurable on an interval scale, nonparametric tests waste (by ignoring) this knowledge about intervals. By wasting data, such tests gain the advantage of not having to make parametric assumptions, but sacrifice power in terms of using all available information to reject a false null hypothesis. Nonparametric tests need more observations than parametric tests to achieve the same size of Type I and Type II errors. For example a nonparametric test using ranked data requires about 105 observations to be as powerful as a $t$-test based on 100 observations of interval measurement data.*

### Matched-pairs Sign Test

An example of a nonparametric test which may be used instead of the $t$-test on $\Delta$ for matched-pairs samples is the *sign test*. It is designed to determine whether significant differences exist between two populations, based on two samples which are related in such a manner that each observation from one sample can be matched with a specific observation from the other sample. For example, one may wish to study the behavior of identical twins under two "treatments," the "before-and-after" effect of a certain drug, or the attitudes of husbands in contrast to the attitudes of their wives. In the sign test, ordinal data is assumed, so that it is meaningful to rank the observations from one sample (e.g., the husbands) only as higher than, equal to, or lower than their corresponding value in the other sample (the wife group). An easy way to record which sample has the higher value for each matched pair (husband vs. wife) is to give each of these pairs a "sign," either a plus $(+)$ sign representing the fact that the first sample has the higher value, or

---

* The test referred to is the Wilcoxon matched-pairs signed-ranks test. For a discussion of nonparametric tests see S. Siegel, *Nonparametric Statistics* (New York: McGraw-Hill Book Company, 1956).

a minus $(-)$ sign representing the fact that the second sample has the higher value. The null hypothesis is usually that the two samples were drawn from populations with the same central tendency, measured by the median. Thus, the probability of a plus sign $(p)$ or a minus sign $(q)$ for each matched pair is $p = q = \frac{1}{2}$. This hypothesis can be tested for small samples by using the binomial distribution.

Suppose that an I.Q. test is given to nine men and their wives to test the null hypothesis that this sample was drawn from a population in which the median I.Q. of a man and his wife do not differ. The alternative hypothesis is that either the husbands or the wives have higher I.Q.'s $(H_a: p \neq \frac{1}{2})$. Table 10.2 gives the data appropriate for use in the sign test. Note that the only relevant fact about these matched scores is whether the husband's score is higher or lower than his wife's score—the I.Q. scores themselves cannot be used unless interval measurement is assumed. The $t$-test cannot be used unless the additional assumption of normality of the populations is made. The binomial test can now be used to test the null hypothesis that the probability that a husband's score will exceed his wife's score (or vice versa) equals $\frac{1}{2}$. If the null hypothesis is true, then $p = q = \frac{1}{2}$; that is, half of the signs should be positive in the entire population of couples. We must find the probability that the sample distribution of signs would occur if $p = \frac{1}{2}$, and compare this to a significance level, say $\alpha = 0.05$. Since we are not hypothesizing whether the husbands or the wives shall have the higher scores, this is an example of a two-sided test, and we must compare the probability of the sample to the value of $\alpha/2$. If the calculated probability is *less* than $\alpha/2$, then we *reject* the null hypothesis and conclude that there is a significant difference in the central location of the distributions of husbands' and wives' I.Q. scores.

Referring to Table 10.2, we see that eight of the nine signs comparing husband's scores and wife's scores are positive. According to the binomial distribution in

**Table 10.2** I.Q. scores in matched-pair samples

| Sample | Wife's score | Husband's score | Sign |
|--------|--------------|-----------------|------|
| 1 | 129 | 115 | + |
| 2 | 110 | 108 | + |
| 3 | 117 | 123 | − |
| 4 | 120 | 104 | + |
| 5 | 114 | 110 | + |
| 6 | 101 | 98 | + |
| 7 | 107 | 106 | + |
| 8 | 125 | 119 | + |
| 9 | 105 | 95 | + |

Table I for $n = 9$ and $p = \frac{1}{2}$,

$$P(x \geqslant 8) = \sum_{x=8}^{9} \binom{9}{x} \left(\frac{1}{2}\right)^x \left(\frac{1}{2}\right)^{9-x} = 0.0196.$$

Since $0.0196 < \alpha/2 = 0.025$, the null hypothesis is rejected. If the alternative hypothesis for this problem had been a one-sided test (e.g., if it had been predicted that the wives would score higher), then $H_0$ could be rejected at the 0.0196 level of significance, or higher.

As in the binomial test, the null hypothesis in the sign test need not specify that $p = q = \frac{1}{2}$. Consider a problem comparing Product A and Product B; and suppose that we asked each of ten people interviewed to rate these two products on a scale of 100. We might hypothesize that the probability of A being preferred to B is not $\frac{1}{2}$, but some other value, say $\frac{3}{4}$ [that is, $P(A > B) = \frac{3}{4}$]. The alternative to this hypothesis could be that $P(A > B) < \frac{3}{4}$. The sample results of the ten interviews given in Table 10.3 show five positive signs.

**Table 10.3** Scaled values of preference for two products A and B

| Consumer | Product A score | Product B score | Sign |
|---|---|---|---|
| 1 | 75 | 58 | + |
| 2 | 85 | 92 | − |
| 3 | 61 | 69 | − |
| 4 | 55 | 50 | + |
| 5 | 82 | 71 | + |
| 6 | 88 | 84 | + |
| 7 | 45 | 78 | − |
| 8 | 90 | 79 | + |
| 9 | 63 | 69 | − |
| 10 | 71 | 80 | − |

Under the null hypothesis with $n = 10$ and $p = \frac{3}{4}$, the probability of five or less $+$'s is (from Table I):

$$P(x \leqslant 5) = \sum_{x=0}^{5} \binom{10}{x} \left(\frac{3}{4}\right)^x \left(\frac{1}{4}\right)^{5-x} = 0.078.$$

It is not possible, on the basis of this sample, to reject the null hypothesis $H_0 \colon p = \frac{3}{4}$ at conventional levels of significance such as $\alpha = 0.01$ or $0.05$. We conclude that it remains an acceptable view that $\frac{3}{4}$ of the population may prefer product A to B.

**Use of Normal Approximation for Large Samples**

When the number of matched pairs $n$ is large and the hypothesized value of $p$ is not extremely close to 0 or 1, then the normal approximation to the binomial may be applied to this sign test (as discussed in Section 5.6). The mean and variance of a binomial random variable are $\mu_x = np$ and $\sigma_x^2 = npq$. Thus, an approximately standardized normal variable may be determined when $npq > 3$ by:

$$z = \frac{x - np}{\sqrt{npq}}$$

where $x$ is the number of positive signs. To make the approximation more accurate, the *correction for continuity* should be made by reducing by 0.5 the difference between the observed number of positive signs and the expected number under the null hypothesis. Thus, a suitable test statistic for the sign test in large samples is:

$$z = \frac{(x \pm 0.5) - np}{\sqrt{npq}}. \tag{10.5}$$

The plus sign is used when $x$ is less than $np$; the negative sign is used when $x$ exceeds $np$.

For example, suppose we test whether $\frac{3}{4}$ of the population prefers product A to B by asking $n = 180$ consumers to rate each product on a scale of 100. If the comparative results contain 120 positive signs, then we calculate

$$z = \frac{(x + 0.5) - np}{\sqrt{npq}} = \frac{120.5 - 135}{\sqrt{33.75}} = -2.496.$$

Since the probability $P(z \leqslant -2.496) = 0.0063$ is smaller than $\alpha = 0.01$, we reject the null hypothesis and conclude that less than $\frac{3}{4}$ of the population prefer product A to B.

## 10.5  CHI-SQUARE TEST FOR INDEPENDENCE

In the chi-square test of Section 8.8, a set of observed values classified into $c$ categories according to a *single* attribute were tested for goodness-of-fit against a set of expected values. At this time we extend that analysis by assuming that more than one attribute is under investigation, and we want to determine whether or not these attributes are independent. For example, instead of investigating car sales relative to the single attribute "months of the year," we might wish to construct a test to determine if the attributes "car model" (such as sedans vs. hardtops) and the attribute "months of the year" are independent in their effect on sales. Similarly, in our supermarket example of Section 4.5, we might have been interested in determining whether or not the pattern of arrivals per minute is independent of the attribute "days of the week."

In examples of this type, a direct extension of the chi-square test of Section 8.8 can be used to test the null hypothesis that the attributes under investigation are independent. For such problems there is generally no "theory" available to use in determining the expected frequency for each category. However, we can use the observed data to calculate expected frequencies under the assumption that the null hypothesis (of independence) is true. For example, suppose that, in our car-sales example, 20 percent of the total number of sales fall in the month of June. If the null hypothesis is true, namely, that car model (sedan, hardtop) and months of the year are independent, then we would expect about 20 percent of all sedan sales to fall in June, and 20 percent of all hardtop sales to fall in June. The same relationship should hold for all six months under study. If the alternative hypothesis is true (i.e., these attributes are not independent), then we would expect to find differences in the proportion of sedan and hardtop sales across the six months.

In the above example, let's say that of the 150 cars the dealer sold in the six-month period, 50 were sedans and 100 were hardtops. Since June sales represented 20% of this total, if $H_0$ is true, he would thus "expect" 20% of the 50 sedan sales to occur in June, which would be 0.20(50) = 10 cars. Similarly, he would "expect" 20% of the 100 hardtop sales to occur in June, or 0.20(100) = 20 cars. (Note that he expects to sell 10 sedans plus 20 hardtops, for a total of 30 cars, which is exactly 20% of the 150 cars sold.) Also, note that he expects to sell hardtops and sedans in June in a ratio of 2 to 1. This ratio exactly agrees with his total sales ratio of 100 hardtops to 50 sedans (or 2 to 1). The entire set of expected frequencies $(E_{ij})$ and observed frequencies $(O_{ij})$ for this problem are given in Table 10.4 (in

**Table 10.4** Observed and expected sales of automobiles

| Models | Jan. | Feb. | Mar. | Apr. | May | June | Total |
|--------|------|------|------|------|-----|------|-------|
| Sedans | 9 / 3 | 6 / 3 | 5 / 4 | 8 / 12 | 12 / 16 | 10 / 12 | 50 |
| Hardtops | 18 / 24 | 12 / 15 | 10 / 11 | 16 / 12 | 24 / 20 | 20 / 18 | 100 |
| Totals | 27 | 18 | 15 | 24 | 36 | 30 | 150 |

each cell the expected frequency is in the upper left corner and the observed frequency is in the lower right corner).

The chi-square test for this type of problem is exactly as before, except that we now sum over all cells in two rows rather than just one:

$$\chi^2_{(c-1)} = \sum_{i=1}^{2} \sum_{j=1}^{6} \frac{(O_{ij} - E_{ij})^2}{E_{ij}},$$

$$\chi^2_5 = \frac{(3-9)^2}{9} + \frac{(3-6)^2}{6} + \cdots + \frac{(18-20)^2}{20}$$

$$= 14.15.$$

The number of degrees of freedom for this problem is 5, for if the marginal totals in Table 10.4 are considered fixed, then only 5 of the 12 cells are free to vary at any one time. The probability $P(\chi^2 > 14.5)$ for $v = 5$ degrees of freedom lies between $\alpha = 0.02$ and $\alpha = 0.01$ (see Table IV). Thus, for any level of significance higher than 0.02 we can reject the null hypothesis that the attributes "car model" and "months of the year" are independent. That is, we conclude that the proportion of sedans to hardtops does vary from month to month.

**Generalizing the Chi-square Test**

The chi-square test illustrated above can be generalized to include problems involving any number of categories for each attribute. Let's designate the two attributes as A and B, where attribute A is assumed to have $r$ categories ($r > 1$) and attribute B is assumed to have $c$ categories ($c > 1$).* Furthermore, assume the total number of observations in the problem is labeled $n$. A representation of these $n$ observations in matrix form is shown in Fig. 10.2, where $O_{ij}$ represents the observation in the $i$th row and the $j$th column. A matrix in the form of Fig. 10.2 is called a *contingency table*.

The dots in the column and row totals in the matrix indicate that these numbers represent the sum of a particular set of values. For example, the number $O_{.1}$ represents the sum of all the observed values in the first column, while $O_{1.}$ represents the sum of all the observed frequencies in the first row. The symbol $O_{..}$ represents the sum over all rows and columns, hence $O_{..}$ must equal $n$, the total number of observations.

Calculating the expected frequency $E_{ij}$ for each cell in a contingency table involves multiplying the *proportion* of the total number of observations falling in

---

* The reader should recognize that this use of $r$ has no connection with the standard statistical notation of $r$ representing a correlation coefficient, as in Chapters 11 and 12. The number of categories here are designated $r$ and $c$ to correspond to the number of *rows* and *columns* in a contingency table, such as Fig. 10.2.

Attribute B

|   | 1 | 2 | 3 | $\cdots$ | $j$ | $\cdots$ | $c$ | Totals |
|---|---|---|---|---|---|---|---|---|
| 1 | $O_{11}$ | $O_{12}$ | $O_{13}$ | $\cdots$ | $O_{1j}$ | $\cdots$ | $O_{1c}$ | $O_{1.}$ |
| 2 | $O_{21}$ | $O_{22}$ | $O_{23}$ | $\cdots$ | $O_{2j}$ | $\cdots$ | $O_{2c}$ | $O_{2.}$ |
| 3 | $O_{31}$ | $O_{32}$ | $O_{33}$ | $\cdots$ | $O_{3j}$ | $\cdots$ | $O_{3c}$ | $O_{3.}$ |
| . | . | . | . | | . | | . | . |
| . | . | . | . | | . | | . | . |
| . | . | . | . | | . | | . | . |
| $i$ | $O_{i1}$ | $O_{i2}$ | $O_{i3}$ | $\cdots$ | $O_{ij}$ | $\cdots$ | $O_{ic}$ | $O_{i.}$ |
| . | . | . | . | | . | | . | . |
| . | . | . | . | | . | | . | . |
| . | . | . | . | | . | | . | . |
| $r$ | $O_{r1}$ | $O_{r2}$ | $O_{r3}$ | $\cdots$ | $O_{rj}$ | $\cdots$ | $O_{rc}$ | $O_{r.}$ |
| Totals | $O_{.1}$ | $O_{.2}$ | $O_{.3}$ | $\cdots$ | $O_{.j}$ | $\cdots$ | $O_{.c}$ | $O_{..} = n$ |

(row label *Attribute A* at left)

**Fig. 10.2.** Contingency table.

the $j$th category for Attribute B (which is $O_{.j}/n$) times the *number* of observations falling in the $i$th category of Attribute A (which is $O_{i.}$).*

*Expected frequency in $i$th row, $j$th column:*

$$E_{ij} = \left(\frac{O_{.j}}{n}\right)(O_{i.}) = \frac{O_{i.}O_{.j}}{n}. \tag{10.6}$$

After completing the expected frequency for each of the cells in the contingency table [using Formula (10.6)], the calculated value of $\chi^2$ can be determined by the following formula:

$$\chi^2_{(r-1)(c-1)} = \sum_{i=1}^{r} \sum_{j=1}^{c} \frac{(O_{ij} - E_{ij})^2}{E_{ij}}. \tag{10.7}$$

The number of degrees of freedom for this $\chi^2$-statistic can be determined by noting that, in calculating the expected frequency for each cell, we must assume that the marginal totals ($O_{i.}$ and $O_{.j}$) are fixed quantities. This means that one degree of

---

* This expectation is a direct result of the relationship presented in Chapter 2, which says that two discrete events $A_i$ and $B_j$ are independent if and only if $P(A_i \cap B_j) = P(A_i)P(B_j)$. Since our estimates of $P(A_i)$ and $P(B_j)$ are $O_{i.}/n$ and $O_{.j}/n$, respectively, the product $(O_{i.}/n)(O_{.j}/n)$ is our estimate of the joint probability. Multiplying this product by the total number of observations ($n$) gives Formula (10.6).

freedom is lost for each row and each column, so that the total number of degrees of freedom is $(r - 1)(c - 1)$.

## A Sample Problem

To illustrate the use of Formula (10.7), consider the problem of trying to determine whether the prices of certain stocks on the New York Stock Exchange are independent of the industry to which they belong. Assume that four categories of industries are investigated (labeled I, II, III, and IV), and that stock prices in these industries are classified into one of three categories ("high-priced," "middle-priced," or "low-priced.") The data from such an analysis might look like the values shown in Table 10.5, where once again the expected values are in the upper left of each cell, and the observed values in the lower right.

**Table 10.5** Frequencies for the stock example

|  | $E_{ij}$ |
|--|--|
|  | $O_{ij}$ |

|  | Industry | | | | |
|--|--|--|--|--|--|
| Stock prices | I | II | III | IV | Total |
| High | 13.8 / 15 | 10.4 / 8 | 8.7 / 10 | 12.1 / 12 | 45 |
| Med. | 18.5 / 20 | 13.9 / 16 | 11.5 / 12 | 16.1 / 12 | 60 |
| Low | 7.7 / 5 | 5.7 / 6 | 4.8 / 3 | 6.8 / 11 | 25 |
| Total | 40 | 30 | 25 | 35 | 130 |

To illustrate the calculation of expected frequencies, note that the expected frequency of high-priced stocks in Industry I is found by multiplying the proportion of stock in Industry I to the total number of observations, which is 40/130, by the number of observations in the high-priced category (45). This product is

$$(40/130)(45) = 13.8,$$

which is shown in the first cell. Similarly, the expected frequency for high-priced stocks in Industry II is $(30/130)(45) = 10.4$. Note that, for each row and column, the sum of the expected frequencies must be the same as the sum of the observed frequencies. The number of degrees of freedom for this problem is $(r - 1)(c - 1) = (3)(2) = 6$, and the calculated value of $\chi^2$ is

$$\chi_6^2 = \frac{(15 - 13.8)^2}{13.8} + \frac{(8 - 10.4)^2}{10.4} + \cdots + \frac{(11 - 6.8)^2}{6.8}$$
$$= 6.264.$$

To be significant at the 0.05 level, the value of $\chi^2$ has to be greater than 12.50 for 6 degrees of freedom (see Table IV). Since the computed value $\chi^2 = 6.264$ is less than this value, the null hypothesis cannot be rejected at the 0.05 level of significance. We thus conclude that the price of stocks is independent of the industry associated with that stock.

### 10.6 ANALYSIS-OF-VARIANCE TEST OF DIFFERENCES AMONG MEANS OF TWO OR MORE POPULATIONS

The tests of hypotheses in Section 10.3 were designed to test for differences between two population means. Often, practical situations may arise where we want to compare more than two populations, such as in comparing the yield from several varieties of corn plants, the gasoline mileage of four automobiles, the smoking habits of five groups of college students, and so forth. In these circumstances one normally does not want to (and usually should not) consider all possible combinations of two populations at a time, and test for differences in each pair. Rather, we want to investigate the differences among the means of all the populations *simultaneously*. The method for performing this simultaneous test is called "ANOVA," which is an abbreviation for *analysis of variance*. The essence of ANOVA is that the total amount of variation in a set of data is broken down into two types, that amount which can be attributed to chance, and that amount which can be attributed to specified causes. ANOVA tests thus involve a comparison between these amounts.

In general, one can investigate any number of factors which are hypothesized to influence the dependent variable. For example, one may wish to investigate the effect on gasoline mileage (the dependent variable) of such things as the speed of the automobile (factor I), the horsepower of the engine (factor II), and the make of the car (factor III). In addition, the categories within each of these factors may have a large number of possible values (e.g., there are an infinite number of car speeds we might investigate). It should not be difficult for even the beginning student of ANOVA to see that the relationship between the factors may become quite complex (for example, how do horsepower and the speed of an automobile interact in their effect on gasoline mileage?). For this reason we present the analysis for only the one-factor model in this text.

## The One-factor Model

In our one-factor model, we will assume that we want to test the null hypothesis that the means of $J$ different populations are all equal. To make this test, we will take a sample from each of the $J$ populations. Analysis-of-variance methods test for differences among the means of the populations by examining the amount of variation *within* each of these samples, relative to the amount of variation *between* the samples. For example, in testing for differences in gasoline mileage attributable to the make of automobile used (the one factor), we could take a sample of $J$ different makes. In ANOVA the $J$ different samples are often called the $J$ "treatments." This terminology stems from early applications of ANOVA to agricultural problems, where, for example, the amount of crop yielded by a certain type of soil was tested by "treating" the soil with various kinds of fertilizer. The first step in building an ANOVA model is to specify the underlying population relationships. To do this, suppose we denote the different treatments by the letter $j$, where $j = 1, 2, 3, \ldots, J$. If we let $N_j$ be the size of the $J$th population, then the values within the $j$th population can be denoted as $i = 1, 2, 3 \ldots, N_j$. Assume, $y_{ij}$ represents the $i$th value of the $j$th population under investigation. If we now denote the mean value of $y$ in the $j$th population as $\mu_j$, and the mean of all values of $y_{ij}$ in all $J$ columns as $\mu$ (or, equivalently, $\mu$ is the mean of the values of $\mu_j$), then the following matrix form represents this situation.

| | | | J Populations (Treatments) | | | | |
|---|---|---|---|---|---|---|---|
| | | 1 | 2 | $3 \cdots j \cdots J$ | | | |
| | 1 | $y_{11}$ | $y_{12}$ | $y_{13} \cdots y_{ij} \cdots y_{1J}$ | | | |
| | 2 | $y_{21}$ | $y_{22}$ | $y_{23} \cdots y_{2j} \cdots y_{2J}$ | | | |
| | 3 | $y_{31}$ | $y_{32}$ | $y_{33} \cdots y_{3j} \cdots y_{3J}$ | | | |
| | . | . | . | .    .    . | | | |
| Values within | . | . | . | .    .    . | | | |
| each population | . | . | . | .    .    . | | | |
| | $i$ | $y_{i1}$ | $y_{i2}$ | $y_{i3} \cdots y_{ij} \cdots y_{iJ}$ | | | |
| | . | . | . | .    .    . | | | |
| | . | . | . | .    .    . | | | |
| | . | . | . | .    .    . | | | |
| Mean of the $j$th population | | $\mu_1$ | $\mu_2$ | $\mu_3 \cdots \mu_{j.} \cdots \mu_J$ | | $\mu$ = Grand mean | |
| Population size | | $N_1$ | $N_2$ | $N_3 \cdots N_j \cdots N_J$ | | | |

As we indicated above, in ANOVA we are interested in determining the variation within and between the populations. In terms of the variation within a given population, we will assume that the values of $y_{ij}$ differ from the mean of this population ($\mu_j$) only because of random effects. That is, there are influences on $y_{ij}$ which are unexplainable (i.e., "random") in terms of our one-factor model. The difference between $y_{ij}$ and $\mu_j$ is usually denoted by the symbol $\varepsilon_{ij}$ (where $\varepsilon$ is the Greek letter epsilon). Thus,

$$\varepsilon_{ij} = y_{ij} - \mu_j$$

or

$$y_{ij} = \varepsilon_{ij} + \mu_j \tag{10.8}$$

for $(j = 1, 2, \ldots, J)$ and $(i = 1, 2, \ldots, N_j)$.

In examining differences *between* populations, we will assume that the difference between the mean of the $j$th population ($\mu_j$) and the grand mean ($\mu$) is attributable to what is called a "treatment effect." That is, $\mu_j$ is not exactly equal to $\mu$ because of the effect of the $j$th treatment. Suppose we label this treatment effect as $\tau_j$ ($\tau$ is the Greek letter tau), where

$$\tau_j = \mu_j - \mu$$

or

$$\mu_j = \tau_j + \mu \tag{10.9}$$

for $j = 1, 2, \ldots, J$.

The one-factor ANOVA model can be formulated by substituting Formula (10.9) into Formula (10.8):

> *One-factor ANOVA model:*
>
> $$y_{ij} = \mu + \tau_j + \varepsilon_{ij} \qquad \begin{array}{l} (j = 1, 2, \ldots, J) \\ (i = 1, 2, \ldots, N_j). \end{array} \tag{10.10}$$

This model says that the value of $y_{ij}$ is composed of three components (or effects): a common effect ($\mu$) plus a treatment effect ($\tau_j$) plus a random effect ($\varepsilon_{ij}$).

Before we can proceed with our ANOVA test it is necessary to make a number of assumptions about the effects which comprise the model. First, we must assume that the grand mean ($\mu$) is a fixed constant, and that the treatment effects ($\tau_j$) are also fixed constants. We must also assume that the random error terms ($\varepsilon_{ij}$) are independent, and that for each of the $J$ treatments (populations), the errors $\varepsilon_{ij}$ are normally distributed with a mean of zero, and have a variance ($\sigma^2$) which is the same for all populations.

### Sums of Squares in ANOVA

The null hypothesis in our one-factor model is that the treatment effects are all zero. This hypothesis can be stated either as $H_0: \tau_j = 0$ $(j = 1, 2, 3, \ldots, J)$, or, equivalently as $H_0: \mu_1 = \mu_2 = \cdots = \mu_J$. The alternative hypothesis is that the treatment effects are not all equal to zero. To test these hypotheses we assume that a sample of size $n_j$ has been taken from each of the $J$ populations. For each sample we calculate the mean value $\bar{y}_j$ for $j = 1, 2, \ldots, J$, and we also calculate the grand sample mean, which is denoted as $\bar{y}$.

To illustrate these concepts, suppose we measure the gasoline mileage ($y$) of three different makes of compact automobiles. Four cars of each make are sampled, and each car selected has a standard transmission, four-cylinder engine, and no power equipment. All cars are run on the same trip (city and country driving), using the same drivers, fuel, and weather conditions. In other words, all possible influences except for the brand are controlled as carefully as possible; of course, there may be factors which are not (or cannot be) controlled, such as traffic conditions and tire conditions, but these factors are assumed to be random in their effect. Now, assume the sample results for this test are those shown in Table 10.6.

**Table 10.6** Gasoline mileage test

| Observa-tion ($i$) | Sample number($j$) | Make of Car | | |
|---|---|---|---|---|
| | | 1 | 2 | 3 |
| 1 | | 28 | 24 | 31 |
| 2 | | 25 | 23 | 32 |
| 3 | | 27 | 18 | 37 |
| 4 | | 28 | 27 | 24 |
| Average | | $\bar{y}_1 = 27$ | $\bar{y}_2 = 23$ | $\bar{y}_3 = 31$     $\bar{y} = 27$ |

If $H_0$ is true (i.e., $\mu_1 = \mu_2 = \mu_3$), then the observed differences between $\bar{y}_1$, $\bar{y}_2$, and $\bar{y}_3$ in Table 10.6 can be attributed to random effects. The alternative hypothesis in this case is $H_a$: $\mu_1$, $\mu_2$, and $\mu_3$ are not all equal. Note that we can use the column means in Table 10.6 to make estimates of the treatment effects (our estimates are denoted as $\hat{\tau}_j$):

$$\begin{array}{l} \text{Estimated} \\ \text{treatment} \\ \text{effects} \end{array} \begin{cases} \hat{\tau}_1 = \bar{y}_1 - \bar{y} = 27 - 27 = 0, \\ \hat{\tau}_2 = \bar{y}_2 - \bar{y} = 23 - 27 = -4, \\ \hat{\tau}_3 = \bar{y}_3 - \bar{y} = 31 - 27 = 14. \end{cases}$$

In effect, the ANOVA test is concerned with determining whether the estimated values of $\tau_j$ are large enough to convince us that $H_0$ is not, in fact, true.

Whenever $H_0$ is true, we would expect the variability between the $J$ means to be the same as the variability within each sample, since in this case the random effects ($\varepsilon_{ij}$) are the only source of variation. If the treatment effects are not all zero, then the variability between samples should be larger than the variability within the samples. Our measure of variability in ANOVA is similar to that used in calculating variances; i.e., first we calculate the sum of the squared deviations about the mean.

The variation *within* the $J$ samples is calculated by first summing the squared deviations of $y_{ij}$ about $\bar{y}_j$ for each sample, which is $\sum_{i=1}^{n_j} (y_{ij} - \bar{y}_j)^2$. If we now sum this variation over all $J$ samples, the result is called the *sum of squares within* (abbreviated SSW):

$$\text{Sum of squares within:} \quad \text{SSW} = \sum_{j=1}^{J} \sum_{i=1}^{n_j} (y_{ij} - \bar{y}_j)^2. \quad (10.11)$$

We will need the amount of variation *between* samples, which is called *sum of squares between* (SSB). In this case, we first take the squared deviation of the $J$th column mean and the grand mean, which is $(\bar{y}_j - \bar{y})^2$. This deviation must then be multiplied by (i.e., weighted by) the number of observations in the $J$th sample, and then summed over all values of $J$. That is,

$$\text{Sum of squares between:} \quad \text{SSB} = \sum_{j=1}^{J} n_j(\bar{y}_j - \bar{y})^2. \quad (10.12)$$

There is one other variation in ANOVA, which is the total variation among all observations in the sample. This variation, denoted as SST (sum of squares total), is the sum of squared deviations of all values of $y_{ij}$ about the grand mean $\bar{y}$:

$$\text{Sum of squares total:} \quad \text{SST} = \sum_{j=1}^{J} \sum_{i=1}^{n_j} (y_{ij} - \bar{y})^2.$$

A fundamental equation of ANOVA states that *total variation equals the sum of the between and within variations:* i.e.,

$$\text{SST} = \text{SSB} + \text{SSW}.$$

We can illustrate the calculation of these three measures of variation using the data in Table 10.6. First, applying Formula (10.11), we find SSW to be the sum of the squared deviations of the first column (which is 6) plus the variation in column 2 (which is 42) plus the variation in column 3 ($=86$). Hence

$$\text{SSW} = \sum_{i=1}^{4} (y_{i1} - \bar{y}_1)^2 + \sum_{i=1}^{4} (y_{i2} - \bar{y}_2)^2 + \sum_{i=1}^{4} (y_{i3} - \bar{y}_3)^2$$
$$= 6 + 42 + 86 = 134.$$

In using Formula (10.12) to calculate the variation between the column means, we see that $n_1 = n_2 = n_3 = 4$, and

$$\text{SSB} = n_1(\bar{y}_1 - \bar{y})^2 + n_2(\bar{y}_2 - \bar{y})^2 + n_3(\bar{y}_3 - \bar{y})^2$$
$$= 4(0) + 4(16) + 4(16) = 128.$$

Finally, we know that SST = SSB + SSW = 128 + 134 = 262. The reader may wish to verify this value, using Table 10.6, and the formula for SST.

### Mean Squares and the ANOVA Table

As we have seen above, the first step in ANOVA is to calculate SSB and SSW. In order to compare the variability within samples to the variability between samples we need to divide these sums by their respective degrees of freedom (for the same reason that $\sum(x_i - \bar{x})^2$ is divided by its d.f. $(n - 1)$, in calculating the sample variance $s^2$). The d.f. for SSB is always one less than the number of populations, or $J - 1$. Similarly, for SST the number of d.f. is one less than the total sample size, which is $\sum_{j=1}^{J} n_j - 1$. For SSW, the expression, $\sum_{j=1}^{J} n_j - J$, gives the degrees of freedom. Note that these d.f. sum in the same manner as do the sums of squares:

*Sums of squares:*       SST     =     SSB    +     SSW

$$d.f.: \quad \sum_{j=1}^{J} n_j - 1 = (J - 1) + \sum_{j=1}^{J} n_j - J.$$

A sum of squares divided by its degrees of freedom is called a *mean square* (abbreviated MS). Hence,

*Mean square between:*     $\text{MSB} = \text{SSB}/(J - 1);$

*Mean square within:*      $\text{MSW} = \text{SSW}\Big/\left(\sum_{j=1}^{J} n_j - J\right).$

The various components necessary for ANOVA are usually presented in what is called an "analysis-of-variance table." The general format of such a table is shown in Table 10.7.

**Table 10.7** Analysis-of-variance table

| Source of variation | SS | d.f. | MS = SS/d.f. |
|---|---|---|---|
| Between samples (treatments) | $\sum_{j=1}^{J} n_j(\bar{y}_j - \bar{y})^2$ | $J - 1$ | $\text{SSB}/(J - 1)$ |
| Within samples | $\sum_{j=1}^{J}\sum_{i=1}^{n_j}(y_{ij} - \bar{y}_j)^2$ | $\sum_{j=1}^{J} n_j - J$ | $\text{SSW}\Big/\left(\sum_{j=1}^{J} n_j - J\right)$ |
| Total | $\sum_{j=1}^{J}\sum_{i=1}^{n_j}(y_{ij} - \bar{y})^2$ | $\sum_{j=1}^{J} n_j - 1$ | |

**Table 10.8** ANOVA for gasoline mileage test

| Source of variation | SS | d.f. | MS |
|---|---|---|---|
| Between samples (brands) | 128 | 2 | 64 |
| Within samples | 134 | 9 | 14.9 |
| Total | 262 | 11 | |

We can illustrate this type of table with our gasoline mileage example. These data are shown in Table 10.8.

In ANOVA we test the null hypothesis by comparing the value of MSB to the value of MSW. If the amount of variability between samples (MSB) is small relative to the variability within samples (MSW), then we should conclude that $H_0$ cannot be rejected. On the other hand, if MSB is large relative to MSW, then we want to reject $H_0$. We can determine whether the size of MSB to MSW is large enough to reject $H_0$ by using a test based on a probability distribution that we have not yet presented in this text. This continuous probability density function is called the *F*-distribution.

### 10.7   THE *F*-DISTRIBUTION

The *F*-distribution is named in honor of R. A. Fisher, who first studied it in 1924. This distribution is usually defined in terms of the ratio of two independent $\chi^2$ variables, each divided by its degrees of freedom. Suppose that $A_1$ is a $\chi^2$-variable, with $v_1$ degrees of freedom, and $A_2$ is another $\chi^2$-variable with $v_2$ degrees of freedom. The ratio of $A_1$ divided by $v_1$ to $A_2$ divided by $v_2$ is a random variable which is denoted by the letter $F$:

$$F_{(v_1,\, v_2)} = \frac{A_1/v_1}{A_2/v_2}. \tag{10.13}$$

This random variable has $v_1$ degrees of freedom in the numerator, and $v_2$ d.f. in the denominator. The value of $F$ must always be positive or zero since $A_1$ and $A_2$ are squares and can never assume negative values. On the other hand, there is no upper limit to the value of $F$, so this distribution, like the $\chi^2$-distribution, is skewed to the right. A typical *F*-distribution is shown in Fig. 10.3. Tables VIII (a) and (b) list the critical values of $P(F > F)$ for selected values of $v_1$ and $v_2$ corresponding to the two most commonly employed significance levels, $\alpha = 0.05$ and $\alpha = 0.01$.* For example, from Table VIII (a) it can be seen that, when

---

* Notice that Table VIII is not a cumulative distribution function ($P(F < F)$), but rather gives the values in the upper tail of the pdf, $P(F > F)$.

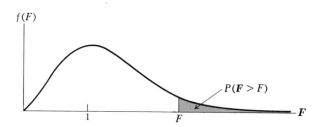

**Fig. 10.3.** A typical **F**-distribution.

$v_1 = 10$ and $v_2 = 6$, the probability that the random variable $F$ will exceed 4.06 is 0.05; that is, $P(F > 4.06) = 0.05$. Similarly, from Table VIII (b), the probability is 0.01 that $F$ is larger than 7.87 for these same degrees of freedom. The reader should note that only values of $F$ which are greater than 1.0 are presented in Table VIII, in order to save on the space necessary to present these values. Also, remember that the only inconvenience this causes is that it is always necessary to assume that the *larger* value of $A/v$ occurs in the *numerator* of the $F$ ratio.

**The *F*-test in ANOVA**

In ANOVA, the values of SSB and SSW can both be shown to follow $\chi^2$-distributions, with $(J - 1)$ and $(\sum n_j - J)$ degrees of freedom respectively. Thus, in Formula (10.13) if we let $A_1 = \text{SSB}$ with $v_1 = (J - 1)$, and $A_2 = \text{SSW}$ with $v_2 = (\sum n_j - J)$, then

$$F_{(J-1,\,\Sigma\,n_j - J)} = \frac{\text{SSB}/(J - 1)}{\text{SSW}/(\sum n_j - J)}$$
$$= \frac{\text{MSB}}{\text{MSW}}. \qquad (10.14)$$

Formula (10.14) therefore gives us a method for testing the size of MSB relative to MSW. Note that for this test the critical region lies in the *upper* tail of the *F*-distribution since $H_0$ is rejected only for large values of $F$. That is, when MSB is large relative to MSW, we reject $H_0$ because it appears that the treatment effects are not all equal to zero. On the other hand, when the treatment effects are all equal to zero (that is, $\mu_1 = \mu_2 = \cdots = \mu_J$), then we would expect the ratio of MSB to MSW to be relatively small, which means that $H_0$ is not rejected for small values of $F$.

Returning to our gasoline mileage test, we can now show how the comparison between MSB and MSW can be made by using the *F*-distribution. First recall, from Table 10.8, that MSB $= 64$ and MSW $= 14.9$, and $v_1 = 2, v_2 = 9$. The value of $F$ is thus

$$F_{(2, 9)} = \frac{MSB}{MSW} = \frac{64}{14.9} = 4.30.$$

Referring to Table VIII (a) for 2 and 9 degrees of freedom, we see that the critical value for $\alpha = 0.05$ is 4.26. Since our calculated value is larger than this number, we conclude that the treatment effects are not all equal to zero; i.e., that a difference does exist in the gas mileage among the three makes of automobiles.

**An Easier Calculational Procedure**

The reader should note that not all samples need to have the same number of observations, and that the number of populations may be two or more. If $J = 2$, then this $F$-test is equivalent to the one-sided $t$-test described in Section 10.3. To illustrate the case when the sample sizes are not equal, consider the problem of an agricultural advisor who is interested in the yield of fruit from orchards located near each other and similar in all ways except that different treatments of fungicides and pesticides are applied. A random selection of fruit trees is made and the output recorded, as shown in Table 10.9.

**Table 10.9** Agricultural experiment yields

| Treatment 1 | Treatment 2 | Treatment 3 |
|---|---|---|
| 130 | 126 | 131 |
| 126 | 127 | 128 |
| 128 | 129 | 133 |
| 132 | 124 | 130 |
| 126 | 126 | 130 |
|  | 130 | 132 |
|  | 125 |  |
|  | 126 |  |
|  | 128 |  |
|  | 126 |  |

A useful step to reduce computational drudgery in ANOVA is to subtract an arbitrary value from all observations to make the numbers easier to manipulate. (This process does not affect the variations between or within columns since the same number is subtracted from all $y_{ij}$.) Table 10.10 shows the same data as Table 10.9 after subtracting the arbitrary value 120 from each item. It also shows the sample sums and means and the estimated treatment effects for the new coded data.

**Table 10.10** Agricultural experiment, using coded yields

| Treatment 1 | Treatment 2 | Treatment 3 | |
|---|---|---|---|
| 10 | 6 | 11 | |
| 6 | 7 | 8 | |
| 8 | 9 | 13 | |
| 12 | 4 | 10 | |
| 6 | 6 | 10 | |
| | 10 | 12 | |
| | 5 | | |
| | 6 | | |
| | 8 | | |
| | 6 | | |
| $n_1 = 5$ | $n_2 = 10$ | $n_3 = 6$ | $n = 21$ |
| $\sum_{i=1}^{5} y_{i1} = 42$ | $\sum_{i=1}^{10} y_{i2} = 67$ | $\sum_{i=1}^{6} y_{i3} = 64$ | $\sum_{j=1}^{3}\sum_{i=1}^{n_j} y_{ij} = 173$ |
| $\bar{y}_1 = 8.4$ | $\bar{y}_2 = 6.7$ | $\bar{y}_3 = 10.7$ | $\bar{y} = 8.2$ |
| $\hat{\tau}_1 = 0.2$ | $\hat{\tau}_2 = -1.5$ | $\hat{\tau}_3 = 2.5$ | |

While the formulas given previously help describe the different measures used in ANOVA, they are computationally tedious to apply when the number of observations is large or when the sample means have several decimals. The following computational formulas are much easier to use, although they may appear more complex.

$$\text{C.F. (correction factor)} = \left(\sum_{j}^{J}\sum_{i}^{n_j} y_{ij}\right)^2 \Big/ n$$

$$\text{SST} = \sum_{i}^{n_j}\sum_{j}^{J} y_{ij}^2 - \text{C.F.},$$

$$\text{SSB} = \sum_{j=1}^{J}\left[\frac{\sum_{i=1}^{n_j}(y_{ij})^2}{n_j}\right] - \text{C.F.},$$

(10.15)

and

$$\text{SSW} = \text{SST} - \text{SSB}.$$

We now apply these formulas to the data of Table 10.10. The correction factor, using the sum of all observations, is

$$\text{C.F.} = \frac{(173)^2}{21} = 1425.19.$$

**Table 10.11** ANOVA for agricultural experiment

| Source of variation | SS | d.f. | MS |
|---|---|---|---|
| Due to treatments | SSB = 59.18 | 2 | 29.59 |
| Due to error | SSW = 72.63 | 18 | 4.04 |
| Total | SST = 131.81 | 20 | |

The sum of squares of all observations is $\sum_i^{n_j} \sum_j^J y_{ij}^2 = 1557$; hence,

$$SST = 1557 - 1425.19 = 131.81.$$

The square of sums of entries in each column is used to obtain:

$$SSB = \frac{1764}{5} + \frac{4489}{10} + \frac{4096}{6} - 1425.19 = 59.18.$$

Then, by subtraction, SSW = 131.81 − 59.18 = 72.63. Forming an ANOVA table, we obtain Table 10.11.

Using the **F**-ratio MSB/MSW yields

$$F_{(2, 18)} = \frac{29.59}{4.04} = 7.32.$$

Since the critical value for $\alpha = 0.01$ is $F_{(2, 18)} = 6.01$, we conclude that the null hypothesis of no difference in yield due to treatments is rejected. There *is* a significant difference among the means of the populations that were subjected to the different treatments. Referring to the sample means in Table 10.10, it appears that treatment 3 is most effective in increasing yield.

## REVIEW PROBLEMS

1. Two types of new car are tested for gas mileage. One group, consisting of 36 cars, had an average gas mileage of 24 miles per gallon, with a standard deviation of 1.5 miles, while the corresponding figures for the other group, consisting of 72 cars, were 22.5 and 2.0. Use these values to test the null hypothesis that there is no difference between the two types of car with respect to gas mileage, at the 0.01 level of significance.

2. Suppose that a random sample is drawn from two populations, both normally distributed and having equal variances. The mean of the first sample of 11 observations is 25.0, with a variance of 300. The second sample mean, resulting from 30 observations, is $\bar{x}_2 = 20.0$, with a variance of $s_2^2 = 100$.

   a) At the 0.05 level of significance, can the null hypothesis $H_0: \mu_1 - \mu_2 = 0$ be rejected in favor of the alternative hypothesis $H_a: \mu_1 - \mu_2 \neq 0$?

   b) At what level of significance can the null hypothesis $H_0: \mu_1 - \mu_2 = 0$ be rejected in favor of $H_a: \mu_1 - \mu_2 > 0$?

3. Suppose an underground newspaper is sold on two college campuses. A random selection of weekly sales figures provides the following data:

| Sample | Sample size | Mean | Standard deviation |
|---|---|---|---|
| Campus A | 10 | 123 | 15 |
| Campus B | 6 | 108 | $\sqrt{185}$ |

Based on sales averages, test whether the newspaper sells more at A than at B. Use alpha = 0.05.

4. A particular student tells you that you get more ice cream in a 20¢ cone at the Dairy Bar than at the IC-Shoppe. You buy cones at both locations at random times during a three-week period, and measure the ice-cream content as follows:

| | Dairy Bar | IC-Shoppe |
|---|---|---|
| Number of cones | 8 | 10 |
| Average (oz.) | 7 | 5.5 |
| Standard deviation (oz.) | 1 | $\sqrt{1.7}$ |

Make a test, at the 0.01 level of significance, that is appropriate to determine whether the student's claim seems correct.

5. Two sections of a statistics course took the same final examination. A sample of 9 was randomly drawn from Section A, and a sample of 4 was randomly drawn from Section B. These scores are arranged in ascending order:

| Section A: | 65, | 68, | 72, | 75, | 82, | 85, | 87, | 91, | 95 |
|---|---|---|---|---|---|---|---|---|---|
| Section B: | 50, | 59, | 71, | 80 | | | | | |

a) What assumptions about the parent population are necessary if we wish to use the $t$-test to test for significant differences between these samples?

b) At the 0.025 level of significance, can $H_0: \mu_1 - \mu_2 = 0$ be rejected in favor of $H_a$: $\mu_1 - \mu_2 \neq 0$?

6. Two swimmers are recognized throughout the world as among the best in a certain event. In a series of independent practice trials, they set the following times in seconds. On the basis of these trials can you detect, with 99% confidence, any difference between the performances of these two swimmers?

| Swimmer A | Swimmer B |
|---|---|
| 30.7 | 31.1 |
| 31.2 | 31.2 |
| 31.3 | 31.4 |
| 30.9 | 31.6 |

7. The monthly sales of automatic feeding birdcages in two different sales districts is recorded for five months selected at random. Test at the 5% significance level to see if sales are significantly higher, on the average, in District B.

| District A | District B |
|------------|------------|
| 12 | 10 |
| 16 | 20 |
| 10 | 16 |
| 14 | 18 |
| 8 | 16 |

8. A student has a new automobile and is trying to determine statistically which of two different gasolines he should use. He measures his mileage on eight consecutive tankfuls, using Brands A and B alternately. He computes the mileage difference $(B - A)$ after every two tankfuls. The average difference is 2.5 mpg and the standard deviation of the differences is 2.0 mpg. Since the dealer selling Brand A is a friend, he will switch to Brand B only if he is 90% sure that his mileage will be at least one mile per gallon better. What should he do?

9. You are considering chartering a boat for some deep-sea fishing. You usually charter with Capt. Mike Ketchum, but you have heard that Capt. Joe Hookum is better at finding the big ones during their feeding hour. You go down to the docks and observe the daily catch from the two different boats. In comparing the six largest fish caught on each boat, you find that the fish from Capt. Joe's boat are on the average 9 pounds heavier with a standard deviation (from this average difference) of 3 pounds. Test, with 95% confidence, whether or not you should switch from your old salty buddy, Capt. Mike, in order to catch heavier fish.

10. A county agent experiments with eight acres of land. Half the acreage is treated with fertilizer $x$ and half is treated with fertilizer $y$. The average difference in yield between paired acres was 10 bushels. The standard deviation of differences was 4. Would you conclude that there is a significant difference of yield between the two differently fertilized tracts? You should have 99% confidence, if you do find a difference.

11. Distinguish between parametric and nonparametric statistical tests. Under what circumstances is each type of test most appropriate? Give several specific examples of problems where a nonparametric test would be more appropriate than a parametric test.

12. Identify each of the following numbers as representing measurement on either nominal, ordinal, interval, or ratio scales:
   a) the numbers designating years (e.g., 1971, 1972, etc.),
   b) the numbers on football players' jerseys,
   c) the numbers representing golf scores in a tournament,
   d) social security numbers,
   e) the numbers representing the order of finish in a horse race.

13. Two groups of overweight men were matched in pairs according to age, weight, occupation, and a number of other criteria; one-half of these men were put on one weight-

reducing program, the other half on another program. The weight losses were as follows:

| Pair | First plan | Second plan |
|------|-----------|-------------|
| 1 | 25 lb | 15 lb |
| 2 | 29 lb | 22 lb |
| 3 | 21 lb | 30 lb |
| 4 | 48 lb | 12 lb |
| 5 | 8 lb | 0 lb |

Use a sign test with $\alpha = 0.10$ to determine whether the weight loss under the two plans differs.

14. a) How is the $F$-distribution related to the chi-square distribution?
   b) What are the parameters of the $F$-distribution?
   c) What is the probability a value of the $F$-distribution will exceed 3.37 when $v_1 = 10$ and $v_2 = 20$? What is $P(F \geqslant 2.77)$ for 20 degrees of freedom in the numerator and 10 in the denominator?

15. Explain the advantage, if any, of a comparison of $J$ means by an analysis of variance and an $F$-test, over the practice of carrying out a $t$-test separately for each pair of means.

16. All restaurants are rated for cleanliness and quality by three state health inspectors. A sample of six restaurants gives the ratings as shown below. Determine at the $\alpha = 0.05$ level of significance whether the inspectors differ significantly in their average rating score, or whether the variation in the average scores can be attributed to sampling error.

| Inspector | Restaurant ratings | | | | | |
|-----------|----|----|----|----|----|----|
|           | 1  | 2  | 3  | 4  | 5  | 6  |
| A | 99 | 90 | 66 | 75 | 85 | 92 |
| B | 95 | 49 | 48 | 71 | 80 | 93 |
| C | 97 | 62 | 60 | 76 | 90 | 88 |

17. Suppose there are four classes in a statistics course, each using the same text and materials but having different teachers. We wish to determine, at $\alpha = 0.01$ level of significance, whether there is a difference in the common exam results due to the different instruction. For this purpose, samples of size five were drawn at random from each class; their scores on a 10-point exam are given below.

| Instructors | Grade observations | | | | |
|-------------|----|----|---|---|----|
| A | 4 | 3 | 1 | 2 | 4 |
| B | 3 | 7 | 6 | 3 | 6 |
| C | 8 | 8 | 5 | 7 | 8 |
| D | 9 | 10 | 8 | 8 | 10 |

Construct an ANOVA table and determine the test result.

18. The numbers of students using a computer terminal, by class, over a selected sample of three weeks, are as follows:

| Week | Freshmen | Sophomores | Juniors | Seniors | Graduates |
|------|----------|------------|---------|---------|-----------|
| 1 | 114 | 171 | 147 | 151 | 167 |
| 2 | 120 | 166 | 134 | 179 | 177 |
| 3 | 150 | 143 | 121 | 156 | 199 |

Determine, at $\alpha = 0.05$ level of significance, whether the average weekly use of the terminal differs among these student classes.

19. The records at a large metropolitan hospital listed the following births during the three shifts of the day during a two-week period.

|  | 7 A.M.–3 P.M. | 3 P.M.–11 P.M. | 11 P.M.–7 A.M. |
|---------|---------------|----------------|----------------|
| Males | 15 | 5 | 10 |
| Females | 5 | 10 | 15 |

a) Determine the expected frequency in each cell. What null hypothesis is being tested with a $\chi^2$-test?
b) How many degrees of freedom are there? At what level of significance can $H_0$ be rejected?

## EXERCISES

20. An I.Q. test was given to two different groups of high-school students. The first group resulted in a mean I.Q. of 112 with a standard deviation of 6, and the second group had an average I.Q. of 114 with a standard deviation of 4.

a) If the first group was of size $n = 60$ and the second group of size $n = 40$, at what level of significance can the null hypothesis $H_0:\mu_1 - \mu_2 = 0$ be rejected in favor of $H_a:\mu_1 - \mu_2 \neq 0$? At what level can $H_0$ be rejected in favor of $H_a:\mu_1 - \mu_2 < 0$?
b) Will your answers to part (a) change if the sample size for both groups is $n = 100$?

21. A sample of 10 fibers treated with a standard technique has average strength of 10, with standard deviation 3.2. Another sample of 17 fibers treated with a new technique has an average of 20 and standard deviation of 3.0. Test at the 0.05 level of significance to see whether the new technique can be expected to produce a population of fibers with average strength *at least 5* greater than the population of fibers treated with the standard technique.

22. Two groups of students were given a written test on driving skills. There were ten students in the first group (A) and eight in the second (B). Their scores were as follows.

| A | B |
|---|---|
| 25 | 45 |
| 30 | 49 |
| 42 | 62 |
| 44 | 63 |
| 58 | 68 |
| 59 | 69 |
| 75 | 69 |
| 79 | 71 |
| 87 | |
| 90 | |

Suppose that the national average for this test is 60. Test the hypothesis that the frequency of students scoring above and below the national average is not the same for these two groups. (Use a $\chi^2$-test.)

23. The hours worked during the summer by college students at various full-time jobs is recorded for a sample of six students in each job. The data follow:

| Lifeguard | Production worker | Construction | Retail clerk |
|---|---|---|---|
| 490 | 525 | 475 | 527 |
| 450 | 506 | 460 | 507 |
| 478 | 473 | 525 | 492 |
| 510 | 526 | 420 | 505 |
| 504 | 502 | 499 | 530 |
| 482 | 505 | 472 | 555 |

Determine whether the average hours worked differ significantly ($\alpha = 0.01$) among the various occupations.

24. Suppose ten applicants for graduate programs are rated by two members of the graduate faculty, on a scale from 1 to 10 (with "ten" indicating excellence). Test whether the ratings (at $\alpha = 0.05$) are significantly different between the two professors.

| Rating of graduate applicants | | | | | | | | | |
|---|---|---|---|---|---|---|---|---|---|
| Professor A    8 | 5 | 2 | 7 | 6 | 9 | 4 | 5 | 7 | 5 |
| Professor B    6 | 5 | 4 | 7 | 8 | 10 | 7 | 6 | 6 | 8 |

25. See if you can prove, without consulting the text, that

$$\text{SS Total} = \text{SS Between} + \text{SS Within}.$$

26. Fifty terminally ill patients in a hospital were randomly assigned to either a control group or an experimental group. The experimental group was given a new drug being tested for this particular illness. The control group was kept on the medicine they had been receiving. The number of deaths in these two groups during the next year is given in the following chart:

|  | Living | Deaths |
|---|---|---|
| Experimental | 9 | 16 |
| Control | 1 | 24 |

    a)  What null hypothesis should be tested here? What level of $\alpha$ would you recommend?
    b)  At what level of significance can $H_0$ be rejected?

27. The ten students in group A in Exercise 22 took a road test for driving skills as well as a written test. Their scores on both tests are as follows:

| Student | Written test | Road test |
|---|---|---|
| 1 | 25 | 38 |
| 2 | 30 | 36 |
| 3 | 42 | 50 |
| 4 | 44 | 45 |
| 5 | 58 | 30 |
| 6 | 59 | 78 |
| 7 | 75 | 77 |
| 8 | 79 | 85 |
| 9 | 87 | 65 |
| 10 | 90 | 76 |

Use a sign test to test the null hypothesis that a student's score on the road test is not different from his score on the written test. At what level of significance can $H_0$ be rejected for a two-sided test? At what level can $H_0$ be rejected if it was hypothesized that the road test scores would be higher than the written test scores?

28. If chi-square is a known probability distribution given its degrees of freedom, and if nonparametric tests involve distribution-free statistics, explain why chi-square tests are so prevalent in nonparametric statistics.

29. In an experiment on oligopoly behavior, an individual producer makes a series of price decisions. For each decision the producer's profit position in the previous period is known. Consider the following observations of the behavior of one producer.

| | Outcome | | |
| Action | Profits increased | Profits same | Profits decreased |
| --- | --- | --- | --- |
| Raise price | 12 | 5 | 20 |
| Price same | 15 | 16 | 2 |
| Lower price | 8 | 4 | 18 |

Test at the 0.05 level, whether the response pattern of price changes is different among the different classifications of profit change.

# Simple Regression and Correlation Analysis

## 11.1  INTRODUCTION

In the past several chapters we have discussed the process of using sample information to make inferences, test hypotheses, or modify beliefs about the characteristics of a population. In this chapter and the next we turn to a related problem, involving two or more variables—making inferences about how changes in one set of variables are related to changes in another set. A description of the *nature* of the relationship between two or more variables is called *regression analysis*, while investigation into the *strength* of such relationships is called *correlation analysis*.

Sir Francis Galton, an English expert on heredity in the late 1800's, was one of the first researchers to work with the problem of describing one variable on the basis of one or more other variables. Galton's work centered on the heights of fathers compared to the heights of their sons. He found that there was a tendency toward the mean—exceptionally short fathers tended to have sons of more average height (i.e., taller than their father), while just the opposite was true for unusually tall fathers. Galton said that the heights of the sons "regressed" or reverted to the mean, thus originating the term *regression*. Nowadays the term "regression" much more generally means the description of the nature of the relationship between two or more variables.

Regression analysis is concerned with the problem of describing or estimating the value of one variable, called the *dependent* variable, on the basis of one or more other variables, called *independent* variables. Suppose, for example, that a businessman is trying to predict his sales for next month (the dependent variable) on the basis of indexes of disposable income, price levels, or any of numerous other independent variables; or perhaps he is trying to predict the performance of one of his products, under certain conditions of stress or at various temperatures; similarly he may be using one or more of a battery of tests in trying to evaluate the ability of prospective employees for new jobs. In these cases regression analysis is being used in an attempt to *predict* or estimate the value of an unknown dependent variable on the basis of the known value of one or more independent variables. In other cases regression may be used to *describe* the relationship between known values of two or more variables. An economist may use it for this purpose as an aid in understanding the relationship between historical observations over a specified time span, such as the relation of consumption to current and past levels of income and wealth, or the relationship between any one or more of a number of economic indicators and prices, or profits, or sales in a given industry.

No matter whether regression analysis is used for descriptive or predictive purposes, one cannot expect to be able to forecast or describe the *exact* value of sales, or profits, or consumption, or any other dependent variable. There may be many factors which could cause variations in the dependent variable for a given value of the independent variables, such as fluctuations in the stock market, changes in the weather, a passing fad, or just differences in human ability and motivation. Because of these possible variations, we shall be interested in determining the *average* relationship between the dependent variable and the independent variables. That is, we will want to be able to estimate the mean value of a dependent variable for any given values of the independent variables. Although regression analysis can involve one or more independent variables, in this chapter we will confine our analysis to the case of *simple* linear regression—i.e., only *one* independent variable. *Multiple* linear regression is presented in Chapter 12.

### The Regression Model

For most regression analysis the average population relationship between the dependent variable (which is usually denoted by the letter $y$) and the independent variable (denoted by the letter $x$) is assumed to be linear. A linear function is used because it is mathematically simple, and yet still provides an approximation to the real-world relationship which is sufficient for most practical purposes.

Since we are interested in determining the mean value of $y$ for a given value of $x$, we are thus interested in the conditional expectation $E[y \mid x]$. Another symbol often used to denote this expectation is $\mu_{y \cdot x}$, which is read as "the mean of the $y$-values for a given $x$-value." By assuming that $y$ and $x$ are linearly related we are saying that all possible conditional means $(E[y \mid x] = \mu_{y \cdot x})$ which might be calculated (one for each possible value of $x$) must be on a *single straight line*. This line is called the *population regression line*. To specify this line (or any straight line), we need to know its *slope* and *intercept*. Suppose we let $\alpha$ be the $y$-intercept and $\beta$ be the slope of the line. This line is thus written as follows:*

*Population regression line:*

$$\mu_{y \cdot x} = E[y \mid x] = \alpha + \beta x. \tag{11.1}$$

To illustrate this population model, suppose $y$ represents the quantity of beef purchased per month by a household, and $x$ represents the retail price of beef. The value of $\mu_{y \cdot x}$ is thus the mean quantity of beef purchased per month for some given price of beef. When some exact value of $x$ is specified, it is customary to denote this value as $x_i$, and to let $\mu_{y \cdot x_i}$ represent the mean of the $y$-values for the specific value of $x$. For example, $x_i$ might be a price of beef, such as $x_i = \$2.79/\text{lb}$, and

$$\mu_{y \cdot x_i} = \mu_{y \cdot \$2.79} = \alpha + \beta(2.79)$$

---

* The meaning of the Greek letters $\alpha$ and $\beta$ as used in this equation has no relationship to the meaning of these same letters used in Chapter 8 to describe the probability of Type I and Type II errors.

would be the mean quantity purchased per month when the price is \$2.79/lb. Thus, when speaking of a specific value of $x$, we merely substitute $x_i$ for $x$ in Formula (11.1).

In addition to estimating the mean value $\mu_{y \cdot x_i}$, we would like to make statements about an *observed* value of $y$ for a given value $x_i$. This value is denoted as $y_i$. For example, we might want to make a statement about the quantity of beef the "John Doe" family would purchase per month when the price of beef is $x_i =$ \$2.79/lb. As we indicated above, an *observed* value $y_i$ for the given value $x_i$ is usually not equal to $\mu_{y \cdot x_i}$. The difference between $y_i$ and $\mu_{y \cdot x_i}$ depends upon the accuracy of the regression model in depicting the real-world situation, and by the accuracy with which the variables $x$ and $y$ are measured. It also depends on the predictability of the underlying behavior of the persons, businesses, governments, etc., involved in the model. Any changes in human or institutional behavior could also cause such differences.

The point of the above discussion is that the difference between $y_i$ and $\mu_{y \cdot x_i}$ is the unpredictable element in regression analysis. For this reason, this difference is usually called the random "error," and denoted by the symbol epsilon, $\varepsilon_i$. That is,

$$\varepsilon_i = y_i - \mu_{y \cdot x_i} \quad \text{or} \quad y_i = \mu_{y \cdot x_i} + \varepsilon_i. \tag{11.2}$$

Suppose the price of beef is $x_i =$ \$2.79, and the average quantity purchased per month by *all* households at this price is

$$\mu_{y \cdot \$2.79} = 5 \text{ lbs.}$$

Should the John Doe family purchase $y_i = 3.2$ lbs per month, then the error in this case would be

$$\varepsilon_i = 3.2 - 5.0 = -1.8 \text{ lbs.}$$

We can now use Formula (11.2) to describe what is called the "population regression model." This model consists of all the terms which, when added together, sum to $y_i$. Substituting $y_i = \mu_{y \cdot x_i} + \varepsilon_i$ into (11.1) we get:

$$\boxed{\textit{Population regression model:} \quad y_i = \alpha + \beta x_i + \varepsilon_i.} \tag{11.3}$$

An example of this model is shown in Fig. 11.1. When values of $x$ and $y$ are plotted in this fashion, the diagram is called a *scatter diagram*.

An as illustration of a problem in which regression might be used to estimate an unknown population relationship, consider the task facing the director of admissions in most graduate schools. Because of limited resources, admission can be granted only to a select group of students, usually those predicted to be the most successful in graduate school. The process of deciding which students to admit is a difficult one, in which most directors presumably attempt to relate

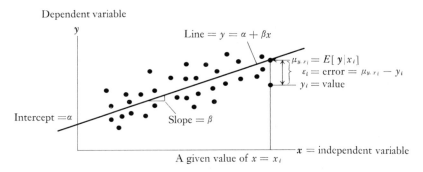

**Fig. 11.1.** The population regression model.

"success" in graduate school to a large variety of factors such as admission test scores, previous grades, recommendations, age, etc. In general, the only way to determine the nature of such a (population) relationship is on the basis of past data (sample information). If it can be determined that certain of these factors have been related to some definition of success in the past, then this information may be helpful in separating the potentially successful applicants from those less likely to be successful. Of considerable interest to most admissions directors, for example, is the relationship between an applicant's score on the Graduate Record Examination and his subsequent performance in graduate school, as measured by his grade point average.

Let us denote by $x$ the Graduate Record Examination scores (GRE), and let $y$ represent the grade-point averages (GPA). The population in this case might be considered to be all possible candidates for admission during a specified time span, such as 1965 through 1980. Once again it must be pointed out that use of the term "the population relationship" does not imply that there is necessarily an exact relationship, which always holds true, between two variables. The variables GRE and GPA are obviously not exactly related, as a higher GRE score doesn't *always* lead to a higher GPA; still, one would expect to find a positive relationship between GRE scores and the *mean* GPA for students with various GRE scores. It is therefore meaningful to attempt to determine how changes in the independent variable influence the mean value of a dependent variable. The methods of this chapter enable us to estimate the parameters of the population regression model by using sample data.

We can illustrate the characteristics of a population regression model by assuming (for the moment) that the population parameters for this GRE–GPA example are known quantities. For example, let's assume $\alpha = 0.95$ and $\beta = 0.0039$. This means that the population regression line is

$$\textit{Population regression line:} \qquad \mu_{y \cdot x} = 0.95 + .0039x,$$

and the population regression model is:

*Population regression model:*
$$y_i = 0.95 + 0.0039x_i + \varepsilon_i.$$

This regression model is diagrammed in Fig. 11.2, where each dot represents one observation (i.e., one person) in the population.

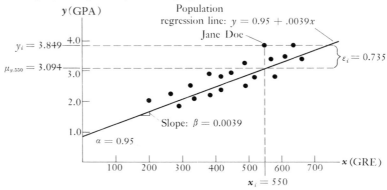

**Fig. 11.2.** Population regression line for GRE–GPA example.

In Fig. 11.2 the population mean value for a GRE of 550 is seen to be

$$E[y \mid 550] = \mu_{y \cdot 550} = 3.094.$$

In other words, the average GPA for all students with GRE scores of 550 is 3.094. This value is easily derived by substituting $x_i = 550$ into the population regression line, as follows:
$$\mu_{y \cdot 550} = 0.95 + 0.0039(550) = 3.094.$$

Now, let's assume one student in this population, Jane Doe, had a GRE of 550, and she earned a GPA of $y_i = 3.849$. For her, the value of the error $\varepsilon_i$ is

$$\varepsilon_i = y_i - \mu_{y \cdot 550} = 3.849 - 3.094 = 0.735.$$

Our assumption that $\alpha$ and $\beta$ are known is, of course, an unrealistic one. Usually $\alpha$ and $\beta$ can only be estimated on the basis of sample data. For example, we might use as sample data all the GRE and GPA scores, from 1965 to the present, of students who actually enrolled and graduated. The discussion in the following section introduces the sample regression model as the basis for estimating $\alpha$ and $\beta$.

**The Sample Regression Model**

In regression analysis it is customary to let $a$ be the best estimate of $\alpha$, $b$ be the best estimate of $\beta$, and to let $\hat{y}$ be the resulting estimate of $\mu_{y \cdot x}$. The values $a$ and $b$ are called the *regression coefficients*, and the line relating $a$, $b$, and $\hat{y}$ is called the *sample regression line.* This line has the same form as the population regression line (Formula (11.1)).

$$\text{Sample regression line:} \quad \hat{y} = a + bx. \tag{11.4}$$

Just as we did for the population regression line, we can add the subscript $i$ to these variables to indicate specific values. Thus, if $x_i$ is a specific value of $x$, then $\hat{y}_i = a + bx_i$ is the equation for finding $\hat{y}_i$, which is the best estimate of $\mu_{y \cdot x_i}$ for this value of $x$. We can also specify a *sample regression model* just as we specified a population regression model. Again we need to define an error term, which in this case is the difference between the predicted value $\hat{y}_i$ and the actual value $y_i$. This error term is denoted as $e_i$, and we can use the sample regression error $e_i$ as an estimate of the population error $\varepsilon_i$. The errors $e_i$ are called "residuals."

$$\text{Residuals:} \quad e_i = y_i - \hat{y}_i \quad \text{or} \quad y_i = \hat{y}_i + e_i.$$

If we use the estimator $\hat{y}_i = y_i - e_i$ in Formula (11.4), we get the sample regression model:

$$\boxed{\text{Sample regression model:} \quad y_i = a + bx_i + e_i.} \tag{11.5}$$

Suppose we illustrate the concept of a sample regression model by assuming that the admissions director in our GRE–GPA example has selected a random sample of students, one of whom is Jane Doe. Figure 11.3 shows these observations in a scatter diagram; the line drawn through these points is the sample regression line (we will explain later how this line was derived). From Fig. 11.3 we see that

$$a = 0.751 \quad \text{and} \quad b = 0.00435,$$

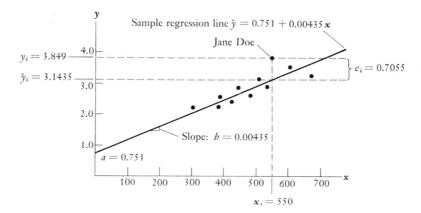

**Fig. 11.3.** Sample scatter diagram of observed values for **y** and x.

and the sample regression line is:

$$\text{Sample regression line:} \quad \hat{y} = 0.751 + 0.00435x.$$

Note that, on the basis of our sample regression line, we predict that the mean GPA for all students with a GRE of 550 will be

$$\hat{y}_i = 0.751 + 0.00435(550) = 3.1435.$$

We can also determine the amount of "error" in predicting Jane Doe's GPA by recalling that her GPA was $y_i = 3.849$. The value of $e_i$ is therefore

$$\begin{aligned} e_i &= y_i - \hat{y}_i \\ &= 3.849 - 3.1435 = 0.7055. \end{aligned}$$

Thus, the value $a = 0.751$ is an estimate of $\alpha = 0.95$; $b = 0.0435$ is an estimate of $\beta = 0.039$; and for Jane Doe, the error $e_i = 0.7055$ is an estimate of the true error $\varepsilon_i = 0.735$.

Now that we have specified the sample and population regression model, we need a procedure for determining values of $a$ and $b$ which provide the "best" estimates of $\alpha$ and $\beta$. The procedure for finding such estimates is called the *method of least squares*. This method will be presented in Section 11.3. Before doing so, however, we would like to specify certain characteristics of the estimators resulting from the least-squares approach.

Basically, the method of least squares is just a curve-fitting technique in which the values of $a$ and $b$ are derived by finding the sample regression line which provides the best fit to the sample data. The advantage of this approach is that if certain assumptions are made about the population, then the resulting estimators $a$ and $b$ can be shown to be unbiased, consistent, and efficient. Furthermore, these assumptions permit us to more easily construct the confidence intervals and test the hypotheses that are so crucial to regression analysis. In other words, the assumptions provide the rationale behind the widespread use of the least-squares approach. In every regression-analysis problem, the researcher must therefore satisfy himself that such assumptions are reasonable.

The assumptions are concerned with the random variable $\varepsilon$, for this variable describes how well $\mu_{y \cdot x_i}$ estimates $y_i$. Making assumptions about the mean and the variance of $\varepsilon$ enables us to make inferences about this variable on the basis of the sample values of $e$.

## 11.2 ASSUMPTIONS AND ESTIMATION

Many possible sets of assumptions about the distribution of the variables in the population regression model could be formulated. One particular group of assumptions has become known as the "ideal" assumptions, in that they yield relatively simple estimators possessing many desirable properties, and they result in test statistics which possess commonly known distributions. In the following section, we describe the five "ideal" assumptions of simple linear regression.

### The Ideal Assumptions

*Assumption* **1.** *The random variable* $\varepsilon$ *is assumed to be independent of the values of* $x$.

This means that the covariance between the values of the independent variable and the corresponding error terms is zero. Such an assumption might be violated when, for example, errors are a *percentage* of the values of $x$ because of measurement error. In this case the values of $\varepsilon$ would tend to get larger as $x$ increases.

*Assumption* **2.** *The random variable* $\varepsilon$ *is assumed to be normally distributed.*

Since $\varepsilon_i$ is a composite of many factors (such as errors of measurement, errors in specifying the model, or irregular errors such as economic, political, social, and business fluctuations), it is reasonable to expect that many of these factors tend to offset each other so that large values of $\varepsilon_i$ are much less likely than small values of $\varepsilon_i$. Indeed, if many of these factors are unrelated, a form of the central-limit theorem guarantees that their effect (represented by $\varepsilon_i$) will be normally distributed. Figure 11.4 illustrates the meaning of this assumption by showing the normal distribution of errors $\varepsilon_i$ about the population straight-line relationship between $y$ and $x$.

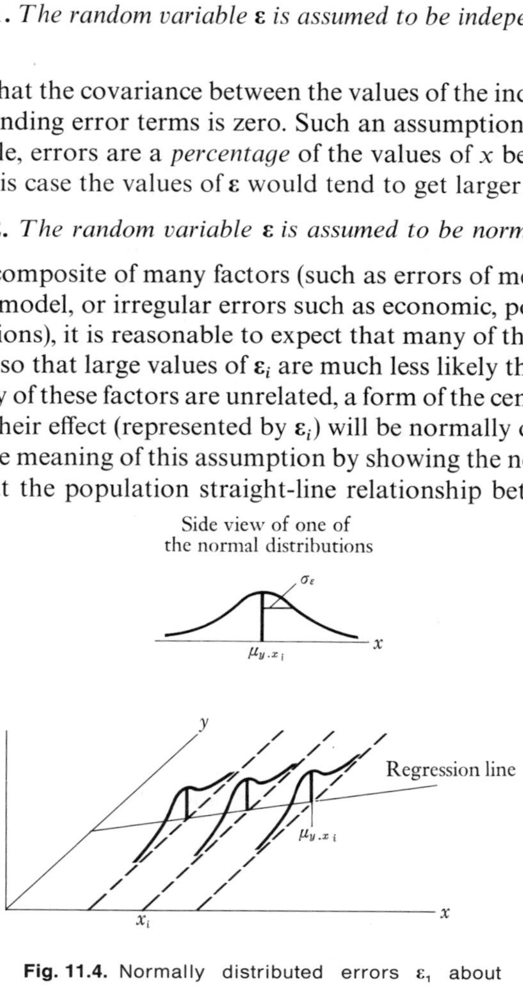

Side view of one of
the normal distributions

**Fig. 11.4.** Normally distributed errors $\varepsilon_1$ about $\mu_{y \cdot x_i}$ for any value $x_i$.

*Assumption* **3.** *The variable* $\varepsilon$ *is assumed to have a mean of zero; that is,*

$$E[\varepsilon] = 0.$$

This means that, for a given $x_i$, the differences between $y_i$ and $\mu_{y \cdot x_i}$ are sometimes positive, sometimes negative, but on the average are zero. Thus, the distribution of $\varepsilon_i$ about the population regression line $\mu_{y \cdot x}$ (as shown in Fig. 11.4) is always centered at the value $\mu_{y \cdot x_i}$ for any given $x_i$.

*Assumption* **4.** *Any two errors, $\varepsilon_k$ and $\varepsilon_j$, are assumed to be independent of each other; that is, their covariance is zero,*

$$C[\varepsilon_k, \varepsilon_j] = 0.$$

This assumption means that the error of one point in the population cannot be related systematically to the error of any other point in the population. In other words, knowledge about the size or sign of one or more errors does not help in predicting the size or sign of any other error. For example, knowing that the error in describing the GPA of one (or more) graduate students is positive does not give you any help in determining whether or not the error for another graduate student will also be positive.

*Assumption* **5.** *The random variable $\varepsilon$ is assumed to have a finite variance $\sigma_\varepsilon^2$ which is constant for all given values of $x_i$.*

This means that the dispersion or variability of points in the population about the population regression line must be constant. In Fig. 11.4, this constant variance of $\varepsilon$ is represented by depicting all the normal distributions about $\mu_{y \cdot x_i}$ as having the same standard deviation. No one distribution is more spread out or more peaked than another for a different value of $x$.*

**Properties of the Regression Coefficients**

As we indicated previously, the five ideal assumptions presented above provide the rationale for the widespread use of the least-squares procedure. Given assumption 2, it can be shown that the estimators of $\alpha$, $\beta$, and $\mu_{y \cdot x_i}$ obtained by using the least-squares criterion are *identical* to the estimators which would result using the principle of maximum-likelihood estimation. Such estimators have a number of desirable properties (such as consistency), and in addition they are intuitively appealing.

A second important result relating to the least squares estimators is called the *Gauss–Markov theorem.* This classical result of linear estimation was formulated by the German mathematician and astronomer, Karl F. Gauss (1777–1855), in his early works published in 1807 and 1821. Since these involved applications in physics and planetary motion, they generally remained unknown to social scientists and businessmen until they were restated in a more modern context by

---

* We might point out that when Assumption 1 is violated because of proportional errors in measuring $x$, this means that Assumption 5 (constant variance) will also be violated.

A. A. Markov in 1912, in a study of linear processes. In the 1930's the work of Markov was extended and applied directly to least-squares estimation in several ways, and the Gauss–Markov Theorem assumed the identity it has today:

> **Gauss–Markov Theorem.** *If Assumptions* 1, 3, 4, *and* 5 *hold true, then the estimators of* α, β *and* $\mu_{y \cdot x}$ *determined by the least-squares criterion are Best Linear Unbiased Estimators* (*i.e., BLUE*).

In the above context, the term *linear* means that the estimators are straight-line functions of the values of the dependent variable *y*. They are *unbiased* because their expected value is equal to the population value (given that assumptions 1 and 3 are true). They are *best* in the sense of being *efficient* (if assumptions 4 and 5 are true). That is, the least-squares estimators have a variance which is smaller than that of any other linear unbiased estimator. Thus, the importance of the Gauss–Markov theorem is that if assumptions 1, 3, 4, and 5 hold, then the least-squares estimators have the desirable properties of unbiasedness and efficiency.

## 11.3 ESTIMATING THE VALUES OF α AND β BY LEAST SQUARES

A first step in finding a sample regression line of best fit is to plot the data in a scatter diagram. Such a plot allows us to visually determine whether a straight-line approximation to the data appears reasonable and to make rough estimates of *a* and *b*. Although this approach often yields fairly satisfactory results, there are at least two reasons for having a more systematic approach to finding the "best" straight-line fit to the data. First, different people are likely to find slightly different values for *a* and *b* by the freehand drawing method. Secondly, the freehand estimation procedure provides no way of measuring the sampling errors, which are always important in forming confidence intervals or doing tests of hypotheses on population parameters.

What we need is a mathematical procedure for determining the sample regression line that best fits the sample data. The difficulty in establishing such a mathematical procedure to give the line of best fit is in determining the criterion to use in defining "best fit." Perhaps the most reasonable criterion is to find values *a* and *b* so that the resulting values of $\hat{y}$ (in the equation $\hat{y} = a + bx$) are as close as possible to the observed values $y_i$. That is, we want to minimize the values of $e_i$, where

$$e_i = y_i - \hat{y}_i$$

(which in Fig. 11.3 is seen to be the vertical distance from $\hat{y}_i$ to $y_i$).

Unfortunately, we can't find the line of best fit merely by finding a line for which the sum of all the residuals is equal to zero; that is,

$$\sum e_i = \sum (y_i - \hat{y}_i) = 0.$$

All that such a procedure accomplishes is to ensure that the positive residuals will exactly balance the negative residuals. For example, if we set $a = 0$ and $b = \bar{y}$, the resulting line ($\hat{y} = \bar{y}$) has the property $\sum e_i = 0$, but this line usually is not a good fit to the data (it's just a horizontal straight line).

The approach almost universally adopted to find the line of best fit in regression analysis is to determine the values of $a$ and $b$ which minimize the sum of the *squared residuals*. This procedure is known as the *method of least squares*. Since $e_i = y_i - \hat{y}_i$, this method is defined as follows:

*Method of least-squares estimation:*

$$\text{Minimize } \sum_{i=1}^{n} e_i^2 = \sum_{i=1}^{n} (y_i - \hat{y}_i)^2. \tag{11.6}$$

The sample regression line determined by minimizing $\sum e_i^2$ is called *the least-squares regression line*. Since $\hat{y}_i = a + bx_i$, minimizing

$$\sum e_i^2 = \sum (y_i - \hat{y}_i)^2.$$

is equivalent to minimizing

$$\sum_{i=1}^{n} [y_i - (a + bx_i)]^2.$$

Finding the value of the two unknowns which minimize this function is a problem which is solvable by the methods of calculus.* The reader should take care to understand that the two unknowns are the estimators $a$ and $b$ (not $y$ and $x$), since it is $a$ and $b$ which must be chosen from among an infinite possible set of values

---

* For convenience, let's denote the function to be minimized as $G = \sum_{i=1}^{n} [y_i - a - bx_i]^2$. Since this function is to be minimized with respect to $a$ and $b$, it is necessary to take the partial derivatives of $G$ with respect to these two variables, set each of these partials equal to zero, and then solve the resulting two equations simultaneously. The partial derivatives are:

$$\frac{\partial G}{\partial a} = \sum_{i=1}^{n} 2(y_i - a - bx_i)(-1),$$

$$\frac{\partial G}{\partial b} = \sum_{i=1}^{n} 2(y_i - a - bx_i)(-x_i).$$

Setting these equal to zero yields the following two equations (called *the normal equations*) which can be solved to obtain Formula (11.7):

$$\sum_{i=1}^{n} y_i = na + b \sum_{i=1}^{n} x_i,$$

$$\sum_{i=1}^{n} x_i y_i = a \sum_{i=1}^{n} x_i + b \sum_{i=1}^{n} x_i^2.$$

($y$ and $x$ are values given by the sample data). By minimizing $\sum e_i^2$, the values $a$ and $b$ that provide the line of best fit, according to the least-squares criterion, are found to be:

---

*Slope of least-squares line:*

$$b = \frac{\dfrac{1}{n-1} \sum\limits_{i=1}^{n} (x_i - \bar{x})(y_i - \bar{y})}{\dfrac{1}{n-1} \sum\limits_{i=1}^{n} (x_i - \bar{x})^2} \; ;$$

*Intercept of least-squares line:*

$$a = \bar{y} - b\bar{x}.$$

(11.7)

---

Perhaps the formula for $b$ in Formula (11.7) will be a bit easier to remember if we point out that its denominator is the *sample variance of* $x$, since

$$s_x^2 = \frac{1}{n-1} \sum_{i=1}^{n} (x_i - \bar{x})^2.$$

The numerator of the expression for $b$ is the *sample covariance* of $x$ and $y$. The concept of a covariance was introduced in Chapter 3, where we defined

$$C[x, y] = \frac{1}{N} \sum (x_i - \mu_x)(y_i - \mu_y).$$

The symbol $s_{xy}$ is usually used to denote an unbiased estimate of $C[x, y]$, which is

$$s_{xy} = \frac{1}{n-1} \sum_{i=1}^{n} (x_i - \bar{x})(y_i - \bar{y}).$$

Thus,

$$b = \frac{s_{xy}}{s_x^2} = \frac{\left(\begin{array}{c}\text{Sample covariance between}\\ \text{independent and dependent variables}\end{array}\right)}{\text{Sample variance of independent variable}}.$$

(11.8)

Two features of the sample regression line should be noted from Formula (11.7). First, note that if we substitute the intercept $a = \bar{y} - b\bar{x}$ into the line

$$\hat{y} = a + bx,$$

we get $\hat{y} = \bar{y} - b(x - \bar{x})$. This means that, whenever $x = \bar{x}$, then $\hat{y} = \bar{y}$; hence, the regression line always goes through the point $(\bar{x}, \bar{y})$. Secondly, in order to minimize $\sum e_i^2$, it is necessary to set $\sum_{i=1}^{n} e_i = 0$ (see footnote on page 406). Thus, our line estimating the average relationship between $y$ and $x$ passes through the point of averages $(\bar{x}, \bar{y})$, and splits the scatter diagram of observed points so that the positive residuals (underestimates of the true point) always exactly cancel the negative residuals (overestimates of the true points). Such a sample regression line therefore unbiasedly estimates the population regression line.

To illustrate the technique of finding a least-squares regression line, consider again our GPA–GRE problem. Suppose we now use the method of least squares to determine $a$ and $b$ for a random sample of eight observations (i.e., eight students).* Columns (1) and (2) in Table 11.1 give the hypothetical data for these students, ordered from the lowest GRE to the highest score. We will use the rest of the data in Table 11.1 in a moment.

**Table 11.1** Sample points for GRE and GPA and calculations of their means, variations, and covariations

| (1) GRE($x_i$) | (2) GPA($y_i$) | (3) $(x_i - \bar{x})$ | (4) $(y_i - \bar{y})$ | (5) $(y_i - \bar{y})(x_i - \bar{x})$ | (6) $(x_i - \bar{x})^2$ |
|---|---|---|---|---|---|
| 480 | 2.70 | $-60$ | $-0.40$ | 24 | 3,600 |
| 490 | 2.90 | $-50$ | $-0.20$ | 10 | 2,500 |
| 510 | 3.30 | $-30$ | $+0.20$ | $-6$ | 900 |
| 510 | 2.90 | $-30$ | $-0.20$ | 6 | 900 |
| 530 | 3.10 | $-10$ | $0.00$ | 0 | 100 |
| 550 | 3.00 | $+10$ | $-0.10$ | $-1$ | 100 |
| 610 | 3.20 | $+70$ | $+0.10$ | 7 | 4,900 |
| 640 | 3.70 | $+100$ | $+0.60$ | 60 | 10,000 |
| Sum     4320 | 24.80 | 0 | 0 | 100.00 | 23,000 |
| Mean  $\bar{x} =$  540 | $\bar{y} =$  3.10 | | | | |

Our first step in analyzing the data in the first two columns of Table 11.1 is to construct a scatter diagram, to see if the assumption of linearity is a reasonable one in this case. Figure 11.5 indicates that it is.

The sums in columns (5) and (6) of Table 11.1 give the information necessary to calculate $a$ and $b$. Using Formula (11.7), we can determine $b$ as follows:

$$b = \frac{\dfrac{1}{n-1}\displaystyle\sum_{i=1}^{n}(y_i - \bar{y})(x_i - \bar{x})}{\dfrac{1}{n-1}\displaystyle\sum_{i=1}^{n}(x_i - \bar{x})^2}$$

$$= \frac{100/7}{23,000/7} = 0.00435.$$

* For most practical purposes a sample size of eight students would not be sufficient. We assume it here only for computational ease.

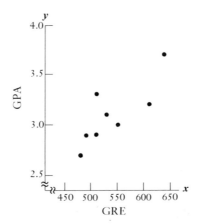

**Fig. 11.5.** Scatter diagram for GRE– GPA sample values.

Using this value of $b$ and the means of $x$ and $y$ shown in columns (1) and (2) of Table 11.1, we obtain the following value of $a$:

$$a = \bar{y} - b\bar{x} = 3.10 - 0.00435(540) = 0.751.$$

Hence, the least-squares regression line for this example is

$$\hat{y} = 0.751 + 0.00435x. \tag{11.9}$$

Figure 11.6 illustrates this sample regression line. Since the line in Fig. 11.6 was determined by the method of least squares, there is no other line which could be drawn such that the sum of the squared residuals between the points and the line (measured in a vertical direction) could be smaller than for this line. The residuals and the estimated values of $y_i$ for all eight sample points are given in Table 11.2. The sum of the residuals in this case $(-0.003)$ differs from zero only because of rounding error.

At this time the reader should check his understanding of the least-squares regression line by verifying that the point of means $(\bar{x}, \bar{y})$ does lie on the line, and by examining the corresponding values of the residuals in Table 11.2. Note from Fig. 11.6 that a positive residual such as $e_3 = 0.33$ means that the point $(x_3, y_3)$ lies above the line and therefore, $\hat{y}_3$ underestimates $y_3$. On the other hand, a negative residual such as $e_7 = -0.205$ corresponds to an overestimation of $y_7$ by $\hat{y}_7$, in that the point $(x_7, y_7)$ lies below the regression line.

**Further Discussion of the Estimators a and b**

Formula (11.7) provides a convenient method for calculating the regression slope $b$ when the values of $x$ and $y$ are integers. When they are not integers, an alternative

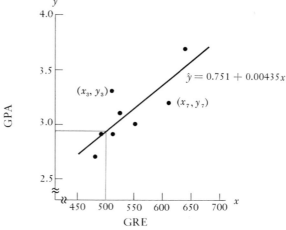

**Fig. 11.6.** The least-squares regression line **y** = 0.751 + 0.00435x.

**Table 11.2** Observed, estimated, and residual values for the least-squares regression shown in Fig. 11.6

| $x_i$ | Observed value ($y_i$) | Predicted value ($\hat{y}_i = 0.751 + 0.00435x_i$) | Residual $e_i = (y_i - \hat{y}_i)$ |
|---|---|---|---|
| 480 | 2.70 | 2.839 | −.139 |
| 490 | 2.90 | 2.883 | .017 |
| 510 | 3.30 | 2.970 | .330 |
| 510 | 2.90 | 2.970 | −.070 |
| 530 | 3.10 | 3.057 | .043 |
| 550 | 3.00 | 3.144 | −.144 |
| 610 | 3.20 | 3.405 | −.205 |
| 640 | 3.70 | 3.535 | .165 |
| Sum  4320 | 24.80 | 24.80 | −.003 |

formula for $b$ may often prove to be computationally easier. This formula is:

$$b = \frac{\sum x_i y_i - \frac{1}{n} \sum x_i \sum y_i}{\sum x_i^2 - \frac{1}{n}\left(\sum x_i\right)^2}. \tag{11.10}$$

We will illustrate this formula in terms of our GRE–GPA example. Before doing so, however, we should point out that since the value of $b$ depends on the relative values of $x$ and $y$, not on their absolute size, subtracting a constant from each value of $x$ and/or $y$ will not affect the regression slope. The advantage of such subtractions is that the arithmetic of Formula (11.10) may then become much easier. To illustrate this fact, suppose we subtract some convenient number from each value of $x$ and $y$ in Table 11.1, and then use Formula (11.10) to determine $b$. We (arbitrarily) select 500 to be subtracted from each value of $x$, and 3.00 from each value of $y$. These values are shown in Table 11.3.

**Table 11.3** GRE–GPA data repeated, with 500 subtracted from each $x$ value, and 3.00 from each $y$ value

|  | $x$ | $y$ | $xy$ | $x^2$ | $y^2$ |
|---|---|---|---|---|---|
|  | $-20$ | $-.30$ | 6.0 | 400 | 0.09 |
|  | $-10$ | $-.10$ | 1.0 | 100 | 0.01 |
|  | 10 | .30 | 3.0 | 100 | 0.09 |
|  | 10 | $-.10$ | $-1.0$ | 100 | 0.01 |
|  | 30 | .10 | 3.0 | 900 | 0.01 |
|  | 50 | 0 | 0 | 2500 | 0 |
|  | 110 | .20 | 22.0 | 12100 | 0.04 |
|  | 140 | .70 | 98.0 | 19600 | 0.49 |
| Sum | 320 | .80 | 132.0 | 35,800 | 0.74 |

Using these sums and (11.10) we can calculate $b$:

$$b = \frac{(132.0) - \frac{1}{8}(320)(0.80)}{(35,800) - \frac{1}{8}(320)^2} = 0.00435.$$

This is the same value of $b$ calculated previously. In using this computational formula, one must remember to return to the original data in order to calculate the value of $a$, for an incorrect $a$ value will result from the coded data.

The least-squares regression line described by Formula (11.9) serves several purposes. First, the regression coefficients provide point estimates of the population parameters, 0.751 being an estimate of $\alpha$, and 0.00435 an estimate of $\beta$. These point estimates serve a variety of research needs. Knowing the regression line also enables us to use the values of $\hat{y}$ to estimate the conditional mean of the dependent variable, $\mu_{y \cdot x}$, for specific values of $x$. Perhaps most importantly, the regression line permits prediction of the *actual* values of the dependent variable given a value of the independent variable.

For example, suppose a student has a Graduate Record Examination score of $x_i = 500$ and we would like to predict the grade-point average, $y_i$, that this person will earn. The best estimate of $y_i$ using Formula (11.9) is given by $\hat{y}_i$, which in this case is

$$\hat{y}_i = 0.751 + 0.00435(500) = 2.926.$$

In using a regression line to make predictions about the dependent variable, special care must be taken when the value of the independent variable falls outside the range of past experience (historical data), since it may be that these values cannot be represented by the same equation. The regression equation described above, for example, predicts that the GPA for students with a GRE of 100 will be

$$\hat{y}_i = 0.751 + 0.00435(100) = 1.21.$$

But since students scoring this low are usually not admitted to graduate school, there is really no way of knowing what GPA they might achieve. Similarly, the equation we derived predicts the GPA for a student with a test score of 800 to be

$$\hat{y}_i = 0.751 + 0.00435(800) = 4.22,$$

which is not even possible. Clearly special care must be taken in this example when attempting to predict outside a range of GRE scores running from about 400 to 700.

Several aspects of the method of least squares need to be emphasized at this point. First, this method is just a curve-fitting technique, and as such it requires no assumptions about the distributions of $x$ or $y$ or $\varepsilon$. It is only when probability statements about the parameters of the population are desired that we need such assumptions, because then $a$ and $b$ become random variables. Economists, for example, use regression analysis in an attempt to form interval estimates and to test various assumptions about population parameters in economic models, such as the marginal propensity to consume, the price elasticity of demand, and the factor shares of labor and capital in production. Also, they study the effect of changes in one variable (e.g., a tax change) on one or more other variables such as employment, consumption, and prices.

Secondly, the method of least squares can be adapted to apply to nonlinear populations. Although the formulas necessary for applying the method of least squares to a nonlinear relationship will naturally differ from those used for linear relationships, the objective in fitting the curve is the same in the two cases—to find the line minimizing the sum of the squared residuals. Nonlinear regression models, which are often quite complex, will not be discussed in this text.

As we indicated earlier in this chapter, an important aspect of regression analysis is concerned with how well a given sample regression line estimates the population regression line. Since such an analysis is based on how good a "fit" the sample observations are to the sample regression line, this topic is often referred to as "goodness of fit."

## 11.4 MEASURES OF GOODNESS OF FIT

In this section we will present two measures of goodness of fit. The first is a measure of the *absolute* fit of the sample points to the sample regression line, called the *standard error of the estimate*. The second measure is an index of the *relative* goodness of fit of a sample regression line, called the *coefficient of determination*. Our presentation of these measures will be easier if we first present some of the components of variability in regression analysis.

In regression analysis, the difference between $y_i$ and the mean of the $y$ values ($\bar{y}$) is often called the *total deviation of y*; that is, it represents the total amount the $i$th observation deviates from the mean of all $y$ values. By a mathematical identity, this total deviation can be written as the *sum* of two other deviations, one of which is $(y_i - \hat{y})$, and the other is $(\hat{y}_i - \bar{y})$. Since the deviation $(\hat{y}_i - \bar{y})$ is that part of the total deviation which is "explained" (i.e. accounted for by the regression lines), this term is called *explained deviation*. Since the deviation $(y_i - \hat{y})$ is the residual for the $i$th sample observation, and we have no basis for explaining why it occurred, this term is called *unexplained deviation*. That is,

$$\begin{array}{ccc} \text{Total} & \text{Unexplained} & \text{Explained} \\ \text{deviation} & = \text{deviation} & + \text{deviation} \end{array}$$

$$(y_i - \bar{y}) = (y_i - \hat{y}_i) + (\hat{y}_i - \bar{y}) \tag{11.11}$$

This relationship is illustrated in Fig. 11.7 in the context of our GRE–GPA example.

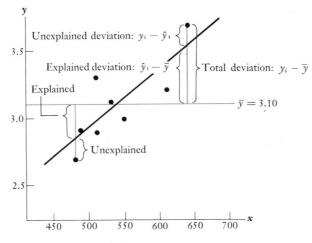

**Fig. 11.7.** Explained and unexplained variation.

Because the two parts of the total deviation, shown in Formula (11.11), are independent, it can be shown that this same relationship holds when we *square* each deviation, and sum over all $n$ observations. That is,

$$\sum_{i=1}^{n} (y_i - \bar{y})^2 = \sum_{i=1}^{n} (y_i - \hat{y}_i)^2 + \sum_{i=1}^{n} (\hat{y}_i - \bar{y})^2. \qquad (11.12)$$

The lefthand side of Formula (11.12) is referred to as the *total variation*, or as the *sum of squares total* (which is abbreviated as SST). The first term on the right is the *unexplained variation*, or equivalently, the *sum of squares error* (SSE). You should recognize SSE as $\sum e_i^2$, the term which was minimized in finding the least-squares regression line. The last term in (11.12), which is the sum of squares of the values of $\hat{y}$ about $\bar{y}$, is called the *explained variation*, or the *sum of squares regression* (SSR). Thus, we can rewrite (11.12) as follows:

$$
\begin{array}{ccccc}
\text{Total} & = & \text{Unexplained} & + & \text{Explained} \\
\text{variation} & & \text{variation} & & \text{variation}
\end{array}
$$

$$\sum_{i=1}^{n} (y_i - \bar{y})^2 = \sum_{i=1}^{n} (y_i - \hat{y}_i)^2 + \sum_{i=1}^{n} (\hat{y}_i - \bar{y})^2 \qquad (11.13)$$

$$
\boxed{\quad \text{SST} \quad = \quad \text{SSE} \quad + \quad \text{SSR} \quad}
$$

The advantage of breaking total variation into these two components is that we can now talk about goodness of fit in terms of the size of SSE. For example, if the line is a perfect fit to the data, then SSE $= 0$. Usually, however, the line is not a perfect fit, hence, SSE $\neq 0$.

We can calculate SST, SSE, and SSR for our GRE–GPA example from the data in Tables 11.1 and 11.2. First, SST can be derived by squaring the values in column 4 of Table 11.1. These values are shown in column 1 of Table 11.4. The

**Table 11.4** Calculation of SST, SSE, and SSR

| $(y_i - \bar{y})^2$ | $(y_i - \hat{y}_i)^2$ | $(\hat{y}_i - \bar{y})$ | $(\hat{y}_i - \bar{y})^2$ |
|---|---|---|---|
| 0.16 | 0.019 | $2.839 - 3.10$ | 0.068 |
| 0.04 | 0.000 | $2.883 - 3.10$ | 0.047 |
| 0.04 | 0.109 | $2.970 - 3.10$ | 0.017 |
| 0.04 | 0.005 | $2.970 - 3.10$ | 0.017 |
| 0.00 | 0.002 | $3.054 - 3.10$ | 0.002 |
| 0.01 | 0.021 | $3.144 - 3.10$ | 0.002 |
| 0.01 | 0.042 | $3.405 - 3.10$ | 0.093 |
| 0.36 | 0.027 | $3.535 - 3.10$ | 0.189 |
| SST $= 0.66$ | SSE $= 0.225$ | | SSR $= 0.435$ |

value of SSE is derived by squaring the errors $e_i = (y_i - \hat{y}_i)$ shown in the final column of Table 11.2. These squares (rounded to three decimals) are shown in column 2 of Table 11.4. Finally, we can calculate

$$\text{SSR} = \sum (\hat{y}_i - \bar{y})^2$$

by subtracting $\bar{y} = 3.10$ from each value of $\hat{y}$ in column 3 of Table 11.2, and then squaring these differences. These values are shown in column 4 of Table 11.4.

We thus see that

$$\text{SST} = \text{SSE} + \text{SSR}$$
$$0.660 = 0.225 + 0.435.$$

In practice, there are more efficient ways of calculating SST, SSR, and SSE. Since $V[y] = (1/(n - 1))\sum(y_i - \bar{y})^2$, the term $\sum(y_i - \bar{y})^2$ can be found more easily by using the computational formula for variance. Similar to the terms in the denominator of Formula (11.10), we can write:

$$\sum_{i=1}^{n} (y_i - \bar{y})^2 = \text{SST} = \sum_{i=1}^{n} y_i^2 - \frac{1}{n}\left(\sum_{i=1}^{n} y_i\right)^2. \tag{11.14}$$

Next, SSR can be calculated without first calculating each value of $\hat{y}_i$. And then once SSR is known, SSE can be derived by letting $\text{SSE} = \text{SST} - \text{SSR}$. A formula for calculating SSR is given below:

$$\text{SSR} = b\sum_{i=1}^{n} (x_i - \bar{x})(y_i - \bar{y}) = b\left[\sum_{i=1}^{n} x_i y_i - \frac{1}{n}\left(\sum_{i=1}^{n} x_i\right)\left(\sum_{i=1}^{n} y_i\right)\right]. \tag{11.15}$$

In Table 11.1 the value of $\Sigma(x - \bar{x})(y - \bar{y})$ was shown to be 100.0. Since $b = 0.00435$,

$$\text{SSR} = 0.00435(100.0) = 0.435,$$

which agrees with the value given previously.

Our first measure of goodness of fit (the standard error of the estimate) is based on the value of SSE; the second measure (the coefficient of determination) is based on the size of SSE relative to SST. These measures are described below.

### Standard Error of the Estimate

Our measure of the absolute size of the goodness of fit is the value of SSE divided by its degrees of freedom. Since SSE is the sum of the squared residuals, this measure is thus the sample variance of the values of $e_i$.* The number of degrees of

---

* Recall from Chapter 7 that an unbiased sample variance is calculated by dividing the sum of squared deviations by the degrees of freedom.

freedom in this measure is $n - 2$, because *two* sample statistics ($a$ and $b$) must be calculated before the values of $\hat{y}$ can be computed (since $\hat{y} = a + bx$). Hence,

*Variance of residuals:*

$$V[e] = \frac{1}{n - 2} \sum_{i=1}^{n} (y_i - \hat{y}_i)^2 = \frac{1}{n - 2} \text{(SSE)}.$$

Since standard deviations are usually easier to interpret than variances, the absolute measure of goodness of fit for regression analysis is the square root of $V[e]$, which is called the *standard error of the estimate* (denoted by the symbol $s_e$).

---

*Standard error of estimate:*

$$s_e = \sqrt{\frac{1}{n - 2} \sum_{i=1}^{n} (y_i - \hat{y}_i)^2} = \sqrt{\frac{\text{SSE}}{n - 2}}. \qquad (11.16)$$

---

To illustrate the calculation of $s_e$ for the GRE–GPA example, we need to recall from column 2 of Table 11.4 that

$$\sum e_i^2 = \text{SSE} = 0.225.$$

Since $n = 8$ for that example, the value of $s_e$ is:

$$s_e = \sqrt{\frac{\text{SSE}}{n - 2}} = \sqrt{\frac{0.225}{6}} = 0.1936.$$

The value of $s_e$ can be interpreted in a manner similar to the sample standard deviation of the values of $x$ about $\bar{x}$. That is, given that Assumption 2 holds (i.e., the $\varepsilon_i$ are normal, with mean of zero), then approximately 68.3 percent of the observations will fall within $\pm 1s_e$ units of the regression line, 95.4 percent will fall within $\pm 2s_e$ units of this line, and 99.7 percent will fall within $\pm 3s_e$ units of it.* Using this information gives one a good indication of the fit of the regression line to the sample data. In our example, a range of $\pm 3s_e$ would be

$$\pm 3(0.1936) = 0.5808,$$

which means the potential error in estimating GPA on the basis of GRE will be about 0.60 grade points. This is a rather large potential error, since the average GPA $= 3.10$, and the limits on GPA are zero and 4.0; but we must remember

---

* Technically, this interpretation of $s_e$ should be used only when the sample size is relatively large, as it gives only an approximation to the correct interval (see Section 11.9). For present purposes, however, we will ignore this distinction between the $z$ and $t$ distributions.

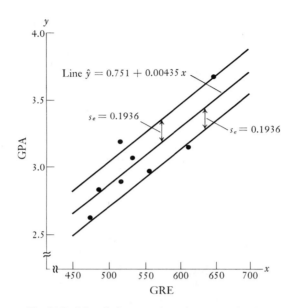

**Fig. 11.8.** A band of one $s_e$ about the regression line.

that our sample size was very small ($n = 8$). On the other hand, over two-thirds (68.3%) of the actual values of $y$ will fall within $\pm s_e = \pm 0.1936$ grade points of the estimated values $\hat{y}$. A band of $\pm 1 s_e = \pm 0.1936$ about the regression line is illustrated in Fig. 11.8. In this graph we see that six of the eight "errors" are less than 0.1936.

### The Coefficient of Determination

Our second measure of goodness of fit, which is useful in interpreting the *relative* amount of the variation that has been explained by the sample regression line, is called the *coefficient of determination*. This measure, which is denoted by the symbol $r^2$, is derived from a commonsense use of the same terms we have been discussing, SST, SSR, and SSE.

Suppose that, in the relationship SST = SSE + SSR, we divide each term by SST, as follows:

$$\frac{\text{SST}}{\text{SST}} = 1.0 = \frac{\text{SSE}}{\text{SST}} + \frac{\text{SSR}}{\text{SST}}.$$

Since SSE is the unexplained variation in $y$, the ratio SSE/SST is the *proportion* of total variation that is unexplained by the regression relation; similarly, the ratio SSR/SST is the *proportion* of total variation that is *explained* by the regression line. This last ratio, SSR/SST, is the *relative* measure of goodness of fit we sought, and is called the *coefficient of determination.*

---

*Coefficient of determination:*

$$r^2 = \frac{\text{SSR}}{\text{SST}} = \frac{\text{Variation explained}}{\text{Total variation}}. \qquad (11.17)$$

---

If the regression line perfectly fit all the sample points, then all residuals would be zero; then SSE would be zero, and hence

$$\left(\frac{\text{SSR}}{\text{SST}}\right) = r^2 = 1.0.$$

In other words, a perfect straight-line fit always results in a value of $r^2 = 1$. As the level of fit becomes less accurate, less and less of the variation in $y$ is explained by the relation with $x$ (i.e., SSR decreases), which means that $r^2$ must decrease. The lowest value of $r^2$ is 0, which will occur whenever SSR $= 0$. A value of $r^2 = 0$ thus means that none of the variation in $y$ is explained by $x$.

Once the value of $r^2$ is calculated in a regression analysis, we have a measure of goodness of fit. For example, if $r^2 = 0.70$, this means that 70% of the total variation in the $y$ values is explained by the regression. Similarly, if $r^2 = 0.5$, then 50 percent of the variation in $y$ has been explained. To further illustrate this concept, let's calculate $r^2$ in our GRE–GPA example. In this case we already know that SSR $= 0.435$ and SST $= 0.660$. Hence,

$$r^2 = \frac{\text{SSR}}{\text{SST}} = \frac{0.435}{0.660} = 0.659.$$

The interpretation of this result is that 65.9% of the total sample variation in GPA is explained by the values of GRE.* The remaining 34.1% of the variation in GPA is still unexplained. Probably some other factors omitted from our regression model could help determine some additional portion of the variation. If these other factors could be measured and included as additional independent variables, we would have a *multiple* regression model. Such an extension is considered in Chapter 12.

### 11.5 CORRELATION ANALYSIS

Although we have presented the relative measure of goodness of fit ($r^2$) in terms of the regression relationship between $y$ and $x$, the strength or closeness of the

---

* The similarity in size between SST and $r^2$ is purely coincidental.

linear relation between two variables can be measured *without* estimating the population regression line. The measurement of how well two (or more) variables vary together is called *correlation analysis.*

One measure of the population relationship between two random variables is their covariance, which is

$$C[x, y] = E[(x - \mu_x)(y - \mu_y)].$$

Although the covariance has many important statistical uses, this measure in general is *not* a good indicator of the relative strength of the relationship between two variables, because its magnitude depends so highly on the *units* used to measure the variables. For example, the covariance between two measures of length $x$ and $y$ will be much smaller if $x$ is scaled in feet than if $x$ is scaled in inches. For this reason, it is necessary to "standardize" the covariance of two variables in order to have a good measure of fit. This standardization is accomplished by dividing $C[x, y]$ by $\sigma_x$ and $\sigma_y$. The resulting measure is called the *population correlation coefficient*, and is denoted by the Greek letter $\rho$ (rho):

$$\rho = \frac{\text{Covariance of } x \text{ and } y}{(\text{Std. dev. of } x)(\text{Std. dev. of } y)} = \frac{C[x, y]}{\sigma_x \sigma_y}. \tag{11.18}$$

Three values of $\rho$ serve as benchmarks for interpretation of a correlation coefficient. First, let's consider the population where the values of $x$ and $y$ all fall on a single straight line with a positive slope. In this case, which is referred to as a "perfect positive linear relationship" between $x$ and $y$, the value of $C[x, y]$ will exactly equal the value of $\sigma_x$ times $\sigma_y$; hence, $\rho$ will equal $+1$.

When the relationship between $x$ and $y$ is a perfect *negative* linear relationship, all values of $x$ and $y$ lie on a straight line with a negative slope. This situation results in a value of $C[x, y]$ which exactly equals $-(\sigma_x)(\sigma_y)$. Thus, in this case $\rho$ will equal $-1$.

If $x$ and $y$ are not linearly related (i.e., they are independent random variables), then the value of the correlation coefficient will be zero, since in this case $C[x, y] = 0$, which means that

$$\rho = \frac{C[x, y]}{\sigma_x \sigma_y} = 0.$$

Thus, $\rho$ measures the strength of the linear association between $x$ and $y$. Values of $\rho$ close to zero indicate a weak relation; values close to $+1.0$ indicate a strong "positive" correlation, and values close to $-1.0$ indicate a strong "negative" correlation. Figure 11.9 illustrates some representations of values of $\rho$ for selected scatter diagrams.

Note, from Figs. 11.9(b) and 11.9(c), that two populations which appear quite different can have the same correlation coefficient. Figures 11.9(c) and 11.9(d) show the difference between positive and negative correlation. The last two diagrams

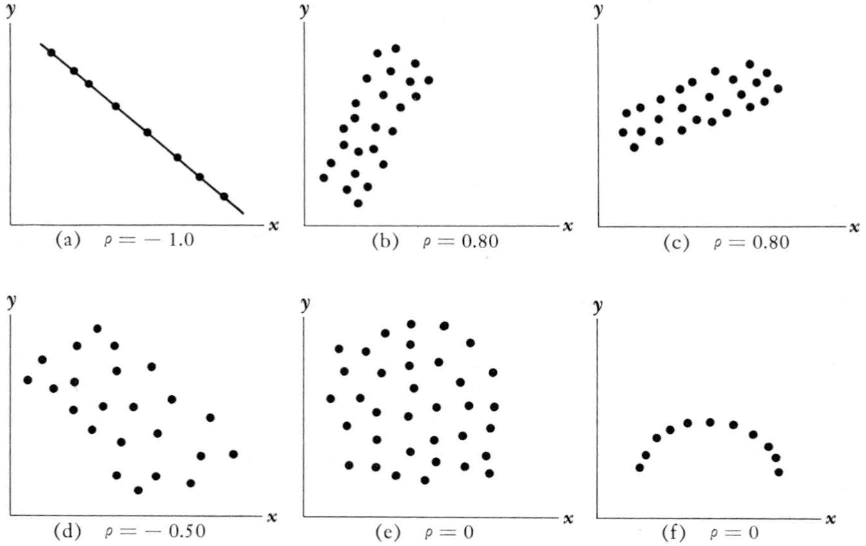

**Fig. 11.9.** The population correlation coefficient.

show different examples of a population with zero correlation. In Fig. 11.9(f), $x$ and $y$ are related in a nonlinear fashion, yet still $\rho = 0$, which emphasizes the fact that $\rho$ measures the strength of the *linear* relationship.

### The Sample Correlation Coefficient

As in all estimation problems, we use sample data to estimate the population parameter $\rho$. In this case, the sample statistic is called the *sample correlation coefficient*, and denoted by the letter $r$. The value of $r$ is defined in the same way as $\rho$, except that we substitute for each population parameter its best estimate based on the sample data. For instance, the best estimate of $C[x, y]$ in Formula (11.18) is the sample covariance, which is denoted by the symbol $s_{xy}$, where

$$s_{xy} = \frac{1}{n-1} \sum_{i=1}^{n} (x_i - \bar{x})(y_i - \bar{y}).$$

Similarly, the best estimate of $\sigma_x^2$ is the sample variance

$$s_x^2 = \frac{1}{n-1} \sum_{i=1}^{n} (x_i - \bar{x})^2,$$

and the best estimate of $\sigma_y^2$ is the sample variance

$$s_y^2 = \frac{1}{n-1} \sum_{i=1}^{n} (y_i - \bar{y})^2.$$

Substituting these estimates in Formula (11.18) yields the following formula for $r$:

---

*Sample correlation coefficient:*

$$r = \frac{s_{xy}}{s_x s_y} = \frac{\text{Covariance of } x \text{ and } y}{(\text{Std. dev. of } x)(\text{Std. dev. of } y)}$$

$$= \frac{\dfrac{1}{n-1} \sum (x_i - \bar{x})(y_i - \bar{y})}{\sqrt{\dfrac{1}{n-1} \sum_{i=1}^{n} (x_i - \bar{x})^2} \sqrt{\dfrac{1}{n-1} \sum_{i=1}^{n} (y_i - \bar{y})^2}}. \tag{11.19}$$

---

A sample correlation coefficient is interpreted in the same manner as $\rho$, except that it measures the strength of the *sample* data rather than the population values. For example, when $r = \pm 1$, there is a perfect straight-line fit between the sample values of $x$ and $y$; hence, they are said to have a perfect correlation. If the sample values of $x$ and $y$ have little or no relationship, then $r$ will be close to zero.

To illustrate the determination of the sample correlation coefficient, once again consider the data for our GRE–GPA example. The values for these variables are repeated in Table 11.5, followed by the calculations necessary for determining $r$.

We see that the sample data in this example has a correlation of 0.81, indicating a fairly strong linear relationship. This result agrees with our goodness-of-fit analysis of the same data. In fact, we will show, in the remainder of this chapter, that there are many ties between regression and correlation analysis. For example, we can explore the connection between the value of $r$ and the value of the slope $b$ by comparing Formulas (11.8) and (11.19), which are reproduced below:

$$b = \frac{s_{xy}}{s_x^2} \quad \text{and} \quad r = \frac{s_{xy}}{s_x s_y}.$$

Because $s_x^2 = s_x s_x$, by substitution we have

$$r = b \frac{s_x}{s_y}. \tag{11.20}$$

Since $s_x$ and $s_y$ are never negative, a positive correlation must always correspond to a regression line with a positive slope, and a negative $r$ corresponds to a negative slope. That is, the sign of $r$ and $b$ will always be the same.

**Table 11.5**

| x(GRE) | y(GPA) | $(x - \bar{x})$ | $(y - \bar{y})$ | $(x - \bar{x})(y - \bar{y})$ | $(x - \bar{x})^2$ | $(y - \bar{y})^2$ |
|---|---|---|---|---|---|---|
| 480 | 2.70 | $-60$ | $-0.40$ | 24 | 3,600 | 0.16 |
| 490 | 2.90 | $-50$ | $-0.20$ | 10 | 2,500 | 0.04 |
| 510 | 3.30 | $-30$ | $+0.20$ | $-6$ | 900 | 0.04 |
| 510 | 2.90 | $-30$ | $-0.20$ | 6 | 900 | 0.04 |
| 530 | 3.10 | $-10$ | 0.00 | 0 | 100 | 0.00 |
| 550 | 3.00 | $+10$ | $-0.10$ | $-1$ | 100 | 0.01 |
| 610 | 3.20 | $+70$ | 0.10 | 7 | 4,900 | 0.01 |
| 640 | 3.70 | $+100$ | $+0.60$ | 60 | 10,000 | 0.36 |
| Sum  4320 | 24.80 | 0 | 0.00 | 100 | 23,000 | 0.66 |

$$r = \frac{\dfrac{1}{n-1}\sum_{i=1}^{n}(x_i - \bar{x})(y_i - \bar{y})}{\sqrt{\dfrac{1}{n-1}\sum(x_i - \bar{x})^2}\sqrt{\dfrac{1}{n-1}\sum(y_i - \bar{y})^2}} = \frac{100/7}{\sqrt{23,000/7} \cdot \sqrt{0.66/7}} = 0.81$$

Now consider the connection between the correlation coefficient $r$ and the coefficient of determination. Recall that $r^2 = \text{SSR/SST}$. It is not difficult to show that

$$\frac{\text{SSR}}{\text{SST}} = \frac{s_{xy}^2}{s_x^2 s_y^2},$$

which exactly equals the square of the correlation coefficient (Formula (11.19)). The reader can now understand the use of the letter $r$ for the correlation coefficient and the symbol $r^2$ for the coefficient of determination. The latter is indeed the square of the former. In most cases, $r^2$ is much easier to interpret than $r$. For instance, a sample correlation coefficient of $r = 0.50$ means that 25 percent of the sample variation has been explained by $x$, since $r^2 = 0.25$. Similarly, if $r = 0.70$, 49 percent of the sample variation has been explained. In the GRE–GPA example, the correlation coefficient was $r = 0.81$, which implies that about 66 percent of the variability of graduate grades in this sample can be explained on the basis of differences in GRE scores (since $r^2 = 0.81^2 = 0.66$). About 34 percent remains unexplained.

Just as Formula (11.10) presented a computational formula for calculating $b$, so can we present a formula for calculating $r$ that simplifies the calculation of this statistic when $\bar{x}$ and $\bar{y}$ are not integers.

> *Computational formula for r:*
>
> $$r = \frac{\sum x_i y_i - \frac{1}{n}\left(\sum x_i\right)\left(\sum y_i\right)}{\sqrt{\sum x_i^2 - \frac{1}{n}\left(\sum x_i\right)^2}\sqrt{\sum y_i^2 - \frac{1}{n}\left(\sum y_i\right)^2}}.$$ (11.21)

We can now show that this formula yields the same value of $r$ for our GRE–GPA example. Since subtracting a constant from every value of $y$ and $x$ will not change the value of $r$ (just as it didn't change the value of $b$), we will use the coded data from Table 11.3 to find $r$.

$$r = \frac{(132) - \frac{1}{8}(320)(0.80)}{\sqrt{(35,800) - \frac{1}{8}(320)^2}\sqrt{(0.74) - \frac{1}{8}(0.80)^2}}$$

$$= \frac{100}{\sqrt{23,000}\sqrt{0.66}} = 0.81.$$

This value agrees with the $r$ we calculated previously.

It is important to note at this point that the value of the correlation coefficient does not depend on which variable is designated as $x$ and which as $y$. The distinction *is* important in regression analysis, however, for the conditional distribution of $y$, given $x$, results in a different regression line than the conditional distribution of $x$, given $y$. Formula (11.20) holds only if the numerator is $b$ multiplied by the standard deviation of the *independent* variable.

A note of caution must be added to anyone attempting to infer cause and effect from correlation or regression analysis, since a high correlation or a good fit to a regression line does *not* imply that $x$ is "causing" $y$. It does not even imply that $x$ will provide a good estimate of $y$ in the future, or for any other set of sample observations. For example the weekly Dow–Jones stock index was reported to have a 0.84 correlation with the number of points scored by a New York City basketball team in its weekend game, and liquor consumption in the U.S.A. is supposed to be highly correlated with teachers' salaries. In this latter case the high correlation undoubtedly results because of the presence of one or more additional influences on both variables, such as increases in the general economic well-being.

The above discussion should not be interpreted to mean that one cannot, or should not, draw inferences or conclusions from regression or correlation analysis, but only that care must be taken in assuming cause and effect. Most graduate schools, for example, believe that admission test scores are one means for estimating academic success; they presumably are basing their opinion not on the fact that higher GRE scores "cause" higher grades, but rather on the fact that whatever these tests measure (memory, intelligence, vocabulary) *does* have an influence on graduate grades. As another example, economists make frequent use of regression techniques in attempts to determine cause-and-effect relationships, especially those which might prove useful in predicting the future of the economy. These techniques form the basis for the field of econometrics.

### 11.6  TEST ON THE SIGNIFICANCE OF THE SAMPLE REGRESSION LINE

We have presented a way of estimating the best regression line fitting the linear relation between $y$ and $x$, and we have discussed measures of the strength of the linear relationship. However, we have not given any rules or guidelines to help determine whether knowledge of the independent variable $x$ is useful in predicting the values of $y_i$. Suppose the population relationship is such that $\beta = 0$. This means that the population regression line must be a horizontal straight line, where $\hat{y} = \bar{y}$. Since $\hat{y}$ is a constant whenever $\beta = 0$, the values of $x$ are of no use in predicting $y$. If $\beta$ is not equal to zero, then the values of $x$ *are* meaningful in predicting $y$. Thus, to determine whether or not the estimation of the $y$ values is improved by using the regression line, we need to test the null hypothesis that $\beta = 0$.

#### Test on the Slope

Suppose the null hypothesis is
$$H_0: \beta = 0.$$
Rejecting $H_0$ in this case means concluding, on the basis of the sample information given, that $\beta$ does not equal zero, and hence that the regression line improves our estimate of the dependent variable. The alternative hypothesis for this test might take on a number of different forms. For example, we could use the one-sided alternative that the slope is greater than zero ($H_a: \beta > 0$), or the one-sided alternative that the slope is less than zero ($H_a: \beta < 0$). A two-sided alternative would be to hypothesize merely that the slope does not equal zero ($H_a: \beta \neq 0$).

Although the above forms of alternative hypotheses are by far the most commonly used, the hypothesized value need not be zero. To illustrate a problem where $\beta$ is not assumed to be equal to zero, suppose that it is desirable to test the null hypothesis that the slope of the regression line relating personal income and consumption (i.e., marginal propensity to consume) has not deviated from some historical value, such as $\beta = 0.90$. If we assume that the alternative hypothesis is two-sided, then the appropriate test is $H_0: \beta = 0.90$ against $H_a: \beta \neq 0.90$. The assumed value of $\beta$ in such a test is usually denoted as $\beta_0$; a two-sided alternative would thus be
$$H_0: \beta = \beta_0 \qquad \text{vs.} \qquad H_a: \beta \neq \beta_0.$$

A $t$-test is used to test the null hypothesis $H_0: \beta = \beta_0$. This test is very similar to the $t$-test about a population mean, since we are again testing a mean ($\beta$), the population is assumed to be normal (due to assumption 2 about the normality of the $\varepsilon_i$'s), and the population standard deviation is unknown. In the present case the sample statistic is $b$ (rather than $\bar{x}$), the hypothesized population value is $\beta_0$ (rather than $\mu_0$), and the sample standard error is $s_b$ (rather than $s_{\bar{x}}$), where $s_b$ is defined as follows;

*Standard error of the regression coefficient b:*

$$s_b = s_e \sqrt{\frac{1}{\sum\limits_{i=1}^{n} (x_i - \bar{x})^2}}. \tag{11.22}$$

The value $s_b$ is a measure of the amount of sampling error in the regression coefficient $b$, just as $s_{\bar{x}}$ was a measure of the sampling error of $\bar{x}$. We can now test the null hypothesis $H_0: \beta = \beta_0$ by subtracting the hypothesized value $\beta_0$ from $b$ and dividing by the standard error of the regression coefficient. The resulting statistic,

$$t_{(n-2)} = \frac{b - \beta_0}{s_b}, \tag{11.23}$$

follows a $t$-distribution with $(n - 2)$ degrees of freedom.*

To illustrate the use of Formula (11.23), suppose in our GRE–GPA example that the null hypothesis $H_0: \beta = 0$ is tested against $H_0: \beta > 0$ (we use the one-sided test here because it is not reasonable to expect an *inverse* relationship between GRE and GPA). First we need to calculate $s_b$ by substituting into Formula (11.23) the previously determined values

$$s_e = 0.1936 \quad \text{and} \quad \sum_{i=1}^{n} (x_i - \bar{x})^2 = 23{,}000;$$

we get:

$$s_b = 0.1936 \sqrt{\frac{1}{\sum (x_i - \bar{x})^2}} = 0.00128.$$

Therefore,

$$t = \frac{b - \beta_0}{s_b} = \frac{0.00435 - 0}{0.00128} = 3.40.$$

---

\* A similar test statistic could be used to test hypotheses about the population $y$-intercept $\alpha$, based on the sample estimate of $a$ and its standard error $(s_a)$:

$$s_a = \sqrt{s_e^2 \sum_{i=1}^{n} x_i^2 \bigg/ \sum_{i=1}^{n} (x_i - \bar{x})^2}.$$

The proper $t$ statistic with $(n - 2)$ degrees of freedom is $t_{(n-2)} = (a - \alpha)/s_a$.

For $n - 2 = 6$ degrees of freedom, the probability that $t$ is larger than 3.40 falls between 0.01 and 0.005 (see Table VII). Thus, it is highly unlikely that a slope of $b = 0.00435$ will occur by chance when $\beta = 0$, and we can conclude that the regression line does seem to improve our ability to estimate the dependent variable (i.e., we reject $H_0$).

In addition to being able to test hypotheses about $\beta$, it is possible to construct a $100(1 - \alpha)$-percent confidence interval for $\beta$. Since the regression coefficient $b$ follows a $t$-distribution with $(n - 2)$ degrees of freedom and standard deviation $s_b$, the desired interval is:

$$b - t_{(\alpha/2,\ n-2)}s_b \leqslant \beta \leqslant b + t_{(\alpha/2,\ n-2)}s_b. \qquad (11.24)$$

A 95-percent confidence interval, given that $n = 8$, $t_{(\alpha/2,\ n-2)} = t_{(0.025,\ 6)} = 2.447$, and $s_b = 0.001278$, would be:

$$0.00435 - (2.447)(0.00128) \leqslant \beta \leqslant 0.00435 + (2.447)(0.00128)$$
$$0.00122 \leqslant \beta \leqslant 0.00748.$$

On the basis of our sample of eight students, we would thus expect an improvement of between 0.122 and 0.748 points in GPA for each 100-point increase in GRE. This is rather a wide interval for any precise forecasting of GPA, but again we must point out that the sample size is very small.

**Test on the Correlation Coefficient**

In addition to the $t$-test presented above for testing $H_0\colon \beta = 0$, there is an equivalent $t$-test based on null hypothesis $H_0\colon \rho = 0$. In order to test this hypothesis, it is necessary to know the sampling distribution of some random variable involving the sample correlation coefficient $r$ (which is our best estimate of $\rho$). The following $t$-distributed random variable, with $(n - 2)$ degrees of freedom, can be used:

$$t_{(n-2)} = \frac{r\sqrt{n - 2}}{\sqrt{1 - r^2}}. \qquad (11.25)$$

To illustrate how this ratio can be used to test the hypothesis $H_0\colon \rho = 0$, we again use our GRE–GPA example, in which $r = 0.81$. For the same reason that we used the alternative $H_a\colon \beta > 0$, we now use the alternate hypothesis $H_a\colon \rho > 0$. To determine the probability that a value such as 0.81 would occur by chance, given that $\rho = 0.81$ and $n = 8$, we use Formula (11.25) to obtain:

$$t = \frac{0.81\sqrt{8 - 2}}{\sqrt{1 - 0.81^2}} = 3.40.$$

Referring to Table VII for six degrees of freedom, we find that the probability that $t$ is equal to or larger than 3.40 is between 0.005 and 0.01, so that it is very unlikely that a sample correlation this high will occur by chance when $\rho = 0$. Thus, we can accept the alternate hypothesis that there is a positive linear relation between $y$ and $x$. The value of $t$ obtained in this analysis, $t = 3.40$, is *exactly* the same result as that obtained when testing $H_0$: $\beta = 0$. This agreement between the two is more than mere coincidence, since the outcome of a simple regression analysis and a correlation analysis on the same data must yield identical results when testing the hypothesis that there is *no* relationship between $x$ and $y$. These tests are not equivalent if the null hypothesis specifies that the slope is equal to some value other than zero, although the previous test on $\beta$ (Formula (11.24)) is still appropriate. In that case, the test on $\beta$ (using Formula (11.24)) is not a test of the significance of the linear relationship, but rather a test on some proposed population parameter $\beta_0$. Such a test would therefore not be equivalent to the $t$-test in this section for the *significance* of $\rho$.

Finally, we should mention that since the alternative hypothesis in our GRE–GPA example was one-sided, we read the value $P(t > 3.40)$ directly from Table VII. If the alternative had been a two-sided alternative, then the probabilities from Table VII would have to be *doubled* (that is, $0.02 \geqslant 2P(t > 3.40) \geqslant 0.01$).

## 11.7 A SAMPLE PROBLEM

Having specified all the concepts and formulas essential for a simple regression and correlation analysis, let us now review them all by applying them. One of the important decisions faced by business managers is the amount of *investment* they should make in new plant and equipment and in maintenance and repair of existing capital goods. Economists are very interested in the level of aggregate investment over all private business in the U.S., since this value is an important factor in determining national income and the potential for growth in an economy.

Suppose a model is specified where the dependent variable $y$ is the level of investment. In such a model, one might want to estimate the amount of investment for a *single* firm by analyzing the relationship between investment and one or more independent variables in a sample of firms. On the other hand, one may be more interested in estimating the *aggregate* investment for all firms, using as the sample data a number of past time periods. For this example, we choose the latter concept, because the necessary data is easily available without a survey of firms. The

Department of Commerce and the Federal Reserve Board, among other agencies, report the key monthly, quarterly, or annual economic aggregates for the United States economy.

Any textbook of basic economic principles suggests some theoretical relationships between investment and other variables. Generally, it is recognized that the level of investment depends on the availability of funds for investment and on the need for expanding production capacity. Thus, any variables which reflect these supply or demand factors might be appropriate in explaining or predicting changes in levels of investment. Such variables as the existing amount of plant and equipment which is depreciating, current and past levels of profits, the interest rate at which funds could be borrowed, indicators of the current general economic conditions, the amount of labor that is unemployed, etc., might be selected as independent variables to help explain variation in investment.

For simplicity, suppose we specify the population regression model to be $y_i = \alpha + \beta x_i + \varepsilon_i$, where $x_i$ represents a composite price index for 500 common stocks during a given time period, and $y_i$ represents the amount of aggregate investment *during the following time period*. We might postulate such a relationship because we believe that stock market prices are indicative of the general level of business expectations for the future. For this example, we will estimate the population regression line on the basis of 20 quarterly observations. The data, shown in Table 11.6, represents investment measured at an annual rate in billions of dollars. The sample statistics needed for computing the least-squares regression line follow Table 11.6.

We know that the line that provides the best least squares fit to this data has the following slope and intercept using Formulas (11.7) and (11.8).

*Sample slope:*

$$b = \frac{\text{Covariance of } x \text{ and } y}{\text{Variance of } x} = \frac{4085.53}{33,090.16} = 0.12347;$$

*Sample intercept:*

$$a = \bar{y} - b\bar{x} = 88.915 - (0.12347)(688.92) = 3.855.$$

The least-squares estimating line is

$$\hat{y}_i = 3.855 + 0.12347x_i.$$

The positive value for $b$ confirms our assumption that investment increases when stock prices increase (we see that, for this data, a 10-unit increase in the stock index $x$ in one quarter is associated with an increase in the annual rate of investment during the next quarter of about 1.23 billion dollars).

To obtain the goodness-of-fit measures, the components of total variation

**Table 11.6** Data for estimating the investment equation

| Observation | Investment $y$ | Stock index $x$ | Observation | Investment $y$ | Stock index $x$ |
|---|---|---|---|---|---|
| 1 | 62.30 | 398.4 | 11 | 84.30 | 581.8 |
| 2 | 71.30 | 452.6 | 12 | 85.10 | 707.1 |
| 3 | 70.30 | 509.8 | 13 | 90.80 | 776.6 |
| 4 | 68.50 | 485.4 | 14 | 97.90 | 875.3 |
| 5 | 57.30 | 445.7 | 15 | 108.70 | 873.4 |
| 6 | 68.80 | 539.8 | 16 | 122.40 | 943.7 |
| 7 | 72.20 | 662.8 | 17 | 114.00 | 830.6 |
| 8 | 76.00 | 620.0 | 18 | 123.00 | 907.5 |
| 9 | 64.30 | 632.2 | 19 | 126.20 | 905.3 |
| 10 | 77.90 | 703.0 | 20 | 137.00 | 927.4 |

$$Mean\ of\ y = \frac{1}{n}\sum_{i=1}^{n} y_i = 88.915 \qquad Mean\ of\ x = \frac{1}{n}\sum_{i=1}^{n} x_i = 688.92$$

$$Variance\ of\ y = \frac{1}{n-1}\sum_{i=1}^{n}(y_i - \bar{y})^2 \qquad Variance\ of\ x = \frac{1}{n-1}\sum_{i=1}^{n}(x_i - \bar{x})^2$$

$$= \frac{11,485}{19} = 604.47 \qquad\qquad = \frac{628,713}{19} = 33,090.16$$

$$Covariance\ of\ x\ and\ y = \frac{1}{n-1}\sum_{i=1}^{n}(y_i - \bar{y})(x_i - \bar{x}) = \frac{77,685}{19} = 4085.53.$$

need to be found. The total variation in $y$ to be explained can be shown to be

$$\text{SST} = \sum (y_i - \bar{y})^2 = 11,485.$$

Its components are:

$$\text{Unexplained variation-SSE} = \sum (y_i - \hat{y}_i)^2 = 1901,$$

and

$$\text{Explained variation-SSR} = \sum (\hat{y}_i - \bar{y})^2 = 9584.$$

The value $\text{SSE} = \sum e_i^2 = 1901$ can be used to find the standard error of the estimate using Formula (11.16):

$$s_e = \sqrt{\sum_{i=1}^{n} e_i^2/(n-2)} = \sqrt{1,901/18} = 10.3.$$

The reader might wish to check to see whether approximately 68 percent of the data points lie within 10.3 vertical units (billions of dollars) above or below the estimating line, by drawing a sketch of the scatter diagram and the line for this example.

Our second measure of goodness of fit is the sample coefficient of determination, $r^2$, given by Formula (11.17).

$$r^2 = \frac{\text{Explained variation (SSR)}}{\text{Total variation (SST)}} = \frac{9{,}584}{11{,}485} = 0.834.$$

This means that 83.4 percent of the variation in investment has been explained by the estimating relationship between it and the stock price index.

Any of the tests or confidence intervals on the population parameters, as described in the preceding sections, can now be easily calculated. For example, a $t$-test on the significance of the slope parameter would be

$$H_0: \beta = 0 \text{ vs. } H_a: \beta > 0.$$

The test statistic with a $t$-distribution and $n - 2 = 18$ degrees of freedom using Formula (11.23) is,

$$t_{18 \text{ d.f.}} = \frac{b - \beta_0}{s_b},$$

where $\beta_0 = 0$ and $s_b = s_e/\sqrt{\Sigma(x_i - \bar{x})^2}$. Using a significance level of $\alpha = 0.005$, we find the critical region, from Table VII, to be all values of $t > t_{(0.005, \; 18)} = 2.878$. The calculated value of $s_e$ is 10.3 and $s_b = 10.3/\sqrt{11.485} = 0.013$; thus the calculated value of $t$ is:

$$t = \frac{b - 0}{s_b} = \frac{0.12347}{0.013} = 9.5.$$

Since $9.5 > 2.878$, we can reject the null hypothesis and conclude that there is a positive linear relationship between current investment and the stock-market index of the previous quarter. If the stock prices do reflect business expectations, then this data supports the theory that business firms are more willing to expand their plant and equipment when they foresee "good times" (higher incomes, greater demands, and more potential sales and profits) ahead.

### *11.8 THE F-TEST

There is still another way of testing the null hypothesis $H_0: \beta = 0$, this one using the measures of unexplained and explained variation. Before the reader bemoans the presentation of one more test to learn, we must point out that the test in this

---

* This section assumes that the reader has studied Section 10.7 on the use of an $F$-distributed random variable in analysis-of-variance tests.

section is particularly important because it can be generalized to problems involving more than just one independent variable (i.e., to the multiple-regression
case). This use is discussed in Chapter 12.

Recall that SST = SSE + SSR. You may also recall that the degrees of freedom associated with SST is $n - 1$ (since only $\bar{y}$ needs to be calculated before SST
can be computed), while the d.f. for SSE is $n - 2$ (both $a$ and $b$ need to be calculated
before computing SSE). Since the d.f. for SST must equal the sum of those for
SSE and SSR, we see by subtraction that the d.f. for SSR = 1 (because $(n - 1) =
(n - 2) + (1)$). Now, a sum of squares divided by its degrees of freedom is called
a *mean square*. The two mean squares we will need are *mean square error* (MSE)
and *mean square regression* (MSR):

$$\text{Mean square error:} \quad MSE = \frac{SSE}{n - 2} = s_e^2$$

$$\text{Mean square regression:} \quad MSR = \frac{SSR}{1}.$$

It is customary to present information about MSE and MSR in what is called
an *analysis-of-variance* table, such as shown in Table 11.7.

**Table 11.7** Analysis-of-variance table for simple regression

| Source of the variation | Sum of squares | Degrees of freedom | Mean square |
|---|---|---|---|
| Regression | SSR | 1 | SSR/1 |
| Error (or residual) | SSE | $n - 2$ | $SSE/(n - 2)$ |
| Total | SST | $n - 1$ | |

One word of caution is necessary here: Although the sums of squares and the
degrees of freedom are additive, it is *not* true that the mean square terms are
additive. Note that, in the analysis-of-variance table, no MS term is given in the row
labeled "Total." The SST and the degrees of freedom total are given in the table,
so that it can be verified that elements in the body of the table do sum to these
values.

To illustrate the use of an analysis-of-variance table, suppose we construct
such a table for the GRE–GPA example. Recall from Section 11.4 that

$$SST = 0.66, \quad SSE = 0.225, \quad \text{and} \quad SSR = 0.435.$$

Hence,

$$MSR = \frac{SSR}{1} = \frac{0.435}{1} = 0.435 \quad \text{and} \quad MSE = \frac{SSE}{n - 2} = \frac{0.225}{6} = 0.0375.$$

Table 11.8 shows the analysis-of-variance table for this example.

**Table 11.8** Analysis-of-variance table for the regression of GPA on GRE

| Source | SS | d.f. | Mean square |
|--------|------|------|-------------|
| Regression | 0.435 | 1 | 0.435 |
| Error | 0.225 | 6 | 0.0375 |
| Total | 0.660 | 7 | |

Now we return to our new test of $H_0: \beta = 0$. If the value of MSR is high relative to the value of MSE, then a large proportion of the total variability of $y$ is being explained by the regression line; this implies that we should reject this null hypothesis. If, however, MSR is *small* relative to the MSE, then the regression line does not explain much of the variability in the sample values of $y$, and the null hypothesis would not be rejected. Thus, the null hypothesis $H_0: \beta = 0$ can be tested by using the ratio of the mean squares, MSR/MSE. This ratio can be shown to have an $F$-distribution with 1 and $(n - 2)$ degrees of freedom:

$$F_{(1,\ n-2,\ \text{d.f.})} = \frac{\text{MSR}}{\text{MSE}}. \tag{11.26}$$

We can now use the data in Table 11.8 to test the hypothesis $H_0: \beta = 0$. If we choose a significance level of $\alpha = 0.05$, the critical value (from Table VIII(a)) is

$$F_{(0.05;\ 1,\ 6\ \text{d.f.})} = 5.99.$$

As shown in Fig. 11.10, the decision rule is to reject $H_0$ (i.e., accept that the regression line does contribute to the explanation of the variation in $y$) if the calculated value of $F$ based on our sample exceeds 5.99.

Using Formula (11.26) and Table 11.8, we calculate the value of $F$ to be

$$F = \frac{\text{MSR}}{\text{MSE}} = \frac{0.435}{0.0375} = 11.6.$$

Since the sample value for $F$ exceeds the critical value of 5.99, we reject the null hypothesis and conclude that the linear relationship *does* help explain the variation in GPA.

We see from the above discussion that the $F$-test on the variation ratio is comparable to the $t$-test on the significance of the slope $b$ or the correlation coef-

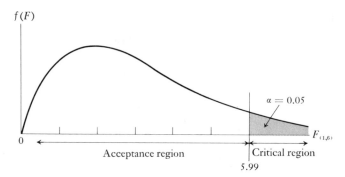

**Fig. 11.10.** The critical region for the analysis-of-variance test on the GRE–GPA regression.

ficient $r$, in that they are all tests on the strength of the relationship between $y$ and $x$ in linear regression. In fact, it can be shown that the $t$-test on $\rho$ and the $F$-test are equivalent tests for the significance of the linear relationship between two variables. The calculated value of $F$ should always equal the square of the calculated value of $t$. In our example,

$$t^2 = (3.40)^2 = 11.56,$$

which differs from $F = 11.6$ only due to rounding errors. The advantage of the $F$-test is that it can be generalized to a test of significance when there is more than one independent variable, while the $t$-test cannot.

## 11.9 CONSTRUCTING A FORECAST INTERVAL

One of the important uses of the sample regression line is to obtain forecasts of the dependent variable, given some value of the independent variable. The estimated value $\hat{y}_i = a + bx_i$ is the best estimate we can make of both $\mu_{y \cdot x_i}$ (the mean value of $y$, given a value $x_i$) and of $y_i$ (the actual value of $y$ that corresponds to the given value $x_i$). Forecasts of both types are frequently desired. Economists may desire to forecast the average or expected level of unemployment, given assumed values of independent variables under policy control. From such forecasts, they might argue which variables should be affected by policy, by how much, and in what direction, so that unemployment can be expected to be reduced toward a certain policy goal, 4% unemployed. In other cases, the forecast of the actual value of the dependent variable may be desired, as, for example, in making predictions of the level of unemployment which will occur in the second quarter of the next year, or of the level of the price of General Motors common stock at the end of this year, or of the total sales this year for Sears.

**Point Estimates of Forecasts**

To obtain the best point estimate for forecasts of both the mean value and the actual value of $y$, the given value of the independent variable (call it $x_g$) is substituted into the estimating equation to obtain the forecast value

$$\hat{y}_g = a + bx_g.$$

Thus, $\hat{y}_g$ is an estimate of *both* $\mu_{y \cdot x_g}$ and $y_g$.

   Suppose that in our GRE–GPA example we wish to obtain the forecast value for the given GRE score of $x_g = 500$. Using the estimated regression coefficients $a = 0.751$ and $b = 0.00435$, we obtain

$$\hat{y}_g = a + bx_g = 0.751 + 0.00435(500) = 2.926.$$

Thus, our best estimate for one person who has a GRE of 500 is $\hat{y}_g = 2.926$.

   Similarly, our estimate for the mean of *all* persons having GRE's of 500 is also $\hat{y}_g = 2.926$. Although these estimates both equal the same value, we must emphasize that they are interpreted differently. This difference will become important when we investigate the process of making interval estimates. Specifically, we will see that the confidence interval for estimating a single value will necessarily be larger than the confidence interval for estimating the mean value because the former will always have a larger standard error.

**Interval Estimates of Forecasts**

Recall that an interval estimate uses a point estimate as its starting point, and then uses the standard error of the point estimate and its probability distribution to find the endpoints of the interval. From the discussion above we know that the starting point is always the same value, $\hat{y}_g$. And, as we pointed out in Section 11.4, the standard error $s_e$ can be used to form the endpoint of the interval when estimating the actual sample value $y_i$ using previous observations, $x_i$. Special care must be taken, however, in estimating forecasts based on *extra* sample values, because then $s_e$ is actually only an approximation to the appropriate standard error. The appropriate standard error is usually called the *standard error of the forecast*, which we will denote as $s_f$. We write the formula for $s_f$ below in terms of $s_e$:

$$s_f = s_e \sqrt{1 + \frac{1}{n} + \frac{(x_g - \bar{x})^2}{\sum\limits_{i=1}^{n} (x_i - \bar{x})^2}}. \tag{11.27}$$

Note that $s_f$ will always be larger than $s_e$ since the term under the square root will always be greater than one. Also, note that $s_f$ depends on the particular value of $x_g$ of interest. Finally, we see that if $n$ is large and if the new sample value of $x_g$ is close to $\bar{x}$ (the mean of the previous sample values) then the term under the square root will be close to 1.0; hence, $s_e$ and $s_f$ will be approximately equal. This result should not be too surprising, for we know that the larger the sample, and the less that a given value $x_g$ deviates from $\bar{x}$, the more faith we have in the sampling results and in the subsequent forecast.

We can now use our point estimate, $y_g$, and the standard error $s_f$, to construct a $100(1 - \alpha)\%$ confidence interval for $y_g$. The appropriate test statistic in this case has the *t*-distribution with $(n - 2)$ degrees of freedom.

---

*Endpoints of a* $100(1 - \alpha)\%$ *forecast interval* $\hat{y}_g$:

$$\hat{y}_g \pm t_{(\alpha/2,\ n-2)}s_f. \tag{11.28}$$

---

Suppose we want to construct, on the basis of our sample of $n = 8$, a 95-percent forecast interval for the GPA of an individual student with a GRE score of 500. From Table VII the value of $t_{(0.025,\ 6)} = 2.447$, and we know from our previous analysis that

$$\hat{y}_{500} = 2.926, \qquad s_e = 0.1936, \qquad \bar{x} = 540, \qquad \text{and} \qquad \sum_{i=1}^{n} (x_i - \bar{x})^2 = 23,000.$$

Substituting these values into Formula (11.28), and using the definition of $s_f$ in Formula (11.27), we get the following endpoints for the forecast interval:

$$2.926 \pm 2.447(0.1936)\sqrt{1 + (1/8) + [(500 - 540)^2/23,000]}$$
$$= 2.926 \pm 2.477(0.1936)(1.09296)$$
$$= 2.926 \pm 0.518$$
$$= 2.408 \quad \text{and} \quad 3.444.$$

We can thus assert, with 95% confidence, that the GPA of an individual with a GRE of 500 will fall between 2.408 and 3.444. Again, this interval is quite wide since we have used a very small sample, $n = 8$, for illustrative purposes.

Now we turn to the problem of constructing a forecast interval for $\mu_{y \cdot x_g}$, the *mean* of the *y*-values. In this case the appropriate standard error is denoted by the symbol $s_{\bar{y}}$, where:

$$s_{\bar{y}} = s_e \sqrt{\frac{1}{n} + \frac{(x_g - \bar{x})^2}{\sum\limits_{i=1}^{n} (x_i - \bar{x})^2}}. \tag{11.29}$$

As was the case for $s_f$, $s_{\bar{y}}$ depends on $n$, $x_g$, and $s_e$. The value of $s_{\bar{y}}$, however, will always be smaller than $s_f$ since $s_f$ contains one additional positive term under the square-root sign. Again, the appropriate test statistic is the *t*-distribution, with

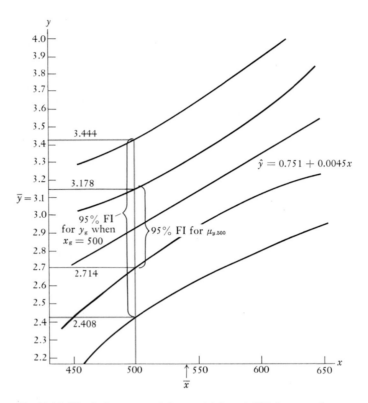

**Fig. 11.11.** Ninety-five percent forecast interval (FI) for $\mu_{y \cdot x_g}$ (narrow band) and $y_g$ (wide band).

$(n - 2)$ degrees of freedom. The endpoints of a $100(1 - \alpha)\%$ interval are thus:

> *Endpoints for a $100(1 - \alpha)\%$ forecast interval on $\mu_{y \cdot x_g}$:*
>
> $$\hat{y}_g \pm t_{(\alpha/2;\ n-2)} s_{\bar{y}}.$$
>
> (11.30)

Substituting the appropriate values in (11.29), and using Formula (11.30) for $s_{\bar{y}}$, we obtain the following endpoints:

$$2.926 \pm 2.447(0.1936)\sqrt{(1/8) + [(500 - 540)^2/23,000]} = 2.926 \pm 0.212$$
$$= 2.714 \quad \text{and} \quad 3.138.$$

The interval 2.714 to 3.138 thus represents a 95% forecast interval for the mean GPA of all students with GRE's of 500 ($\mu_{y \cdot 500}$).

Figure 11.11 shows the forecast interval for values of $x$ from $x_g = 450$ to $x_g = 650$ for both $\mu_{y \cdot x_g}$ (the narrow band) and for $y_g$ (the wide band). As we indicated before, the forecast interval is wider for $y_g$ than for $\mu_{y \cdot x_g}$, because predicting the grade-point average for an individual involves a lot more variability than predicting for an average over many such individuals (i.e., $s_f$ is larger than $s_y$). Note that both bands become increasingly wider for values of $x_g$ further away from $\bar{x}$. This reflects the fact that our least squares estimating line is most accurate at the point of means, $(\bar{x}, \bar{y})$, since it is a representation of the average relationship between $y$ and $x$.

## REVIEW PROBLEMS

1. Explain why a disturbance term $\varepsilon$ is included in the population regression model.

2. Explain what is meant by "best" in a Best Linear Unbiased Estimator (BLUE).

3. Suppose that $y$ = average test score on college entrance boards and $x$ = hundreds of dollars of expenditures per pupil in the student's respective high school. Given that a regression of $y$ on $x$ gives $\hat{y} = 320 + 50x$, what is the interpretation of the value 50 and what value would you predict for $y$ if $x = 50$?

4. Given the following data on five observations for the variables $y$ and $x$, find the slope and intercept for the least-squares estimating line, and write the complete estimating equation.

| $y$: | 3 | 19 | 18 | 22 | 23 |
|---|---|---|---|---|---|
| $x$: | $-2$ | $-1$ | 0 | 1 | 2 |

5. A manufacturing firm bases its sales forecast, for each year, on government estimates for total demand in the industry. The following data give the government estimate for total demand and this firm's sales for the past ten years.

| Government estimate | Sales |
|---|---|
| 200,000 | 5,000 |
| 220,000 | 6,000 |
| 400,000 | 12,000 |
| 330,000 | 7,000 |
| 210,000 | 5,000 |
| 390,000 | 10,000 |
| 280,000 | 8,000 |
| 140,000 | 3,000 |
| 280,000 | 7,000 |
| 290,000 | 10,000 |

   a) Draw a scatter diagram and verify that a linear approximation would be appropriate in this problem.
   b) Find the least-squares regression line.
   c) What sales figure represents the least-squares estimate if the government estimates total demand to be 300,000?

6. The following data pertain to selling prices and number of pages of new statistics books.

| Price | Number of pages |
|-------|-----------------|
| $10   | 400             |
| 12    | 600             |
| 12    | 500             |
| 10    | 300             |
| 8     | 400             |
| 8     | 200             |

Find the regression equation of price (in dollars) on number of pages (in hundreds), using the method of least squares.

7. The following data on production of spark plugs and average cost were collected.

| Average cost (cents) | Production per month (000's) |
|----------------------|------------------------------|
| 3                    | 25                           |
| 9                    | 20                           |
| 40                   | 10                           |
| 15                   | 20                           |
| 33                   | 15                           |

Find the regression equation of cost on production, using the method of least squares.

8. In a simple regression and correlation analysis based on 72 observations, we find $r = 0.8$ and $s_e = 10$.

   a) Find the amount of unexplained variation.
   b) Find the proportion of variation unexplained to the total variation.
   c) Find the total variation of the dependent variable.

9. In a simple regression, explain the importance of $r^2$ and $s_e$, and differentiate between them.

10. Use the data from the following table to compute the regression equation of sales ($y$) on advertising expenditure ($x$). Then compute a measure which describes the proportion of the variation of sales which is explained by the regression.

| Region | Sales, $y$ | Advertising expense, $x$ ($10,000) |
|--------|------------|-------------------------------------|
| A | 31 | 5 |
| B | 40 | 11 |
| C | 25 | 3 |
| D | 30 | 4 |
| E | 20 | 2 |
| F | 34 | 5 |

11. The least-squares estimating line of the number of motel rooms which are rented, ($y$), based on the number of advance reservations made, ($x$), for a certain Holiday Inn is $\hat{y} = 26 + (3/4)x$ with an average $\bar{y} = 60$.

   a) For a particular night, 60 advance reservations are received. Suppose the manager needs one maid for each 9 rooms that need cleaning the following morning. Advise him the minimum number he should employ for this particular case, based on your estimate of the number of rooms that will be occupied and will need cleaning.

   b) Suppose that, for one day, the number of advance reservations received is 36 and the number of rooms occupied turns out to be 55. Find the total, explained, and unexplained deviation for this case.

12. The people of West Overage valley make a living by raising cows and hens. It is finally decided that, for many reasons, the hens are becoming a public nuisance. It is the custom for people to bring cows with them when they come to the valley and to buy hens after arriving. The holdings of the residents are listed below.

| Resident | Cows | Hens |
|----------|------|------|
| Zeke | 15 | 100 |
| Sam | 35 | 50 |
| Sam Jr. | 45 | 30 |
| Sam III | 20 | 70 |
| Zeke Jr. | 10 | 120 |
| Archibald Chauncey | 55 | 20 |

   a) Find the estimating line $\hat{y} = a + bx$, where $y$ is the number of hens and $x$ is the number of cows.

   b) These residents decide to forbid anyone to settle who will probably acquire more than 75 hens. A new settler appears with 40 cows. Decide whether the residents will permit him to stay, on the basis of the number of hens you estimate he will acquire.

   c) Find the proportion of the variance of the number of hens that is explained by the holdings of cows.

13. We wish to explore the relationship between family monthly food consumption, $y$, and family monthly income, $x$, both measured in hundreds of dollars. We are given information for 100 families as follows:

| $n = 100$ | $\bar{y} = \$6$ | $\bar{x} = \$8$ |
|---|---|---|
| $\sum xy = 6000$ | $\sum y^2 = 4500$ | $\sum x^2 = 10,000$ |

a) Find the regression equation $\hat{y} = a + bx$, using the method of least squares.
b) Using the regression equation, make an estimate of consumption for a family with a monthly income of $1500.
c) Compute an absolute measure of goodness of fit of the regression equation and interpret its meaning.

14. In a simple regression it is found that $r^2 = 0.6$ and $s_e^2 = 81$ based on 20 observations. Find the total, explained, and unexplained variation.

15. Given the following data for quiz scores ($y$) and class absences ($x$) for a certain statistics class:

| $\sum y = 1800$ | $\sum xy = 4750$ | $\sum y^2 = 113,000$ |
|---|---|---|
| $\sum x = 90$ | $n = 30$ | $\sum x^2 = 400$ |

a) Find the estimating equation $\hat{y} = a + bx$.
b) Find the coefficient of determination.
c) Determine a student's expected quiz score if he had six absences; and comment on the validity of using the estimating equation for subsequent classes or quizzes.

16. In a regression analysis using 16 observations, the explained variation is 40, out of a total variation of 60. Find the standard error of estimate for the regression equation.

17. Given the information below, compute the measures required:

| $x$ | 2 | 4 | 6 | 8 | 10 | 12 |
|---|---|---|---|---|---|---|
| $y$ | 3 | 4 | 4 | 6 | 6 | 7 |

a) Regression equation of $y$ on $x$.
b) Coefficient of determination.
c) Standard error of estimate.
d) For the fifth entry, determine the total deviation, the explained deviation, and the unexplained deviation.

18. Define or describe briefly each of the following:
a) Covariance
b) Coefficient of determination
c) Method of least squares
d) Normal equations

19. In any simple correlation analysis, state what the logical limits are for values of $r^2$, and explain why.

20. Suppose we have data on number of sheep $x$ (in millions) and production of wool sweaters $y$ (in thousands) for a certain region of the United Kingdom, as follows:

| Year | $y$ | $x$ |
|------|-----|-----|
| 1920 | 2 | 1 |
| 1940 | 5 | 4 |
| 1960 | 8 | 4 |

a) Find the estimating equation for $\hat{y} = a + bx$, by the method of least squares.
b) Find the coefficient of correlation between $y$ and $x$.

21. What is the coefficient of correlation between two variables if:

a) One of the variables is constant?
b) The value of one variable always exceeds the value of the other variable by 100?
c) The unexplained variation is twice the explained variation?

22. Suppose that the following three values represent observations for the random variables $x$ and $y$.

| $x$ | $y$ |
|-----|-----|
| 3 | 4 |
| 0 | 2 |
| 3 | 3 |

a) Compute the sample correlation coefficient.
b) What is the covariance of $x$ and $y$? How much of the variability in $y$ is explained by $x$ if the regression model $\hat{y} = a + bx$ is estimated?

23. What is the difference between regression analysis and correlation analysis? When should each be used? What assumptions about the parent population are made in correlation analysis?

24. A Peace Corps representative works with five Thai farmers in a cooperative shop rebuilding small (United States surplus) gasoline motors for use in water pumps and on sampans. He recognizes a difference in the workers' individual ability to learn the new job and to do it properly without supervision. After several weeks, he records the following information, where $y$ = average weekly output of correctly rebuilt motors, and $x$ = years of education of each of the five Thai workers.

| $x$ | 7 | 5 | 6 | 10 | 4 |
|-----|----|----|----|----|----|
| $y$ | 15 | 7 | 10 | 20 | 8 |

    a)  Find the regression equation of $y$ on $x$.

    b)  Find the correlation coefficient between $y$ and $x$.

25.  In a regression of $y$ on $x$ using 11 observations, the value of the coefficient of determination is 0.36.

    a)  Does this indicate a significant correlation between $y$ and $x$ at the 0.05 significance level?

    b)  What is the proportion of variation left unexplained in this regression?

26.  Suppose that the following ten observations were obtained in a survey to determine the relationship between an individual's educational level and his salary.

| Years of higher education | Income |
|:---:|:---:|
| 3 | $10,000 |
| 4 | 8,000 |
| 7 | 13,000 |
| 9 | 20,000 |
| 1 | 6,000 |
| 0 | 5,000 |
| 2 | 8,000 |
| 1 | 7,000 |
| 8 | 14,000 |
| 5 | 9,000 |

    a)  What is the correlation between years of higher education and income for this sample?

    b)  Use your answer to part (a) to test the null hypothesis

$$H_0: \rho = 0$$

        against the alternative hypothesis

$$H_a: \rho > 0.$$

27.  Given a regression equation, $\hat{y} = 14 + 6x$, based on 12 observations, test the null hypothesis $H_0: \beta = 0$ using a significance level of 0.05. The standard error of $b$ is $s_b = 1.5$.

28.  In a correlation between corporate net investment and long-term interest rate, using quarterly observations from the third quarter of 1966 through the fourth quarter of 1975, a correlation coefficient of $+0.60$ is obtained. Determine whether this is a significant positive correlation, using $\alpha = 0.01$.

29.  Suppose that, in analyzing the relationship between 26 observations of two variables, you find SST equal to 120.0 and SSR equal to 13.2. The slope of the sample regression line is positive.

    a)  What is the coefficient of determination for this problem? What percent of the sample variation has been explained?

    b)  Use the $t$ test described in Section 11.6 to test the null hypothesis $H_0: \rho = 0$ against $H_a: \rho > 0$. Can the null hypothesis of no linear correlation be rejected at the 0.05 level of significance?

c) Compute the *F*-value necessary for testing the null and alternative hypothesis in part (b). Is this value of **F** consistent with the *t*-value calculated in part (b)? Does it lead to acceptance or rejection of the null hypothesis at the 0.05 level of significance?

30. Let *x* represent income payments in Texas (billions of dollars) and let *y* represent retail sales of Texas jewelry stores (millions of dollars.) The regression equation is $\hat{y} = 8.505x - 7.41$. The standard error of the estimate is 3.5 and the correlation coefficient is 0.95.

a) For a year in which income payments are $10.0 billion, what is the best estimate of sales in jewelry stores?

b) What proportion of the variation in the retail jewelry sales is explained by the variation in income payments?

c) This correlation indicates that in Texas, higher retail jewelry sales causes higher incomes. Comment.

31. The following data represent the dollar value of sales and advertising for a retail store.

| Advertising | Sales |
| --- | --- |
| $600 | $5,000 |
| 400 | 4,000 |
| 800 | 7,000 |
| 200 | 3,000 |
| 500 | 6,000 |

a) Draw the scatter diagram for these data. Fit a line by the freehand method. Does the linear approximation seem appropriate?

b) Find the least-squares regression line.

c) What value for sales would you predict if advertising is $700? What value would you predict if advertising is zero?

d) Construct a 95-percent confidence interval for the *mean* value of sales when advertising is $500.

e) Construct a 95-percent confidence interval for actual values of sales when advertising is $500.

f) Test the null hypothesis that the slope of the regression line is zero.

g) Find the sample correlation coefficient.

h) Test the null hypothesis

$$H_0: \rho = 0$$

against the alternative

$$H_a: \rho > 0.$$

At what level of significance can $H_0$ be rejected?

## EXERCISES

32. How would you choose among alternative unbiased estimators of the slope in the population regression model?

33. Explain why least-squares estimates are popular by relating them to the concepts of *maximum likelihood* and BLUE.

34. What assumptions about the parent population are necessary to fit a least-squares regression line to a set of observations? What assumptions about the parent population are necessary to make interval estimates on the basis of a least-squares regression line?

35. Find the least-squares estimator for $b$ in the function $y = bx^3$, based on $n$ sample observations of $y$ and $x$.

36. a) Estimate, on the basis of the following data, the weight of a person who is 71 in. tall.

| Height, in. | Weight, lb |
|---|---|
| 65 | 150 |
| 70 | 170 |
| 75 | 160 |

 b) Estimate the height of a person who weighs 153 lb.
 c) Plot the regression lines you calculated for parts (a) and (b). Why do these lines differ?
 d) Find the sample correlation coefficient. Does the value of the correlation coefficient depend on which variable is dependent and which is independent in the regression equation?

37. If you used some method other than least-squares to get linear unbiased estimates of the coefficients in a regression model, discuss how your values for $s_e$ and $r^2$ would compare to those determined by means of a least-squares regression based on the same sample data.

38. Derive the normal equations for the least-squares estimate of the function $y = \alpha + \beta x^2$.

39. Discuss the following statement: "Cause-and-effect inferences can never be made from regression analysis."

40. If it is true that (under the ideal assumptions) the distribution of the least-squares estimator of a coefficient is *normal*, then why does statistical inference on such a coefficient involve a $t$-distributed test statistic rather than the standardized normal $z$?

41. In a simple regression, the estimated value of the coefficient of variable $x$ is $b = 2.5$ with $s_b = 0.8$. Find a 90% confidence interval on the true parameter $b$ if $n = 22$.

42. Given that, for a sample of 17 pairs of observations on $y$ and $x$, the total variation is 28,416, the SSR is 7,104, and the covariation of $x$ and $y$ is $-42,624$.

 a) Find $s_e$ and $r$ and explain their meaning.
 b) Test the hypothesis that there is no correlation between $y$ and $x$. Should a one- or two-sided alternate hypothesis be used? Let $\alpha = 0.10$.

43. Define or describe briefly each of the following:
 a) Standard error of the estimate;
 b) Standard error of the forecast;
 c) Standard error of the regression coefficient.

44. In a simple regression based on 32 observations, it is found that $r = 0.6$ and $s_e^2 = 100$.

a) Find SST, SSR, and SSE.
b) Do an analysis-of-variance test to determine whether the linear relationship is significant at the 0.01 level.

45. Suppose that a least-squares regression line fitted to 62 observations yields the following
equation: $\hat{y} = 16.9 + 1.225x$. SST for this sample was determined to be 594, while SSE
was found to be 540.

a) Plot the regression line for values of $x$ between zero and three.
b) If $x = 1$, what is the least-squares estimate of $y$? What is the least-squares estimate
of $y$ if $x = 3$?
c) Assume, for this example, that $\sum(x_i - \bar{x})^2 = 36$. Use this information to find the
standard error of the regression coefficient. Test the null hypothesis $H_0: \beta = 0$ against
$H_a: \beta \neq 0$, by means of a $t$-test.
d) Find the standard error of the estimate for this sample. Use this answer to construct
a 95-percent confidence interval for $y_i$ when $x = 1$, and when $x = 3$. Draw, on your
graph for part (a), a band representing a 95-percent confidence interval for $y_i$ when
$0 < x < 3$.
e) Construct a 95-percent confidence interval for the value of $\beta$.
f) From the information given in this example, determine MSR and MSE and use these
values to determine the value of $F$ necessary for testing the hypothesis of *no linear
regression*. How is this value of $F$ related to the value of $t$ you calculated in part (c)
above? Show that they both lead to rejection of the null hypothesis at the same level
of significance.
g) Determine the value of the correlation coefficient $r$.
h) Test the null hypothesis $H_0: \rho = 0$ against $H_a: \rho \neq 0$, by means of a $t$-test. Does this
value agree with the value of $t$ calculated in part (c)?
i) How much of the variation in $y$ is explained by $x$ for these data?

46. A model is estimated to be $\hat{y} = 10 + 3x$, based on observations such that $\bar{y} = 160$ and
$\bar{x} = 50$. Explain whether there is any difference in the precision of forecasts, $\hat{y}$, based on
this model, depending on whether the new value for $x$ is 40 or 100, respectively.

47. Given the following values of $x$ and $y$:

| $x$ | 1.00 | 1.44 | 1.96 | 3.24 | 4.00 | 7.84 |
|---|---|---|---|---|---|---|
| $y$ | 1.0 | 2.0 | 3.0 | 4.0 | 5.0 | 6.0 |

a) Plot these six values of $x$ and $y$ on a graph, and then make a freehand estimate of the
curve relating the two variables. Would a linear function be appropriate in this
circumstance? What type of relationship does $y$ appear to have to the values of $x$?
b) Transform the variable $x$ into a new variable by taking the square root of $x$, and then
plot this new variable against $y$. Is this relationship approximately linear?
c) Use the transformation in part (b) to establish a least-squares regression line for the
relationship between $y$ and $\sqrt{x}$. Calculate a "list of residuals."
d) Find the sample correlation coefficient for the original data and then find it for the

transformed data. Explain how the different values give a hint to the best-fitting form of model relating $y$ to $x$.

48. Determine the least-squares estimate for the following data, assuming that $\hat{y} = ax^b$ (let $\log \hat{y} = \log a + b \log x$). Plot the original data and the least-squares estimate on graph paper.

| $x$ | 1 | 2 | 3 | 4 | 5 |
|-----|-----|-----|-----|-----|-----|
| $y$ | 1.0 | 2.1 | 4.3 | 8.1 | 13.0 |

Determine the least-squares estimate for the model $y = \alpha + \beta x$ and find the residuals. By comparing $s_e$ for both forms of the estimation, determine which model specification provides the best fit.

49. The following data represent the growth pattern of a certain plant life, where $x$ is in months and $y$ is in inches.

| $x$ | 1 | 2 | 3 | 4 | 5 | 6 | 7 |
|-----|-----|-----|-----|-----|-----|-----|-----|
| $y$ | 0.80 | 1.10 | 1.70 | 2.60 | 3.80 | 5.70 | 8.50 |

a) Find the least-squares equation relating $x$ and $y$, of the form $\hat{y} = ab^x$. (Take the logarithm of both sides of this equation, letting $\log \hat{y} = \log a + x \log b$.)

b) Plot the original data and your least-squares estimate, and then use this sketch to find the error of prediction for these seven observations.

# 12

# Extensions of
# Regression Analysis

## 12.1 INTRODUCTION TO MULTIPLE REGRESSION

In Chapter 11 the method of least-squares estimation was found to yield estimates which have the desirable properties of unbiasedness, efficiency, and consistency, given a set of five standard assumptions. In this chapter, we extend that analysis in two ways; first, we discuss problems in which the linear relationship under study involves one dependent variable and *two or more* independent variables. Then, we consider the effect of violations of some of the underlying assumptions. In particular, if a certain assumption is not valid, what is the effect on the properties of the least-squares estimates and on the statistical methods of inference about the parameters in the population model? Also, we discuss some simple graphic methods for discovering whether an assumption seems to be violated. These methods and concepts are often called *residual analysis*, since they deal with the residuals, $e_i = y_i - \hat{y}_i$.

### The Extended Population Model

In most applications many factors may be related to the dependent variable, any of which could help explain its variation. Suppose we assume that there are $m$ independent variables, and that the population model relating these variables to the dependent variable $y$ is given by the following linear relationship.

*Population regression model:*

$$y_i = \alpha + \beta_1 x_{1i} + \beta_2 x_{2i} + \beta_3 x_{3i} + \cdots + \beta_m x_{mi} + \varepsilon_i.$$

As was the case for simple linear regression, the subscript $i$ on each variable represents one of the values in the population. Also, $\alpha$ equals the $y$-intercept, $\beta_1$ equals the slope of the relationship between $y$ and $x_1$, $\beta_2$ equals the slope between $y$ and $x_2$, and so forth. The fact that the relationship is linear means that the relationship between $y$ and *each one* of the independent variables can be approximated by a straight line. In other words, the conditional mean of the dependent variable is given by the following population regression model:

*Population multiple linear regression equation:*

$$\mu_{y \cdot x_1, x_2, \ldots, x_m} = \alpha + \beta_1 x_1 + \beta_2 x_2 + \cdots + \beta_m x_m. \tag{12.1}$$

The coefficients $\beta_1, \beta_2, \ldots, \beta_m$ are called the *partial regression coefficients*, since they indicate the (partial) influence of each independent variable on $y$, with the influence of all the remaining independent variables *held constant*.

For example, we know that the level of GPA for graduate-school students is related not only to a person's GRE scores, but also to a variety of other factors, such as his undergraduate grades (UGG), age, etc. Suppose we assume that there are just two independent variables, GRE($x_1$) and UGG($x_2$), and the relationship between these two variables and the mean GPA ($\mu_{y \cdot x_1, \, x_2}$) is:

$$\mu_{y \cdot x_1, \, x_2} = -1.75 + 0.005x_1 + 0.70x_2.$$

The value $\beta_1 = 0.005$ in this case indicates that, after eliminating or taking into account the influence on GPA of $x_2$ (UGG), a one-unit increase in $x_1$ (GRE) will increase the mean value of $y$ (GPA) by 0.005 units. Similarly, a one-unit increase in $x_2$ (UGG) will increase the mean GPA by 0.70 units (assuming that the influence of GRE is being held constant), since $\beta_2 = 0.70$. Figure 12.1 is a graph of the plane represented by this multiple regression equation.

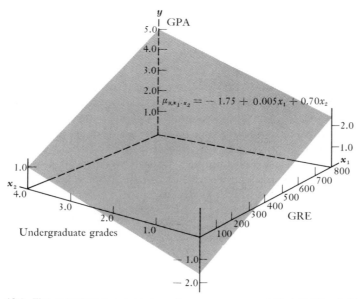

**Fig. 12.1.** The population regression plane $\mu_{y \cdot x_1, x_2} = -1.75 + 0.005x_1 + 0.70x_2$.

The process of using sample information to estimate the parameters of a multiple linear regression equation involves the same techniques used in the simple linear regression case. Suppose that we have a sample consisting of $n$ observations for each of the $m$ variables. The problem is to find the sample regression equation which provides the "best fit" to these data and to use the coefficients of that equation as estimates of the parameters of the population regression equation. For multiple regression the sample equation is:

$$\hat{y} = a + b_1x_1 + b_2x_2 + \cdots + b_mx_m. \tag{12.2}$$

The value of $\hat{y}$ is the estimate of $\mu_{y \cdot x_1, x_2, \ldots, x_m}$; $a$ is the estimate of the intercept $\alpha$; and $b_1, b_2, \ldots, b_m$ are the estimates of the partial regression coefficients $\beta_1, \beta_2, \ldots, \beta_m$. Notice that the multiple regression equation reduces to the simple regression line when $m = 1$.

## 12.2  MULTIPLE LEAST-SQUARES ESTIMATION

The least-squares estimates for multiple regression are again based on the principle of minimizing the squared error (i.e., the sum of the squares of the residuals). As before, each residual $(e_i)$ is the difference between $y_i$ and $\hat{y}_i$, where

$$\hat{y}_i = a + b_1x_1 + b_2x_2 + \cdots + b_mx_m.$$

That is, we want to minimize the function

$$G = \sum_{i=1}^{n} e_i^2 = \sum_{i=1}^{n} (y_i - a_i - b_1x_{1i} - b_2x_{2i} - \cdots - b_mx_{mi})^2.$$

The procedure for minimizing this function is the same as in the simple linear case (shown in a footnote in Section 11.3). In this case, the result is a set of $(m + 1)$ *normal equations* which, when solved simultaneously, yield the $(m + 1)$ estimates $a, b_1, \ldots, b_m$. Although solving for these estimates is not a particularly difficult task, the process usually requires tiresome arithmetic that is prone to computational errors. For this reason, computer programs, based on the techniques of matrix algebra, are generally employed to calculate the sums, sums of squares, and sums of cross products, of the sample observations, and to solve such systems of normal equations. A student who wishes to study more advanced methods and applications of multiple regression or correlation analysis in business and economics is well advised to include a course in matrix algebra in his program of study and to become familiar with some standard computer program for regression analysis. In this chapter, we will not emphasize the calculation of the least-squares estimates, but rather we will emphasize the understanding and interpretation of the results.

### A Sample Problem

To illustrate multiple regression analysis, suppose we extend the example of Section 11.7 by including a second independent variable in our analysis of the dependent variable (investment). Recall that our first variable, which we now label $x_1$, was the price index of 500 common stocks. Our second variable, which is denoted by $x_2$, is the *retained earnings of firms*. Retained earnings are the portion of profits after taxes which is not distributed to owners (stockholders), but is kept within the firm as working capital. Since these retained earnings are often the source of funds used to purchase new land, buildings, and equipment, we presume that a positive relationship exists between $x_2$ and $y$. Both current, past, and expected levels of $x_2$ may influence overall investment levels. In particular, we will attempt to relate the value of retained earnings (in billions of dollars measured at an annual rate) in one quarter with investment in the following quarter. The twenty observations of $x_2$ for the example are shown in Table 12.1. The corresponding values for $y$ and $x_1$ were given in Table 11.6 in Chapter 11.

**Table 12.1** Retained earnings of U.S. firms ($x_2$) in billions of dollars

| Observation | $x_2$ | Observation | $x_2$ | Observation | $x_2$ |
|---|---|---|---|---|---|
| 1 | 16.2 | 8 | 14.3 | 15 | 26.1 |
| 2 | 17.4 | 9 | 10.9 | 16 | 29.0 |
| 3 | 14.8 | 10 | 16.0 | 17 | 24.6 |
| 4 | 14.6 | 11 | 16.2 | 18 | 27.8 |
| 5 | 8.2 | 12 | 16.4 | 19 | 23.3 |
| 6 | 14.9 | 13 | 20.4 | 20 | 21.6 |
| 7 | 15.1 | 14 | 20.5 | | |

Since there are two variables in our model, the population equation to be estimated is

$$y = \alpha + \beta_1 x_1 + \beta_2 x_2.$$

If we solve the normal equations using these twenty observations, the following equation is the least-squares regression line:

$$\hat{y} = 1.677 + 0.07856x_1 + 1.7984x_2. \tag{12.3}$$

The unexplained variation is now

$$\text{SSE} = \sum e_i^2 = \sum (y_i - \hat{y}_i)^2 = 1264.$$

Note how these results compare with those of the analysis involving only $x_1$ and $y$, where the regression line was

$$\hat{y} = 3.855 + 0.12347x_1,$$

and the unexplained variation was

$$\sum e_i^2 = \text{SSE} = 1901.$$

Thus, as this example demonstrates, the introduction of a new variable into the sample regression model usually has several effects:

a) The coefficients of previously included variables change;

b) More of the variation of $y$ is explained;

c) The values of $t$- (or $F$-)distributed statistics change.

In this section we explore in depth the causes and the meaning of the first of these changes. Subsequent sections will deal with changes (b) and (c).

We note that the estimate for the coefficient of $x_1$ (stock index) changes from 0.12347 to 0.07856 when the variable $x_2$ (retained earnings) is included in the estimating equation for $y$ (investment). That is, a 10-point change in the stock index last quarter is now associated with an increase in investment (annually) of only about \$0.79 billion rather than \$1.23 billion. The slope between $y$ and $x_1$ has decreased by 36 percent of its previous value. Obviously, the new estimate $b_1$ must have a different meaning than the estimate $b$ in the simple model. The estimate $b_1 = 0.07856$ is obtained by considering the influence of $x_2$ on $y$, and its relationship with $x_1$, as well as the simple influence of $x_1$ on $y$. The technique of multiple regression is similar to a laboratory-controlled experiment in which one independent variable at a time is varied to examine its influence on the dependent variable, while holding all other controlled factors constant. In this case, the variables included in the model are the only ones being controlled; other factors are subsumed into the error term. The partial regression coefficient measures the influence of one variable on $y$ while holding the influence of the other variables constant.* Thus, $b_1$ measures the partial effect of changes in last quarter's stock prices on investment as if we had controlled the real world in such a way that the amount of *retained earnings* in the previous quarter was constant. The value of $b_1$ depends on the selection of the other factors included in the model. If another factor, such as interest rate, were included in addition to (or instead of) retained earnings, then the value of $b_1$ would change because the controlled environment in which the influence of $x_1$ is being measured would be different. Only if all the explanatory variables in the model are independent of $x_1$ (i.e., have zero covariance with $x_1$) would the estimate of $b_1$ remain unchanged when these variables are included.

Similarly, the value of $b_2 = 1.7984$ represents the partial influence of retained earnings on current investment, where the influence of the index of *stock prices* is held constant. Using a familiar phrase in economics, the effect of a one-billion-dollar increase in retained earnings is an increase of about \$1.8 billion in investment, *ceteris paribus* (holding other things constant, here meaning stock prices).

## 12.3 GOODNESS-OF-FIT MEASURES IN MULTIPLE ANALYSIS

As in the case of simple regression, some goodness-of-fit measures are needed to judge how well the multiple regression equation fits the observed data. Again, an absolute measure and a relative measure are common; they have an interpretation

---

* Students with a knowledge of calculus can interpret a simple regression coefficient of $x_1$ as the derivative $dy/dx_1$, whereas the partial regression coefficient of $x_1$ is the partial derivative, $\partial y/\partial x_1$, obtained by treating $x_2$ as if it were a constant.

completely analogous to those discussed in Section 11.4. Before presenting these measures, we must again present the "ideal" assumptions (about the errors $\varepsilon_i$) that are necessary for interpretating the measures.

### Assumptions for the Multiple Regression Model

Again we must emphasize that the least-squares procedure does not require *any* assumptions about the population, since this procedure is merely a curve-fitting technique. However, in order to be able to test the goodness of fit of a sample regression equation, it is once more necessary to make certain "ideal" assumptions about the error term ($\varepsilon$) in the population regression model. The first five of these assumptions are parallel to those specified in Section 11.2 for the simple regression model. We repeat them below for the multiple regression case:

*Assumption* **1.** *The error term* $\varepsilon$ *is independent of each of the m independent variables* $x_1, x_2, \ldots, x_m$.

*Assumption* **2.** *The errors for all possible sets of given values of* $x_1, x_2, \ldots, x_m$ *are normally distributed.*

*Assumption* **3.** *The expected value of the errors is zero for all possible sets of given values* $x_1, x_2, \ldots, x_m$.

*Assumption* **4.** *The variance of the errors is finite, and is the same for all possible sets of given values* $x_1, x_2, \ldots, x_m$.

*Assumption* **5.** *Any two errors* $\varepsilon_i$ *and* $\varepsilon_j$ *are independent*; *i.e., their covariance is zero.*

In addition to these five assumptions, two additional conditions are necessary to obtain least-squares estimates in the multiple regression equation.

*Assumption* **6.** *None of the independent variables is an exact linear combination of the other independent variables.*

This means that no one variable $x_i$ is an exact multiple of any other independent variable. Further, if $m \geqslant 2$, this assumption means that no one variable $x_i$ can be written as

$$x_i = a_1 x_1 + a_2 x_2 + \cdots + a_{i-1} x_{i-1} + a_{i+1} x_{i+1} + \cdots + a_m x_m,$$

where the $a$'s are constants. This assumption is a weak condition, since it requires only that the variables not be *perfectly* related to each other in a linear function. In practice, the independent variables are often only partially linearly related to each other, or sometimes perfectly related to each other in some nonlinear way. Although least-squares estimators can be calculated in these situations, problems sometimes arise in their interpretation.

*Assumption* 7. *The number of observations* ($n$) *must exceed the number of independent variables* ($m$) *by at least two* (*that is,* $n \geqslant m + 2$).

Since, in the multiple regression equation, there are $m + 1$ parameters to be estimated, the number of degrees of freedom is $n - (m + 1)$. Thus, this assumption merely specifies that there be at least one degree of freedom (that is, $n - (m + 1) \geqslant 1$). In practice, the sample size needs to be quite a bit larger than this value; otherwise, any measure of goodness of fit may be more a consequence of having only a small amount of data, rather than of having a correctly specified model.

### Multiple Standard Error of Estimate

The standard error of the estimate for the multiple regression equation is defined as

$$s_e = \sqrt{\left( \frac{\text{Unexplained variation}}{\text{Degrees of freedom}} \right)},$$

just as it is for simple regression. However, since $(m + 1)$ parameters must be estimated before a residual from the multiple regression equation can be calculated, the degrees of freedom in this statistic are $n - (m + 1)$. Thus, we have,

*Multiple standard error of estimate:*

$$s_e = \sqrt{\frac{\text{SSE}}{n - m - 1}} = \sqrt{\frac{1}{n - m - 1} \sum_{i=1}^{n} e_i^2}. \tag{12.4}$$

As before, we know that about 68 percent of all sample points should lie within one standard error of the estimated values of $y_i$; about 95 percent should lie within two standard errors. To illustrate the calculation of $s$, recall that for our multiple regression example the value of $\sum e_i^2 = 1264$, $m = 2$ (two independent variables), and $n = 20$ (see page 451). Thus,

$$s_e = \sqrt{\frac{1264}{20 - 2 - 1}} = \sqrt{\frac{1264}{17}} = \$8.62 \text{ billion}.$$

Comparing this value of $s_e$ with that obtained using the simple regression model (Section 11.7), where the standard error was \$10.3 billion, we observe that the value of $s_e$ has decreased. Since the amount of variation explained in a regression model (SSR) can never be reduced by the addition of another variable, $s_e$ will usually decrease. If a weakly related independent variable is added, however, the reduction in unexplained variation could be so small that it would not compensate for the loss of one degree of freedom due to its inclusion. In this case, $s_e$ would increase, and the inclusion of the new variable in the model would not be worthwhile.

**Multiple Coefficient of Determination**

In the multiple regression case, the relative measure of goodness of fit is designated by the symbol $R^2$, to differentiate it from the simple coefficient of determination $r^2$. This *multiple coefficient of determination*, $R^2$, is the ratio of the variation explained by the multiple regression equation (SSR) to the total variation of $y$ (SST). The only difference between $R^2$ and $r^2$ is that, in the former case, the explained variation results from $m$ independent variables rather than from the single independent variable used in the latter case. It is customary to write the multiple coefficient of determination as $R^2_{y \cdot x_1, x_2, \ldots, x_m}$, where the dependent variable is specified before the dot and the independent variables are listed after the dot.

*Multiple coefficient of determination:*

$$R^2_{y \cdot x_1, x_2, \ldots, x_m} = \frac{\text{Variation explained by all } x_i\text{'s (SSR)}}{\text{Total variation of } y \text{ (SST)}}. \tag{12.5}$$

To illustrate the calculation of $R^2$, recall that, in our investment example, the unexplained variation with $x_1$ and $x_2$ in the analysis is SSE $= 1264$. From Section 11.8 we know that the total variation in $y$ is

$$\text{SST} = \sum (y_i - \bar{y})^2 = 11,485.$$

Since the explained variation due to regression (SSR) equals the difference between the total variation and the unexplained variation, we know that

$$\text{SSR} = \sum (\hat{y}_i - \bar{y})^2 = 11,485 - 1264 = 10,221.$$

Therefore,

$$R^2_{y \cdot x_1, x_2} = \frac{10,221}{11,485} = 0.89.$$

This means that 89 percent of the variation in investment is explained by the linear relationship between investment, stock prices, and retained earnings. In comparing the value of $R^2 = 0.89$ to the value $r^2 = 0.835$ obtained from the simple model (using stock prices alone), we see that the addition of variable $x_2$ to the analysis explains an additional 5.5 percent of the variation in investment. Since the explained variation (SSR) can never decrease by adding another independent variable, $R^2$ will always either increase, or remain the same, as more variables are included in the model.

The fact that $x_1$ explains 83.5 percent of the variability in $y$, and $x_2$ only an additional 5.5 percent, does not imply that stock prices ($x_1$) are *better* predictors of investment than are retained earnings ($x_2$). If retained earnings had been the variable considered first and then controlled during the addition of stock prices to the analysis, then retained earnings would appear to explain the greater share of total variation. It alone would explain 78 percent of the variation in $y$ and the addition of $x_1$ would explain an extra 11 percent, giving the joint total explained, as before, of 89 percent.

## 12.4  MULTIPLE CORRELATION ANALYSIS

### The Coefficient of Multiple Correlation

Multiple linear correlation bears the same relationship to simple linear correlation as multiple linear regression does to simple linear regression; that is, it represents an extension of the techniques for handling the relationship between *more than two* variables. In multiple linear correlation, the objective is to estimate the *strength* of the relationship between a variable $y$ and a group of $m$ other variables $x_1$, $x_2, \ldots, x_m$. The measure usually used for this purpose is called the *coefficient of multiple correlation*, and is denoted by the symbol $R_{y \cdot x_1, x_2, \ldots, x_m}$. This measure can be interpreted in a manner similar to $r$, since a multiple linear correlation coefficient represents the simple linear correlation coefficient between the sample values of $y$ and estimates of these values provided by the multiple regression equation. However, the value of $R$ is never negative, but rather $0 \leqslant R \leqslant 1$ (this is because the sign of $R$ does *not* indicate the slope of the regression equation, since it is not possible to indicate all the signs of the regression coefficients which relate $y$ to the variables $x_1, x_2, \ldots, x_m$ by a *single* plus or minus sign). As we indicated above, the square of the multiple correlation coefficient ($R^2$) indicates the fraction of the total variation in $y$ accounted for by the regression equation.

For our investment example, the value of $R$ is the square root of $R^2 = 0.89$:

$$R_{y \cdot x_2, x_3} = \sqrt{R^2_{y \cdot x_2, x_3}} = \sqrt{0.89} = 0.943.$$

### Partial Correlation Coefficient

The value of $R$ measures the degree of association between the variable $y$ and *all* of the variables $x_1, x_2, \ldots, x_m$. One may, however, be more interested in the degree of association between $y$ and *one* of the variables $x_1, x_2, \ldots, x_m$, *with the linear effect of all the other variables removed*. A measure of the strength of the relationship between the dependent variable and one other variable, with the linear effect of the rest of the variables eliminated, is called a *partial correlation coefficient*. A partial correlation coefficient is analogous to a partial regression coefficient, in that all other factors are "held constant." Simple correlation, on the other hand, ignores the effect of all other variables, even though these variables might be quite closely related to the dependent variable, or to one another.

Partial correlation measures the strength of the relationship between $y$ and a single independent variable by considering the *relative* amount that the unexplained variation is reduced by including this variable in the regression equation. For instance, in our investment example, we might want to calculate the partial correlation between $y$ and $x_2$, where the linear effect of $x_1$ is held constant (i.e., eliminated). This partial correlation is denoted by the symbol $r_{y, x_2 \cdot x_1}$, where the variables before the dot indicate those whose correlation is being measured ($y$ and $x_2$), and the variables after the dot indicate those whose influence is being held constant ($x_1$). Thus, $r_{y, x_2 \cdot x_1}$ would be a measure of the strength of the relationship between investment ($y$) and retained earnings ($x_2$) while controlling stock prices ($x_1$) to be constant.

As before, the *square* of a correlation coefficient is usually easier to interpret than the coefficient itself. In the case of a partial correlation coefficient this square is called a *partial coefficient of determination*. The partial coefficient of determination measures the proportion of the unexplained variation in $y$ that is *additionally* explained by the variable that is *not* being held constant. Thus, the interpretation of $r_{y, x_2 \cdot x_1}^2$ is as follows:

$$r_{y, x_2 \cdot x_1}^2 = \frac{\begin{pmatrix} \text{Extra variation in } y \text{ explained} \\ \text{by the additional influence of } x_2 \end{pmatrix}}{\text{Variation in } y \text{ unexplained by } x_1 \text{ alone}}. \qquad (12.6)$$

Figure 12.2 illustrates the determination of the value of $r_{y, x_2 \cdot x_1}^2$ for our investment example. The total variation in $y$ (investment) to be explained is

$$\sum (y_i - \bar{y})^2 = 11,485,$$

represented by the area of the entire rectangle. Based on the simple relationship between $y$ and $x_1$ (stock prices) given in Chapter 11, the amount of unexplained variation when $x_i$ is the only independent variable is SSE = 1901. The denominator of Formula (12.6) is thus 1901. Now, recall from Section 12.2 that the amount

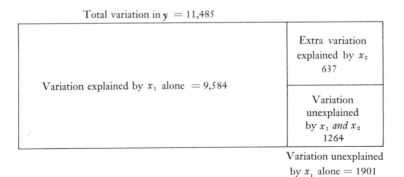

Total variation in $y$ = 11,485

| Variation explained by $x_1$ alone = 9,584 | Extra variation explained by $x_2$ 637 |
| | Variation unexplained by $x_1$ *and* $x_2$ 1264 |

Variation unexplained by $x_1$ alone = 1901

**Fig. 12.2.** The elements of variation used in a partial coefficient of determination, $r_{y.x_2 \cdot x_1}^2$.

of unexplained variation with both $x_1$ and $x_2$ in the analysis is 1264. Thus, the *extra* amount of variation explained by adding $x_2$ to the analysis is

$$1901 - 1264 = 637,$$

which is the value needed for the numerator of Formula (12.6). Thus, the proportion of previously unexplained variation that is explained by the addition of $x_2$ is:

$$r^2_{y,\,x_2 \cdot x_1} = \frac{637}{1901} = 0.335.$$

The square root of this value gives the partial correlation coefficient,

$$r_{y,\,x_2 \cdot x_1} = \sqrt{0.335} = 0.579,$$

between investment and retained earnings, holding stock prices constant.

### *12.5  TESTS FOR THE MULTIPLE ANALYSIS

A variety of test procedures involving the parameters of the multiple regression model or the multiple correlation coefficient have been developed. Not all these will be discussed here, since the complexity is better handled in a more advanced text.† The usual questions of interest in a multiple linear relationship are the overall goodness of the fit, as well as the significance of the partial regression parameters.

#### Analysis-of-Variance Test

The test of the significance of the entire multiple linear regression is equivalent to the simple linear test of the significance of the linear relationship. Hence, the same type of analysis-of-variance table and an **F**-distributed statistic can be utilized (see Table 11.7 and Formula (11.26)).

If the linear regression equation of Formula (12.2) fits the data well, then the amount of variation in $y$ that is explained (SSR) should be large relative to the amount of variation that is left unexplained (SSE). If each of these amounts of variation is divided by its degrees of freedom, then a mean square is obtained. The ratio of the mean square explained (MSR) to the mean square unexplained (MSE) has an **F**-distribution. In multiple regression $(m + 1)$ parameters are estimated on the basis of $n$ observations, so the unexplained variation will have $n - (m + 1)$ degrees of freedom. The degrees of freedom for the explained variation is equal to the number of independent variables $(m)$. Total variation always has $(n - 1)$

---

* This section assumes that the reader has studied Section 11.8 utilizing the **F**-distributed statistic for analysis of variance.
† See James L. Murphy, *Introductory Econometrics* (Homewood, Ill., Richard D. Irwin, 1973), Chapter 11.

**Table 12.2** Analysis-of-variance table for multiple regression

| Source of the variation | Sum of squares | Degrees of freedom | Mean square |
|---|---|---|---|
| Multiple regression | SSR | $m$ | $\text{SSR}/m = \text{MSR}$ |
| Residual | SSE | $n - m - 1$ | $\text{SSE}/(n - m - 1) = \text{MSE}$ |
| Total | SST | $n - 1$ | |

degrees of freedom. Table 12.2 is the analysis-of-variance table for multiple linear regression, analogous to Table 11.7 for simple regression analysis.

The appropriate statistic to test the significance of the entire multiple regression equation follows an *F*-distribution with $m$ and $(n - m - 1)$ degrees of freedom:

$$F_{(m,\ n-m-1)} = \frac{\text{SSR}/m}{\text{SSE}/(n - m - 1)} = \frac{\text{MSR}}{\text{MSE}}. \tag{12.7}$$

We illustrate this test by our investment example. Remember (from Section 12.4) that

$$\text{SST} = 11{,}485, \qquad \text{SSR} = 10{,}221, \qquad \text{and} \qquad \text{SSE} = 1264.$$

In this example, $n = 20$ and $m = 2$. From the analysis-of-variance table (Table 12.4), we find

$$F = \frac{\text{MSR}}{\text{MSE}} = \frac{5110.50}{74.35} = 68.75.$$

This value far exceeds the critical value for $\alpha = 0.01$ with 2 and 17 d.f., which is

$$F_{(0.01;\ 2,\ 17)} = 6.11$$

(from Table VIII(b)); so we conclude that the null hypothesis of no linear relationship can be rejected. Thus, the variables *stock prices* and *retained earnings* do appear to be related to investment.

**Table 12.3** Analysis-of-variance for the multiple regression of investment on stock prices and retained earnings

| Source | SS | d.f. | MS |
|---|---|---|---|
| Regression | $\text{SSR} = 10{,}221$ | $m = 2$ | $\text{MSR} = 5110.50$ |
| Error | $\text{SSE} = 1{,}264$ | $n - m - 1 = 17$ | $\text{MSE} = s_e^2 = 74.35$ |
| Total | $\text{SST} = 11{,}485$ | $n - 1 = 19$ | |

## Tests on a Particular Parameter

To determine the significance of an individual coefficient ($\beta_i$) in the regression model, a test similar to that for the slope in the simple regression equation is used. The null hypothesis, $H_0$: $\beta_i = 0$ means that the variable $x_i$ has no significant linear relationship with $y$, *holding the effect of the other independent variables constant*. The best linear unbiased estimate of $\beta_i$ is the sample partial regression coefficient $b_i$. Under the assumption that the unknown disturbances are normally distributed, then the test for this null hypothesis follows the $t$-distribution with $(n - m - 1)$ degrees of freedom:

$$t_{(n-m-1)} = \frac{b_i - 0}{s_{b_i}}. \tag{12.8}$$

Here $s_{b_i}$ is the estimated standard error of the estimate $b_i$. Calculation of $s_{b_i}$ is quite tedious, but it is always shown in the output of a regression analysis program for a computer. Thus, the determination of $t$ in a practical application is done simply by forming the ratio of the coefficient to its estimated standard error. When the calculated value of $t$ exceeds the critical value $t_{(\alpha; \, n-m-1)}$ determined from Table VII, then the null hypothesis of no significance can be rejected. It is then concluded that the variable $x_i$ does have an important influence on the dependent variable $y$, even after accounting for the influence of all other independent variables included in the model.

For our example, the estimated standard errors of the coefficients $b_1$ and $b_2$ of the variables $x_1$ (stock prices) and $x_2$ (retained earnings) are

$$s_{b_1} = 0.0188 \qquad \text{and} \qquad s_{b_2} = 0.614,$$

respectively. Since $n = 20$ and $m = 2$ in this case, the critical value for a one-sided test on either coefficient (using a significance level of $\alpha = 0.01$) is

$$t_{(\alpha; \, n-m-1)} = t_{(0.01; \, 17)} = 2.567.$$

Thus, the critical region for a one-sided test when $H_0$: $\beta_1 = 0$   (or $H_0$: $\beta_2 = 0$) is *all values of $t$ that exceed* 2.567. We choose a one-sided test because our *a priori* theoretical propositions were that both $x_1$ and $x_2$ were positively related to $y$.

For the test on $\beta_1$, the value of $b_1$ was 0.07856; hence,

$$t = \frac{b_1}{s_{b_1}} = \frac{0.07856}{0.0188} = 4.179.$$

For the test on the significance of $\beta_2$, the value of $b_2$ was 1.7984, so

$$t = \frac{b_2}{s_{b_2}} = \frac{1.7984}{0.614} = 2.929.$$

We conclude that both variables, $x_1 = $ stock prices and $x_2 = $ retained earnings, are significantly related to $y$ = investment. The variable $x_1$ is the more influential of the two, since it has the higher $t$ value.

Now that we have found that both regression coefficients provide a good fit for the data, we can proceed to the next logical task, that of determining the best point forecast based on the previous quarterly observations of stock prices and retained earnings. In doing so, we must remember that even if the regression equation has been shown to fit well, and has all very significant coefficients, such results may not hold for future data. The relationship in a future quarter may differ due to some change in the social, political, or economic environment. However, let's be courageous and assume that the past relationship (given in Formula (12.3)) will hold for the next quarter also. If the given values of the independent variables are $x_1$ = 950 and $x_2$ = 25, then the forecasted value is:

$$\hat{y}_g = a + b_1 x_1 + b_2 x_2$$
$$= 1.677 + 0.07856(950) + 1.7984(25)$$
$$= 121.27.$$

The estimated level of investment for the next quarter is \$121.27 billion.

It is important to recognize that we have presented only part of the analysis possible in our investment model. We could have conducted tests on $y_i$ or on $\mu_{y \cdot x_1 x_2}$; or we might have added additional independent variables, and then performed the same $F$-test (for the overall relationship) and $t$-test (for the individual coefficients) for the new data.

## 12.6 MULTICOLLINEARITY

We now return to a consideration of special problems which arise when one of the assumptions specified in Section 12.3 is violated. This section deals with the violation or near violation of Assumption 6, which specifies that none of the independent variables can be an exact linear combination of the other independent variables. If the independent variables, $x_1, x_2, \ldots, x_m$ are perfectly linearly related to each other, then they are linearly *dependent*. In this case, no estimates of the partial regression coefficients can be obtained, since the normal equations will not be solvable; that is, the method of least squares breaks down and no estimates can be calculated. Perfect dependency seldom occurs in practice, because most investigators are careful not to include in the regression model two or more explanatory variables which represent the same influence on the dependent variable $y$. Indeed, even if an investigator did accidentally include two or more such variables, it is unlikely that the *sample* observations representing measures of these variables would be perfectly related, because some slight errors of measurement and sampling are almost inevitable.

Special problems do sometimes occur, however, when two or more of the independent variables are strongly (but not perfectly) related to one another. This situation is known as *multicollinearity*. When multicollinearity occurs it is possible to calculate least-squares estimates, but the difficulty arises in the interpretation of the strength of the effect of each variable.

To illustrate multicollinearity, recall, in our investment example, that the variable $x_2$ (retained earnings) was correlated with stock price index ($x_1$). Using $x_1$ alone, 83.4% of the variation in $y$ is explained; using $x_2$ alone, 77.7% of the variation in $y$ is explained. However, using both $x_1$ and $x_2$, the combined explained variation is 89% of the total variation. Thus, there is considerable *overlap* in the explanatory roles of the variables $x_1$ and $x_2$, probably since both react to other economic, political, and social factors within the society. The problem of precisely distinguishing the separate influences of the two variables is the problem of multicollinearity. In this example we realize that there is some collinearity because of the large correlation between $x_1$ and $x_2$ (the simple correlation between the two independent variables is $r_{x_1 x_2} = 0.8157$).

### Detection of a Multicollinearity Problem

From the discussion above we see that a high correlation between any pair of explanatory variables $x_i$ and $x_j$ may be used to help identify multicollinearity. It is possible, however, for all independent variables to have relatively small *mutual* correlations and yet to have some multicollinearity among three or more of them. Sometimes it is possible to detect these higher-order associations by using a multiple correlation coefficient dealing only with the explanatory variables. Suppose we use the symbol $R_j$ to denote the multiple correlation coefficient of variable $x_j$ with all the other ($m - 1$) independent variables, $x_1, x_2, \ldots, x_{j-1}, x_{j+1}, \ldots, x_m$. Such a measure could be determined for each of the independent variables. Generally, if one or more of these values, $R_1, R_2, \ldots, R_j, \ldots, R_m$, is approximately the same size as the multiple correlation coefficient $R_{y \cdot x_1 \cdots x_m}$, then multicollinearity is a problem. In other words, if the strength of the association among any of the independent variables is approximately as great as the strength of their combined linear association with the dependent variable, then the amount of overlapping influence may be substantial enough to make the interpretation of the separate influences difficult and imprecise.

To illustrate, suppose a model has four independent variables,

$$y = \alpha + \beta_1 x_1 + \beta_2 x_2 + \beta_3 x_3 + \beta_4 x_4,$$

and the multiple correlation coefficient for this model is

$$R_{y \cdot x_1 x_2 x_3 x_4} = 0.90.$$

To check for multicollinearity, one would first calculate the six simple correlations between pairs of independent variables

$$r_{x_1x_2}, \quad r_{x_1x_3}, \quad r_{x_1x_4}, \quad r_{x_2x_3}, \quad r_{x_2x_4}, \quad r_{x_3x_4}.$$

If one of these is close to unity, then imprecise estimation will result. The next step would be to calculate the multiple correlation coefficients of each independent variable with the other three: that is,

$$R_{x_1 \cdot x_2x_3x_4}, \quad R_{x_2 \cdot x_1x_3x_4}, \quad R_{x_3 \cdot x_1x_2x_4}, \quad \text{and} \quad R_{x_4 \cdot x_1x_2x_3}.$$

If any of these are as large as

$$R_{y \cdot x_1x_2x_3x_4} = 0.90,$$

then the problem of multicollinearity may be substantial. There is really no statistical method for testing whether these values indicate high multicollinearity or not, since this is not a problem of statistical inference about the population, but merely a property of the sample observations.

**Effects of Multicollinearity**

When multicollinearity occurs, the least-squares estimates are still unbiased and efficient. The problem is that the estimated standard error of the coefficient (say, $s_{b_i}$ for the coefficient $b_i$) tends to be inflated. This standard error tends to be larger than it would be in the absence of multicollinearity because the estimates are very sensitive to any changes in the sample observations or in the model specification. In other words, including or excluding a particular variable or certain data points may greatly change the estimated partial coefficient. When $s_{b_i}$ is larger than it should be, then the $t$-value for testing the significance of $\beta_i$ is smaller than it should be (Formula (12.8)). Thus, one is likely to conclude that a variable $x_i$ is not important in the relationship when it really is.

If one is interested primarily in the forecasts of $y_i$, or $\mu_{y \cdot x_1, x_2, \ldots, x_m}$, rather than in the significance of the separate coefficients $b_1, b_2, \ldots, b_m$, then multicollinearity may not be a problem. Suppose the combined fit for the regression equation is very good. If the observed linear relationships among all the independent variables can be expected to remain true for some new observations, then the regression model should also give a close fit for the new sample values even if multicollinearity is present.

**Correction of Multicollinearity**

When multicollinearity in a regression model is severe and more precise estimates of the coefficients are desired, one common procedure is to select the independent variable *most seriously involved* in the multicollinearity and remove it from the model. The difficulty with this approach is that the model may now not correctly represent the population relationship and all estimated coefficients would contain a specification *bias*. It would be better to try to replace the multicollinear variable with another that is less collinear but may still measure the same theoretical construct. For example, if the theoretical variable is "business expectations" and it was measured by a stock price index which is highly collinear with retained earnings, then it may be possible to replace the stock index with some other measure, perhaps an index of business expectations obtained by surveying executives in the 500 largest corporations. In this way, the multicollinearity may be reduced while still retaining the theoretical base for the model.

## 12.7 HETEROSCEDASTICITY AND AUTOCORRELATION

Recall that Assumption 4 states that each $\varepsilon_i$ has the same variance ($V[\varepsilon_i] = \sigma_\varepsilon^2$), and Assumption 5 states that the covariance between any two disturbance variables $\varepsilon_k$ and $\varepsilon_j$ is zero,

$$C[\varepsilon_k, \varepsilon_j] = 0.$$

We mentioned that these two assumptions are crucial in obtaining simple least-squares estimates of the regression coefficients which are *efficient*. This means that these estimators have a smaller variance than any other linear unbiased estimator that might be devised. If one or both of these assumptions is violated, then the estimator calculated by the method of least-squares would not have the smallest variance; some *other* estimator which uses more information would be the efficient one. This loss in efficiency occurs whenever either one of two problems is encountered: *heteroscedasticity* or *autocorrelation*. These terms are defined below:

*Heteroscedasticity occurs when Assumption 4 is violated.*
It means that the variance of the disturbances $\varepsilon_i$ is not constant, but changing.

*Autocorrelation occurs when Assumption 5 is violated.*
It means that there is a correlation between the error terms.

The effect of either of these problems is a least-squares estimate of the regression coefficient for which the standard error of the coefficient is not minimized. Thus, tests of hypothesis or confidence intervals based on this property will not be correct.

### Detecting these Violations

Some statistical tests have been developed in which the residuals $[e_i = y_i - \hat{y}_i]$ of the least-squares equation can be used to infer whether or not Assumptions 4 or 5 are violated. A description of such tests is not presented here, but some graphical methods of residual analysis are discussed.*

---

* Texts on econometrics generally have rather complete discussions of these tests, such as the Durbin–Watson test, Theil–Nagar test, Goldfeld–Quandt test, and Tukey test.

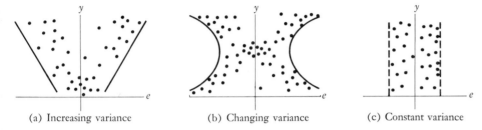

(a) Increasing variance        (b) Changing variance        (c) Constant variance

**Fig. 12.3.** Plotting residuals against *y* to detect heteroscedasticity.

To help detect a situation where the variance of the errors is not constant, it is often useful to plot each $y$ value against its corresponding residual ($y_i - \hat{y}$). For example, the V-shaped slope of the boundary lines for the scatter of points in Fig. 12.3(a) indicates an increasing variance of the residuals as the value of $y$ increases. Such a plot is an indication that the fit of the model is not uniform and that the disturbances may not have a constant variance. A changing variance would also be indicated if the boundary lines approximated an inverted V or if they were close together at some points and wider apart at others, as, for example, in Fig. 12.3(b). Assumption 4 of constant variance would *not* seem to be violated if the boundary lines are approximately parallel, as in Fig. 12.3(c).

For our example, the values of $y$ and the values of the residuals based on the estimating equation in Formula (12.3) are given in Table 12.4. A plot of these values is shown in Fig. 12.4. There seems to be a fairly even spread of errors, except for observation number 20, as indicated by the dotted boundary line. Despite this one sample point, we can probably conclude that the assumption of constant variance is not seriously violated in this example.

**Table 12.4** Values of investment (*y*) and the residuals (*e*) based on Formula (12.3)

| Observation | $y$ | Residual $e$ | Observation | $y$ | Residual $e$ |
|---|---|---|---|---|---|
| 1 | 62.3 | 0.191 | 11 | 84.3 | 7.783 |
| 2 | 71.3 | 2.775 | 12 | 85.1 | −1.621 |
| 3 | 70.3 | 1.957 | 13 | 90.8 | −8.574 |
| 4 | 68.5 | 2.433 | 14 | 97.9 | −9.408 |
| 5 | 57.3 | 5.862 | 15 | 108.7 | −8.530 |
| 6 | 68.8 | −2.080 | 16 | 122.4 | −5.568 |
| 7 | 72.2 | −8.702 | 17 | 114.0 | 2.830 |
| 8 | 76.0 | −0.101 | 18 | 123.0 | 0.034 |
| 9 | 64.3 | −6.645 | 19 | 126.2 | 11.500 |
| 10 | 77.9 | −7.780 | 20 | 137.0 | 23.621 |

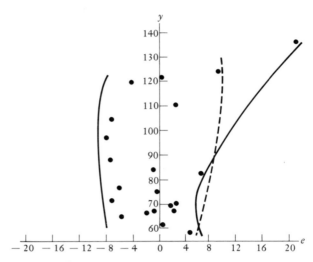

**Fig. 12.4** Plot of *y* against *e* using values from Table 12.6.

Frequently, the assumption of constant variance is not seriously violated when using economic or business data *measured over time* unless some significant structural change occurred to affect the observations, such as a new law, a war, a revolution, or some natural disaster. More often, the problem of heteroscedasticity arises when cross-sectional data, *at a given point in time, is used*, such as employment or production data across firms, or tax and revenue data across states. In these cases, the disturbances may not have constant variances because of differing factors related to the size or the legal code of the different cross-sectional entities. For example, large corporations have different structures and operate under different tax laws from those that concern small business firms. Thus, one would expect a specified model to better represent one of these types over the other. The variance of disturbances for the one type that it fits best will be smaller than the variance of disturbances for observations of the other type.

Violations of Assumption 5, on the other hand, tend to occur most frequently when the observations for the variables are taken at periodic intervals *over time*. If some underlying factors not specified in the model exert an influence on the fit of the model over several time periods, then the disturbances tend to be correlated to each other. Consider a change in corporate tax laws that might affect both the amount of investment (due to investment tax credits or depreciation write-offs) and the amount of retained earnings (due to taxes on profits). This legal factor may not be represented in the model, but its effect may be seen on the errors of the regression equation. For example, the average relationship estimated may give values too high before the tax-law change and too low afterwards. The residuals would then tend to all be negative for observations in time before the tax change and all be positive for observations occurring after the change. Thus, they would not be occurring at random but would systematically be related to each other, i.e., autocorrelated.

The most frequently considered form of autocorrelation is the linear association of successive residuals. If we denote a residual at time period $t$ by $e_t$ and the previous residual by $e_{t-1}$, then first-order autocorrelation refers to the simple linear correlation of $e_t$ with $e_{t-1}$ over the entire set of observations, $t = 2, 3, 4,$ $\ldots, n$. A measure of this correlation is given by the correlation coefficient between these two variables, $r_{e_t e_{t-1}}$. A geometric representation can be obtained by a scatter diagram of the points corresponding to each pair $(e_t, e_{t-1})$. Figure 12.5 illustrates three cases: positive autocorrelation (a); negative autocorrelation (b); and no autocorrelation (c). When the points are predominantly in the positive quadrants (a), this means that successive residuals tend to have the same sign. If the points lie in the negative quadrants (b), then successive residuals tend to have opposite signs. If the scatter of points is spread over all quadrants (c), successive residuals tend to be independent, or not correlated.

If we use the residuals from our investment example (Table 12.4), the plot of $e_t$ against $e_{t-1}$ would tend to look more like Fig. 12.5(a), indicating positive autocorrelation. The reader can make such a sketch and note that the signs of the residuals tend to be grouped over time, first positive and then mostly negative until the final four residuals, which are all positive. Thus, the least-squares estimates for this model, as given in Formula (12.3), are probably not efficient due to the problem of autocorrelation.

### Improving the Estimates

When either Assumption 4 or 5 is violated and the least-squares estimates are not efficient, an improved estimator can be determined that is efficient. This improved

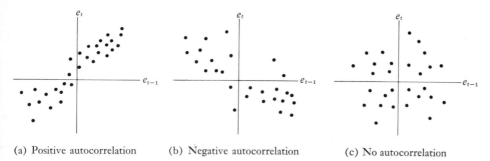

(a) Positive autocorrelation        (b) Negative autocorrelation        (c) No autocorrelation

**Fig. 12.5.** Plots of successive residuals.

estimator, called a *generalized* least-squares estimator, involves some complex calculations; hence, the formulas are not given here.* The new complexity arises from the need to incorporate into the estimates the extra information about the changing variances or the autocorrelation of the disturbances. In other words, if something is known about the way the variance increases or decreases (as in Fig. 12.3(a) or (b)), then this information should be used to weight the observations before determining the least-squares estimators. In general, those observations for which the variance of the errors is smallest should be the more reliable, and therefore should be weighted most heavily. Similarly, if some pattern of autocorrelation among the errors is known, then this information should be used in finding better estimates.

## 12.8 NONPARAMETRIC CORRELATION MEASURES

All the correlation measures presented thus far have been based on measurements of variables for which a mean, variance, and covariance can be determined. Sometimes, it is desirable to be able to calculate correlations among variables where such measures are not obtainable. In general, the determination of means, variances, and covariances requires *interval measurement,* which means that the difference (i.e., interval) between any two observations is meaningful. Frequently, data may be observed which has only *ordinal measurement.* In such a case, only the *relative* ranking of any two observations has meaning. For example, a stock analyst may list a ranking of the ten best common stocks for purchase by investors interested primarily in safety and income. The rank of any two stocks in such a list has meaning, but not the difference between ranks. For example, we know that stock 4 is ranked better than stock 7, and stock 2 is ranked higher than stock 5. However, we do *not* know that the difference in desirability between stocks 4 and 7 is identical to the difference between stocks 2 and 5, even though this difference is 3 rank positions in both instances. Now, if one had two such rankings of stocks (by different analysts), a measure of correlation between the two rank orderings might be desired. In such a situation, the previous correlation measures are not appropriate because of the ordinal nature of the data. Fortunately there are other measures of correlation, called nonparametric measures, which can be used.

### Spearman's Rank Correlation Coefficient

Research published by C. Spearman in 1904 led to development of what is perhaps the most widely used nonparametric measure of correlation. This measure, usually denoted either by the letter $r_s$, or by the Greek word rho, has thus become known

---

* They are most easily presented in terms of matrix and vector notation. See James L. Murphy, *Introductory Econometrics*, Chapter 13, for a more complete discussion of the methods of regaining efficient estimators.

as *Spearman's rho.* Spearman's rank correlation coefficient $r_S$ is very similar to the ordinary correlation coefficient we have studied thus far, except that now ranks are used as the data. A perfect positive correlation ($r_S = +1$) means that the two samples rank each object identically, while a perfect negative correlation ($r_S = -1$) means that the ranks of the two samples have an exactly *inverse* relationship. Values of $r_S$ between $-1$ and $+1$ denote less than perfect correlation. To measure correlation by Spearman's method, we first take the difference between the rank of an object in one sample and its rank in the second sample, and we then square this difference. If this squared difference is denoted as $d_i^2$ for the $i$th pair of observations, then the sum of these squared differences over a set of $n$ pairs of observations is

$$\sum_{i=1}^{n} d_i^2 .$$

The value of $r_S$ is derived from $\sum d_i^2$ as follows:

$$\text{Spearman's rank correlation coefficient:} \quad r_S = 1 - \frac{6 \sum_{i=1}^{n} d_i^2}{n^3 - n}. \quad (12.9)$$

As an example of the use of Formula (12.9), suppose that we calculate the rank correlation between the nation's top five football teams for a given year, as reflected in polls by the Associated Press (AP) and the United Press International (UPI). Assume that the teams in the top five are identical (teams A, B, C, D, and E), but they occur in a different order, as given in Table 12.5. We see in Table 12.5 that $\sum d_i^2 = 10$. Thus, the rank correlation between the AP and UPI ratings is

$$r_S = 1 - \frac{6(10)}{5^2 - 5} = 1 - \frac{60}{120} = 0.50.$$

**Table 12.5** Spearman's rho for rank orderings

| Team | AP Rank | UPI Rank | $d_i$ | $d_i^2$ |
|------|---------|----------|-------|---------|
| A | 1 | 1 | 0 | 0 |
| B | 2 | 4 | $-2$ | 4 |
| C | 3 | 2 | 1 | 1 |
| D | 4 | 5 | $-1$ | 1 |
| E | 5 | 3 | 2 | 4 |
| | $n = 5$ | | | Sum 10 |

### Kendall's Correlation Coefficient

An alternative method for determining a rank correlation coefficient is to calculate Kendall's correlation coefficient. This statistic, developed by the statistician M.G. Kendall, is denoted by the Greek letter $\tau$ (tau), and called Kendall's tau. Although Kendall's tau is suitable for determining the rank correlation of the same type of data for which Spearman's rho is useful, the two methods use different techniques for determining this correlation, so their values will not normally be the same. Spearman's rho is perhaps more widely used, but Kendall's tau has the advantage of being generalizable to a partial correlation coefficient.

The rank correlation coefficient $\tau$ is determined by first calculating an index which indicates how the ranks of one set of observations, *taken two at a time*, differ from the ranks of the other set of observations. The easiest way to determine the value of this index is to arrange the two sets of rankings so that one of them, say the first sample, is in ascending order, from the lowest score (rank) to the highest score (rank). The other set, representing the second sample, will not be in ascending order unless the ranks of the two samples agree perfectly. Now, consider all possible combinations of the $n$ ranks in this second sample, taken two at a time (i.e., all pairs); assign a value of $+1$ to each pair in which the two ranks are in the same (ascending) *order* as they are in the first sample, and assign $-1$ to each pair in which the two ranks are *not* in the same order as they are in the first sample. The sum of these $+1$ and $-1$ values is an indication of how well the second set of rankings agrees with the first set. Since there are $\binom{n}{2}$ combinations of $n$ objects taken two at a time, this sum (or index) can assume any value between $+\binom{n}{2}$ and $-\binom{n}{2}$. Kendall's tau is defined as the ratio of the computed value of this index to the maximum value it can assume (which is $\binom{n}{2}$).

---

*Kendall's rank correlation coefficient:*

$$\tau = \frac{\text{Computed index}}{\text{Maximum index}}.$$    (12.10)

---

Note that when there is perfect positive correlation, $\tau$ will equal $+1$, since the computed index and the maximum index will both equal $\binom{n}{2}$; if there is a perfect negative correlation, the computed index will equal $-\binom{n}{2}$ and $\tau$ will equal $-1$.

The value of Kendall's tau can be determined for the set of data concerned with the relationship between UPI and AP football rankings. Let's use the AP ranking as the base for comparison, since these ranks are already in ascending order. Now the UPI rankings in Table 12.5 must be compared, two at a time, in order to determine the number of pairs in the correct order ($+1$) and the number in the incorrect order ($-1$). First we compare Team A with each of the other four teams. Since Team A is ranked ahead of B in both polls, we score a $+1$ for the A–B comparison. Similarly, both polls rank A ahead of teams C, D, and E. Hence, we score a $+1$ for each of the comparisons A–C, A–D, and A–E. Now we need to compare Team B with the other three teams (C, D, E). We see that B is ahead of C in the AP ranking, but the ordering is reversed in the UPI ranking; hence, the B–C comparison is given a $-1$ score. All such paired comparisons for this example are shown in Table 12.6.

**Table 12.6** Kendall's tau for rankings in Table 12.5

| Pair | Value | Pair | Value | Pair | Value |
|------|-------|------|-------|------|-------|
| A vs B | +1 | B vs C | −1 | C vs E | +1 |
| A vs C | +1 | B vs D | +1 | D vs E | −1 |
| A vs D | +1 | B vs E | −1 | | |
| A vs E | +1 | C vs D | +1 | | |

Sum of values = 4

The maximum index in this example is the number of paired comparisons, which is

$$\binom{n}{2} = \binom{5}{2} = 10.$$

The computed index is the number of $+1$ scores (7), minus the number of $-1$ scores (3), which is

$$7 - 3 = 4.$$

Thus, Kendall's tau is

$$\tau = \frac{4}{10} = 0.40.$$

Note that Kendall's tau (0.40) is less than the comparable value of Spearman's rho (0.50). Both coefficients, however, utilize the same amount of information about the association between two variables, and for a given set of observations both will reject the null hypothesis that two variables are unrelated in the population at the same level of significance. For small samples, tables are available for determining the probability of a given value of $\tau$ or $r_S$ under this null hypothesis. For large samples, methods of statistical inference involving $r_S$ and $\tau$ can be constructed utilizing the $t$-distribution and the normal distribution respectively.*

---

* See any text on nonparametric procedures, such as S. Siegel, *Nonparametric Statistics* (New York: McGraw-Hill Book Co., 1956).

## REVIEW PROBLEMS

1. Discuss the usefulness and value of the extension of regression analysis to include more than one explanatory factor.

2. Explain the difference in meaning between the simple regression coefficient, and a partial regression coefficient in a multiple regression analysis.

3. Explain in what way, if any, the assumptions for multiple regression analysis differ from the underlying assumptions for simple regression analysis.

4. In a multiple regression analysis of changes in annual average U.S. interest rates ($y$) on three explanatory variables ($x_1$, $x_2$, and $x_3$), the following results are found:

$$\sum (y - \bar{y})^2 = 600, \qquad \sum e^2 = 150,$$

and the variation explained by $x_1$ and $x_2$ is 350.

   a) Find the multiple coefficient of determination, and explain its meaning.
   b) Find $r^2_{yx_3 \cdot x_1x_2}$, and interpret its meaning.

5. Suppose the variation in $y$ is 500 units and the model

$$\hat{y} = a + b_1x_1 + b_2x_2$$

   leaves 240 units unexplained (based on 15 observations). Extending the model to include variable $x_3$ explains 80 more units of variation in $y$. Find $R^2_{y \cdot x_1x_2x_3}$ and $r^2_{y \cdot x_3 \cdot x_1x_3}$.

6. Suppose that, in a multiple regression of $y$ on variables $x_1$, $x_2$, and $x_3$, we obtain SST $= 1000$, SSE $= 200$, and the variation explained by only $x_2$ and $x_3$ is 400. Find $R^2_{y \cdot x_1x_2x_3}$ and $r^2_{yx_1 \cdot x_2x_3}$.

7. Discuss whether each of the following statements is true or false.

   a) If $s_e = s_y$, then $b = 0$.
   b) If $R_{y \cdot x_1x_2x_3} = 1$, then $r_{yx_1 \cdot x_2x_3} = 0$.
   c) If $R_{y \cdot x_1x_2x_3} = R_{y \cdot x_1x_2}$, then $r_{yx_3 \cdot x_1x_2} = 0$.
   d) $R^2_{y \cdot x_1x_2x_3} \geqslant R^2_{y \cdot x_1x_2}$.
   e) $r^2_{yx_1} + r^2_{yx_2} = R^2_{y \cdot x_1x_2}$.

8. In a multiple regression of $y$ on the variables $x$, $z$, and $w$, the total variation is 200, the residual variation is 20, and the variation explained by only variables $z$ and $w$ is 120 units.

   a) Find $R^2_{y \cdot xzw}$.
   b) Find $r^2_{yx \cdot zw}$.

9. Given a multiple regression model, $\hat{y} = a + b_1x_1 + b_2x_2 + b_3x_3$ based on 24 observations on each variable. Suppose the total variation, SST, is $= 300$, the unexplained variation is 60, and the amount of variation explained by variables $x_1$ and $x_2$ together is 160.

   a) Calculate the value of the multiple coefficient of determination and interpret its meaning.
   b) Prepare a diagram similar to Fig. 12.2 to explain the meaning and value of $r^2_{yx_3 \cdot x_1x_2}$.
   c) Complete an analysis-of-variance table to make a test on the significance of the linear relation.

10. In a multiple regression analysis:

   a) What measures are used to determine whether the equation fits the data well and may be useful for forecasting?

   b) How can you determine which of the explanatory factors included in the model has the most significance in explaining the variation of the dependent variable $y$?

11. Given the following results from a multiple regression analysis, using 53 observations (standard errors are in parenthesis).

$$\hat{y} = 6 + 3x_1 + 10x_2 - 4x_3.$$
$$\quad\;\; (1.5) \quad (2) \quad\;\; (4) \quad\;\; (0.8)$$

   a) Calculate the values of the $t$-statistic for making one-sided tests of the significance of each individual estimate of the regression coefficients, and find the critical value for such tests if alpha $= 0.05$.

   b) Explain which independent variable is most important and which is least important, in explaining the variation in the dependent variable.

   d) Suppose $x_1$, $x_2$, and $x_3$ are policy variables which can be manipulated. If $x_1$ and $x_2$ are each increased by 20 units while $x_3$ is increased by 50 units, what is your best estimate of the change in $y$?

12. Given the following information on a linear multiple-regression model:

   $y$ = average yield of corn per acre on an Iowa farm in bushels;
   $x_2$ = amount of summer rainfall, District 3 weather station, Iowa;
   $x_3$ = average daily use of tractors on the farm in machine hours;
   $x_4$ = amount of fertilizer, type XS80, used per acre.

   The sample includes observations for ten crop years.

   *Results:*

$$\hat{y} = 16 + 75x_1 + 6x_2 + 48x_3 \qquad \text{Regression equation}$$
$$\quad\;\; (10) \quad (25) \quad (4) \quad\;\; (8) \qquad \text{Standard errors of}$$
$$\text{regression coefficients}$$

   $n = 10$, $\quad s_e = 20$ bushels, $\quad s_y = 40$ bushels, $\quad r^2_{yx_1 \cdot x_2x_3} = 0.60.$

   Answer the following questions:

   a) What are the degrees of freedom for $t$-distributed test statistics concerned with this regression?

   b) Explain which variable appears to be most important in explaining the variation of yield.

   c) From the regression results, is it proper to argue that more machine hours of use of tractors causes more yield, or that more yield requires more machine hours use of tractors? Explain.

   d) Find the coefficient of multiple correlation, $R^2_{y \cdot x_1x_2x_3}$.

e) Account for the different values of

$$R^2_{y \cdot x_1 x_2 x_3} \quad \text{and} \quad r^2_{yx_1 \cdot x_2 x_3}$$

by explaining the different meanings of the two coefficients.

13. Using 12 observations, a model $y = \alpha + \beta_1 x_1 + \beta_2 x_2 + \beta_3 x_3 + \beta_4 x_4$ is estimated by the method of least squares. Here SST = 400, SSE = 170, and the amount of variation explained jointly by $x_1$, $x_2$, and $x_3$ is 200.

a) Find $r^2_{yx_4 \cdot x_1 x_2 x_3}$ and explain what it means.
b) Do an ANOVA test with $\alpha = 0.05$ to determine whether this linear relation is significant.

14. In estimating a multiple-regression model, what is the significance of obtaining a high value of $R^2$, but small values of the $t$-statistics associated with each partial-regression coefficient?

15. Explain the meaning of multicollinearity, and specify one of its effects that you think is important.

16. Suppose the values of $y$ are related to both variables $x$ and $z$. The observations for $x$, $y$, and $z$ are:

| $y$ | $x$ | $z$ |
|-----|------|-----|
| 1.0 | 1.00 | 5.0 |
| 2.0 | 1.44 | 3.5 |
| 3.0 | 1.96 | 3.0 |
| 4.0 | 3.24 | 4.0 |
| 5.0 | 4.00 | 1.0 |
| 6.0 | 7.84 | 2.0 |

a) Plot the relationship between $y$ and $x$, and between $y$ and $z$. Is the relationship approximately linear in both cases? If not, what would be the problem in fitting an equation of the form $\hat{y} = a + bx + cz$?
b) Find the simple correlation coefficient of variables $x$ and $z$. Do you think multicollinearity might be a problem in using $x$ and $z$ as copredictors of $y$?

17. The following ten observations represent the price movement of a certain common stock over a ten-year period.

| $x$ (Year) | $y$ (Price) | $x$ (Year) | $y$ (Price) |
|------------|-------------|------------|-------------|
| 1 | 100 | 6 | 35 |
| 2 | 120 | 7 | 60 |
| 3 | 75 | 8 | 75 |
| 4 | 50 | 9 | 80 |
| 5 | 40 | 10 | 70 |

a) Sketch the relationship between $x$ and $y$. What type of function does this relationship seem to follow for the given ten years?

b) Use the method of least squares to fit the equation $\hat{y} = a + bx$.

c) Determine the residuals for these ten observations, and check to see whether autocorrelation may be a problem in this estimation.

18. The five major colleges in a certain state are rated by two Councils of Education on the national eminence of their respective faculties. In a national listing, these five schools are placed as follows:

| School | First ranking | Second ranking |
|--------|---------------|----------------|
| 1 | 25 | 15 |
| 2 | 39 | 22 |
| 3 | 21 | 30 |
| 4 | 48 | 12 |
| 5 | 8 | 10 |

Find and interpret Spearman's rho and Kendall's tau.

19. A group of ten students are scored in their ability to perform two sets of tasks emphasizing strength and coordination, respectively. Using the data below, compute and compare the values of Spearman's rho and Kendall's tau. Interpret your results.

| Student | Task 1 | Task 2 |
|---------|--------|--------|
| 1 | 25 | 38 |
| 2 | 30 | 36 |
| 3 | 42 | 50 |
| 4 | 44 | 45 |
| 5 | 58 | 30 |
| 6 | 59 | 78 |
| 7 | 75 | 77 |
| 8 | 79 | 85 |
| 9 | 87 | 65 |
| 10 | 90 | 76 |

**EXERCISES**

20. a) List all the assumptions underlying statistical inference based on a least-squares estimation of a multiple-regression model.

b) Indicate, by a short statement, the essential value of each assumption to the analysis.

c) Suggest one type of situation or type of data for which a particular assumption is likely to be violated.

21. Given the following model, where the variables are $y$, $x$ and $w$:

$$y_i = \alpha + \beta_1 x_i + \beta_2 x_i^2 + \beta_3 x_i w_i + \varepsilon_i.$$

Write an expression in terms of these variables for what needs to be minimized in order to apply the method of least squares for estimating the coefficients of the model.

22. I have a linear model with two independent variables, $x_1$ and $x_2$. How should I make a best selection of a third variable to include in my model from among three obvious candidates, say $z_1$, $z_2$, and $z_3$?

23. In a multiple regression based on 40 observations, the following results are obtained:

$$\hat{y} = 10 + 4x_1 + 6x_2 - 2x_3$$

with standard errors:            (1.2)    (5.0)    (0.4)

a) Explain the meaning of the coefficient for $x_2$.
b) Using some test statistic, explain which of the independent variables is the most significant.
c) Explain one effect of dropping variable $x_2$ from the regression model and re-estimating.

24. Consider a linear-regression model, $y = \alpha + \beta_1 x_1 + \beta_2 x_2 + \varepsilon$, where

$y$ = learning by grade 12, as measured by an academic test score composite, with mean 300 and standard deviation 150, for the entire population of 12th-graders;
$x_1$ = school expenditures per pupil during 3 years of high school (in hundreds of dollars);
$x_2$ = an index of socioeconomic status of the individual, with mean at 10 and standard deviation of 2, for the entire population of 12th-graders.

Based on a sample of 25 twelfth-grade-level individuals who were arrested on drug possession charges, the following results are obtained. Analyze and interpret, and explain the results in the way you think most appropriate and meaningful.

| Variable | Mean | Standard deviation |
|---|---|---|
| $y$ | 306.67 | 175.98 |
| $x_1$ | 12.58 | 9.31 |
| $x_2$ | 11.17 | 8.95 |

Correlations:
$$r_{yx_1} = 0.83; \qquad r_{yx_2} = 0.35; \qquad r_{x_1 x_2} = 0.10.$$

| Coefficient | Estimate | Standard error | $t$-value |
|---|---|---|---|
| $a$ | 10.16 | 11.9 | 0.85 |
| $b_1$ | 17.6 | 0.62 | 28.3 |
| $b_2$ | 4.3 | 2.90 | 1.48 |

Multiple $R = 0.92$;    $s_e = 3.015$.

| Analysis of variance | SS | d.f. | Mean square |
|---|---|---|---|
| Regression | 1090 | 2 | 545 |
| Residual | 200 | 22 | 9.09 |

25. Give an argument why any one of the standard assumptions for regression analysis would be violated, in each of the following situations, for a simple model of the form $y = \alpha + \beta x + \varepsilon$. Also, suggest how the violation of this assumption would affect the properties of the ordinary least-squares (OLS) estimators.

   a) $y$ measures wealth of an individual and $x$ measures his age, and $V[x]$ probably increases with age.
   b) Observations on $y$ and $x$ are daily stock averages and volume of trading respectively.

26. Explain the meaning of each of the following assumptions, where $\varepsilon$ is a random term in a linear-regression model. Give an example of some specified model which might violate each assumption, and explain why.

   a) $E[\varepsilon_i, \varepsilon_j] = 0$ for $i \neq j$;
   b) $V[\varepsilon_i] = \sigma_\varepsilon^2$ for all $i$.

27. The application of OLS treats all observations as equally important. State one situation in which this may be an inappropriate procedure, and explain the general principle or method of a better procedure.

28. Consider a simple model, $y_i = \alpha + \beta x_i + \varepsilon_i$, for which it is known that $\varepsilon_i = 0.3\varepsilon_{i-1} + v_i$, where the $v_i$ are normally and independently distributed with constant variance.

   a) Explain the problem of using ordinary least squares in this situation.
   b) Construct the appropriate expression that will minimize the sum of squares, if the variables are transformed to correct for the problem.

29. Given the following multiple-regression results, where $n = 12$:

| Variable | Mean | Standard deviation | | |
|---|---|---|---|---|
| $y$ | 306.67 | 175.98 | $r_{yx_1} = 0.9348$ |
| $x_1$ | 12.58 | 9.31 | $r_{yx_2} = 0.3501$ |
| $x_2$ | 11.17 | 8.95 | $r_{x_1 x_2} = 0.0096$ |

| Coefficient | Estimator | Standard deviation | |
|---|---|---|---|
| $b_1$ | 17.61 | 0.623 | $R_{y \cdot x_1 x_2} = 0.995$ |
| $b_2$ | 6.70 | 0.647 | $s_e = 19.220$ |
| $a$ | 10.17 | 11.972 | |

| Observation | Observed $y$ | Residual |
|:-----------:|:------------:|:--------:|
| 1  | 650 | $-1.63$  |
| 2  | 80  | 11.99    |
| 3  | 120 | $-18.46$ |
| 4  | 180 | $+32.34$ |
| 5  | 360 | 17.79    |
| 6  | 140 | $-21.07$ |
| 7  | 450 | 7.11     |
| 8  | 550 | 11.43    |
| 9  | 280 | $-14.37$ |
| 10 | 300 | $-16.30$ |
| 11 | 350 | $-17.43$ |
| 12 | 220 |          |

a) Which independent variable is most important in determining $y$?

b) Find the residual for the 12th observation if the observed values for $x_1$ and $x_2$ are 8 and 9, respectively.

c) What is the number of degrees of freedom for the $F$-test on the multiple linear association represented by this estimated model?

d) What percentage of the total variation in $y$ has been explained by this regression?

e) Using a plot, comment on the validity of the assumption of homoscedasticity in this model.

f) Using a plot, comment on the validity of the assumption of nonautocorrelation in this model.

g) Do you think multicollinearity is a problem in this estimation? Explain your answer.

h) If new observations on $x_1$ and $x_2$ are obtained, a corresponding value of $y$ can be predicted using the regression equation. Discuss the accuracy of such a prediction if $x_1$ and $x_2$ are 20 and 30, respectively, as compared to a prediction of $y$ if $x_1$ and $x_2$ are 10 and 12, respectively.

30. The power-efficiency of the Spearman rank-correlation coefficient is 91% of the Pearson correlation coefficient. Explain the meaning of this statement.

# Time Series and Index Numbers

## 13.1 INTRODUCTION TO TIME SERIES

In Chapter 11 we studied methods for describing the nature of the relationship between two variables. In this chapter we turn to what is in some sense a subset of this type problem, where the independent variable under investigation is time.

Recording observations of a variable which is a function of time results in a set of numbers called a *time series*. Most data in business and economic publications take the form of a time series—e.g., the monthly sales receipts in a retail store, Gross National Product (G.N.P.) of the U.S. by year, and indexes of consumer and wholesale prices, to name just a few. The analysis of time-series data in such circumstances usually focuses on two types of problems:

1) Attempting to estimate the factors (or components as they are called) which produce the pattern in the series; and

2) Using these estimates in forecasting the future behavior of the series.

In this book we shall concentrate our attention mainly on estimating the components of the time series, rather than studying the forecasting implications provided by these estimates. Our approach will emphasize economic time series because of their central importance in the planning function performed by many business and governmental agencies.

### Components of a Time Series

In general, the fluctuations in an economic time series are assumed to result from four different components: trend ($T$), seasonal variation ($S$), cyclical variation ($C$), and irregular or random variation ($I$). *Trend* is the long-term movement in a time series. (G.N.P., for example, has grown at a rate of approximately three to four percent a year over the past 20 years. The tendency toward a decreasing work week and increasing price levels over the past several decades also illustrates long-term movements or trends.) *Seasonal variation* represents fluctuations which repeat themselves within a fixed period of time less than one year; e.g., a series based on daily fluctuations repeats itself every week, while a series based on monthly fluctuations repeats itself every year. (The increased sale of ski equipment in the winter and gardening supplies in the spring are quarterly fluctuations; heavy sales receipts in supermarkets on Saturdays represent daily fluctuations.) The *cyclical* component of a time series represents a pattern repeated over time periods of differing length, usually longer than one year. (Business cycles, with their stages of prosperity, recession, and recovery are important examples of such cyclical movements.)

The movements in a time series generated by trend, seasonal and cyclical variation are assumed to be based on systematic causes; that is, these movements do not occur merely by chance, but reflect factors whose influence is more or less regular. Exactly the opposite holds true for random or *irregular variation*, which

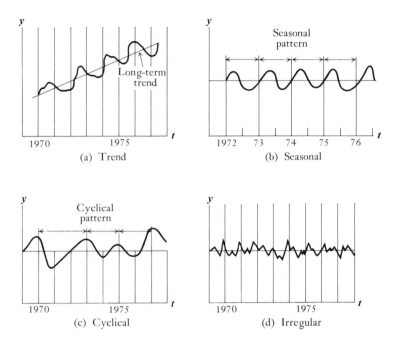

**Fig. 13.1.** The four components of a time series.

is, by definition, those fluctuations which are unpredictable, or take place at various points in time, by chance or randomly. (Floods, strikes, and fashion fads illustrate the irregular component of a time series.) Figure 13.1 presents a graphical view of the four components of a time series. In each case the dependent variable $y$ is expressed as a function of the independent variable $t$ (time).

**The Time-series Model**

Now that we have specified the components of a time series, the problem becomes one of estimating each one of these components for a given series. That is, we want to be able to estimate what portions of the value of $y$ for any given year are attributable respectively to trend, seasonal factors, cyclical factors, or random or irregular variation. In order to separate these values, we need to make some assumptions about how these components are related in the population under investigation; these assumptions are referred to as the *time-series model*.

Time-series models usually fall into one of two categories, depending on whether their components are expressed as sums or products. The first of these, called the *additive model*, assumes that the value of $y$ equals the *sum* of the four components, or that $y = T + S + C + I$. By assuming that the components of a time series are additive we are, in effect, assuming that these components are independent of one another. Thus, for example, trend can affect neither seasonal nor cyclical variation, nor can these components affect trend. The other major type of relationship between the components expresses $y$ in the form $y = T \times S \times C \times I$ and is called the *multiplicative model*. A model of this form assumes that the four components are related to one another, yet still allows for the components to result from different basic causes. Note that one can transform a multiplicative model into a linear (i.e. additive) model by taking the logarithm of both sides,

$$\log y = \log T + \log S + \log C + \log I.$$

Other models exist in addition to the additive and multiplicative ones, although these models usually take the form of combinations of additive and multiplicative elements, such as

$$y = S + T \times C \times I,$$

or

$$y = C + T \times S \times I.$$

For the purpose of estimating each of the components of a time series, models normally treat $S$, $C$, and $I$ as deviations from the trend. In other words, trend is usually estimated first, and the variation which can be attributed to trend is eliminated from the values of $y$. The variation remaining in $y$ must then be due to either seasonal, cyclical, or irregular factors. Each of these components of the time series can then be isolated by other statistical techniques. While some of these methods are excessively tedious, the basic concepts are similar to the simple methods to be discussed in this chapter. Once all the components of a time series are estimated, then forecasts of the value of the time series at some future point in time can be made by estimating first the value of the trend component at that point, and then modifying this trend value by an adjustment which takes the seasonal and cyclical components into account.

### Some Purposes and Problems of Forecasts

The primary purpose of making a forecast of a time series is for planning. A businessman wants to forecast the trend in his sales in order to make long-range planning decisions about investment in more plant capacity and new equipment. He may need to forecast cyclical movements in order to take advantage of lower interest-rate periods to conduct a bond sale or a period of high investor expectations to release a new issue of common stocks. Seasonal variation is important for short-run planning of inventories and employment levels, as he needs to be prepared for periods of high demand for his products and services. As consumers, we recognize these seasonal components also, and plan to make purchases at times of special sales when prices are lower. Of course, the marketing experts use the seasonal components to plan advertising campaigns to entice buyers to purchase during the high season when the product or service is most desirable. We are sure that any salesman would prefer to handle ski equipment in the fall and winter, and air conditioners in the spring and summer, rather than vice versa.

The most obvious problem inherent in forecasting future values of a time series is the potential size of the (unpredictable) irregular component. This component is a random variable whose size is a result of a large number of independent factors affecting the economic variable $y$. Psychological and sociological variables of individual and group behavior, as well as other economic and business considerations, could affect the forecast. In addition, the irregular component might consist of one single occurrence in the time period being forecast, such as a flood, a major political event, or an energy crisis. Neither the size nor the direction of the irregular component can be predicted.

There is also an important problem in trying to forecast the more regular parts of the time series, the trend, cyclical, and seasonal components. In forecasting, it is necessary to assume that no change occurs in the fundamental causes underlying these regular patterns. For this reason, it is always very dubious to make forecasts far beyond the range of presently observable values of the time series. In most cases, this means that the forecaster must not try to predict very far into the future, or he is apt to be greatly embarrassed.

## 13.2  LINEAR TREND

The first step in analyzing a time series is usually estimation of the trend component $T$. As was true in regression analysis, the first decision usually must be whether or not the trend can be assumed to be linear. Several linear methods will be discussed briefly, before we discuss the methods for estimating trend when the relationship is not linear.

The linear trend is written $y = a + bx$, where $x$ is a measure of time. Often a graph of the time-series data indicates quite well whether or not this linear relationship provides a good approximation to the long-term movement of the series. If a linear relationship is appropriate, then there are several methods for roughly approximating $T$. A fairly accurate approximation can often be obtained by merely drawing the line which, by observation, seems to best represent the long-run movement of the points. A more systematic approach is to use what is called the *method of semi-averages*. In this approach the data is divided into two equal parts, one representing the values associated with the first half of the years under investigation, and the second representing the remaining years. The average value of the independent (time) and dependent variables is calculated for each of these parts, and then the points representing these averages are connected by a straight line representing the trend line. For data containing an odd number of observations, the middle observation may be left out when dividing the data into two equal parts. Similarly, it may be advisable to eliminate one or two observations when calculating the mean value of $y$, if these observations are clearly atypical of the rest of the series and if their inclusion will disturb the whole trend line.

The method of least squares, developed in Chapter 11, represents the most popular method of fitting a trend line to time-series data. Recall that, in order to determine the values of $a$ and $b$ in a least-squares analysis, it is necessary to solve the following two normal equations:

$$\sum_{i=1}^{n} y_i = na + b \sum_{i=1}^{n} x_i,$$

$$\sum_{i=1}^{n} x_i y_i = a \sum_{i=1}^{n} x_i + b \sum_{i=1}^{n} x_i^2. \tag{13.1}$$

We will describe the use of these equations shortly.

### Scaling and Interpreting the Time Variable

In any time-series analysis it is important to carefully define the units of the time variable $x$, and to scale this variable so that it is easy to manipulate. Since time is continuous, it really makes little difference if the units used to express time are years, months, weeks, days, or any other desirable period. Also, because the point in time which is selected as $x = 0$ has no influence on the analysis, any period can be assigned this value. For whatever period is used, however, it is usually convenient to let $x = 0$ represent the *middle* of that period. For example, if $x$ is in months, then $x = 0$ would be the middle of one of the months; similarly, if $x$ is in periods six months long (half-years), then $x = 0$ would represent the middle of this period (i.e., after 3 months). A simple coding rule which is often used in assigning the values of $x$ is:

a) If the number of time periods is odd, let $x$ be in the same units as the observed time periods (years, months, etc.) and assign $x = 0$ to that time period falling in the exact center. Then let time periods before $x = 0$ be denoted by $\ldots, -3, -2, -1$, and future time periods be denoted by $+1, +2, +3, \ldots$

b) If the number of time periods is even, let $x$ be in units one-half as large as the observed time periods (half-years, and half-months, etc.), and assign $x = 0$ to the midpoint in time between the two middle observations. Then denote time periods before $x = 0$ as $\ldots -5, -3, -1$, and denote future periods as $+1, +3, +5, \ldots$

This coding procedure has the advantage that the mean of the $x$-values will always be $\bar{x} = 0$, and also that the middle of each time period is represented by an integer value of $x$. This means that the solution of the least-squares trend

**Table 13.1** Monthly observations of employment in a firm

| $y$ Employment | Time period | Value of $x$ |
|---|---|---|
| 16 | March, 1975 | $-5$ |
| 17 | April, 1975 | $-3$ |
| 18 | May,1975 | $-1$ |
| 20 | June, 1975 | $+1$ |
| 18 | July, 1975 | $+3$ |
| 24 | August, 1975 | $+5$ |

equation is computationally easier, and the interpretation and use of the trend
equation is simpler. We will illustrate both of these two types of scaling. First,
suppose the time-series data we have available concerns the variable $y$ = employment in a firm, measured over the six-month period shown in Table 13.1. Since
the number of time periods is even, we use coding procedure (b).

The units of $x$ in Table 13.1 are expressed in units of half-months, with $x = 0$
corresponding to the date, midnight May 31, 1975. The values of $x$ now represent
the number of half-months before or after the end of May, 1975. For example,
the value $x = 5$ represents August 15, 1975, which is 5 half-months after May 31st.
Note that the sum of the $x$-values is zero, which means that $\bar{x} = 0$. To forecast
a trend value for November, 1975, the value $x = 11$ would be substituted into the
trend equation. To forecast a trend value at the *end* of the year (December 31,
1975), the value $x = 14$ would be used.

To take another example, consider the sales data ($y$) in Table 13.2. This data
uses $x$ measured in years since the number of time periods is odd, with the origin
$x = 0$ corresponding to July 1, 1975. The value $x = -1$ thus corresponds to July 1,
1974, and $x = 2$ is July 1, 1977. Again note that $\sum x_i = 0$, and $\bar{x} = 0$. We will use
the data in the remaining columns of Table 13.2 shortly.

**Table 13.2** Sales (in thousands of
dollars)

| Year | $x$ | $y$ | $x^2$ | $xy$ |
|---|---|---|---|---|
| 1973 | $-2$ | $\$2$ | 4 | $-4$ |
| 1974 | $-1$ | 6 | 1 | $-6$ |
| 1975 | 0 | 10 | 0 | 0 |
| 1976 | 1 | 13 | 1 | 13 |
| 1977 | 2 | 16 | 4 | 32 |
| Sum | 0 | 47 | 10 | 35 |

### Using Simple Regression to Find the Trend Line

By using coding rules outlined above the solution of the normal equations in Formula (13.1) for time series is relatively simple. Since $\sum_{i=1}^{n} x_i = 0$ under these coding rules, if we substitute $\sum x = 0$ into the two normal equations, the result is the following two "reduced" equations.

$$\text{Reduced equations:} \qquad \sum_{i=1}^{n} y_i = na \quad \text{or} \quad a = \frac{\sum_{i=1}^{n} y_i}{n},$$

$$\sum_{i=1}^{n} x_i y_i = b \sum x_i^2 \quad \text{or} \quad b = \frac{\sum_{i=1}^{n} x_i y_i}{\sum_{i=1}^{n} x_i^2}. \tag{13.2}$$

Let us now substitute the appropriate values from Table 13.2 into these reduced equations in order to find the least-squares regression line for the sales data. The value of $a$ equals $47/5 = 9.4$, while the value of $b$ is $35/10 = 3.5$. Hence, the line of best fit is

$$\hat{y} = 9.4 + 3.5x$$

with $x$ in units of a year and $x = 0$ at midyear 1975. An estimate of the value of sales for 1980 ($x = 5$) based on trend factors only, would thus be

$$\hat{y} = 9.4 + 3.5(5)$$
$$= 26.9.$$

The trend line based on the data of Table 13.2 is shown in Fig. 13.2. A good exercise for the reader would be to determine and graph the trend line for the data in Table 13.1.

### 13.3 NONLINEAR TRENDS

The problem of fitting a trend line to a nonlinear time series is essentially the same problem we mentioned in Chapter 11 concerning nonlinear regression—that of finding an equation which best describes the relationship between an independent variable (time, in this case) and the dependent variable (the time-series values). As is true in fitting a regression line, it is not sufficient merely to find an equation which provides a good fit to the data, but it is necessary to find a model which is justifiable in terms of the underlying economic nature of the series. In estimating the trend in a time series, there are a number of nonlinear equations which can be justified under a wide variety of circumstances.

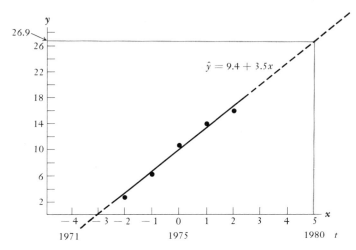

**Fig. 13.2.** Trend line for data of Table 13.2. Forecast value of sales for mid-1980 = 26.9.

## Exponential Curve

Time series are often used to describe data that increase or decrease at a constant proportion over time, such as population growth, the sales of a new product, or the spread of a highly communicable disease. Data taking this form can be approximated by an equation referred to as the *exponential curve*:

$$\text{Exponential curve:} \quad y = ab^x. \qquad (13.3)$$

The form of the exponential curve depends on the values of $a$ and $b$. If $b$ is between zero and one, then the value of $y$ will decrease as $x$ increases. When $b$ is larger than one, $y$ will increase as $x$ increases. The value of $a$ gives the $y$-intercept of the curve, as shown in Fig. 13.3.

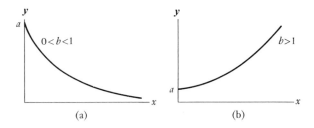

**Fig. 13.3.** The exponential curve $y = ab^x$.

**Table 13.3** Sales

| Year | $x$ | $y$ |
|------|-----|-----|
| 1966 | $-2$ | $1 |
| 1967 | $-1$ | 3 |
| 1968 | 0 | 6 |
| 1969 | 1 | 14 |
| 1970 | 2 | 41 |

Note that by taking the logarithm of both sides of Formula (13.3), we can transform the exponential curve into a linear relationship,

$$\log y = \log (ab^x) = \log a + x \log b.$$

Our model is now linear, and the least-squares approach can be used to find the line of best fit. To illustrate, consider the sales data given in Table 13.3 and graphed in Fig. 13.4. Again, the values of $x$ have been transformed so that $\sum_{i=1}^{n} x_i = 0$. From Fig. 13.4 we see that sales from 1966 to 1970 were not linear, and that an equation of the type shown in Fig. 13.2(b) would not be unreasonable. To use least-squares regression analysis to find the values of $a$ and $b$ in the equation $y = ab^x$, it is necessary to substitute $\log a$ for $a$, $\log b$ for $b$, and $\log y$ for $y$ in the normal equations. Since $\sum_{i=1}^{n} x_i = 0$, the appropriate equations are thus:

$$\log a = \frac{\sum_{i=1}^{n} \log y_i}{n}, \qquad \log b = \frac{\sum_{i=1}^{n} x_i \log y_i}{\sum_{i=1}^{n} x_i^2}. \tag{13.4}$$

To solve these equations we need to transform the data in Table 13.3 by finding the logarithm of $y$ and finding $x \log_{10} y$ (given in Table 13.4). We can now solve

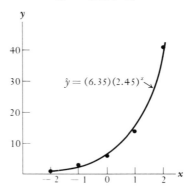

$$\hat{y} = (6.35)(2.45)^x$$

**Figure 13.4**

**Table 13.4** Sales (thousands of dollars)

| $x$ | $y$ | $\log_{10} y$ | $x \log_{10} y$ | $x^2$ |
|---|---|---|---|---|
| $-2$ | 1 | 0.000 | 0.000 | 4 |
| $-1$ | 3 | 0.477 | $-0.477$ | 1 |
| 0 | 6 | 0.778 | 0.000 | 0 |
| 1 | 14 | 1.146 | 1.146 | 1 |
| 2 | 41 | 1.613 | 3.226 | 4 |
| Sum | | 4.014 | 3.895 | 10 |

for $a$ and $b$. Substituting the values calculated in Table 13.4 into Formula (13.4) yields:

$$\log a = 0.8028, \qquad \log b = 0.3895.$$

Taking the antilog of these values yields the least-squares estimates $a = 6.35$ and $b = 2.45$. Thus, the exponential trend equation is

$$\hat{y} = (6.35)(2.45)^x.$$

Using this equation, forecast sales for 1971 ($x = 3$) equal $(6.35)(2.45)^3 = 93.34$. The fit provided by this equation for all values of $x$ from $-2$ to $+2$ is shown in Fig. 13.4.

**Modified Exponential Curve**

In a number of circumstances it is desirable to allow for more flexibility in deciding on the position of the trend line than is provided by the exponential curve, without altering the basic form of this curve. One way to accomplish this objective is to modify the exponential curve by adding a constant to the equation. Suppose we add the constant $c$. The resulting equation is called the *modified exponential curve*:

$$\textit{Modified exponential curve:} \qquad y = c + ab^x. \qquad (13.5)$$

The modified exponential, like the exponential itself, can assume many different forms depending on the values of $a$, $b$, and in this case $c$. Although the addition of the constant $c$ merely serves to shift the exponential curve up or down by a constant amount, such a shift is convenient in describing a time series whose values approach an upper or lower limit, as shown in Fig. 13.5.

Finding the best fit to a modified exponential curve is not as easy as was the case for the exponential curve, as there is no simple transformation which makes the equation linear, and finding the best least-squares fit directly is a fairly difficult task. There is one relatively straightforward method if one is willing to use just three of the observations to determine the equation for estimating trend. With

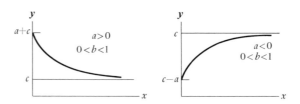

**Fig. 13.5.** The modified exponential curve $y = c + ab^x$.

three observations of $x$ and $y$ we can form three equations in three unknowns, and then solve these equations simultaneously. To illustrate, consider the sales data given in Table 13.5 and graphed in Fig. 13.6.

Suppose that we choose the three observations representing $x = -2$, $x = 0$, and $x = 2$, and substitute these values into Formula (13.5). The result is the following three equations in three unknowns:

$$3 = c + ab^{-2},$$
$$39 = c + ab^0,$$
$$51 = c + ab^2.$$

Solving these three equations simultaneously yields $a = -18.1$, $b = 0.58$, and $c = 57.1$. Thus, our modified exponential curve is

$$\hat{y} = 57.1 - 18.1(0.58)^x.$$

If we want to forecast $y$ for 1971 ($x = 3$) on the basis of these data, then the appropriate predicted value is

$$\hat{y} = 57.1 - 18.1(0.58)^3 = 53.46.$$

Figure 13.6 shows the fit of this equation for values of $x$ from $-2$ to $+2$.

The procedure just described for fitting a trend line to a set of observations is called the "three-point method." When more than just a small number of observations is involved, it is customary to apply this method by dividing the relevant

**Table 13.5** Sales

| Year | $x$ | $y$ |
|------|-----|-----|
| 1966 | $-2$ | $3 |
| 1967 | $-1$ | 27 |
| 1968 | 0 | 39 |
| 1969 | $+1$ | 46 |
| 1970 | $+2$ | 51 |

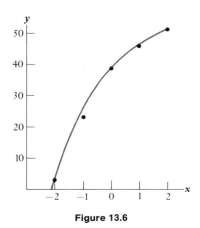

**Figure 13.6**

years into three equal periods. The mean value of $y$ for each of these periods, rather than the value of $y$ from a single year, is used for each of the three points. In either case, the three-point method should be viewed as providing only an approximation to the more precise fitting obtainable by the method of nonlinear least squares (which is not presented here).

**Logistic Curve**

There are a number of additional curves used to estimate trend in a time series, two of which are suitable for mention here. The first is known as a *logistic curve*, and is defined by the following equation:

$$\text{Logistic curve:} \qquad y = \frac{1}{c + ab^x}. \tag{13.6}$$

Note that the logistic curve is just the reciprocal of the modified exponential curve. Its rate of growth or decline is relatively rapid at first, but slows down in the later stages of the series, as shown in Fig. 13.7. Bacterial or population growth sometimes

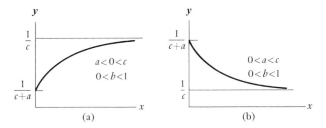

**Fig. 13.7.** The logistic curve $y = 1/(c + ab^x)$.

exhibits such a pattern over time. Solving for the values of $a$, $b$, and $c$ in a logistic curve presents the same problem as solving for these values in the modified exponential curve. The same approach described previously, using three of the time-series values to establish three equations in three unknowns, can be used in determining a logistic curve by letting $1/y$ be the dependent variable, rather than $y$. The resulting equation, $1/y = c + ab^x$, has the same form as the modified exponential.

**Gompertz Curve**

The final trend equation we shall discuss is called the *Gompertz curve*, named after Benjamin Gompertz, who used the curve in the early 1800's in work concerning mortality tables. This curve is similar to the exponential curve except that the constant $a$ is raised to the $b^x$ power instead of to the $x$ power:

$$\textit{Gompertz curve:} \quad y = ca^{b^x}. \tag{13.7}$$

When the value of $b$ in a Gompertz curve is between zero and one, the power to which $a$ is raised will approach zero as $x$ increases; hence the value of $y$ will become closer and closer to $c$, as shown in part (a) of Fig. 13.8. When the value of $b$ is greater than one, the curve will either increase without bound (when $a > 1$) or will approach zero (when $0 < a < 1$) as $x$ gets larger and larger, as shown in part (b) of Fig. 13.8. The Gompertz curve can be fitted to time-series data in essentially the same fashion described for the modified exponential curve, by taking the logarithm of both sides of Formula (13.7), as follows:

$$\log y = \log c + (b^x) \log a. \tag{13.8}$$

Formula (13.8) is a special form of the modified exponential curve, with unknowns $\log a$, $\log b$, and $\log c$, instead of $a$, $b$, and $c$.

The problem of deciding which of the many available nonlinear equations to use in estimating trend in a given circumstance is often a difficult one. It is not

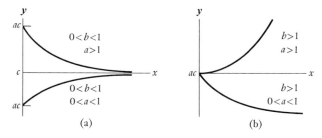

**Fig. 13.8.** The Gompertz curve $y = ca^{b^x}$.

unusual, in fitting the commonly used mathematical curves to a set of time-series observations, to find that two or more equations provide approximately the same closeness of fit, so that for the purpose of merely describing the data there is little to choose between these curves. The problem arises when such curves are used to predict future values of the time series (i.e., extrapolations into the future), for the curves tend to diverge rather quickly, and unacceptably large divergences may result from even a small extrapolation. Thus, as we pointed out earlier in this chapter, it is not sufficient to merely find a trend line providing a "good fit" to the data, but rather fit a curve which can be justified by a general assessment of the underlying nature of the series. It is common, for example, to find particular types of curves in the demand for commodities in certain industries. In other cases, there may be no single persistent relationship between the dependent variable and time.

## 13.4  MOVING AVERAGES TO SMOOTH A TIME SERIES

In the discussion thus far, we have assumed that a trend line can be calculated without first removing, or at least minimizing, the effect of seasonal, cyclical, and random movements in the series. In some circumstances, however, it may be easier to estimate trend if the effects of these fluctuations are removed from the data. Methods for removing these effects are usually referred to as smoothing techniques.

### The Method of Moving Averages

The most commonly used method of smoothing data is called the *method of moving averages*. A moving average is actually a series of averages, where each average is the mean value of the time series over a fixed interval of time, and where all possible averages of this time length are included in the analysis. A 12-month moving average, for example, must include the mean value for each 12-month period in the series. These averages represent a new series which has been smoothed to eliminate fluctuations which occur within a 12-month period. Since a seasonal pattern will repeat itself every 12 months, a 12-month moving average will eliminate fluctuations caused by the seasons of the year. Note that a 12-month moving average will also eliminate any other fluctuations whose pattern repeats itself over an interval of less than a year's duration, such as daily or weekly patterns. Similarly, a five-year moving average could be used to eliminate a pattern or cycle that repeats itself every five years. Such a moving average would also eliminate seasonal patterns, as well as daily and weekly fluctuations.

To illustrate the process of calculating a moving average, suppose we use this approach to smooth the data in Table 13.6. These data appear to fluctuate in a cyclical pattern which repeats itself every five years. In calculating a five-year moving average, we first need to determine the mean value for the initial five years, 1961 through 1965. This mean value, which equals $(1/5)(2 + 4 + 5 + 7 + 8) =$

**Table 13.6** Profits (thousands of dollars)

| Year | $y$ | Centered M.A. | Year | $y$ | Centered M.A. |
|------|-----|---------------|------|-----|---------------|
| 1961 | 2   |               | 1969 | 13  | 11.4          |
| 1962 | 4   |               | 1970 | 14  | 12.6          |
| 1963 | 5   | 5.2           | 1971 | 11  | 14.0          |
| 1964 | 7   | 6.0           | 1972 | 14  | 15.4          |
| 1965 | 8   | 6.8           | 1973 | 18  | 17.2          |
| 1966 | 6   | 8.0           | 1974 | 20  |               |
| 1967 | 8   | 9.2           | 1975 | 23  |               |
| 1968 | 11  | 10.4          |      |     |               |

5.2, is "centered" (i.e., placed) in the middle of the five years being averaged, 1963; similarly, the next moving average, 6.0, is computed from the values corresponding to 1962 through 1966 and centered at the year 1964. This process continues until the last observation is included; the last moving average is 17.2, and this value is centered at the year 1973. The moving average values tend to reduce the variation compared to the original values, and represent a smoothed version of the time series in which the cyclical component has been reduced. If the observation units were one-quarter of a year or less, the moving average technique will smooth out seasonal variations also. The comparison between the original time series and its representation by the moving average values is shown in Fig. 13.9.

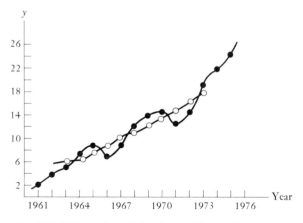

**Fig. 13.9.** Values of annual profits and five-period moving average values (–o–) from Table 13.6.

This moving average procedure gives equal weights to all five observations in calculating each mean value. Sometimes, one may wish to increase the relative importance of one or more observations by using a *weighted moving average*. In calculating a weighted moving average each observation being averaged is given a weight which reflects its relative importance in calculating that average. In a five-year weighted moving average, for example, the weights might be 1, 2, 3, 2, 1, based on the assumption that the middle value in a series of observations should have the largest weight, and the first and fifth observation should have the smallest weights. Using this weighting system on the data in Table 13.6 yields a weighted moving average centered at 1963 equal to

$$\frac{1(2) + 2(4) + 3(5) + 2(7) + 1(8)}{1 + 2 + 3 + 2 + 1} = 5.22.$$

The second weighted moving average would be centered at 1964 and equal to

$$\frac{1(4) + 2(5) + 3(7) + 2(8) + 1(6)}{1 + 2 + 3 + 2 + 1} = 5.67.$$

By continuing this process we could derive the entire series for the weights 1, 2, 3, 2, 1. Had we used some other weighting system, an entirely different moving average would be determined. In general, the weights used depend on the degree to which the analyst wishes to emphasize particular values.

**The Use of Moving Averages for Smoothing**

It is important to note that by smoothing a time series we have not solved the problem of estimating the trend component for that series. Smoothing the series merely serves to eliminate some of the variability not attributable to trend, in the hope that the trend component can then be more easily identified. One method for identifying the appropriate trend curve (e.g., exponential, Gompertz, logistic) is to look at certain characteristics of a curve's moving average. In their book *Mathematical Trend Curves*, for example, Gregg, Hossell, and Richardson advise using a transformation of each curve's slope in deciding among the common trend equations, since dispersions in the projection of these curves will arise largely because of differences in the rate of change of the curve. They suggest finding, for each curve considered, that transformation of the slope which yields a straight line when the transformed values are plotted against time. The estimated slopes of the time series, for each year in the period under consideration, are then subjected to all such transformations. If one of these transformations yields values which approximate a straight line, then the curve corresponding to this transformation may be expected to be a satisfactory predictor.

The following transformations yield straight lines for the curves discussed in Section 13.3.

| Trend curve | Transformation |
| --- | --- |
| Exponential | Slope/moving average |
| Modified exponential | Logarithm of the slope |
| Gompertz | Logarithm of {slope/moving average} |
| Logistic | Logarithm of {slope/(moving average)$^2$} |

The reader is referred to the book mentioned above, *Mathematical Trend Curves*, for a further discussion and examples of the process of fitting trend curves to time-series data.

Although the use of a moving average may help to identify trend, there are several weaknesses to moving average methods as a smoothing device. First of all, this method is often only partially successful in removing all the effects of $S$, $C$, and $I$ from $y$ when determining $T$, because $C$ and $I$ usually do not have regular fluctuations. In addition, the moving-average method tends to introduce spurious cyclical movements into the data being smoothed, so that an analyst trying to remove cyclical effects from his data may, in fact, introduce a nonexistent cycle.

Another weakness of the moving-average method involves the storage space (e.g., in a computer) necessary to compute each average, as the number of pieces of data which need to be kept track of will always equal the number of observations in the moving average. Thus, a company with a large variety of products that wishes to forecast demand for each of its products, with each individual product forecast based on an analysis of past sales data, will have a large storage requirement.

A final problem with the use of moving averages is the arbitrary choice of its length, denoted by $h$, the number of consecutive values used in the averages. The larger the value of $h$, the more the moving average smooths out the original data, but the greater is $(h - 1)$, the number of total observations at the beginning and end of the data for which no moving average value can be determined. Using the five-period moving average, we see in Table 13.6 that a total of four observations are lost in the moving-average series. The size of $h$ must be large enough for smoothing purposes but small enough to retain sufficient observations. Also in smoothing cycles, the value of $h$ should be selected to coincide with the length of the cycle (or an integer multiple of its length). If the cycle length varies, some complex moving-average techniques using different values of $h$ may be most appropriate.

## 13.5 ESTIMATION OF SEASONAL AND CYCLICAL COMPONENTS

Until now we have been concerned with estimating only trend. Often, however, it may be just as (or even more) important to be able to estimate the *seasonal* component in a time series. From a planning point of view, for example, it is often necessary for business managers to take into consideration fluctuations other than trend when financing future operations, purchasing materials or merchandise, establishing employment practices, etc. Many of these fluctuations are of a seasonal nature caused by the weather, such as the increase in swim-wear sales during the summer months, or by social customs, such as the increase in retail sales during the Christmas season.

### Seasonal Index

The seasonal component of a time series is usually expressed by a number, called a seasonal index number, which expresses the value of the seasonal fluctuation in each month as a percent of the trend value expected for that month. A value of 104 thus means that, because of seasonal factors, the time series value is expected to be four percent above the trend value. If the index number is 93, then the time series value is expected to be seven percent less than the trend value.

Because of seasonal variation, special care must be taken in forecasting sales for any given month, or in comparing sales between months. Such forecasts or comparisons are usually made by using a seasonal index to *deseasonalize* the data. Since the base value of a seasonal index is always 100, the value of the seasonal index for each month represents how much above or below the overall monthly average that particular month usually falls. To illustrate this type of index, let's assume that bicycle sales for a given manufacturer have averaged 50,000 units a month for the past five years. However, the average sales in each month is not always equal to 50,000. In February, for example, sales over the past five years has averaged 30,000 units. Similarly, May has averaged 50,000 units, and December, 80,000 units. Since February has averaged only 60% of the overall average sales of 50,000, the seasonal index for February is $S_{Feb.} = 60.0$. The indexes for May and December would be $S_{May} = 100.0$ and $S_{Dec.} = 160$.

In deseasonalizing a time series value, one first divides by the seasonal index, and then multiplies by 100. That is,

$$\text{Seasonally adjusted value} = \frac{\text{Original value} \times 100}{\text{Seasonal index}}. \qquad (13.9)$$

We can illustrate the use of this formula to adjust values in our bicycle problem by assuming that the actual sales for February and December in a given year were 32,000 and 84,000, respectively. Adjusting these values we obtain:

$$\text{Adjusted February sales} = \frac{\$32,000 \times 100}{60} = \$53,000;$$

$$\text{Adjusted December sales} = \frac{\$84,000 \times 100}{160} = \$52,500.$$

Note how this seasonal adjustment makes the comparison of sales between the two months very easy. Here we see that February sales were slightly better than December sales on a seasonally adjusted basis. Because the average seasonal index for these two months exceeds $100\ [(60 + 160)/2 = 110]$, their total adjusted values ($105,500) must be less than the total sales which actually occurred ($116,000). For a seasonal index which has been correctly formulated, the adjusted total for twelve months will always equal the actual total over all twelve months.

### Ratio-to-Trend Method

In order to determine seasonal index numbers, it is necessary to estimate the seasonal component of a time series. Suppose that a given time series can be represented by the multiplicative model $y = T \times S \times C \times I$. One method for estimating the seasonal component in this model is called the *ratio-to-trend method*, which estimates $S$ by removing trend from the series, but does not attempt to remove cyclical and irregular variation. Assume that a value of $T$ has been calculated for each monthly value of $y$ in the series. Trend can be eliminated by dividing $y$ by these monthly trend values, $y/T = S \times C \times I$; the remaining fluctuations are assumed to represent, primarily, seasonal variation. To illustrate the ratio-to-trend method, assume that the data in Table 13.7 represents the monthly sales in a department store over the three-year period from 1973 to 1975.

**Table 13.7** Monthly sales

| Year | Jan. | Feb. | Mar. | Apr. | May | June | July | Aug. | Sept. | Oct. | Nov. | Dec. | Monthly average |
|------|------|------|------|------|-----|------|------|------|-------|------|------|------|-----------------|
| 1973 | 67 | 71 | 72 | 73 | 71 | 70 | 67 | 64 | 66 | 74 | 82 | 99 | 73.0 |
| 1974 | 80 | 86 | 89 | 89 | 84 | 83 | 79 | 76 | 77 | 87 | 98 | 114 | 86.8 |
| 1975 | 92 | 98 | 101 | 103 | 96 | 96 | 89 | 87 | 90 | 99 | 112 | 131 | 99.5 |

The trend component of these observations can be estimated by fitting a least-squares regression line to the data. Rather than fit a line to all 24 points, we can estimate the trend line by using just the three monthly averages shown in the last column of Table 13.7 as follows:

| Year | $x$ | $y$ | $x^2$ | $xy$ |
|------|-----|------|-------|-------|
| 1973 | $-1$ | 73.0 | 1 | $-73.0$ |
| 1974 | 0 | 86.8 | 0 | 0.0 |
| 1975 | 1 | 99.5 | 1 | 99.5 |
| Sum | 0 | 259.3 | 2 | 26.5 |

Since $\sum_{i=1}^{n} x_i = 0$, we can use the reduced form of the normal equations, as in Formula (13.2),

$$a = \sum_{i=1}^{n} y_i/n, \quad \text{and} \quad b = \sum_{i=1}^{n} x_i y_i \bigg/ \sum_{i=1}^{n} x_i^2.$$

Substituting the appropriate values into these equations yields a value of $a$ equal to $259.3/3 = 86.4$ and $b$ equal to $26.5/2 = 13.3$. These values determine the following trend line:

$$\hat{y} = 86.4 + 13.3x, \tag{13.10}$$

with $x$ in units of one year and $x = 0$ at mid-1974. Since it is monthly trend changes we want to estimate, this equation will be slightly easier to work with if $x$ is expressed in months, so we need merely to divide the slope of the line by 12.* The resulting equation,

$$\hat{y} = 86.4 + 1.108x, \tag{13.11}$$

indicates that $y$ will increase 1.108 each month due to trend, accumulating an annual increase of 13.3, as in Formula (13.10). When $x = 0$ the value of $\hat{y}$ is 86.4 for mid-1974. The difficulty with this value is that the value of $\hat{y}$ on July 1, 1974, is really not representative of the entire month of July. A more representative date for July would be July 15th, or $\frac{1}{2}$ month later. The value of $y$ estimated by the trend line for July 15 can be determined by substituting $x = \frac{1}{2}$ into equation (13.11), so that $86.4 + 1.108\,(\frac{1}{2}) = 87.0$. To find the trend value for preceding months, approximately 1.1 is subtracted successively from 87.0. These values are shown in Table 13.8.

**Table 13.8** Monthly trend

| Year | Jan. | Feb. | Mar. | Apr. | May | June | July | Aug. | Sept. | Oct. | Nov. | Dec. |
|------|------|------|------|------|-----|------|------|------|-------|------|------|------|
| 1973 | 67.2 | 68.3 | 69.4 | 70.5 | 71.6 | 72.7 | 73.8 | 74.9 | 76.0 | 77.1 | 78.2 | 79.3 |
| 1974 | 80.4 | 81.5 | 82.6 | 83.7 | 84.8 | 85.9 | 87.0 | 88.1 | 89.2 | 90.3 | 91.4 | 92.5 |
| 1975 | 93.6 | 94.7 | 95.8 | 96.9 | 98.0 | 99.1 | 100.2 | 101.3 | 102.4 | 103.5 | 104.6 | 105.7 |

We have now determined a value of $T$ for each month in the series. The next step is to divide each month's sales value shown in Table 13.7 by the corresponding monthly trend value given in Table 13.8. The results of this division are shown in the first three rows of Table 13.9. Since there are three different index numbers for each month in Table 13.9, we combine these values by taking their average,

---

* To change from one trend equation to an equivalent one with different units for $x$, simply multiply the slope by the ratio of the size of the new unit for $x$ to the size of the original unit for $x$. Since one year has 12 months, the appropriate ratio here is 1/12.

**Table 13.9** Seasonal index numbers

| Year | Jan. | Feb. | Mar. | Apr. | May | June | July | Aug. | Sept. | Oct. | Nov. | Dec. |
|------|------|------|------|------|-----|------|------|------|-------|------|------|------|
| 1973 | 99.7 | 103.9 | 103.7 | 103.5 | 99.1 | 96.2 | 90.7 | 85.4 | 86.8 | 95.9 | 104.8 | 124.8 |
| 1974 | 99.5 | 105.5 | 107.7 | 106.3 | 99.0 | 96.6 | 90.8 | 86.2 | 86.3 | 96.3 | 107.2 | 123.2 |
| 1975 | 98.2 | 103.4 | 105.4 | 106.2 | 97.9 | 96.8 | 88.8 | 85.8 | 87.8 | 95.6 | 107.0 | 123.9 |
| Average | 99.1 | 104.3 | 105.6 | 105.3 | 98.7 | 96.5 | 90.1 | 85.8 | 87.0 | 95.9 | 106.3 | 124.0 |

shown in the bottom row of Table 13.9. To avoid seasonal adjustments which bias the data, these averages should have a grand average of 100.0. Aside from rounding errors, this procedure guarantees such a result unless some extreme values were dropped from the calculation due to a specific disturbance. The values shown in the bottom row of Table 13.9 are the monthly seasonal index values, and may be used to deseasonalize observations, as discussed before. The deseasonalized data can be more accurately examined for trend and cycles than the original data, since any confusion of seasonal movements is eliminated.

Perhaps Fig. 13.10 will help summarize some of the material in this section. The lower portion of this figure shows the original sales data from Table 13.7 and the trend component of this data (Table 13.8). The seasonal index at the top of the figure was derived by dividing each of the trend values into the sales data, and then averaging, for each month, over the three years 1973, 1974, 1975.

### Ratio-to-Moving-Average Method

The ratio-to-trend procedure we have just described produces a seasonal index which still contains $C$ and $I$. Although the process is considerably more cumbersome, there is a way to calculate an index in which these fluctuations have been removed, called the *ratio-to-moving-average method*. The first step in this approach is to smooth the data by using a 12-month moving average. If we disregard the irregular component whose causes and occurrences are unknown, the smoothed series will contain fluctuations attributable only to $T$ and $C$, since a 12-month moving average will remove seasonal variations. In terms of the multiplicative model we have, in effect, divided $y$ by $S$, so that the new series is $y/S = T \times C$. If the *original* time-series values, $y$, are now divided by this newly calculated series containing only $T$ and $C$, the result is $y/(T \times C) = (T \times S \times C)/(T \times C) = S$. Thus, by a rather roundabout route, we have isolated $S$ and determined a seasonal index in which the variations attributable to $T$ and $C$ have been removed. To illustrate how this process works, we again use the data of Table 13.7.

First, a 12-month moving average is taken over all months in the series. This moving average is shown in columns 3, 6, and 9 of Table 13.10. Now, to associate the moving-average values with the fifteenth rather than the first day of each

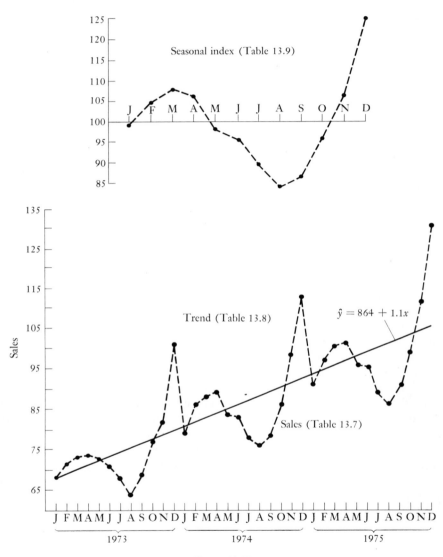

**Figure 13.10**

**Table 13.10**

| (1) Month | (2) 1973 Sales | (3) 12-month M.A. | (4) Centered 12-month M.A. | (5) 1974 Sales | (6) 12-month M.A. | (7) Centered 12-month M.A. | (8) 1975 Sales | (9) 12-month M.A. | (10) Centered 12-month M.A. |
|---|---|---|---|---|---|---|---|---|---|
| Jan. | 67 | | | 80 | | 80.8 | 92 | | 93.6 |
| | | | | | 81.3 | | | 94.0 | |
| Feb. | 71 | | | 86 | | 81.8 | 98 | | 94.5 |
| | | | | | 82.3 | | | 94.9 | |
| Mar. | 72 | | | 89 | | 82.8 | 101 | | 95.5 |
| | | | | | 83.3 | | | 96.0 | |
| Apr. | 73 | | | 89 | | 83.8 | 103 | | 96.5 |
| | | | | | 84.3 | | | 97.0 | |
| May | 71 | | | 84 | | 85.0 | 96 | | 97.6 |
| | | | | | 85.7 | | | 98.2 | |
| June | 70 | | | 83 | | 86.3 | 96 | | 98.9 |
| | | 73.0 | | | 86.9 | | | 99.6 | |
| July | 67 | | 73.6 | 79 | | 87.4 | 89 | | |
| | | 74.1 | | | 87.9 | | | | |
| Aug. | 64 | | 74.7 | 76 | | 88.4 | 87 | | |
| | | 75.3 | | | 88.9 | | | | |
| Sept. | 66 | | 76.1 | 77 | | 89.4 | 90 | | |
| | | 76.8 | | | 89.9 | | | | |
| Oct. | 74 | | 77.5 | 87 | | 90.6 | 99 | | |
| | | 78.1 | | | 91.2 | | | | |
| Nov. | 82 | | 78.7 | 98 | | 91.7 | 112 | | |
| | | 79.2 | | | 92.1 | | | | |
| Dec. | 99 | | 79.8 | 114 | | 92.7 | 131 | | |
| | | 80.3 | | | 93.2 | | | | |

month, these values are "centered" by taking the average of each two adjacent months. The centered moving-average values are shown in columns 4, 7, and 10 and represent $T \times C$ and some parts of $I$.

We have now eliminated $S$ and some remaining parts of $I$ from the series. To find a seasonal index we must divide the original time-series values in Table 13.7 by the 12-month centered moving-average values in Table 13.10. The result of this division, shown in Table 13.11, is two index numbers for each month of the year.

Since these index numbers may differ, some irregular component variation $I$ is still included. By averaging the monthly $S \times I$ estimates for each calendar month, some more of the irregular component is eliminated. In a long time series with at least five values of $S \times I$ for each month, this averaging virtually isolates

**Table 13.11** Seasonal index numbers

| Year | Jan. | Feb. | Mar. | Apr. | May | June | July | Aug. | Sept. | Oct. | Nov. | Dec. |
|---|---|---|---|---|---|---|---|---|---|---|---|---|
| 1973 | | | | | | | 91.0 | 85.7 | 86.9 | 95.5 | 104.1 | 124.0 |
| 1974 | 99.0 | 105.1 | 107.1 | 106.1 | 98.8 | 96.2 | 90.4 | 86.0 | 86.2 | 96.0 | 106.7 | 122.9 |
| 1975 | 98.4 | 103.9 | 105.9 | 106.8 | 98.5 | 97.0 | | | | | | |
| Aver. | 98.7 | 104.5 | 106.5 | 106.5 | 98.7 | 96.6 | 90.7 | 85.9 | 86.6 | 95.8 | 105.4 | 123.5 |

the pure seasonal index component $S$, with $T$, $C$, and $I$ removed. The averages of these two values are the seasonal indexes found in the bottom row of Table 13.11, which are quite similar to those in Table 13.9, where the ratio-to-trend method was used.

Dividing a time series by the appropriate seasonal index values, such as those shown in Table 13.11, "seasonally adjusts" a time series so that the changes in the series are not caused merely by changes in the season. Quarterly GNP figures, for example, are usually seasonally adjusted to remove seasonal fluctuations.

**Determining the Cyclical Variation**

In some analyses, it is desirable to isolate the cyclical component of a time series so that turning points and peaks and troughs may be studied. If a stable cyclical pattern, such as a 3.4-year business cycle or a 20-year housing construction cycle could be isolated, it would greatly aid economists in determining the underlying causes and in forecasting future movements. Quite a wide variety of complicated methods have been developed for studying the cyclical component of a time series. However, we can use a roundabout method again, to find $C$ in a simple way.

Given a time series $y = T \times C \times S \times I$, the seasonal component $S$ can be determined by the ratio-to-moving-average method. The values of $y$ can then be divided by the seasonal index $S$ to obtain $y \times 100/S = T \times C \times I$. Next, we would eliminate the trend component from $T \times C \times I$ by a least-squares fit, using the most appropriate linear or nonlinear curve. Dividing $T \times C \times I$ by the values of $\hat{y}$ obtained from this least-squares analysis would thus leave only the components $C \times I$. To eliminate $I$, a weighted moving average of three or five periods can be determined, using the $C \times I$ data. If a longer moving average were used, the cycles might be smoothed out too much and partially eliminated also. The best type of weights to use are large weights for the center values and much smaller weights for the values farther away from the center. If the weights are constructed so that their sum is equal to one, then no division by the sum of the weights is necessary. For example in a five-period moving average, the weights might be $-0.1$, $0.3$, $0.6$, $0.3$, and $-0.1$. The advantage of this extra computation using weights is that it produces a smoother curve for the cycle component $C$ while it is also more sensitive to the original fluctuations because it preserves the amplitude of the cycles more faithfully. Using the odd-period moving average is convenient for centering the resulting values (which are the component $C$). For example, based on monthly data, the cyclical component for April would be:

$$C_{\text{April}} = -0.1(C \times I)_{\text{Feb.}} + 0.3(C \times I)_{\text{March}} + 0.6(C \times I)_{\text{April}}$$
$$+ 0.3(C \times I)_{\text{May}} - 0.1(C \times I)_{\text{June}}.$$

Values of $C$ for other months would be found similarly.

## 13.6  INDEX NUMBERS

Earlier in this chapter, index numbers were used to express the seasonal components of a time series. Our objective at that point and in the following discussion is to develop a measure which summarizes the characteristics of large masses of data. An index number accomplishes this purpose by relating the values of a dependent variable, such as sales or price, to an independent variable, such as time, income, or occupation. Although index numbers are used in many areas of the behavioral and social sciences, their main application involves describing business and economic activity, such as changes in price, production, wages, and employment, over a period of time.

### Uses of Index Numbers

The primary purpose of an index number is to provide a value useful for comparing magnitudes of aggregates of related variables to each other and to measure their changes over time. Consequently, many different index numbers have been developed for special uses. Let us briefly mention some of their common uses.

First, index numbers are useful summary measures for policy guides. The Federal Reserve Board may use index numbers on interest rates or employment or consumer credit, as inputs to discussions on appropriate open-market transactions. The Cost-of-Living Council may use price and income indexes in setting wage and price guidelines.

Second, many index numbers have been developed to be indicators of business conditions, such as the *Forbes*, *Fortune*, or *Business Week* indexes. The National Bureau of Economic Research, the Bureau of Labor Statistics, and the Commerce Department all publish various index numbers for this same purpose.

Some of these indexes are commonly used in a third way for making comparisons of changes among different sectors of the economy. Growth in the agricultural or mining sector might be compared to growth in the manufacturing sector. Levels of state and local government expenditures might be analyzed and compared among regions by using index numbers.

A fourth use of certain special indexes, such as wage, productivity, and cost-of-living indexes is in wage contracts and labor-management bargaining. Management often likes to tie wage increases to productivity increases, while labor unions like to relate the need for wage increases to cost-of-living increases. Similar relations among indexes are used in adjusting insurance coverage, changing retirement and social security benefits, etc.

Finally, a fifth and very common use of an index number is as a deflator. Price or cost indexes are divided into certain measures of economic activity, in order to obtain the *real* or constant dollar value of these measures. That is, an adjustment is made for the changing value of the dollar, so that more meaningful comparisons over time can be made.

## Using an Index as a Deflator

The use of an index as a deflator is so common that it deserves some examples. Suppose a person's income increases from $10,000 to $15,000 over a given period, while the consumer price index (C.P.I.) increases from 100 to 130. The nominal increase in income of $5000 is offset by general inflation indicated by the 30-percent increase in the C.P.I. If this index is assumed to be relevant to the consumer purchases of such a person, it could be used to "deflate" the air out of his income increase and find his "real" increase in income. Since the base value of any index number is 100, the real income is obtained by the rule:

$$\text{Real income} = \frac{(\text{Nominal income}) \times 100}{\text{Consumer price index}}.$$

The real income at the beginning and at the end of the period being studied, respectively, is:

$$\frac{10{,}000 \times 100}{100} = \$10{,}000 \quad \text{(Beginning real income)},$$

$$\frac{15{,}000 \times 100}{130} = \$11{,}538 \quad \text{(End real income)}.$$

With the extra $5000 income, this person could buy only an extra $1,538 worth of goods and services because prices increased. Real income is usually called the "purchasing power" of the money income. In comparing incomes, wages, or rents of individuals, or gross national product, or personal income per capita of different countries it is common to use an appropriate deflator. In this way, the true standard of living is more easily recognized.

As another example, suppose a state government allocated $25 million for highway construction in 1970 when costs of labor, materials, land, equipment, etc., were measured by a construction cost index to be 125 (relative to a base period equal to 100, in 1967). Suppose that in 1975 some highway projects receive funding of $30 million. However, the index of costs in 1975 has risen to 180. We find the real value of the available money in terms of constant base-year costs in each case by deflating by the cost index:

$$\text{Real value in 1970} = \frac{\$25 \text{ million} \times 100}{125} = \$20 \text{ million};$$

$$\text{Real value in 1975} = \frac{\$30 \text{ million} \times 100}{180} = \$16.67 \text{ million}.$$

The 1975 allocation will build only 83 percent as much as the lower 1970 allocation would have permitted five years earlier.

### Constructing an Index

Having discussed the nature and uses of index numbers, we turn in the next section to a discussion of constructing a price index. While this is only one type of index, it shares many problems of construction with indexes of all other types. The reader should be ready to recognize the following usual problems.

Each index number must have a base period for which the value of the index is 100.0. The selection of a typical base year is somewhat arbitrary but very important. If a particularly exceptional period were selected when the values of the variables were quite extreme, then all other values of the index would be affected.

A second arbitrary choice involves the selection of the type of index to use (simple, relative, weighted, etc.) and of the appropriate weights to use in combining the values of different items included in the aggregate measure. Incidentally, the choice of items to include is often very debatable. An attempt should be made to include the most common and representative items. An index for new car prices in the United States may not include prices for all possible cars that could be purchased, but it surely must include the most popular cars of the largest auto-makers. Similarly, an index of United States wages per man-hour in manufacturing must include a sample of representative types of workers across a representative group of industries selected to be representative of all regions of the country. The problems inherent in such selections are numerous. The final result may be some representative index number that does not suit anyone in particular.

The most compelling practical problem in constructing an index is obtaining sufficient and accurate data for all the items included. To obtain a consumer price index, one might sample consumers to find out how much they paid and what quantity they bought of a large list of products, or one might sample storekeepers to find out their sales and prices on all the different items. Once a set of quantities and prices is obtained, the calculation of the index is essentially a problem involving weighted averages.

Effectively, an index number is simply the ratio of two numbers expressed as

**Table 13.12** Civilian labor force index

| (1) Year | (2) Civilian labor force (millions) | (3) Index with 1954 = 100 |
|------|------|------|
| 1950 | 63 | 98.4 |
| 1952 | 63 | 98.4 |
| 1954 | 64 | 100.0 |
| 1956 | 68 | 106.3 |
| 1958 | 71 | 110.9 |
| 1960 | 72 | 112.5 |

a percentage. For example, consider the data in Table 13.12. Seven values of the civilian labor force for selected years are shown in column 2. If the year 1954 is chosen as the base year, then the index number for year $i$, as shown in column 3, is obtained as follows:

$$\text{(Index number)}_i = \frac{y_i \times 100}{y_{\text{Base}}}.$$

For example, the index number for 1960 is $(72 \times 100)/64 = 112.5$. This indicates a 12.5-percent increase in 1960 relative to the base year 1954.

## 13.7  PRICE INDEX NUMBERS

Since prices play a major role in every economy, there has been great interest in developing appropriate price indexes. In this section, several different types of price index are defined, and some of the procedures for determining their value are outlined. Most of these use the simple *price relative* as a cornerstone. It is an index number for prices of a single item comparable to the index number for civilian employment in Table 13.12. A price relative is the ratio of the price of a certain commodity in a given period ($p_n$) to the price of that same commodity in some base period ($p_0$). If we let $I_n$ represent the index number for a given year relative to the base year, then $I_n = (p_n/p_0) \times 100$. Following this convention, the index for the base year is 100, since $I_0 = (p_0/p_0) \times 100$. In general, a price relative shows the percentage increase or decrease in the price of a commodity from the base period to a given period.

To illustrate the construction of a price relative, assume the price of eggs for the years from 1972 to 1974 to be 60, 68, and 80 cents a dozen, respectively. With 1972 as the base period or year ($1972 = 100$), the price relatives for the three years are:

$$1972 \text{ price relative} = \left(\frac{p_0}{p_0}\right) \times 100 = \frac{60}{60} \times 100 = 100;$$

$$1973 \text{ price relative} = \left(\frac{p_1}{p_0}\right) \times 100 = \frac{68}{60} \times 100 = 113;$$

$$1974 \text{ price relative} = \left(\frac{p_2}{p_0}\right) \times 100 = \frac{80}{60} \times 100 = 133$$

Thus the price of eggs increased 13 percent from 1972 to 1973 and 33 percent from 1972 to 1974.

### Simple Indexes

Rather than describe changes in a single commodity, one may wish to describe changes in the general price level. To do this we combine a representative number

Table 13.13

| Commodity | Quantity units | Prices 1970 ($p_0$) | Prices 1973 ($p_3$) | Price relatives $\left(\dfrac{p_3}{p_0}\right)$ |
|---|---|---|---|---|
| Milk | 1 pt. | 20 | 25 | 1.25 |
| Bread | 8 oz. | 15 | 20 | 1.33 |
| Cheese | 4 oz. | 40 | 45 | 1.13 |
| Butter | 1 lb. | 55 | 65 | 1.18 |
| Oranges | 1 doz. | 60 | 65 | 1.08 |
| Total | | 190 cents | 220 cents | 5.97 |

of commodities, sometimes called a *market basket*, into an aggregate price index. The prices of these commodities in a given year are compared to the prices of the same market basket in a base year. One such index, called the *simple aggregate price index*, can be expressed as follows:

$$\text{Simple aggregate price index:} \qquad I_n = \frac{\sum p_n}{\sum p_0} \times 100.$$

Suppose that we use the data in Table 13.13 to compute a simple aggregate price index for 1973, based on 1970 prices. The sum of the prices in 1973 ($p_3$) is 220 cents, while the sum of the prices in 1970 ($p_0$) is 190 cents; the ratio of these two numbers times 100 gives the simple aggregate price index

$$I_3 = \frac{\sum p_3}{\sum p_0} \times 100 = \frac{220 \text{ cents}}{190 \text{ cents}} \times 100 = 116.$$

The simple aggregate price index has two main disadvantages. First, it fails to consider the relative importance of the commodities; secondly, the use of different units to measure each commodity affects the price index. The second disadvantage can be overcome merely by changing to a *simple average of price relatives* index. As the name suggests, this index equals the average of all price relatives. Since $\sum(p_n/p_0)$ equals the sum of these relatives, and $N$ equals the total number of commodities, then:

$$\text{Simple average of price relatives:} \qquad I_n = \frac{\sum (p_n/p_0)}{N} \times 100.$$

A simple average of price relatives index constructed for the data given in Table 13.13 can be determined by dividing the sum total of the price relatives, 5.97, by the number of commodities, 5 (and then multiplying by 100):

$$I_3 = \frac{5.97}{5} \times 100 = 119.$$

**Weighted Price Indexes**

Weighted index numbers overcome the first disadvantage of the simple index. They permit consideration of the relative importance of the commodities in the market basket, which is usually measured in terms of the total amount of money spent on each commodity during a year. The product of the price of a commodity and the quantity consumed in a given year represents the total amount of money consumers spend on that particular commodity. If $q_i$ represents the quantity consumed in year $i$, then $p_0q_0$ equals the total amount spent on a particular commodity in the base year, and $p_nq_n$ represents the amount spent some other year. A *weighted average of price relatives* index weights the various price relatives in a market basket by the total amount spent on that commodity. This index may take on two different forms, depending on whether base-year weights ($p_0q_0$) or the weights of some given year ($p_nq_n$) are used in constructing the index.

*Weighted average of price relatives:*

$$\text{Base-year weights: } I_n = \frac{\sum \left(\frac{p_n}{p_0} p_0q_0\right)}{\sum p_0q_0} \times 100;$$

(13.12)

$$\text{Given-year weights: } I_n = \frac{\sum \left(\frac{p_n}{p_0} p_nq_n\right)}{\sum p_nq_n} \times 100.$$

A weighted average of price relatives index can be constructed from the data in Table 13.13 if the quantities purchased for each commodity are known. Suppose that the yearly purchases of a typical consumer are those shown in Table 13.14. The weighted price relative for milk, for instance, using base-year weights of

**Table 13.14**

| Commodity | Quantities | | $p_0q_0$ | $p_3q_0$ | $p_0q_3$ | $p_3q_3$ |
|---|---|---|---|---|---|---|
| | 1970 ($q_0$) | 1973 ($q_3$) | | | | |
| Milk | 30 | 35 | $6.00 | $7.50 | $7.00 | $8.75 |
| Bread | 25 | 20 | $3.75 | $5.00 | $3.00 | $4.00 |
| Cheese | 20 | 30 | $8.00 | $9.00 | $12.00 | $13.50 |
| Butter | 10 | 5 | $5.50 | $6.50 | $2.75 | $3.25 |
| Oranges | 15 | 20 | $9.00 | $9.75 | $12.00 | $13.00 |
| Total | | | $32.25 | $37.75 | $36.75 | $42.50 |

$p_0 q_0 = \$6.00$, would be

$$\frac{p_3}{p_0} p_0 q_0 = (1.25)(\$6.00) = \$7.50.$$

The weighted price relative values given in Table 13.15 are calculated in the same fashion.

**Table 13.15**

| Commodity | Base-year weights $\dfrac{p_3}{p_0} p_0 q_0$ | Given-year weights $\dfrac{p_3}{p_0} p_3 q_3$ |
|---|---|---|
| Milk | $7.50 | $10.94 |
| Bread | $7.50 | $5.33 |
| Cheese | $9.00 | $15.26 |
| Butter | $6.50 | $3.86 |
| Oranges | $9.75 | $4.04 |
| Total | $37.75 | $49.43 |

We can now use this information to calculate the two types of weighted average of price relatives. Using base-year weights $p_0 q_0$:

$$\frac{\sum \left(\dfrac{p_3}{p_0} p_0 q_0\right)}{\sum p_0 q_0} \times 100 = \frac{\$37.75}{\$32.25} \times 100 = 117.$$

Using given-year weights $p_3 q_3$:

$$\frac{\sum \left(\dfrac{p_3}{p_0} p_3 q_3\right)}{\sum p_3 q_3} \times 100 = \frac{\$49.43}{\$42.50} \times 100 = 116.$$

In this example, the two methods give results which happen to be quite close together.

Rather than weighted price relatives, a popular means of constructing a weighted price index is to weight each price directly by multiplying it by the quantity of that commodity consumed in either the base year or some other year. The value $p_n q_0$, for example, represents the price of a commodity in year $n$ weighted by the quantity of the commodity consumed in the base year. The total theoretical value of the commodities in year $n$ is thus $\sum p_n q_0$. If we now take the ratio of this value to the actual value of these goods in the base year, $\sum p_0 q_0$, the resulting index is called a *Laspeyres price index*:

$$\text{Laspeyres price index:} \quad I_n = \frac{\sum p_n q_0}{\sum p_0 q_0} \times 100. \quad (13.13)$$

If, instead of using $q_0$ to weight prices, the quantity consumed in the given year $n$ is used as a weight, the resulting index is called a *Paasche price index*:

$$\text{Paasche price index:} \quad I_n = \frac{\sum p_n q_n}{\sum p_0 q_n} \times 100. \quad (13.14)$$

Note that the Laspeyres price index is equivalent to the weighted average of price relatives index using base-year weights, since

$$\frac{\sum \left(\frac{p_n}{p_0} p_0 q_0\right)}{\sum p_0 q_0} \times 100 = \frac{\sum p_n q_0}{\sum p_0 q_0} \times 100.$$

Similarly, the Paasche price index equals the weighted average of price relatives index using given year weights as seen by comparing Formulas (13.12) and (13.14). The Laspeyres and Paasche price indexes can be constructed from the data in Table 13.14 by using the information in the last four columns with the same result as for the weighted average of price relatives indexes:

$$\text{Laspeyres price index:} \quad \frac{\sum p_3 q_0}{\sum p_0 q_0} \times 100 = \frac{\$37.75}{\$32.25} \times 100 = 117,$$

$$\text{Paasche price index:} \quad \frac{\sum p_3 q_3}{\sum p_0 q_3} = \frac{\$42.50}{\$36.75} \times 100 = 116.$$

Since construction of the Paasche index requires determination of new weights each year, while the Laspeyres index does not, the Laspeyres index is used much more often. Furthermore, the Laspeyres index has the added advantage that indexes obtained by this formula may be compared from year to year, while indexes obtained by using the Paasche formula can be compared easily only with the base year. Both indexes, however, tend to reflect a slight bias when reporting price changes. Under the usual conditions of a downward sloping demand schedule, when people tend to purchase more of low-priced items and less of high-priced items, the numerator of the Laspeyres index will be somewhat higher than it should be, resulting in an overestimation of price increases. At the same time, the numerator of the Paasche index will tend to be lower than it should be, resulting in an underestimation of price increases. These biases are only slightly obvious in the results of the preceding examples, wherein the Laspeyres index is 117 while the Paasche index is 116.

## 13.8 ECONOMIC INDEXES AND THEIR LIMITATIONS

We have presented just a brief introduction to the topic of index numbers. In practice, the process of constructing and updating an index can become quite involved. Before leaving the subject, we shall describe a number of the more widely used indexes and some of their limitations.

The *Consumer Price Indexes* (C.P.I.) and the *Wholesale Price Index* (W.P.I.), both published by the Department of Labor, Bureau of Labor Statistics, are the two best-known price indexes. These two indexes represent excellent examples of attempts to summarize large masses of data in a single price index: Approximately 400 items comprise the "market basket" of the C.P.I., while the W.P.I. bases its series on over 2,000 items. The C.P.I. reflects the value of the market basket of goods and services purchased by urban clerical workers in the retail market. An indication of price movements in all markets other than the retail market is given by the W.P.I. The graph of the consumer price index for commodities and services for the years 1968 through 1973 is shown in Fig. 13.11. Another price index, the *Daily Spot Market Price Index*, reflects price movements in the primary (raw materials) market. This index combines 22 basic commodities by taking the geometric average of their price relatives.

There are two major economic quantity indexes: (1) the *Index of Industrial Production* (I.I.P.), published by the Federal Reserve Board, and (2) the Export and Import Indexes, published by the U.S. Department of Commerce. The Export and Import Indexes calculate quantity changes in both exports and imports. The I.I.P., which is a weighted average of quantity relatives index, measures changes in manufacturing output.

The limitations of economic indexes follow directly from the problems suggested earlier in the construction of an index. Since the useful and meaningful application of an index often depends on what year is selected as the base, it is important to use as a base a period with "normal" or "average" economic activity. Sometimes the base year can be the average price or quantity of several years, rather than a single year. Exactly which items are included in the index also has an important bearing on the validity of the index. In the case of the C.P.I., every few years the Bureau of Labor Statistics undertakes a detailed study of the consumer "market basket" in order to determine which old items to replace and which new items to add. Since, over a period of time, the quality of products can change, some items fall out of use completely, and others appear that had no earlier counterpart, it is difficult, if not impossible, to compare the prices of many of

Index, 1967 = 100

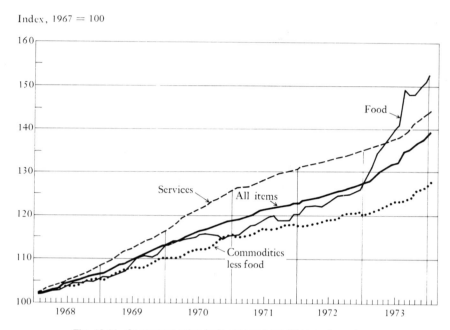

**Fig. 13.11.** Consumer price indexes: commodities and services.

today's goods with those of an earlier period. For example, it is unrealistic to compare the price of an old icebox with the price of today's modern refrigerator, and the price of a television set cannot be compared with the price of a model which didn't even exist 35 years ago.

This inability to compare goods, and consequently prices, over time has led to charges that price indexes exaggerate the real increase in prices, and in the case of the C.P.I., exaggerates the change in the cost of living, because they do not take into account the quality improvement of goods. Even though the Bureau of Labor Statistics attempts to factor out all of the quality changes that can be detected in the 400-odd items used in computing the C.P.I., it is extremely difficult to measure the worth of these increases to consumers. Lloyd G. Reynolds, an economist, illustrates this problem quite well in the following situation: Give a family a 1955 Sears Roebuck Catalog and a 1975 Sears Catalog and $1,000, and allow them to make up an order list from either the 1955 or the 1975 catalog, but not from both. Most families would probably use the 1975 catalog, which must mean that they consider the higher 1975 prices more than offset by new and improved quality products, implying that there was no real increase in consumer prices between 1955 and 1975 even though the C.P.I. indicated a sizable increase.

Because there are so many difficulties in constructing meaningful price indexes, they should be used only as guides and indicators of price movements, and should not be quoted as indisputable facts. Also one should remember that they represent prices to some "average" person with "average" tastes and preferences. Whether or not they are relevant to a particular person or group must always be questioned before they are applied.

### REVIEW PROBLEMS

1. What is a time series? Why is it necessary to distinguish time-series analysis from regression analysis?

2. Personal consumption expenditures ($y$) for the years 1956–1960 in the United States were as shown in the table.

| $y$ | Years |
|-----|-------|
| 270 | 1956 |
| 290 | 1957 |
| 295 | 1958 |
| 310 | 1959 |
| 335 | 1960 |

   a) Fit a straight-line trend line for $y$ by the method of least squares. Be sure to indicate the origin.
   b) Give the amount of total deviation, explained deviation, and unexplained deviation for the 1959 observation.

3. Suppose the trend line for average annual wages of members of a local electricians' union from 1950–1968 is $\hat{y} = 6500 + 300x$, with origin at mid-1954 and $x$ in units of one year.

   a) What is the average annual increase in wages over this period?
   b) Suppose a local nonunion electrician made $9000 in 1963. How much did his income differ from the trend value in 1963 for union members?

4. A study is being made to determine the growth of annual honey production ($y$) over time ($x$) measured in years before and after 1973. The following data is given ($x = 0$ is 1973):

| $x$ | $-2$ | $-1$ | 0 | 1 | 2 |
|-----|------|------|---|---|---|
| $y$ | 6 | 9 | 12 | 11 | 15 |

   a) Find the regression equation of $y$ on $x$.
   b) Find the estimated value of honey production in 1976.

5. The value of building permits issued in a certain town over an eight-year period is given below.

| x (years) | Value (thousands) | x (years) | Value (thousands) |
|-----------|-------------------|-----------|-------------------|
| 1965 | $300 | 1969 | $310 |
| 1966 | 150 | 1970 | 490 |
| 1967 | 210 | 1971 | 380 |
| 1968 | 400 | 1972 | 400 |

a) Plot the above data and make a freehand estimate of the trend line.
b) Use the method of semi-averages to construct a trend line.
c) Use the method of least squares to determine a trend line.
d) Compare your estimates of the trend line for parts (a), (b), and (c). What value would you estimate for 1973 under each of these methods?

6. The following values represent sales data for the years 1965–70.

| Year | Sales (units) |
|------|---------------|
| 1965 | 18,000 |
| 1966 | 19,000 |
| 1967 | 23,000 |
| 1968 | 24,000 |
| 1969 | 26,000 |
| 1970 | 28,000 |

a) Plot the values and then estimate a linear trend line by the freehand method.
b) Use the method of semi-averages to estimate a trend line.
c) Construct a least-squares regression line to estimate trend.

7. The sales of a new product had the following growth pattern:

| x (years) | Sales (thousands) | x (years) | Sales (thousands) |
|-----------|-------------------|-----------|-------------------|
| 1 | 1.6 | 4 | 4.6 |
| 2 | 2.7 | 5 | 5.1 |
| 3 | 3.9 | 6 | 5.4 |

a) Fit a Gompertz curve to these data.
b) Fit a logistic curve.
c) Graph both of the curves derived above, and then indicate which model appears to give the better fit.

8. Describe one advantage and one disadvantage of using a moving-average smoothing method in time-series analysis.

9. Find a three-year moving average for the following data.

| Year | Sales (thousands) |
|------|-------------------|
| 1962 | $18 |
| 1963 | 20 |
| 1964 | 22 |
| 1965 | 19 |
| 1966 | 21 |
| 1967 | 24 |
| 1968 | 21 |
| 1969 | 23 |
| 1970 | 27 |

10. Given the following amounts of money (to the nearest thousand dollars) lost each year by a certain businessman by gambling in Las Vegas during his vacations:

| 1965 | 1966 | 1967 | 1968 | 1969 | 1970 | 1971 | 1972 | 1973 |
|------|------|------|------|------|------|------|------|------|
| 2 | 7 | 5 | 16 | 12 | 23 | 29 | 68 | 2 |

Find the four-period moving average for the amounts lost.

11. a) Compute the trend equation for the following series by the method of least squares.
    b) Compare the trend value for 1970 with the three-period moving average value for 1970.

| Year | 1965 | 1966 | 1967 | 1968 | 1969 | 1970 | 1971 |
|------|------|------|------|------|------|------|------|
| Tons | 30 | 44 | 50 | 42 | 51 | 68 | 65 |

12. a) State two reasons for estimating the seasonal component of a time series.
    b) State two reasons for estimating the secular trend equation for a time series.
    c) State two disadvantages to the use of a (simple) moving average as a fit of trend.

13. Assume the trend line (annual total equation) for suits is $\hat{y} = 3600 + 480x$ with origin at October, 1972. The seasonal index for April is 80. Estimate the seasonally adjusted output for April, 1976.

14. The operating season for a certain tomato cannery is from June to October. Suppose that the following observations represent monthly sales, in thousands, from this cannery from 1973 to 1975.

| Year | June | July | August | September | October |
|------|------|------|--------|-----------|---------|
| 1973 | 75 | 86 | 102 | 105 | 90 |
| 1974 | 83 | 89 | 110 | 115 | 92 |
| 1975 | 84 | 95 | 113 | 118 | 89 |

a) Use the ratio-to-trend method to find a seasonal index for sales for the months June to October.
b) Find a seasonal index using the ratio-to-moving-average method, using a 5-month moving average.
c) Assume that the multiplicative model $y = T \times S \times C \times I$ holds, and that there is no irregular variation. Decompose the index for August of 1974 into the component parts $T$, $S$, and $C$, using your answer to part (a).
d) Repeat the above process, using your answer to part (b).

15. Explain how to find the cyclical component of a time series.

16. Explain the difference between cyclical and seasonal variations in a time series.

17. The Business Research Department of the Carolina Corporation forecasts sales for next year of $12 million, based on a trend projection. It is expected that no sharp cyclical fluctuations will occur during the year, that the effect of trend *within* the year will be negligible, and that the past pattern of quarterly seasonal variation will continue. The pattern is as follows:

| Quarter | 1st | 2nd | 3rd | 4th |
|---|---|---|---|---|
| Seasonal index | 130 | 90 | 75 | 105 |

Prepare a forecast of quarterly sales from the above information for the 1st and 2nd quarters of next year.

18. In finding the cyclical component of a time series, explain how to obtain $C$ once the $CI$ component has been isolated.

19. a) The unadjusted index of sales of Company C-B is 102 for January, 1968. The seasonally adjusted index for the same month is 133. Find the seasonal index for January.
b) The annual sales for 1976 are forecast to be $240,000. The seasonal index for March is computed to be 90. Give a reasonable forecast of the sales for March, 1976.

20. The index of seasonal variation for cement production for selected months is 60 for January, 70 for March, 122 for August, and 100 for November.

a) In which of these months is cement production the greatest?
b) In which of these months is the production most typical of the monthly average?
c) Production in a certain region increased from 6,060,000 barrels in January, 1972, to 11,590,000 barrels in August, 1972. What was the percentage change in cement production after making allowance for seasonal variation?

21. Suppose a budget request for a new university building was $3.5 million in 1965, and a similar request is $4.5 million in 1973. Also, an index of building and construction costs (assumed to be applicable to this situation) with base year $1967 = 100$, is equal to 95 in 1965 and 140 in 1973. Compare the relative real value of the building which could have resulted from the two budget requests.

22. A popular smoothing model not discussed in this text is called the *exponential smoothing model*. Find a book describing exponential smoothing (e.g., *Introduction to Statistical Methods*, by D. L. Harnett) and give a summary of this method.

23. Go to a library or to your local newspaper and find the current value of the *Consumer Price Index* (C.P.I.) and the *Wholesale Price Index* (W.P.I.).

    a) Determine how the value of these indexes has changed over the past one year, as well as over the past ten years. How has the composition of the items included in these indexes changed over the past ten years?
    b) Find as many examples as you can from the news media illustrating current uses of these indexes (e.g., labor–management negotiations, etc.).

24. The Bureau of Labor Statistics consumer price index for services increased from 107 to 115 during a certain period.

    a) If a consumer budgeted $20 per month for services at the beginning of this period, how much does he need to budget for the same level of services at the end of the period?
    b) Suppose a person's salary increased from $500 to $552 per month during this period. What is the real change he experiences in purchasing power for services?

25. In the past year, the GNP of a nation has risen from $200 billion to $212 billion, which represents a 6% growth rate. However, the price index used in calculating the national product has also risen from 125 to 130. What is the real growth rate of this nation's GNP after accounting for inflation?

26. If an index of money wages of workers (1951 = 100) was 250 in 1970 and an appropriate index of living costs (1951 = 100) was 200 in the same year, what has been the percent increase from 1951 to 1970 in:

    a) money wages of workers?
    b) real wages of workers?

27. Suppose a man's income increases from $12,000 to $15,000 over a period of years when the consumer price index increases from 100 to 120. What is the change in his "real" income?

28. Suggest two important practical difficulties in determining a Laspeyres price index.

29. Explain how a Laspeyres price index differs from a simple arithmetic average of price relatives.

30. Suppose that the following market-basket values were observed in 1955 and 1956:

|  | 1955 | | 1956 | |
|---|---|---|---|---|
| Item | Price | Quantity | Price | Quantity |
| Milk | 15 cents qt. | 25 qt. | 15 cents qt. | 20 qt. |
| Eggs | 50 cents doz. | 10 doz. | 45 cents doz. | 15 doz. |
| Bread | 15 cents loaf | 30 loaves | 20 cents loaf | 35 loaves |

Construct the following:

a) price relatives for each commodity,
b) a simple aggregate price index,

c) a simple average of price relatives index,
d) a weighted average of price relatives index, using both base-year and given-year weights.
e) a Laspeyres and a Paasche price index.

31. Given the following data:

| Year | Price A | Quantity A | Price B | Quantity B |
|------|---------|-----------|---------|-----------|
| 1967 | 25 cents | 10 | 10 cents | 30 |
| 1968 | 30 cents | 20 | 20 cents | 30 |
| 1969 | 30 cents | 40 | 25 cents | 20 |
| 1969 | 40 cents | 45 | 20 cents | 25 |

a) Construct a Laspeyres price index using 1967 as the base year;
b) Construct a Paasche price index for these years.

32. Suppose price and quantity data (in relevant units) for sales of major appliances are collected in order to construct a durable consumer-goods price index. From the data, determine a Laspeyres price index for 1975 relative to the base period, 1970, and interpret its meaning.

| | 1970 | | 1975 | |
|------|-----|-----|-----|-----|
| Item | $p$ | $q$ | $p$ | $q$ |
| Range | 1.5 | 8 | 2 | 12 |
| Refrigerator | 3 | 10 | 3.8 | 12 |
| Air conditioner | 1 | 6 | 2 | 15 |

33. Given the following data of monthly purchases by a certain family:

| | 1950 | | 1960 | |
|-----------|-----|-----|-----|-----|
| Commodity | $p$ | $q$ | $p$ | $q$ |
| Eggs (doz.) | 0.40 | 10 | 0.60 | 9 |
| Sugar (lbs.) | 0.10 | 35 | 0.20 | 30 |
| Butter (lbs.) | 0.50 | 5 | 0.40 | 6 |

a) Compute a weighted aggregate Laspeyre's price index, using 1950 as the base year.
b) Assume that your index applied in general for all foods. Also assume a certain family spends $1200 for food in 1950 and $1750 in 1960. In which year would they have had more to eat for their money?

34. The following data on student expenses is dug up by the campus newspaper for an editorial on the "good ole days."

| | 1963 | | 1964 | | 1965 | |
|---|---|---|---|---|---|---|
| Item | Price (cents) | Quan-tity | Price (cents) | Quan-tity | Price (cents) | Quan-tity |
| Lunch | 30 | 6 | 50 | 6 | 60 | 6 |
| Gasoline | 25 | 1 | 30 | 2 | 35 | 3 |
| Movies | 50 | 4 | 90 | 2 | 100 | 3 |

a) Compute the Laspeyres price index for the three years, using 1963 as the base.
b) Assuming that these data are representative of all commodities and all students, ascertain, by deflation, the real level of living, relative to 1963, of a student who spent $500 per semester in 1963, $600 in 1964, and $700 in 1965.

## EXERCISES

35. a) What are the four components of a time series? Give an example of a time series in which you would expect all four components to arise, explaining why you would anticipate them.
    b) Can you think of any type of time series which has only three, or two, of the four components? Are there any other components which might be present in a time series? Give specific examples.

36. Describe the additive and the multiplicative models used in time-series analysis. Which of these models do you think is more realistic for most economic time series? Explain why. Give examples where each model would be appropriate.

37. Assume that the sales volume in a certain industry can be described by the multiplicative model $y = T \times S \times C \times I$. One month last year the trend estimate of sales was 44,000 units and actual sales were 55,000 units. If we assume that the seasonal index was $S = 95$, and the index for cyclical movement was $C = 119$, what index value must be associated with the irregular movement, $I$?

38. The estimate of trend accounted for $180,000 of a department store's sales last October. Assuming a multiplicative model, no irregular variation, and a cyclical index of 110, find the seasonal index for this month. Actual sales were $210,000.

39. Given a trend equation, $\hat{y} = 37.50 + 6x$ with origin at July 1, 1965 and $x$ in units of one year, where $y$ is truck sales in thousands, convert the equation to monthly units with origin at March 1, 1967.

40. Given the following trend equation: $\hat{y} = 137.50 + 8x$. Origin is July 1, 1965, $x$ unit: 1 year; $y$ is fruit sales in thousands. Find the monthly trend value for February 15, 1967.

41. A producer of one-man portable helicopters had sales over the five years, 1971–1975, of 1, 2, 3, 5, and 9 units, respectively.

a) Find the trend line for sales.
b) Find the standard error of estimate for this trend line and explain its meaning.
c) What is your estimate for sales in 1977?
d) For the second observation (year 1972), give the total deviation, the explained deviation, and the unexplained deviation.

42. a) Work Exercise 49, Chapter 11, part (a).
  b) Fit a modified exponential curve to the data in (a).
  c) Does your answer to part (a) or part (b) provide a better fit to the data? Explain why one model is better than the other.

43. a) Fit an exponential curve of the form $y = ab^x$ to the following data on population growth.

| Year | Population |
|------|------------|
| 1971 | 100 |
| 1972 | 400 |
| 1973 | 700 |
| 1974 | 1600 |
| 1975 | 3900 |

  b) Plot this time series and the curve you calculated above on a graph. What value would you predict for population in 1977?
  c) Fit a modified exponential curve to these data.
  d) Plot on a graph both the exponential approximation and the modified exponential approximation to these data. Which gives the better approximation? Explain why one model is better than the other.

44. A company has determined a seasonal index and a trend line for their monthly sales. The seasonal index for December is 140. The trend line for monthly sales is $\hat{y} = 163{,}250 + 4{,}520x$, with $x$ in units of months, with origin at mid-April, 1974. Make the forecast of company sales for December, 1976.

45. Given the following data for sales of boating supplies:

| 1975 Quarter | Actual sales | Trend values | Seasonal index |
|------|------|------|------|
| 1 | 100,000 | 90,000 | 80 |
| 2 | 150,000 | 95,000 | 130 |
| 3 | 120,000 | 100,000 | 110 |
| 4 | 110,000 | 105,000 | 80 |

  a) Give one representation of the trend equation for sales of boating supplies.
  b) Find seasonally adjusted sales for the four quarters of 1975.
  c) Assuming no change in the trend or seasonal pattern and assuming other factors remain constant, give a forecast for second-quarter sales, 1977.

46. The secular trend of sales for the Jones Department Store is accurately described by the equation $\hat{y} = 120{,}000 + 1000x$, where $x$ represents a period of one month and has a value of zero in December, 1967. The seasonal indexes for the company's sales are:

| J | F | M | A | M | J | J | A | S | O | N | D |
|---|---|---|---|---|---|---|---|---|---|---|---|
| 100 | 80 | 90 | 120 | 115 | 95 | 75 | 70 | 90 | 95 | 120 | 150 |

a) Ignoring cyclical and random influences, forecast sales for February, 1969, May, 1972, and December, 1970.

b) What factors could cause these estimates to be incorrect?

c) What might be done to compensate for inaccuracies as they become apparent?

47. The following data are the quasimeans of ratios of original data to the 12-month moving averages for the sales of a retail store. They are based on ten years.

a) Compute the index of seasonal variation from these for the months of March, April, and August. Be sure to adjust so the indexes average 100.

| Month | Modified means of ratios |
|---|---|
| Jan. | 56 |
| Feb. | 60 |
| Mar. | 100 |
| Apr. | 110 |
| May | 105 |
| June | 102 |
| July | 80 |
| Aug. | 72 |
| Sept. | 88 |
| Oct. | 105 |
| Nov. | 120 |
| Dec. | 145 |
| Total | 1143 |

b) Suppose the total sales for next year is estimated in December of this year to be 120 million. What would be the best estimate of sales for April and August?

48. In the library, find a 10-year time series on annual household income for the United States, a region, your state, or for some occupation, race, or sex subgroup. Find a time series on consumer prices which is relevant for this group of households. Calculate the adjusted annual household income, corrected for price changes, and interpret your results in terms of growth of real income over the period.

49. Given the following information

| Year | Import price $y$ | Quantity imported, $x$ |
|---|---|---|
| 1965 | 2 | 6 |
| 1966 | 3 | 5 |
| 1967 | 6 | 4 |
| 1968 | 5 | 5 |
| 1969 | 4 | 7 |
| 1970 | 3 | 10 |
| 1971 | 5 | 9 |
| 1972 | 7 | 7 |
| 1973 | 8 | 8 |
| 1974 | 7 | 9 |

a) For the data above, find the estimating line of import price on the quantity imported.

b) Find the correlation coefficient for this data.

c) Find the percentage of variation in import price that is explained by the variation in the quantity imported. What is the name given to this measure?

d) Suppose the trend line for import price is $\hat{y} = 2.8 + 0.5x$, where $x = 0$ for 1965. Is the import price for these years above or below the trend estimates?

e) Do you think it would be a good idea to remove seasonal variation from this data, as well as trend, before studying the cyclical relatives? Give a reason.

f) Find the values of the three-year moving average of quantity imported for the years 1967 and 1970.

50. Using the library, find the price and quantity sold of shares of ten selected stocks on the last day of the previous four years. Using one year as the base period, construct your own:

a) simple price index of the stocks;

b) Laspeyres price index of these stocks;

c) Find the Dow-Jones stock index at year-end for these same years and, using the same base period, construct an index of the Dow-Jones index to compare with your index in part (b).

d) Explain how your price index of stocks differs in construction and meaning from a typical stock index.

# Selected
# Bibliography

## BUSINESS AND ECONOMIC STATISTICS

CLELLAND, RICHARD, and others, *Basic Statistics with Business Applications*. New York: Wiley, 1966.

FREUND, JOHN E., and FRANK J. WILLIAMS, *Elementary Business Statistics: The Modern Approach*. Englewood Cliffs, New Jersey: Prentice Hall, 1964.

HAYS, WILLIAM L., and ROBERT L. WINKLER, *Statistics: Probability, Inference, and Decision*, New York: Holt, Rinehart, and Winston, 1970.

HUGHES, ANN, and DENNIS GRAWOIG, *Statistics: A Foundation for Analysis*. Reading, Mass.; Addison-Wesley, 1971.

JOHNSTON, J., *Econometric Methods*. New York: McGraw-Hill, 1963.

MILLS, FREDERICK C., *Statistical Methods*, 3rd ed. New York: Holt, Rinehart, and Winston, 1955.

NETER, JOHN, and WILLIAM WASSERMAN, *Fundamental Statistics for Business and Economics*, 3rd ed. Boston: Allyn and Bacon, 1966.

PETERS, WILLIAM S., and GEORGE W. SUMMERS, *Statistical Analysis for Business Decisions*. Englewood Cliffs, New Jersey: Prentice Hall, 1968.

SASAKI, KYOHEI, *Statistics for Modern Business Decision Making*. Belmont, California: Wadsworth, 1968.

SPURR, WILLIAM A., and CHARLES P. BONINI, *Statistical Analysis for Business Decisions*. Homewood, Illinois: Irwin, 1967.

SCHLAIFER, ROBERT, *Probability and Statistics for Business Decisions*. New York: McGraw-Hill, 1959.

SCHLAIFER, ROBERT, *Analysis of Decisions Under Uncertainty*. New York: McGraw-Hill, 1969.

WONNACOTT, THOMAS H., and RONALD J. WONNACOTT, *Introductory Statistics*. New York: Wiley, 1969.

# APPENDIX A

# Subscripts and Summations

Throughout this book we will use certain symbols to distinguish between the numbers in a set of data, and to indicate the sum of such numbers. For example, we may wish to distinguish between the monthly sales of a certain business, and then sum these monthly sales to get the yearly sales. To do this, suppose we let the symbol $x$ denote the monthly sales of this firm. Furthermore, we will add a subscript to this symbol to denote which month is being represented. Thus, $x_1 = $ sales in first month, $x_2 = $ sales in second month, and so forth, with $x_{12} = $ sales in the twelfth month. To illustrate, if sales in the sixth month were 120 units, then we would write $x_6 = 120$. The notation $x_i$ thus stands for "sales in the $i$th month," where $i$ can be any number from 1 to 12; that is, $i = 1, 2, \ldots, 12$. The dots in this last expression are used to indicate "and so on."

Now, assume that we want to sum the sales for all 12 months in a year, which is

$$x_1 + x_2 + \cdots + x_{12}.$$

Another way of writing this sum is to use the Greek letter $\sum$ (capital sigma). This symbol is read as "take the sum of." At the bottom of this $\sum$ sign we usually place the first value of $i$ which is to be included in the sum. The last value of $i$ to be summed is usually placed at the top of the sum sign. Thus,

$$\sum_{i=1}^{12} x_i$$

is read as "sum the values of $x_i$ starting from $i = 1$ and ending with $i = 12$." That is,

$$\sum_{i=1}^{12} x_i = x_1 + x_2 + \cdots + x_{12}.$$

Similarly, suppose we wanted the sum of only the last seven months in the year. This sum would be written as follows:

$$\sum_{i=6}^{12} x_i = x_6 + x_7 + \cdots + x_{12}.$$

In statistics we will often not know in advance what the final value in a summation will be. For example, we know that we want to sum a set of sales values, but we don't know how many values there are to be summed. To designate this situation, we will let the symbol $n$ represent the last number in the sum (where $n$ can be any integer value, such as $1, 2, 3, \ldots$). The notation

$$\sum_{i=1}^{n} x_i = x_1 + x_2 + \cdots + x_n$$

is thus read as "the sum of $n$ numbers, where the first number is $x_1$, the second is $x_2$, and the last is $x_n$." In summing monthly sales over a year, we would thus let $n = 12$, so that $\sum_{i=1}^{n} x_i = \sum_{i=1}^{12} x_i$.

We should perhaps mention that, in some chapters in this book, we will sometimes omit the limits of summation, and simply write $\sum x_i$. This notion should be interpreted to mean "sum all relevant values of $x_i$." In these instances we will make sure that the reader always knows what the relevant values of $x_i$ are. Also, we might point out that the choice of symbols in designating a sum of numbers is often quite arbitrary. For example, we might have used the letter $y$ to denote monthly sales (instead of $x$), and used the letter $j$ as a subscript (instead of $i$). In this case $\sum_{j=1}^{12} y_j$ would denote the sum of the twelve monthly values.

### Double Summations

In a number of chapters in this book we will find it convenient to use *two* subscripts instead of just one. In these instances the first subscript indicates one characteristic under study, and the second subscript some other characteristic. For example, suppose we let $x_{ij}$ = sales in the $i$th month by the $j$th salesman. The notation $x_{6,2} = 15$ would indicate that in the sixth month ($i = 6$), salesman number 2 ($j = 2$) sold 15 units. Using the same procedure as described above, we can denote the total sales over 12 months by the $j$th salesman as the sum of $x_{1j}$ (sales in the 1st month by the $j$th salesman) plus $x_{2j}, \ldots,$ plus $x_{12,j}$ (sales in the 12th month by the $j$th salesman). That is,

*Total sales by salesman $j$:*

$$\sum_{i=1}^{12} x_{ij} = x_{1j} + x_{2j} + \cdots + x_{12,j}.$$

As another example of a similar type of sum, suppose we want to denote the sum of sales in the $i$th month (where $i$ is some number between 1 and 12) over all the salesmen in the company. If we let $m$ = total number of salesmen, then this sum is $x_{i1}$ (sales in month $i$ by salesman #1) plus $x_{i2}, \ldots,$ plus $x_{im}$ (sales in month $i$ by salesman $m$). That is,

*Total sales in month $i$:*

$$\sum_{j=1}^{m} x_{ij} = x_{i1} + x_{i2} + \cdots + x_{im}.$$

Finally, we might wish to sum over all months ($i = 1, 2, \ldots, 12$) and all salesmen ($j = 1, 2, \ldots, m$). This sum could be written as:

*Total sales over all months and all salesmen:*

$$\sum_{\text{All } j} \sum_{\text{All } i} x_{ij} = \left\{ \begin{array}{l} x_{11} \quad + x_{12} + \cdots \quad + x_{1m} \\ + x_{21} \quad + x_{22} + \cdots \quad + x_{2m} \\ \quad \vdots \\ + x_{12,1} + x_{12,2} + \cdots + x_{12,m} \end{array} \right\}.$$

# APPENDIX B

# Tables

# Table I  Binomial Distribution

From Robert Schlaifer, *Analysis of Decisions Under Uncertainty* (New York: McGraw-Hill Book Co., Inc., Preliminary Edition, Volume II, 1967) by specific permission of the President and Fellows of Harvard College, who hold the copyright.

The following table gives values of the binomial mass function defined by

$$p(x) = \binom{n}{x} p^x (1-p)^{n-x}$$

$$= \frac{n!}{x!(n-x)!}\, p^x (1-p)^{n-x}.$$

This is the probability of exactly $x$ successes in $n$ independent Bernoulli trials with probability of success on a single trial equal to $p$. The values of $x$ at the left of any section are to be used in conjunction with the values of $p$ at the top of that section; the values of $x$ at the right of any section are to be used in conjunction with the values of $p$ at the bottom of that section.

*Example:* To evaluate $p(x)$ for $n = 5$, $x = 3$, and $p = 0.83$, locate the section of the table for $n = 5$, the column for $p = 0.83$, and the row for $x = 3$, and read

$$p(x) = 0.1652.$$

## $n = 1$

| $x$ | 01 | 02 | 03 | 04 | 05 | 06 | 07 | 08 | 09 | 10 | $x$ |
|---|---|---|---|---|---|---|---|---|---|---|---|
| 0 | 9900 | 9800 | 9700 | 9600 | 9500 | 9400 | 9300 | 9200 | 9100 | 9000 | 1 |
| 1 | 0100 | 0200 | 0300 | 0400 | 0500 | 0600 | 0700 | 0800 | 0900 | 1000 | 0 |
| $p$ | 99 | 98 | 97 | 96 | 95 | 94 | 93 | 92 | 91 | 90 | $p$ |

| $x$ | 11 | 12 | 13 | 14 | 15 | 16 | 17 | 18 | 19 | 20 | $x$ |
|---|---|---|---|---|---|---|---|---|---|---|---|
| 0 | 8900 | 8800 | 8700 | 8600 | 8500 | 8400 | 8300 | 8200 | 8100 | 8000 | 1 |
| 1 | 1100 | 1200 | 1300 | 1400 | 1500 | 1600 | 1700 | 1800 | 1900 | 2000 | 0 |
| $p$ | 89 | 88 | 87 | 86 | 85 | 84 | 83 | 82 | 81 | 80 | $p$ |

| $x$ | 21 | 22 | 23 | 24 | 25 | 26 | 27 | 28 | 29 | 30 | $x$ |
|---|---|---|---|---|---|---|---|---|---|---|---|
| 0 | 7900 | 7800 | 7700 | 7600 | 7500 | 7400 | 7300 | 7200 | 7100 | 7000 | 1 |
| 1 | 2100 | 2200 | 2300 | 2400 | 2500 | 2600 | 2700 | 2800 | 2900 | 3000 | 0 |
| $p$ | 79 | 78 | 77 | 76 | 75 | 74 | 73 | 72 | 71 | 70 | $p$ |

| $x$ | 31 | 32 | 33 | 34 | 35 | 36 | 37 | 38 | 39 | 40 | 50 | $x$ |
|---|---|---|---|---|---|---|---|---|---|---|---|---|
| 0 | 6900 | 6800 | 6700 | 6600 | 6500 | 6400 | 6300 | 6200 | 6100 | 6000 | 5000 | 1 |
| 1 | 3100 | 3200 | 3300 | 3400 | 3500 | 3600 | 3700 | 3800 | 3900 | 4000 | 5000 | 0 |
| $p$ | 69 | 68 | 67 | 66 | 65 | 64 | 63 | 62 | 61 | 60 | 50 | $p$ |

## $n = 2$

| $x$ | 01 | 02 | 03 | 04 | 05 | 06 | 07 | 08 | 09 | 10 | $x$ |
|---|---|---|---|---|---|---|---|---|---|---|---|
| 0 | 9801 | 9604 | 9409 | 9216 | 9025 | 8836 | 8649 | 8464 | 8281 | 8100 | 2 |
| 1 | 0198 | 0392 | 0582 | 0768 | 0950 | 1128 | 1302 | 1472 | 1638 | 1800 | 1 |
| 2 | 0001 | 0004 | 0009 | 0016 | 0025 | 0036 | 0049 | 0064 | 0081 | 0100 | 0 |
| $p$ | 99 | 98 | 97 | 96 | 95 | 94 | 93 | 92 | 91 | 90 | $p$ |

| $x$ | 11 | 12 | 13 | 14 | 15 | 16 | 17 | 18 | 19 | 20 | $x$ |
|---|---|---|---|---|---|---|---|---|---|---|---|
| 0 | 7921 | 7744 | 7569 | 7396 | 7225 | 7056 | 6889 | 6724 | 6561 | 6400 | 2 |
| 1 | 1958 | 2112 | 2262 | 2408 | 2550 | 2688 | 2822 | 2952 | 3078 | 3200 | 1 |
| 2 | 0121 | 0144 | 0169 | 0196 | 0225 | 0256 | 0289 | 0324 | 0361 | 0400 | 0 |
| $p$ | 89 | 88 | 87 | 86 | 85 | 84 | 83 | 82 | 81 | 80 | $p$ |

| $x$ | 21 | 22 | 23 | 24 | 25 | 26 | 27 | 28 | 29 | 30 | $x$ |
|---|---|---|---|---|---|---|---|---|---|---|---|
| 0 | 6241 | 6084 | 5929 | 5776 | 5625 | 5476 | 5329 | 5184 | 5041 | 4900 | 2 |
| 1 | 3318 | 3432 | 3542 | 3648 | 3750 | 3848 | 3942 | 4032 | 4118 | 4200 | 1 |
| 2 | 0441 | 0484 | 0529 | 0576 | 0625 | 0676 | 0729 | 0784 | 0841 | 0900 | 0 |
| $p$ | 79 | 78 | 77 | 76 | 75 | 74 | 73 | 72 | 71 | 70 | $p$ |

| $x$ | 31 | 32 | 33 | 34 | 35 | 36 | 37 | 38 | 39 | 40 | $x$ |
|---|---|---|---|---|---|---|---|---|---|---|---|
| 0 | 4761 | 4624 | 4489 | 4356 | 4225 | 4096 | 3969 | 3844 | 3721 | 3600 | 2 |
| 1 | 4278 | 4352 | 4422 | 4488 | 4550 | 4608 | 4662 | 4712 | 4758 | 4800 | 1 |
| 2 | 0961 | 1024 | 1089 | 1156 | 1225 | 1296 | 1369 | 1444 | 1521 | 1600 | 0 |
| $p$ | 69 | 68 | 67 | 66 | 65 | 64 | 63 | 62 | 61 | 60 | $p$ |

| $x$ | 41 | 42 | 43 | 44 | 45 | 46 | 47 | 48 | 49 | 50 | $x$ |
|---|---|---|---|---|---|---|---|---|---|---|---|
| 0 | 3481 | 3364 | 3249 | 3136 | 3025 | 2916 | 2809 | 2704 | 2601 | 2500 | 2 |
| 1 | 4838 | 4872 | 4902 | 4928 | 4950 | 4968 | 4982 | 4992 | 4998 | 5000 | 1 |
| 2 | 1681 | 1764 | 1849 | 1936 | 2025 | 2116 | 2209 | 2304 | 2401 | 2500 | 0 |
| $p$ | 59 | 58 | 57 | 56 | 55 | 54 | 53 | 52 | 51 | 50 | $p$ |

(Continued)

# Table I (Continued)

## n = 3

| x | 01 | 02 | 03 | 04 | 05 | 06 | 07 | 08 | 09 | 10 |
|---|----|----|----|----|----|----|----|----|----|----|
| 0 | 9703 | 9412 | 9127 | 8847 | 8574 | 8306 | 8044 | 7787 | 7536 | 7290 |
| 1 | 0294 | 0576 | 0847 | 1106 | 1354 | 1590 | 1816 | 2031 | 2236 | 2430 |
| 2 | 0003 | 0012 | 0026 | 0046 | 0071 | 0102 | 0137 | 0177 | 0221 | 0270 |
| 3 | 0000 | 0000 | 0000 | 0001 | 0001 | 0002 | 0003 | 0005 | 0007 | 0010 |
| p | 99 | 98 | 97 | 96 | 95 | 94 | 93 | 92 | 91 | 90 |
|   | 3 | 2 | 1 | 0 | | | | | | x |

| x | 11 | 12 | 13 | 14 | 15 | 16 | 17 | 18 | 19 | 20 |
|---|----|----|----|----|----|----|----|----|----|----|
| 0 | 7050 | 6815 | 6585 | 6361 | 6141 | 5927 | 5718 | 5514 | 5314 | 5120 |
| 1 | 2614 | 2788 | 2952 | 3106 | 3251 | 3387 | 3513 | 3631 | 3740 | 3840 |
| 2 | 0323 | 0380 | 0441 | 0506 | 0574 | 0645 | 0720 | 0797 | 0877 | 0960 |
| 3 | 0013 | 0017 | 0022 | 0027 | 0034 | 0041 | 0049 | 0058 | 0069 | 0080 |
| p | 89 | 88 | 87 | 86 | 85 | 84 | 83 | 82 | 81 | 80 |
|   | 3 | 2 | 1 | 0 | | | | | | x |

| x | 21 | 22 | 23 | 24 | 25 | 26 | 27 | 28 | 29 | 30 |
|---|----|----|----|----|----|----|----|----|----|----|
| 0 | 4930 | 4746 | 4565 | 4390 | 4219 | 4052 | 3890 | 3732 | 3579 | 3430 |
| 1 | 3932 | 4014 | 4091 | 4159 | 4219 | 4271 | 4316 | 4355 | 4386 | 4410 |
| 2 | 1045 | 1133 | 1222 | 1313 | 1406 | 1501 | 1597 | 1693 | 1791 | 1890 |
| 3 | 0093 | 0106 | 0122 | 0138 | 0156 | 0176 | 0197 | 0220 | 0244 | 0270 |
| p | 79 | 78 | 77 | 76 | 75 | 74 | 73 | 72 | 71 | 70 |
|   | 3 | 2 | 1 | 0 | | | | | | x |

| x | 31 | 32 | 33 | 34 | 35 | 36 | 37 | 38 | 39 | 40 |
|---|----|----|----|----|----|----|----|----|----|----|
| 0 | 3285 | 3144 | 3008 | 2875 | 2746 | 2621 | 2500 | 2383 | 2270 | 2160 |
| 1 | 4428 | 4439 | 4444 | 4443 | 4436 | 4424 | 4406 | 4382 | 4354 | 4320 |
| 2 | 1989 | 2089 | 2189 | 2289 | 2389 | 2488 | 2587 | 2686 | 2783 | 2880 |
| 3 | 0298 | 0328 | 0359 | 0393 | 0429 | 0467 | 0507 | 0549 | 0593 | 0640 |
| p | 69 | 68 | 67 | 66 | 65 | 64 | 63 | 62 | 61 | 60 |
|   | 3 | 2 | 1 | 0 | | | | | | x |

| x | 41 | 42 | 43 | 44 | 45 | 46 | 47 | 48 | 49 | 50 |
|---|----|----|----|----|----|----|----|----|----|----|
| 0 | 2054 | 1951 | 1852 | 1756 | 1664 | 1575 | 1489 | 1406 | 1327 | 1250 |
| 1 | 4282 | 4239 | 4191 | 4140 | 4084 | 4024 | 3961 | 3894 | 3823 | 3750 |
| 2 | 2975 | 3069 | 3162 | 3252 | 3341 | 3428 | 3512 | 3594 | 3674 | 3750 |
| 3 | 0689 | 0741 | 0795 | 0852 | 0911 | 0973 | 1038 | 1106 | 1176 | 1250 |
| p | 59 | 58 | 57 | 56 | 55 | 54 | 53 | 52 | 51 | 50 |
|   | 3 | 2 | 1 | 0 | | | | | | x |

## n = 4

| x | 01 | 02 | 03 | 04 | 05 | 06 | 07 | 08 | 09 | 10 |
|---|----|----|----|----|----|----|----|----|----|----|
| 0 | 9606 | 9224 | 8853 | 8493 | 8145 | 7807 | 7481 | 7164 | 6857 | 6561 |
| 1 | 0388 | 0753 | 1095 | 1416 | 1715 | 1993 | 2252 | 2492 | 2713 | 2916 |
| 2 | 0006 | 0023 | 0051 | 0088 | 0135 | 0191 | 0254 | 0325 | 0402 | 0486 |
| 3 | 0000 | 0000 | 0001 | 0002 | 0005 | 0008 | 0013 | 0019 | 0027 | 0036 |
| 4 | 0000 | 0000 | 0000 | 0000 | 0000 | 0000 | 0000 | 0000 | 0001 | 0001 |
| p | 99 | 98 | 97 | 96 | 95 | 94 | 93 | 92 | 91 | 90 |
|   | 4 | 3 | 2 | 1 | 0 | | | | | x |

| x | 11 | 12 | 13 | 14 | 15 | 16 | 17 | 18 | 19 | 20 |
|---|----|----|----|----|----|----|----|----|----|----|
| 0 | 6274 | 5997 | 5729 | 5470 | 5220 | 4979 | 4746 | 4521 | 4305 | 4096 |
| 1 | 3102 | 3271 | 3424 | 3562 | 3685 | 3793 | 3888 | 3970 | 4039 | 4096 |
| 2 | 0575 | 0669 | 0767 | 0870 | 0975 | 1084 | 1195 | 1307 | 1421 | 1536 |
| 3 | 0047 | 0061 | 0076 | 0094 | 0115 | 0138 | 0163 | 0191 | 0222 | 0256 |
| 4 | 0001 | 0002 | 0003 | 0004 | 0005 | 0007 | 0008 | 0010 | 0013 | 0016 |
| p | 89 | 88 | 87 | 86 | 85 | 84 | 83 | 82 | 81 | 80 |
|   | 4 | 3 | 2 | 1 | 0 | | | | | x |

| x | 21 | 22 | 23 | 24 | 25 | 26 | 27 | 28 | 29 | 30 |
|---|----|----|----|----|----|----|----|----|----|----|
| 0 | 3895 | 3702 | 3515 | 3336 | 3164 | 2999 | 2840 | 2687 | 2541 | 2401 |
| 1 | 4142 | 4176 | 4200 | 4214 | 4219 | 4214 | 4201 | 4180 | 4152 | 4116 |
| 2 | 1651 | 1767 | 1882 | 1996 | 2109 | 2221 | 2331 | 2439 | 2544 | 2646 |
| 3 | 0293 | 0332 | 0375 | 0420 | 0469 | 0520 | 0575 | 0632 | 0693 | 0756 |
| 4 | 0019 | 0023 | 0028 | 0033 | 0039 | 0046 | 0053 | 0061 | 0071 | 0081 |
| p | 79 | 78 | 77 | 76 | 75 | 74 | 73 | 72 | 71 | 70 |
|   | 4 | 3 | 2 | 1 | 0 | | | | | x |

| x | 31 | 32 | 33 | 34 | 35 | 36 | 37 | 38 | 39 | 40 |
|---|----|----|----|----|----|----|----|----|----|----|
| 0 | 2267 | 2138 | 2015 | 1897 | 1785 | 1678 | 1575 | 1478 | 1385 | 1296 |
| 1 | 4074 | 4025 | 3970 | 3910 | 3845 | 3775 | 3701 | 3623 | 3541 | 3456 |
| 2 | 2745 | 2841 | 2933 | 3021 | 3105 | 3185 | 3260 | 3330 | 3396 | 3456 |
| 3 | 0822 | 0891 | 0963 | 1038 | 1115 | 1194 | 1276 | 1361 | 1447 | 1536 |
| 4 | 0092 | 0105 | 0119 | 0134 | 0150 | 0168 | 0187 | 0209 | 0231 | 0256 |
| p | 69 | 68 | 67 | 66 | 65 | 64 | 63 | 62 | 61 | 60 |
|   | 4 | 3 | 2 | 1 | 0 | | | | | x |

| x | 41 | 42 | 43 | 44 | 45 | 46 | 47 | 48 | 49 | 50 |
|---|----|----|----|----|----|----|----|----|----|----|
| 0 | 1212 | 1132 | 1056 | 0983 | 0915 | 0850 | 0789 | 0731 | 0677 | 0625 |
| 1 | 3368 | 3278 | 3185 | 3091 | 2995 | 2897 | 2799 | 2700 | 2600 | 2500 |
| 2 | 3511 | 3560 | 3604 | 3643 | 3675 | 3702 | 3723 | 3738 | 3747 | 3750 |
| 3 | 1627 | 1719 | 1813 | 1908 | 2005 | 2102 | 2201 | 2300 | 2400 | 2500 |
| 4 | 0283 | 0311 | 0342 | 0375 | 0410 | 0448 | 0488 | 0531 | 0576 | 0625 |
| p | 59 | 58 | 57 | 56 | 55 | 54 | 53 | 52 | 51 | 50 |
|   | 4 | 3 | 2 | 1 | 0 | | | | | x |

## n = 5

| x | 01 | 02 | 03 | 04 | 05 | 06 | 07 | 08 | 09 | 10 |
|---|----|----|----|----|----|----|----|----|----|----|
| 0 | 9510 | 9039 | 8587 | 8154 | 7738 | 7339 | 6957 | 6591 | 6240 | 5905 |
| 1 | 0480 | 0922 | 1328 | 1699 | 2036 | 2342 | 2618 | 2866 | 3086 | 3280 |
| 2 | 0010 | 0038 | 0082 | 0142 | 0214 | 0299 | 0394 | 0498 | 0610 | 0729 |
| 3 | 0000 | 0001 | 0003 | 0006 | 0011 | 0019 | 0030 | 0043 | 0060 | 0081 |
| 4 | 0000 | 0000 | 0000 | 0000 | 0000 | 0001 | 0001 | 0002 | 0003 | 0004 |
| p | 99 | 98 | 97 | 96 | 95 | 94 | 93 | 92 | 91 | 90 |
|   | 5 | 4 | 3 | 2 | 1 | 0 | | | | x |

*(Continued)*

# Table I (Continued)

## n = 5

| x | .11 | .12 | .13 | .14 | .15 | .16 | .17 | .18 | .19 | .20 | |
|---|---|---|---|---|---|---|---|---|---|---|---|
| 0 | 5584 | 5277 | 4984 | 4704 | 4437 | 4182 | 3939 | 3707 | 3487 | 3277 | 5 |
| 1 | 3451 | 3598 | 3724 | 3829 | 3915 | 3983 | 4034 | 4069 | 4089 | 4096 | 4 |
| 2 | 0853 | 0981 | 1113 | 1247 | 1382 | 1517 | 1652 | 1786 | 1919 | 2048 | 3 |
| 3 | 0105 | 0134 | 0166 | 0203 | 0244 | 0289 | 0338 | 0392 | 0450 | 0512 | 2 |
| 4 | 0007 | 0009 | 0012 | 0017 | 0022 | 0028 | 0035 | 0043 | 0053 | 0064 | 1 |
| 5 | 0000 | 0000 | 0000 | 0001 | 0001 | 0001 | 0002 | 0002 | 0003 | 0003 | 0 |
| | .89 | .88 | .87 | .86 | .85 | .84 | .83 | .82 | .81 | .80 | x |
| | | | | | | | | | | *p* | |

| x | .21 | .22 | .23 | .24 | .25 | .26 | .27 | .28 | .29 | .30 | |
|---|---|---|---|---|---|---|---|---|---|---|---|
| 0 | 3077 | 2887 | 2707 | 2536 | 2373 | 2219 | 2073 | 1935 | 1804 | 1681 | 5 |
| 1 | 4090 | 4072 | 4043 | 4003 | 3955 | 3898 | 3834 | 3762 | 3685 | 3601 | 4 |
| 2 | 2174 | 2297 | 2415 | 2529 | 2637 | 2739 | 2836 | 2926 | 3010 | 3087 | 3 |
| 3 | 0578 | 0648 | 0721 | 0798 | 0879 | 0962 | 1049 | 1138 | 1229 | 1323 | 2 |
| 4 | 0077 | 0091 | 0108 | 0126 | 0146 | 0169 | 0194 | 0221 | 0251 | 0283 | 1 |
| 5 | 0004 | 0005 | 0006 | 0008 | 0010 | 0012 | 0014 | 0017 | 0021 | 0024 | 0 |
| | .79 | .78 | .77 | .76 | .75 | .74 | .73 | .72 | .71 | .70 | x |
| | | | | | | | | | | *p* | |

| x | .31 | .32 | .33 | .34 | .35 | .36 | .37 | .38 | .39 | .40 | |
|---|---|---|---|---|---|---|---|---|---|---|---|
| 0 | 1564 | 1454 | 1350 | 1252 | 1160 | 1074 | 0992 | 0916 | 0845 | 0778 | 5 |
| 1 | 3513 | 3421 | 3325 | 3226 | 3124 | 3020 | 2914 | 2808 | 2700 | 2592 | 4 |
| 2 | 3157 | 3220 | 3275 | 3323 | 3364 | 3397 | 3423 | 3441 | 3452 | 3456 | 3 |
| 3 | 1418 | 1515 | 1613 | 1712 | 1811 | 1911 | 2010 | 2109 | 2207 | 2304 | 2 |
| 4 | 0319 | 0357 | 0397 | 0441 | 0488 | 0537 | 0590 | 0646 | 0706 | 0768 | 1 |
| 5 | 0029 | 0034 | 0039 | 0045 | 0053 | 0060 | 0069 | 0079 | 0090 | 0102 | 0 |
| | .69 | .68 | .67 | .66 | .65 | .64 | .63 | .62 | .61 | .60 | x |
| | | | | | | | | | | *p* | |

| x | .41 | .42 | .43 | .44 | .45 | .46 | .47 | .48 | .49 | .50 | |
|---|---|---|---|---|---|---|---|---|---|---|---|
| 0 | 0715 | 0656 | 0602 | 0551 | 0503 | 0459 | 0418 | 0380 | 0345 | 0313 | 5 |
| 1 | 2484 | 2376 | 2270 | 2164 | 2059 | 1956 | 1854 | 1755 | 1657 | 1562 | 4 |
| 2 | 3452 | 3442 | 3424 | 3400 | 3369 | 3332 | 3289 | 3240 | 3185 | 3125 | 3 |
| 3 | 2399 | 2492 | 2583 | 2671 | 2757 | 2838 | 2916 | 2990 | 3060 | 3125 | 2 |
| 4 | 0834 | 0902 | 0974 | 1049 | 1128 | 1209 | 1293 | 1380 | 1470 | 1562 | 1 |
| 5 | 0116 | 0131 | 0147 | 0165 | 0185 | 0206 | 0229 | 0255 | 0282 | 0312 | 0 |
| | .59 | .58 | .57 | .56 | .55 | .54 | .53 | .52 | .51 | .50 | x |
| | | | | | | | | | | *p* | |

## n = 6

| x | .01 | .02 | .03 | .04 | .05 | .06 | .07 | .08 | .09 | .10 | |
|---|---|---|---|---|---|---|---|---|---|---|---|
| 0 | 9415 | 8858 | 8330 | 7828 | 7351 | 6899 | 6470 | 6064 | 5679 | 5314 | 6 |
| 1 | 0571 | 1085 | 1546 | 1957 | 2321 | 2642 | 2922 | 3164 | 3370 | 3543 | 5 |
| 2 | 0014 | 0055 | 0120 | 0204 | 0305 | 0422 | 0550 | 0688 | 0833 | 0984 | 4 |
| 3 | 0000 | 0002 | 0005 | 0011 | 0021 | 0036 | 0055 | 0080 | 0110 | 0146 | 3 |
| 4 | 0000 | 0000 | 0000 | 0000 | 0001 | 0002 | 0003 | 0005 | 0008 | 0012 | 2 |
| 5 | 0000 | 0000 | 0000 | 0000 | 0000 | 0000 | 0000 | 0000 | 0000 | 0001 | 1 |
| | .99 | .98 | .97 | .96 | .95 | .94 | .93 | .92 | .91 | .90 | 0 |
| | | | | | | | | | | *p* | x |

| x | .11 | .12 | .13 | .14 | .15 | .16 | .17 | .18 | .19 | .20 | |
|---|---|---|---|---|---|---|---|---|---|---|---|
| 0 | 4970 | 4644 | 4336 | 4046 | 3771 | 3513 | 3269 | 3040 | 2824 | 2621 | 6 |
| 1 | 3685 | 3800 | 3888 | 3952 | 3993 | 4015 | 4018 | 4004 | 3975 | 3932 | 5 |
| 2 | 1139 | 1295 | 1452 | 1608 | 1762 | 1912 | 2057 | 2197 | 2331 | 2458 | 4 |
| 3 | 0188 | 0236 | 0289 | 0349 | 0415 | 0486 | 0562 | 0643 | 0729 | 0819 | 3 |
| 4 | 0017 | 0024 | 0032 | 0043 | 0055 | 0069 | 0086 | 0106 | 0128 | 0154 | 2 |
| 5 | 0001 | 0001 | 0002 | 0003 | 0004 | 0005 | 0007 | 0009 | 0012 | 0015 | 1 |
| 6 | 0000 | 0000 | 0000 | 0000 | 0000 | 0000 | 0000 | 0000 | 0000 | 0001 | 0 |
| | .89 | .88 | .87 | .86 | .85 | .84 | .83 | .82 | .81 | .80 | x |
| | | | | | | | | | | *p* | |

| x | .21 | .22 | .23 | .24 | .25 | .26 | .27 | .28 | .29 | .30 | |
|---|---|---|---|---|---|---|---|---|---|---|---|
| 0 | 2431 | 2252 | 2084 | 1927 | 1780 | 1642 | 1513 | 1393 | 1281 | 1176 | 6 |
| 1 | 3877 | 3811 | 3735 | 3651 | 3560 | 3462 | 3358 | 3251 | 3139 | 3025 | 5 |
| 2 | 2577 | 2687 | 2789 | 2882 | 2966 | 3041 | 3105 | 3160 | 3206 | 3241 | 4 |
| 3 | 0913 | 1011 | 1111 | 1214 | 1318 | 1424 | 1531 | 1639 | 1746 | 1852 | 3 |
| 4 | 0182 | 0214 | 0249 | 0287 | 0330 | 0375 | 0425 | 0478 | 0535 | 0595 | 2 |
| 5 | 0019 | 0024 | 0030 | 0036 | 0044 | 0053 | 0063 | 0074 | 0087 | 0102 | 1 |
| 6 | 0001 | 0001 | 0001 | 0002 | 0002 | 0003 | 0004 | 0005 | 0006 | 0007 | 0 |
| | .79 | .78 | .77 | .76 | .75 | .74 | .73 | .72 | .71 | .70 | x |
| | | | | | | | | | | *p* | |

| x | .31 | .32 | .33 | .34 | .35 | .36 | .37 | .38 | .39 | .40 | |
|---|---|---|---|---|---|---|---|---|---|---|---|
| 0 | 1079 | 0989 | 0905 | 0827 | 0754 | 0687 | 0625 | 0568 | 0515 | 0467 | 6 |
| 1 | 2909 | 2792 | 2673 | 2555 | 2437 | 2319 | 2203 | 2089 | 1976 | 1866 | 5 |
| 2 | 3267 | 3284 | 3292 | 3290 | 3280 | 3261 | 3235 | 3201 | 3159 | 3110 | 4 |
| 3 | 1957 | 2061 | 2162 | 2260 | 2355 | 2446 | 2533 | 2616 | 2693 | 2765 | 3 |
| 4 | 0660 | 0727 | 0799 | 0873 | 0951 | 1032 | 1116 | 1202 | 1291 | 1382 | 2 |
| 5 | 0119 | 0137 | 0157 | 0180 | 0205 | 0232 | 0262 | 0295 | 0330 | 0369 | 1 |
| 6 | 0009 | 0011 | 0013 | 0015 | 0018 | 0022 | 0026 | 0030 | 0035 | 0041 | 0 |
| | .69 | .68 | .67 | .66 | .65 | .64 | .63 | .62 | .61 | .60 | x |
| | | | | | | | | | | *p* | |

(Continued)

Table I (Continued)

**n = 6**

| x | 41 | 42 | 43 | 44 | 45 | 46 | 47 | 48 | 49 | 50 | x |
|---|----|----|----|----|----|----|----|----|----|----|---|
| 0 | 0422 | 0381 | 0343 | 0308 | 0277 | 0248 | 0222 | 0198 | 0176 | 0156 | 6 |
| 1 | 1759 | 1654 | 1552 | 1454 | 1359 | 1267 | 1179 | 1095 | 1014 | 0937 | 5 |
| 2 | 3055 | 2994 | 2928 | 2856 | 2780 | 2699 | 2615 | 2527 | 2436 | 2344 | 4 |
| 3 | 2831 | 2891 | 2945 | 2992 | 3032 | 3065 | 3091 | 3110 | 3121 | 3125 | 3 |
| 4 | 1475 | 1570 | 1666 | 1763 | 1861 | 1958 | 2056 | 2153 | 2249 | 2344 | 2 |
| 5 | 0410 | 0455 | 0503 | 0554 | 0609 | 0667 | 0729 | 0795 | 0864 | 0937 | 1 |
| 6 | 0048 | 0055 | 0063 | 0073 | 0083 | 0095 | 0108 | 0122 | 0138 | 0156 | 0 |
| p | 59 | 58 | 57 | 56 | 55 | 54 | 53 | 52 | 51 | 50 | x |

**n = 7**

| x | 01 | 02 | 03 | 04 | 05 | 06 | 07 | 08 | 09 | 10 | x |
|---|----|----|----|----|----|----|----|----|----|----|---|
| 0 | 9321 | 8681 | 8080 | 7514 | 6983 | 6485 | 6017 | 5578 | 5168 | 4783 | 7 |
| 1 | 0659 | 1240 | 1749 | 2192 | 2573 | 2897 | 3170 | 3396 | 3578 | 3720 | 6 |
| 2 | 0020 | 0076 | 0162 | 0274 | 0406 | 0555 | 0716 | 0886 | 1061 | 1240 | 5 |
| 3 | 0000 | 0003 | 0008 | 0019 | 0036 | 0059 | 0090 | 0128 | 0175 | 0230 | 4 |
| 4 | 0000 | 0000 | 0000 | 0001 | 0002 | 0004 | 0007 | 0011 | 0017 | 0026 | 3 |
| 5 | 0000 | 0000 | 0000 | 0000 | 0000 | 0000 | 0000 | 0001 | 0001 | 0002 | 2 |
| p | 99 | 98 | 97 | 96 | 95 | 94 | 93 | 92 | 91 | 90 | x |

| x | 11 | 12 | 13 | 14 | 15 | 16 | 17 | 18 | 19 | 20 | x |
|---|----|----|----|----|----|----|----|----|----|----|---|
| 0 | 4423 | 4087 | 3773 | 3479 | 3206 | 2951 | 2714 | 2493 | 2288 | 2097 | 7 |
| 1 | 3827 | 3901 | 3946 | 3965 | 3960 | 3935 | 3891 | 3830 | 3756 | 3670 | 6 |
| 2 | 1419 | 1596 | 1769 | 1936 | 2097 | 2248 | 2391 | 2523 | 2643 | 2753 | 5 |
| 3 | 0292 | 0363 | 0441 | 0525 | 0617 | 0714 | 0816 | 0923 | 1033 | 1147 | 4 |
| 4 | 0036 | 0049 | 0066 | 0086 | 0109 | 0136 | 0167 | 0203 | 0242 | 0287 | 3 |
| 5 | 0003 | 0004 | 0006 | 0008 | 0012 | 0016 | 0021 | 0027 | 0034 | 0043 | 2 |
| 6 | 0000 | 0000 | 0000 | 0000 | 0001 | 0001 | 0001 | 0002 | 0003 | 0004 | 1 |
| p | 89 | 88 | 87 | 86 | 85 | 84 | 83 | 82 | 81 | 80 | x |

| x | 21 | 22 | 23 | 24 | 25 | 26 | 27 | 28 | 29 | 30 | x |
|---|----|----|----|----|----|----|----|----|----|----|---|
| 0 | 1920 | 1757 | 1605 | 1465 | 1335 | 1215 | 1105 | 1003 | 0910 | 0824 | 7 |
| 1 | 3573 | 3468 | 3356 | 3237 | 3115 | 2989 | 2860 | 2731 | 2600 | 2471 | 6 |
| 2 | 2850 | 2935 | 3007 | 3067 | 3115 | 3150 | 3174 | 3186 | 3186 | 3177 | 5 |
| 3 | 1263 | 1379 | 1497 | 1614 | 1730 | 1845 | 1956 | 2065 | 2169 | 2269 | 4 |
| 4 | 0336 | 0389 | 0447 | 0510 | 0577 | 0649 | 0724 | 0803 | 0886 | 0972 | 3 |
| 5 | 0054 | 0066 | 0080 | 0097 | 0115 | 0137 | 0161 | 0187 | 0217 | 0250 | 2 |
| 6 | 0005 | 0006 | 0008 | 0010 | 0013 | 0016 | 0020 | 0024 | 0030 | 0036 | 1 |
| 7 | 0000 | 0000 | 0000 | 0000 | 0001 | 0001 | 0001 | 0001 | 0002 | 0002 | 0 |
| p | 79 | 78 | 77 | 76 | 75 | 74 | 73 | 72 | 71 | 70 | x |

**n = 7**

| x | 31 | 32 | 33 | 34 | 35 | 36 | 37 | 38 | 39 | 40 | x |
|---|----|----|----|----|----|----|----|----|----|----|---|
| 0 | 0745 | 0672 | 0606 | 0546 | 0490 | 0440 | 0394 | 0352 | 0314 | 0280 | 7 |
| 1 | 2342 | 2215 | 2090 | 1967 | 1848 | 1732 | 1619 | 1511 | 1407 | 1306 | 6 |
| 2 | 3156 | 3127 | 3088 | 3040 | 2985 | 2922 | 2853 | 2778 | 2698 | 2613 | 5 |
| 3 | 2363 | 2452 | 2535 | 2610 | 2679 | 2740 | 2793 | 2838 | 2875 | 2903 | 4 |
| 4 | 1062 | 1154 | 1248 | 1345 | 1442 | 1541 | 1640 | 1739 | 1838 | 1935 | 3 |
| 5 | 0286 | 0326 | 0369 | 0416 | 0466 | 0520 | 0578 | 0640 | 0705 | 0774 | 2 |
| 6 | 0043 | 0051 | 0061 | 0071 | 0084 | 0098 | 0113 | 0131 | 0150 | 0172 | 1 |
| 7 | 0003 | 0003 | 0004 | 0005 | 0006 | 0008 | 0009 | 0011 | 0014 | 0016 | 0 |
| p | 69 | 68 | 67 | 66 | 65 | 64 | 63 | 62 | 61 | 60 | x |

| x | 41 | 42 | 43 | 44 | 45 | 46 | 47 | 48 | 49 | 50 | x |
|---|----|----|----|----|----|----|----|----|----|----|---|
| 0 | 0249 | 0221 | 0195 | 0173 | 0152 | 0134 | 0117 | 0103 | 0090 | 0078 | 7 |
| 1 | 1211 | 1119 | 1032 | 0950 | 0872 | 0798 | 0729 | 0664 | 0604 | 0547 | 6 |
| 2 | 2524 | 2431 | 2336 | 2239 | 2140 | 2040 | 1940 | 1840 | 1740 | 1641 | 5 |
| 3 | 2923 | 2934 | 2937 | 2932 | 2918 | 2897 | 2867 | 2830 | 2786 | 2734 | 4 |
| 4 | 2031 | 2125 | 2216 | 2304 | 2388 | 2468 | 2543 | 2612 | 2676 | 2734 | 3 |
| 5 | 0847 | 0923 | 1003 | 1086 | 1172 | 1261 | 1353 | 1447 | 1543 | 1641 | 2 |
| 6 | 0196 | 0223 | 0252 | 0284 | 0320 | 0358 | 0400 | 0445 | 0494 | 0547 | 1 |
| 7 | 0019 | 0023 | 0027 | 0032 | 0037 | 0044 | 0051 | 0059 | 0068 | 0078 | 0 |
| p | 59 | 58 | 57 | 56 | 55 | 54 | 53 | 52 | 51 | 50 | x |

**n = 8**

| x | 01 | 02 | 03 | 04 | 05 | 06 | 07 | 08 | 09 | 10 | x |
|---|----|----|----|----|----|----|----|----|----|----|---|
| 0 | 9227 | 8508 | 7837 | 7214 | 6634 | 6096 | 5596 | 5132 | 4703 | 4305 | 8 |
| 1 | 0746 | 1389 | 1939 | 2405 | 2793 | 3113 | 3370 | 3570 | 3721 | 3826 | 7 |
| 2 | 0026 | 0099 | 0210 | 0351 | 0515 | 0695 | 0888 | 1087 | 1288 | 1488 | 6 |
| 3 | 0001 | 0004 | 0013 | 0029 | 0054 | 0089 | 0134 | 0189 | 0255 | 0331 | 5 |
| 4 | 0000 | 0000 | 0001 | 0002 | 0004 | 0007 | 0013 | 0021 | 0031 | 0046 | 4 |
| 5 | | | | | | | | | | | 3 |
| p | 99 | 98 | 97 | 96 | 95 | 94 | 93 | 92 | 91 | 90 | x |

| x | 11 | 12 | 13 | 14 | 15 | 16 | 17 | 18 | 19 | 20 | x |
|---|----|----|----|----|----|----|----|----|----|----|---|
| 0 | 3937 | 3596 | 3282 | 2992 | 2725 | 2479 | 2252 | 2044 | 1853 | 1678 | 8 |
| 1 | 3892 | 3923 | 3923 | 3897 | 3847 | 3777 | 3691 | 3590 | 3477 | 3355 | 7 |
| 2 | 1684 | 1872 | 2052 | 2220 | 2376 | 2518 | 2646 | 2758 | 2855 | 2936 | 6 |
| 3 | 0416 | 0511 | 0613 | 0723 | 0839 | 0959 | 1084 | 1211 | 1339 | 1468 | 5 |
| 4 | 0064 | 0087 | 0115 | 0147 | 0185 | 0228 | 0277 | 0332 | 0393 | 0459 | 4 |

(Continued)

A–10

# Table I (Continued)

## n = 9

| p | 01 | 02 | 03 | 04 | 05 | 06 | 07 | 08 | 09 | 10 | x |
|---|----|----|----|----|----|----|----|----|----|----|---|
| 0 | 9135 | 8337 | 7602 | 6925 | 6302 | 5730 | 5204 | 4722 | 4279 | 3874 | 9 |
| 1 | 0830 | 1531 | 2116 | 2597 | 2985 | 3292 | 3525 | 3695 | 3809 | 3874 | 8 |
| 2 | 0034 | 0125 | 0262 | 0433 | 0629 | 0840 | 1061 | 1285 | 1507 | 1722 | 7 |
| 3 | 0001 | 0006 | 0019 | 0042 | 0077 | 0125 | 0186 | 0261 | 0348 | 0446 | 6 |
| 4 | 0000 | 0000 | 0001 | 0003 | 0006 | 0012 | 0021 | 0034 | 0052 | 0074 | 5 |
| 5 | 0000 | 0000 | 0000 | 0000 | 0000 | 0001 | 0002 | 0003 | 0005 | 0008 | 4 |
| 6 | 0000 | 0000 | 0000 | 0000 | 0000 | 0000 | 0000 | 0000 | 0000 | 0001 | 3 |
| p | 99 | 98 | 97 | 96 | 95 | 94 | 93 | 92 | 91 | 90 | x |

| p | 11 | 12 | 13 | 14 | 15 | 16 | 17 | 18 | 19 | 20 | x |
|---|----|----|----|----|----|----|----|----|----|----|---|
| 0 | 3504 | 3165 | 2855 | 2573 | 2316 | 2082 | 1869 | 1676 | 1501 | 1342 | 9 |
| 1 | 3897 | 3884 | 3840 | 3770 | 3679 | 3569 | 3446 | 3312 | 3169 | 3020 | 8 |
| 2 | 1927 | 2119 | 2295 | 2455 | 2597 | 2720 | 2823 | 2908 | 2973 | 3020 | 7 |
| 3 | 0556 | 0674 | 0800 | 0933 | 1069 | 1209 | 1349 | 1489 | 1627 | 1762 | 6 |
| 4 | 0103 | 0138 | 0179 | 0228 | 0283 | 0345 | 0415 | 0490 | 0573 | 0661 | 5 |
| 5 | 0013 | 0019 | 0027 | 0037 | 0050 | 0066 | 0085 | 0108 | 0134 | 0165 | 4 |
| 6 | 0001 | 0002 | 0003 | 0004 | 0006 | 0008 | 0012 | 0016 | 0021 | 0028 | 3 |
| 7 | 0000 | 0000 | 0000 | 0000 | 0000 | 0001 | 0001 | 0001 | 0002 | 0003 | 2 |
| p | 89 | 88 | 87 | 86 | 85 | 84 | 83 | 82 | 81 | 80 | x |

| p | 21 | 22 | 23 | 24 | 25 | 26 | 27 | 28 | 29 | 30 | x |
|---|----|----|----|----|----|----|----|----|----|----|---|
| 0 | 1199 | 1069 | 0952 | 0846 | 0751 | 0665 | 0589 | 0520 | 0458 | 0404 | 9 |
| 1 | 2867 | 2713 | 2558 | 2404 | 2253 | 2104 | 1960 | 1820 | 1685 | 1556 | 8 |
| 2 | 3049 | 3061 | 3056 | 3037 | 3003 | 2957 | 2899 | 2831 | 2754 | 2668 | 7 |
| 3 | 1891 | 2014 | 2130 | 2238 | 2336 | 2424 | 2502 | 2569 | 2624 | 2668 | 6 |
| 4 | 0754 | 0852 | 0954 | 1060 | 1168 | 1278 | 1388 | 1499 | 1608 | 1715 | 5 |
| 5 | 0200 | 0240 | 0285 | 0335 | 0389 | 0449 | 0513 | 0583 | 0657 | 0735 | 4 |
| 6 | 0036 | 0045 | 0057 | 0070 | 0087 | 0105 | 0127 | 0151 | 0179 | 0210 | 3 |
| 7 | 0004 | 0005 | 0007 | 0010 | 0012 | 0016 | 0020 | 0025 | 0031 | 0039 | 2 |
| 8 | 0000 | 0000 | 0001 | 0001 | 0001 | 0001 | 0002 | 0002 | 0003 | 0004 | 1 |
| p | 79 | 78 | 77 | 76 | 75 | 74 | 73 | 72 | 71 | 70 | x |

| p | 31 | 32 | 33 | 34 | 35 | 36 | 37 | 38 | 39 | 40 | x |
|---|----|----|----|----|----|----|----|----|----|----|---|
| 0 | 0355 | 0311 | 0272 | 0238 | 0207 | 0180 | 0156 | 0135 | 0117 | 0101 | 9 |
| 1 | 1433 | 1317 | 1206 | 1102 | 1004 | 0912 | 0826 | 0747 | 0673 | 0605 | 8 |
| 2 | 2576 | 2478 | 2376 | 2270 | 2162 | 2052 | 1941 | 1831 | 1721 | 1612 | 7 |
| 3 | 2701 | 2721 | 2731 | 2729 | 2716 | 2693 | 2660 | 2618 | 2567 | 2508 | 6 |
| 4 | 1820 | 1921 | 2017 | 2109 | 2194 | 2272 | 2344 | 2407 | 2462 | 2508 | 5 |

## n = 8

| p | 11 | 12 | 13 | 14 | 15 | 16 | 17 | 18 | 19 | 20 | x |
|---|----|----|----|----|----|----|----|----|----|----|---|
| 5 | 0006 | 0009 | 0014 | 0019 | 0026 | 0035 | 0045 | 0058 | 0074 | 0092 | 3 |
| 6 | 0000 | 0001 | 0001 | 0002 | 0003 | 0003 | 0005 | 0006 | 0009 | 0011 | 2 |
| 7 | 0000 | 0000 | 0000 | 0000 | 0000 | 0000 | 0000 | 0000 | 0001 | 0001 | 1 |
| p | 89 | 88 | 87 | 86 | 85 | 84 | 83 | 82 | 81 | 80 | x |

| p | 21 | 22 | 23 | 24 | 25 | 26 | 27 | 28 | 29 | 30 | x |
|---|----|----|----|----|----|----|----|----|----|----|---|
| 0 | 1517 | 1370 | 1236 | 1113 | 1001 | 0899 | 0806 | 0722 | 0646 | 0576 | 8 |
| 1 | 3226 | 3092 | 2953 | 2812 | 2670 | 2527 | 2386 | 2247 | 2110 | 1977 | 7 |
| 2 | 3002 | 3052 | 3087 | 3108 | 3115 | 3108 | 3089 | 3058 | 3017 | 2965 | 6 |
| 3 | 1596 | 1722 | 1844 | 1963 | 2076 | 2184 | 2285 | 2379 | 2464 | 2541 | 5 |
| 4 | 0530 | 0607 | 0689 | 0775 | 0865 | 0959 | 1056 | 1156 | 1258 | 1361 | 4 |
| 5 | 0113 | 0137 | 0165 | 0196 | 0231 | 0270 | 0313 | 0360 | 0411 | 0467 | 3 |
| 6 | 0015 | 0019 | 0025 | 0031 | 0038 | 0047 | 0058 | 0070 | 0084 | 0100 | 2 |
| 7 | 0001 | 0002 | 0002 | 0003 | 0004 | 0005 | 0006 | 0008 | 0010 | 0012 | 1 |
| 8 | 0000 | 0000 | 0000 | 0000 | 0000 | 0000 | 0000 | 0000 | 0001 | 0001 | 0 |
| p | 79 | 78 | 77 | 76 | 75 | 74 | 73 | 72 | 71 | 70 | x |

| p | 31 | 32 | 33 | 34 | 35 | 36 | 37 | 38 | 39 | 40 | x |
|---|----|----|----|----|----|----|----|----|----|----|---|
| 0 | 0514 | 0457 | 0406 | 0360 | 0319 | 0281 | 0248 | 0218 | 0192 | 0168 | 8 |
| 1 | 1847 | 1721 | 1600 | 1484 | 1373 | 1267 | 1166 | 1071 | 0981 | 0896 | 7 |
| 2 | 2904 | 2835 | 2758 | 2675 | 2587 | 2494 | 2397 | 2297 | 2194 | 2090 | 6 |
| 3 | 2609 | 2668 | 2717 | 2756 | 2786 | 2805 | 2815 | 2815 | 2806 | 2787 | 5 |
| 4 | 1465 | 1569 | 1673 | 1775 | 1875 | 1973 | 2067 | 2157 | 2242 | 2322 | 4 |
| 5 | 0527 | 0591 | 0659 | 0732 | 0808 | 0888 | 0971 | 1058 | 1147 | 1239 | 3 |
| 6 | 0118 | 0139 | 0162 | 0188 | 0217 | 0250 | 0285 | 0324 | 0367 | 0413 | 2 |
| 7 | 0015 | 0019 | 0023 | 0028 | 0033 | 0040 | 0048 | 0057 | 0067 | 0079 | 1 |
| 8 | 0001 | 0001 | 0001 | 0002 | 0003 | 0004 | 0005 | 0007 | 0005 | 0007 | 0 |
| p | 69 | 68 | 67 | 66 | 65 | 64 | 63 | 62 | 61 | 60 | x |

| p | 41 | 42 | 43 | 44 | 45 | 46 | 47 | 48 | 49 | 50 | x |
|---|----|----|----|----|----|----|----|----|----|----|---|
| 0 | 0147 | 0128 | 0111 | 0097 | 0084 | 0072 | 0062 | 0053 | 0046 | 0039 | 8 |
| 1 | 0816 | 0742 | 0672 | 0608 | 0548 | 0493 | 0442 | 0395 | 0352 | 0312 | 7 |
| 2 | 1985 | 1880 | 1776 | 1672 | 1569 | 1469 | 1371 | 1275 | 1183 | 1094 | 6 |
| 3 | 2759 | 2723 | 2679 | 2627 | 2568 | 2503 | 2431 | 2355 | 2273 | 2187 | 5 |
| 4 | 2397 | 2465 | 2526 | 2580 | 2627 | 2665 | 2695 | 2717 | 2730 | 2734 | 4 |
| 5 | 1332 | 1428 | 1525 | 1622 | 1719 | 1816 | 1912 | 2006 | 2098 | 2187 | 3 |
| 6 | 0463 | 0517 | 0575 | 0637 | 0703 | 0774 | 0848 | 0926 | 1008 | 1094 | 2 |
| 7 | 0092 | 0107 | 0124 | 0143 | 0164 | 0188 | 0215 | 0244 | 0277 | 0312 | 1 |
| 8 | 0008 | 0010 | 0012 | 0014 | 0017 | 0020 | 0024 | 0028 | 0033 | 0039 | 0 |
| p | 59 | 58 | 57 | 56 | 55 | 54 | 53 | 52 | 51 | 50 | x |

(Continued)

**Table I** (Continued)

$n = 10$

| x | .11 | .12 | .13 | .14 | .15 | .16 | .17 | .18 | .19 | .20 | | x |
|---|---|---|---|---|---|---|---|---|---|---|---|---|
| 5 | 0023 | 0033 | 0047 | 0064 | 0085 | 0111 | 0141 | 0177 | 0218 | 0264 | | 5 |
| 6 | 0002 | 0004 | 0006 | 0009 | 0012 | 0018 | 0024 | 0032 | 0043 | 0055 | | 4 |
| 7 | 0000 | 0000 | 0000 | 0001 | 0001 | 0002 | 0003 | 0004 | 0006 | 0008 | | 3 |
| 8 | 0000 | 0000 | 0000 | 0000 | 0000 | 0000 | 0000 | 0000 | 0001 | 0001 | | 2 |
| | .89 | .88 | .87 | .86 | .85 | .84 | .83 | .82 | .81 | .80 | p | x |

| x | .21 | .22 | .23 | .24 | .25 | .26 | .27 | .28 | .29 | .30 | | x |
|---|---|---|---|---|---|---|---|---|---|---|---|---|
| 0 | 0947 | 0834 | 0733 | 0643 | 0563 | 0492 | 0430 | 0374 | 0326 | 0282 | | 10 |
| 1 | 2517 | 2351 | 2188 | 2030 | 1877 | 1730 | 1590 | 1456 | 1330 | 1211 | | 9 |
| 2 | 3011 | 2984 | 2942 | 2885 | 2816 | 2735 | 2646 | 2548 | 2444 | 2335 | | 8 |
| 3 | 2134 | 2244 | 2343 | 2429 | 2503 | 2563 | 2609 | 2642 | 2662 | 2668 | | 7 |
| 4 | 0993 | 1108 | 1225 | 1343 | 1460 | 1576 | 1689 | 1798 | 1903 | 2001 | | 6 |
| 5 | 0317 | 0375 | 0439 | 0509 | 0584 | 0664 | 0750 | 0839 | 0933 | 1029 | | 5 |
| 6 | 0070 | 0088 | 0109 | 0134 | 0162 | 0195 | 0231 | 0272 | 0317 | 0368 | | 4 |
| 7 | 0011 | 0014 | 0019 | 0024 | 0031 | 0039 | 0049 | 0060 | 0074 | 0090 | | 3 |
| 8 | 0001 | 0002 | 0002 | 0003 | 0004 | 0005 | 0007 | 0009 | 0011 | 0014 | | 2 |
| 9 | 0000 | 0000 | 0000 | 0000 | 0000 | 0000 | 0001 | 0001 | 0001 | 0001 | | 1 |
| | .79 | .78 | .77 | .76 | .75 | .74 | .73 | .72 | .71 | .70 | p | x |

| x | .31 | .32 | .33 | .34 | .35 | .36 | .37 | .38 | .39 | .40 | | x |
|---|---|---|---|---|---|---|---|---|---|---|---|---|
| 0 | 0245 | 0211 | 0182 | 0157 | 0135 | 0115 | 0098 | 0084 | 0071 | 0060 | | 10 |
| 1 | 1099 | 0995 | 0898 | 0808 | 0725 | 0649 | 0578 | 0514 | 0456 | 0430 | | 9 |
| 2 | 2222 | 2107 | 1990 | 1873 | 1757 | 1642 | 1529 | 1419 | 1312 | 1209 | | 8 |
| 3 | 2662 | 2644 | 2614 | 2573 | 2522 | 2462 | 2394 | 2319 | 2237 | 2150 | | 7 |
| 4 | 2093 | 2177 | 2253 | 2320 | 2377 | 2424 | 2461 | 2487 | 2503 | 2508 | | 6 |
| 5 | 1128 | 1229 | 1332 | 1434 | 1536 | 1636 | 1734 | 1829 | 1920 | 2007 | | 5 |
| 6 | 0422 | 0482 | 0547 | 0616 | 0689 | 0767 | 0849 | 0934 | 1023 | 1115 | | 4 |
| 7 | 0108 | 0130 | 0154 | 0181 | 0212 | 0247 | 0285 | 0327 | 0374 | 0425 | | 3 |
| 8 | 0018 | 0023 | 0028 | 0035 | 0043 | 0052 | 0063 | 0075 | 0090 | 0106 | | 2 |
| 9 | 0002 | 0002 | 0003 | 0004 | 0005 | 0006 | 0008 | 0010 | 0013 | 0016 | | 1 |
| 10 | 0000 | 0000 | 0000 | 0000 | 0000 | 0000 | 0000 | 0000 | 0001 | 0001 | | 0 |
| | .69 | .68 | .67 | .66 | .65 | .64 | .63 | .62 | .61 | .60 | p | x |

| x | .41 | .42 | .43 | .44 | .45 | .46 | .47 | .48 | .49 | .50 | | x |
|---|---|---|---|---|---|---|---|---|---|---|---|---|
| 0 | 0051 | 0043 | 0036 | 0030 | 0025 | 0021 | 0017 | 0014 | 0012 | 0010 | | 10 |
| 1 | 0355 | 0312 | 0273 | 0238 | 0207 | 0180 | 0155 | 0133 | 0114 | 0098 | | 9 |
| 2 | 1111 | 1017 | 0927 | 0843 | 0763 | 0688 | 0619 | 0554 | 0494 | 0439 | | 8 |
| 3 | 2058 | 1963 | 1865 | 1765 | 1665 | 1654 | 1464 | 1364 | 1267 | 1172 | | 7 |
| 4 | 2503 | 2488 | 2462 | 2427 | 2384 | 2331 | 2271 | 2204 | 2130 | 2051 | | 6 |
| | .59 | .58 | .57 | .56 | .55 | .54 | .53 | .52 | .51 | .50 | p | x |

(Continued)

$n = 9$

| x | .31 | .32 | .33 | .34 | .35 | .36 | .37 | .38 | .39 | .40 | | x |
|---|---|---|---|---|---|---|---|---|---|---|---|---|
| 5 | 0818 | 0904 | 0994 | 1086 | 1181 | 1278 | 1376 | 1475 | 1574 | 1672 | | 4 |
| 6 | 0245 | 0284 | 0326 | 0373 | 0424 | 0479 | 0539 | 0603 | 0671 | 0743 | | 3 |
| 7 | 0047 | 0057 | 0069 | 0082 | 0098 | 0116 | 0136 | 0158 | 0184 | 0212 | | 2 |
| 8 | 0005 | 0007 | 0008 | 0011 | 0013 | 0016 | 0020 | 0024 | 0029 | 0035 | | 1 |
| 9 | 0000 | 0000 | 0000 | 0001 | 0001 | 0001 | 0001 | 0002 | 0002 | 0003 | | 0 |
| | .69 | .68 | .67 | .66 | .65 | .64 | .63 | .62 | .61 | .60 | p | x |

| x | .41 | .42 | .43 | .44 | .45 | .46 | .47 | .48 | .49 | .50 | | x |
|---|---|---|---|---|---|---|---|---|---|---|---|---|
| 0 | 0087 | 0074 | 0064 | 0054 | 0046 | 0039 | 0033 | 0028 | 0023 | 0020 | | 9 |
| 1 | 0542 | 0484 | 0431 | 0383 | 0339 | 0299 | 0263 | 0231 | 0202 | 0176 | | 8 |
| 2 | 1506 | 1402 | 1301 | 1204 | 1110 | 1020 | 0934 | 0853 | 0776 | 0703 | | 7 |
| 3 | 2442 | 2369 | 2291 | 2207 | 2119 | 2027 | 1933 | 1837 | 1739 | 1641 | | 6 |
| 4 | 2545 | 2573 | 2592 | 2601 | 2600 | 2590 | 2571 | 2543 | 2506 | 2461 | | 5 |
| 5 | 1769 | 1863 | 1955 | 2044 | 2128 | 2207 | 2280 | 2347 | 2408 | 2461 | | 4 |
| 6 | 0819 | 0900 | 0983 | 1070 | 1160 | 1253 | 1348 | 1445 | 1542 | 1641 | | 3 |
| 7 | 0244 | 0279 | 0318 | 0360 | 0407 | 0458 | 0512 | 0571 | 0635 | 0703 | | 2 |
| 8 | 0042 | 0051 | 0060 | 0071 | 0083 | 0097 | 0114 | 0132 | 0153 | 0176 | | 1 |
| 9 | 0003 | 0004 | 0005 | 0006 | 0008 | 0009 | 0011 | 0014 | 0016 | 0020 | | 0 |
| | .59 | .58 | .57 | .56 | .55 | .54 | .53 | .52 | .51 | .50 | p | x |

$n = 10$

| x | .01 | .02 | .03 | .04 | .05 | .06 | .07 | .08 | .09 | .10 | | x |
|---|---|---|---|---|---|---|---|---|---|---|---|---|
| 0 | 9044 | 8171 | 7374 | 6648 | 5987 | 5386 | 4840 | 4344 | 3894 | 3487 | | 10 |
| 1 | 0914 | 1667 | 2281 | 2770 | 3151 | 3438 | 3643 | 3777 | 3851 | 3874 | | 9 |
| 2 | 0042 | 0153 | 0317 | 0519 | 0746 | 0988 | 1234 | 1478 | 1714 | 1937 | | 8 |
| 3 | 0001 | 0008 | 0026 | 0058 | 0105 | 0168 | 0248 | 0343 | 0452 | 0574 | | 7 |
| 4 | 0000 | 0000 | 0001 | 0004 | 0010 | 0019 | 0033 | 0052 | 0078 | 0112 | | 6 |
| 5 | 0000 | 0000 | 0000 | 0000 | 0001 | 0001 | 0003 | 0005 | 0009 | 0015 | | 5 |
| 6 | 0000 | 0000 | 0000 | 0000 | 0000 | 0000 | 0000 | 0000 | 0001 | 0001 | | 4 |
| | .99 | .98 | .97 | .96 | .95 | .94 | .93 | .92 | .91 | .90 | p | x |

| x | .11 | .12 | .13 | .14 | .15 | .16 | .17 | .18 | .19 | .20 | | x |
|---|---|---|---|---|---|---|---|---|---|---|---|---|
| 0 | 3118 | 2785 | 2484 | 2213 | 1969 | 1749 | 1552 | 1374 | 1216 | 1074 | | 10 |
| 1 | 3854 | 3798 | 3712 | 3603 | 3474 | 3331 | 3178 | 3017 | 2852 | 2684 | | 9 |
| 2 | 2143 | 2330 | 2496 | 2639 | 2759 | 2856 | 2929 | 2980 | 3010 | 3020 | | 8 |
| 3 | 0706 | 0847 | 0995 | 1146 | 1298 | 1450 | 1600 | 1745 | 1883 | 2013 | | 7 |
| 4 | 0153 | 0202 | 0260 | 0326 | 0401 | 0483 | 0573 | 0670 | 0773 | 0881 | | 6 |

(Continued)

# Table I (Continued)

## $n = 20$

| $p$ | 21 | 22 | 23 | 24 | 25 | 26 | 27 | 28 | 29 | 30 | $x$ |
|---|---|---|---|---|---|---|---|---|---|---|---|
| $x=0$ | 0090 | 0069 | 0054 | 0041 | 0032 | 0024 | 0016 | 0014 | 0011 | 0008 | 20 |
| 1 | 0477 | 0392 | 0321 | 0261 | 0211 | 0170 | 0137 | 0109 | 0087 | 0068 | 19 |
| 2 | 1204 | 1050 | 0910 | 0783 | 0669 | 0569 | 0480 | 0403 | 0336 | 0278 | 18 |
| 3 | 1920 | 1777 | 1631 | 1484 | 1339 | 1199 | 1065 | 0940 | 0823 | 0716 | 17 |
| 4 | 2169 | 2131 | 2070 | 1991 | 1897 | 1790 | 1675 | 1553 | 1429 | 1304 | 16 |
| 5 | 1845 | 1923 | 1979 | 2012 | 2023 | 2013 | 1982 | 1933 | 1868 | 1789 | 15 |
| 6 | 1226 | 1356 | 1478 | 1589 | 1686 | 1768 | 1833 | 1879 | 1907 | 1916 | 14 |
| 7 | 0652 | 0765 | 0883 | 1003 | 1124 | 1242 | 1356 | 1462 | 1558 | 1643 | 13 |
| 8 | 0282 | 0351 | 0429 | 0515 | 0609 | 0709 | 0815 | 0924 | 1034 | 1144 | 12 |
| 9 | 0100 | 0132 | 0171 | 0217 | 0271 | 0332 | 0402 | 0479 | 0563 | 0654 | 11 |
| 10 | 0029 | 0041 | 0056 | 0075 | 0099 | 0128 | 0163 | 0205 | 0253 | 0308 | 10 |
| 11 | 0007 | 0010 | 0015 | 0022 | 0030 | 0041 | 0055 | 0072 | 0094 | 0120 | 9 |
| 12 | 0001 | 0002 | 0003 | 0005 | 0008 | 0011 | 0015 | 0021 | 0029 | 0039 | 8 |
| 13 | 0000 | 0000 | 0001 | 0001 | 0002 | 0002 | 0003 | 0005 | 0007 | 0010 | 7 |
| 14 | 0000 | 0000 | 0000 | 0000 | 0000 | 0000 | 0000 | 0001 | 0001 | 0002 | 6 |
| $p$ | 79 | 78 | 77 | 76 | 75 | 74 | 73 | 72 | 71 | 70 | $x$ |

| $p$ | 31 | 32 | 33 | 34 | 35 | 36 | 37 | 38 | 39 | 40 | $x$ |
|---|---|---|---|---|---|---|---|---|---|---|---|
| $x=0$ | 0006 | 0004 | 0003 | 0002 | 0002 | 0001 | 0001 | 0001 | 0000 | 0000 | 20 |
| 1 | 0054 | 0042 | 0033 | 0025 | 0020 | 0015 | 0011 | 0009 | 0007 | 0005 | 19 |
| 2 | 0229 | 0188 | 0153 | 0124 | 0100 | 0080 | 0064 | 0050 | 0040 | 0031 | 18 |
| 3 | 0619 | 0531 | 0453 | 0383 | 0323 | 0270 | 0224 | 0185 | 0152 | 0123 | 17 |
| 4 | 1181 | 1062 | 0947 | 0839 | 0738 | 0645 | 0559 | 0482 | 0412 | 0350 | 16 |
| 5 | 1698 | 1599 | 1493 | 1384 | 1272 | 1161 | 1051 | 0945 | 0843 | 0746 | 15 |
| 6 | 1907 | 1881 | 1839 | 1782 | 1712 | 1632 | 1543 | 1447 | 1347 | 1244 | 14 |
| 7 | 1714 | 1770 | 1811 | 1836 | 1844 | 1836 | 1812 | 1774 | 1722 | 1659 | 13 |
| 8 | 1251 | 1354 | 1450 | 1537 | 1614 | 1678 | 1730 | 1767 | 1790 | 1797 | 12 |
| 9 | 0750 | 0849 | 0952 | 1056 | 1158 | 1259 | 1354 | 1444 | 1526 | 1597 | 11 |
| 10 | 0370 | 0440 | 0516 | 0598 | 0686 | 0779 | 0875 | 0974 | 1073 | 1171 | 10 |
| 11 | 0151 | 0188 | 0231 | 0280 | 0336 | 0398 | 0467 | 0542 | 0624 | 0710 | 9 |
| 12 | 0051 | 0066 | 0085 | 0108 | 0136 | 0168 | 0206 | 0249 | 0299 | 0355 | 8 |
| 13 | 0014 | 0019 | 0026 | 0034 | 0045 | 0058 | 0074 | 0094 | 0118 | 0146 | 7 |
| 14 | 0003 | 0005 | 0006 | 0009 | 0012 | 0016 | 0022 | 0029 | 0038 | 0049 | 6 |
| 15 | 0001 | 0001 | 0001 | 0002 | 0003 | 0004 | 0005 | 0007 | 0010 | 0013 | 5 |
| 16 | 0000 | 0000 | 0000 | 0000 | 0000 | 0001 | 0001 | 0001 | 0002 | 0003 | 4 |
| $p$ | 69 | 68 | 67 | 66 | 65 | 64 | 63 | 62 | 61 | 60 | $x$ |

*(Continued)*

## $n = 10$

| $p$ | 41 | 42 | 43 | 44 | 45 | 46 | 47 | 48 | 49 | 50 | $x$ |
|---|---|---|---|---|---|---|---|---|---|---|---|
| $x=5$ | 2087 | 2162 | 2229 | 2289 | 2340 | 2383 | 2417 | 2441 | 2456 | 2461 | 5 |
| 6 | 1209 | 1304 | 1401 | 1499 | 1596 | 1692 | 1786 | 1878 | 1966 | 2051 | 4 |
| 7 | 0480 | 0540 | 0604 | 0673 | 0746 | 0824 | 0905 | 0991 | 1080 | 1172 | 3 |
| 8 | 0125 | 0147 | 0171 | 0198 | 0229 | 0263 | 0301 | 0343 | 0389 | 0439 | 2 |
| 9 | 0019 | 0024 | 0029 | 0035 | 0042 | 0050 | 0059 | 0070 | 0083 | 0098 | 1 |
| 10 | 0001 | 0002 | 0002 | 0003 | 0003 | 0004 | 0005 | 0006 | 0008 | 0010 | 0 |
| $p$ | 59 | 58 | 57 | 56 | 55 | 54 | 53 | 52 | 51 | 50 | $x$ |

## $n = 20$

| $p$ | 01 | 02 | 03 | 04 | 05 | 06 | 07 | 08 | 09 | 10 | $x$ |
|---|---|---|---|---|---|---|---|---|---|---|---|
| $x=0$ | 8179 | 6676 | 5438 | 4420 | 3585 | 2901 | 2342 | 1887 | 1516 | 1216 | 20 |
| 1 | 1652 | 2725 | 3364 | 3683 | 3774 | 3703 | 3526 | 3282 | 3000 | 2702 | 19 |
| 2 | 0159 | 0528 | 0988 | 1458 | 1887 | 2246 | 2521 | 2711 | 2828 | 2852 | 18 |
| 3 | 0010 | 0065 | 0183 | 0364 | 0596 | 0860 | 1139 | 1414 | 1672 | 1901 | 17 |
| 4 | 0000 | 0006 | 0024 | 0065 | 0133 | 0233 | 0364 | 0523 | 0703 | 0898 | 16 |
| 5 | 0000 | 0000 | 0002 | 0009 | 0022 | 0048 | 0088 | 0145 | 0222 | 0319 | 15 |
| 6 | 0000 | 0000 | 0000 | 0001 | 0003 | 0008 | 0017 | 0032 | 0055 | 0089 | 14 |
| 7 | 0000 | 0000 | 0000 | 0000 | 0000 | 0001 | 0002 | 0005 | 0011 | 0020 | 13 |
| 8 | 0000 | 0000 | 0000 | 0000 | 0000 | 0000 | 0000 | 0001 | 0002 | 0004 | 12 |
| 9 | 0000 | 0000 | 0000 | 0000 | 0000 | 0000 | 0000 | 0000 | 0000 | 0001 | 11 |
| $p$ | 99 | 98 | 97 | 96 | 95 | 94 | 93 | 92 | 91 | 90 | $x$ |

| $p$ | 11 | 12 | 13 | 14 | 15 | 16 | 17 | 18 | 19 | 20 | $x$ |
|---|---|---|---|---|---|---|---|---|---|---|---|
| $x=0$ | 0972 | 0776 | 0617 | 0490 | 0388 | 0306 | 0241 | 0189 | 0148 | 0115 | 20 |
| 1 | 2403 | 2115 | 1844 | 1595 | 1368 | 1165 | 0986 | 0829 | 0693 | 0576 | 19 |
| 2 | 2822 | 2740 | 2618 | 2466 | 2293 | 2109 | 1919 | 1730 | 1545 | 1369 | 18 |
| 3 | 2093 | 2242 | 2347 | 2409 | 2428 | 2410 | 2358 | 2278 | 2175 | 2054 | 17 |
| 4 | 1099 | 1299 | 1491 | 1666 | 1821 | 1951 | 2053 | 2125 | 2168 | 2182 | 16 |
| 5 | 0435 | 0567 | 0713 | 0868 | 1028 | 1189 | 1345 | 1493 | 1627 | 1746 | 15 |
| 6 | 0134 | 0193 | 0266 | 0353 | 0454 | 0566 | 0689 | 0819 | 0954 | 1091 | 14 |
| 7 | 0033 | 0053 | 0080 | 0115 | 0160 | 0216 | 0282 | 0360 | 0448 | 0545 | 13 |
| 8 | 0007 | 0012 | 0019 | 0030 | 0046 | 0067 | 0094 | 0128 | 0171 | 0222 | 12 |
| 9 | 0001 | 0002 | 0004 | 0007 | 0011 | 0017 | 0026 | 0038 | 0053 | 0074 | 11 |
| 10 | 0000 | 0000 | 0001 | 0001 | 0002 | 0004 | 0006 | 0009 | 0014 | 0020 | 10 |
| 11 | 0000 | 0000 | 0000 | 0000 | 0000 | 0001 | 0001 | 0002 | 0003 | 0005 | 9 |
| 12 | 0000 | 0000 | 0000 | 0000 | 0000 | 0000 | 0000 | 0000 | 0001 | 0001 | 8 |
| $p$ | 89 | 88 | 87 | 86 | 85 | 84 | 83 | 82 | 81 | 80 | $x$ |

*(Continued)*

Table I (Continued)

**n = 20**

| x \ p | 41 | 42 | 43 | 44 | 45 | 46 | 47 | 48 | 49 | 50 |
|---|---|---|---|---|---|---|---|---|---|---|
| 1 | 0004 | 0003 | 0002 | 0001 | 0001 | 0001 | 0001 | 0000 | 0000 | 0000 |
| 2 | 0024 | 0018 | 0014 | 0011 | 0008 | 0006 | 0005 | 0003 | 0002 | 0000 |
| 3 | 0100 | 0080 | 0064 | 0051 | 0040 | 0031 | 0024 | 0019 | 0014 | 0011 |
| 4 | 0295 | 0247 | 0206 | 0170 | 0139 | 0113 | 0092 | 0074 | 0059 | 0046 |
| 5 | 0656 | 0573 | 0496 | 0427 | 0365 | 0309 | 0260 | 0217 | 0180 | 0148 |
| 6 | 1140 | 1037 | 0936 | 0839 | 0746 | 0658 | 0577 | 0501 | 0432 | 0370 |
| 7 | 1585 | 1502 | 1413 | 1318 | 1221 | 1122 | 1023 | 0925 | 0830 | 0739 |
| 8 | 1790 | 1768 | 1732 | 1683 | 1623 | 1553 | 1474 | 1388 | 1296 | 1201 |
| 9 | 1658 | 1707 | 1742 | 1763 | 1771 | 1763 | 1742 | 1708 | 1661 | 1602 |
| 10 | 1268 | 1359 | 1446 | 1524 | 1593 | 1652 | 1700 | 1734 | 1755 | 1762 |
| 11 | 0801 | 0895 | 0991 | 1089 | 1185 | 1280 | 1370 | 1455 | 1533 | 1602 |
| 12 | 0417 | 0486 | 0561 | 0642 | 0727 | 0818 | 0911 | 1007 | 1105 | 1201 |
| 13 | 0178 | 0217 | 0260 | 0310 | 0366 | 0429 | 0497 | 0572 | 0653 | 0739 |
| 14 | 0062 | 0078 | 0098 | 0122 | 0150 | 0183 | 0221 | 0264 | 0314 | 0370 |
| 15 | 0017 | 0023 | 0030 | 0038 | 0049 | 0062 | 0078 | 0098 | 0121 | 0148 |
| 16 | 0004 | 0005 | 0007 | 0009 | 0013 | 0017 | 0022 | 0028 | 0036 | 0046 |
| 17 | 0001 | 0001 | 0001 | 0002 | 0002 | 0003 | 0005 | 0006 | 0008 | 0011 |
| 18 | 0000 | 0000 | 0000 | 0000 | 0000 | 0000 | 0001 | 0001 | 0001 | 0002 |

(lower reading) x: 19 18 17 16 15 14 13 12 11 10 9 8 7 6 5 4 3 2
p: 59 58 57 56 55 54 53 52 51 50

**n = 50**

| x \ p | 01 | 02 | 03 | 04 | 05 | 06 | 07 | 08 | 09 | 10 |
|---|---|---|---|---|---|---|---|---|---|---|
| 0 | 6050 | 3642 | 2181 | 1299 | 0769 | 0453 | 0266 | 0155 | 0090 | 0052 |
| 1 | 3056 | 3716 | 3372 | 2706 | 2025 | 1447 | 0999 | 0672 | 0443 | 0286 |
| 2 | 0756 | 1858 | 2555 | 2762 | 2611 | 2262 | 1843 | 1433 | 1073 | 0779 |
| 3 | 0122 | 0607 | 1264 | 1842 | 2199 | 2311 | 2219 | 1993 | 1698 | 1386 |
| 4 | 0015 | 0145 | 0459 | 0902 | 1360 | 1733 | 1963 | 2037 | 1973 | 1809 |
| 5 | 0001 | 0027 | 0131 | 0346 | 0658 | 1018 | 1359 | 1629 | 1795 | 1849 |
| 6 | 0000 | 0004 | 0030 | 0108 | 0260 | 0487 | 0767 | 1063 | 1332 | 1541 |
| 7 | 0000 | 0001 | 0006 | 0028 | 0086 | 0195 | 0363 | 0581 | 0828 | 1076 |
| 8 | 0000 | 0000 | 0001 | 0006 | 0024 | 0067 | 0147 | 0271 | 0440 | 0643 |
| 9 | 0000 | 0000 | 0000 | 0001 | 0006 | 0020 | 0052 | 0110 | 0203 | 0333 |
| 10 | 0000 | 0000 | 0000 | 0000 | 0001 | 0005 | 0016 | 0039 | 0082 | 0152 |
| 11 | 0000 | 0000 | 0000 | 0000 | 0000 | 0001 | 0004 | 0012 | 0030 | 0061 |
| 12 | 0000 | 0000 | 0000 | 0000 | 0000 | 0000 | 0001 | 0004 | 0010 | 0022 |
| 13 | 0000 | 0000 | 0000 | 0000 | 0000 | 0000 | 0000 | 0001 | 0003 | 0007 |
| 14 | 0000 | 0000 | 0000 | 0000 | 0000 | 0000 | 0000 | 0000 | 0001 | 0002 |
| 15 | 0000 | 0000 | 0000 | 0000 | 0000 | 0000 | 0000 | 0000 | 0000 | 0001 |

(lower reading) x: 50 49 48 47 46 45 44 43 42 41 ... 40 39 38 37 36 35
p: 99 98 97 96 95 94 93 92 91 90

**n = 50**

| x \ p | 11 | 12 | 13 | 14 | 15 | 16 | 17 | 18 | 19 | 20 |
|---|---|---|---|---|---|---|---|---|---|---|
| 0 | 0029 | 0017 | 0009 | 0005 | 0003 | 0002 | 0001 | 0000 | 0000 | 0000 |
| 1 | 0182 | 0114 | 0071 | 0043 | 0026 | 0016 | 0009 | 0005 | 0003 | 0002 |
| 2 | 0552 | 0382 | 0259 | 0172 | 0113 | 0073 | 0046 | 0029 | 0018 | 0011 |
| 3 | 1091 | 0833 | 0619 | 0449 | 0319 | 0222 | 0151 | 0102 | 0067 | 0044 |
| 4 | 1584 | 1334 | 1086 | 0858 | 0661 | 0496 | 0364 | 0262 | 0185 | 0128 |
| 5 | 1801 | 1674 | 1493 | 1286 | 1072 | 0869 | 0687 | 0530 | 0400 | 0295 |
| 6 | 1670 | 1712 | 1674 | 1570 | 1419 | 1242 | 1055 | 0872 | 0703 | 0554 |
| 7 | 1297 | 1467 | 1572 | 1606 | 1575 | 1487 | 1358 | 1203 | 1037 | 0870 |
| 8 | 0862 | 1075 | 1262 | 1406 | 1493 | 1523 | 1495 | 1420 | 1307 | 1169 |
| 9 | 0497 | 0684 | 0880 | 1068 | 1230 | 1353 | 1429 | 1454 | 1431 | 1364 |
| 10 | 0252 | 0383 | 0539 | 0713 | 0890 | 1057 | 1200 | 1309 | 1376 | 1398 |
| 11 | 0113 | 0190 | 0293 | 0422 | 0571 | 0732 | 0894 | 1045 | 1174 | 1271 |
| 12 | 0045 | 0084 | 0142 | 0223 | 0328 | 0453 | 0595 | 0745 | 0895 | 1033 |
| 13 | 0016 | 0034 | 0062 | 0106 | 0169 | 0252 | 0356 | 0478 | 0613 | 0755 |
| 14 | 0005 | 0012 | 0025 | 0046 | 0079 | 0127 | 0193 | 0277 | 0380 | 0499 |
| 15 | 0002 | 0004 | 0009 | 0018 | 0033 | 0058 | 0095 | 0146 | 0214 | 0299 |
| 16 | 0000 | 0001 | 0003 | 0006 | 0013 | 0024 | 0042 | 0070 | 0110 | 0164 |
| 17 | 0000 | 0000 | 0001 | 0002 | 0005 | 0009 | 0017 | 0031 | 0052 | 0082 |
| 18 | 0000 | 0000 | 0000 | 0001 | 0001 | 0003 | 0007 | 0012 | 0022 | 0037 |
| 19 | 0000 | 0000 | 0000 | 0000 | 0000 | 0001 | 0002 | 0005 | 0009 | 0016 |
| 20 | 0000 | 0000 | 0000 | 0000 | 0000 | 0000 | 0000 | 0002 | 0003 | 0006 |
| 21 | 0000 | 0000 | 0000 | 0000 | 0000 | 0000 | 0000 | 0000 | 0001 | 0002 |
| 22 | 0000 | 0000 | 0000 | 0000 | 0000 | 0000 | 0000 | 0000 | 0000 | 0001 |

(lower reading) x: 30 29 28 ...
p: 89 88 87 86 85 84 83 82 81 80

(Continued)

# Table I (Continued)

## n = 50

| x | p | 21 | 22 | 23 | 24 | 25 | 26 | 27 | 28 | 29 | 30 | x |
|---|---|----|----|----|----|----|----|----|----|----|----|---|
| 1 |  | 0001 | 0001 | 0000 | 0000 | 0000 | 0000 | 0000 | 0000 | 0000 | 0000 | 49 |
| 2 |  | 0007 | 0004 | 0002 | 0001 | 0001 | 0000 | 0000 | 0000 | 0000 | 0000 | 48 |
| 3 |  | 0028 | 0018 | 0011 | 0007 | 0004 | 0002 | 0001 | 0000 | 0000 | 0000 | 47 |
| 4 |  | 0088 | 0059 | 0039 | 0025 | 0016 | 0010 | 0006 | 0004 | 0002 | 0001 | 46 |
| 5 |  | 0214 | 0152 | 0106 | 0073 | 0049 | 0033 | 0021 | 0014 | 0009 | 0006 | 45 |
| 6 |  | 0427 | 0322 | 0238 | 0173 | 0123 | 0087 | 0060 | 0040 | 0027 | 0018 | 44 |
| 7 |  | 0713 | 0571 | 0447 | 0344 | 0259 | 0191 | 0139 | 0099 | 0069 | 0048 | 43 |
| 8 |  | 1019 | 0865 | 0718 | 0583 | 0463 | 0361 | 0276 | 0207 | 0152 | 0110 | 42 |
| 9 |  | 1263 | 1139 | 1001 | 0859 | 0721 | 0592 | 0476 | 0375 | 0290 | 0220 | 41 |
| 10 |  | 1377 | 1317 | 1226 | 1113 | 0985 | 0852 | 0721 | 0598 | 0485 | 0386 | 40 |
| 11 |  | 1331 | 1351 | 1332 | 1278 | 1194 | 1089 | 0970 | 0845 | 0721 | 0602 | 39 |
| 12 |  | 1150 | 1238 | 1293 | 1311 | 1294 | 1244 | 1166 | 1068 | 0957 | 0838 | 38 |
| 13 |  | 0894 | 1021 | 1129 | 1210 | 1261 | 1277 | 1261 | 1215 | 1142 | 1050 | 37 |
| 14 |  | 0628 | 0761 | 0891 | 1010 | 1110 | 1186 | 1233 | 1248 | 1233 | 1189 | 36 |
| 15 |  | 0400 | 0515 | 0639 | 0766 | 0888 | 1000 | 1094 | 1165 | 1209 | 1223 | 35 |
| 16 |  | 0233 | 0318 | 0417 | 0529 | 0648 | 0769 | 0885 | 0991 | 1080 | 1147 | 34 |
| 17 |  | 0124 | 0179 | 0249 | 0334 | 0432 | 0540 | 0655 | 0771 | 0882 | 0983 | 33 |
| 18 |  | 0060 | 0093 | 0137 | 0193 | 0264 | 0348 | 0444 | 0550 | 0661 | 0772 | 32 |
| 19 |  | 0027 | 0044 | 0069 | 0103 | 0148 | 0206 | 0277 | 0360 | 0454 | 0558 | 31 |
| 20 |  | 0011 | 0019 | 0032 | 0050 | 0077 | 0112 | 0159 | 0217 | 0288 | 0370 | 30 |
| 21 |  | 0004 | 0008 | 0014 | 0023 | 0036 | 0056 | 0084 | 0121 | 0168 | 0227 | 29 |
| 22 |  | 0001 | 0003 | 0005 | 0009 | 0016 | 0026 | 0041 | 0062 | 0090 | 0128 | 28 |
| 23 |  | 0000 | 0001 | 0002 | 0004 | 0006 | 0011 | 0018 | 0029 | 0045 | 0067 | 27 |
| 24 |  | 0000 | 0000 | 0001 | 0001 | 0002 | 0004 | 0008 | 0013 | 0021 | 0032 | 26 |
| 25 |  | 0000 | 0000 | 0000 | 0000 | 0001 | 0002 | 0003 | 0005 | 0009 | 0014 | 25 |
| 26 |  | 0000 | 0000 | 0000 | 0000 | 0000 | 0001 | 0001 | 0002 | 0003 | 0006 | 24 |
| 27 |  | 0000 | 0000 | 0000 | 0000 | 0000 | 0000 | 0000 | 0001 | 0001 | 0002 | 23 |
| 28 |  | 0000 | 0000 | 0000 | 0000 | 0000 | 0000 | 0000 | 0000 | 0000 | 0001 | 22 |
|  | p | 79 | 78 | 77 | 76 | 75 | 74 | 73 | 72 | 71 | 70 | x |

## n = 50

| x | p | 31 | 32 | 33 | 34 | 35 | 36 | 37 | 38 | 39 | 40 | x |
|---|---|----|----|----|----|----|----|----|----|----|----|---|
| 4 |  | 0001 | 0000 | 0000 | 0000 | 0000 | 0000 | 0000 | 0000 | 0000 | 0000 | 46 |
| 5 |  | 0003 | 0002 | 0001 | 0000 | 0000 | 0000 | 0000 | 0000 | 0000 | 0000 | 45 |
| 6 |  | 0011 | 0007 | 0005 | 0003 | 0002 | 0001 | 0001 | 0000 | 0000 | 0000 | 44 |
| 7 |  | 0032 | 0022 | 0014 | 0009 | 0006 | 0004 | 0002 | 0001 | 0001 | 0000 | 43 |
| 8 |  | 0078 | 0055 | 0037 | 0025 | 0017 | 0011 | 0007 | 0004 | 0003 | 0002 | 42 |
| 9 |  | 0164 | 0120 | 0086 | 0061 | 0042 | 0029 | 0019 | 0013 | 0008 | 0005 | 41 |
| 10 |  | 0301 | 0231 | 0174 | 0128 | 0093 | 0066 | 0046 | 0032 | 0022 | 0014 | 40 |
| 11 |  | 0493 | 0395 | 0311 | 0240 | 0182 | 0136 | 0099 | 0071 | 0050 | 0035 | 39 |
| 12 |  | 0719 | 0604 | 0498 | 0402 | 0319 | 0248 | 0189 | 0142 | 0105 | 0076 | 38 |
| 13 |  | 0944 | 0831 | 0717 | 0606 | 0502 | 0408 | 0325 | 0255 | 0195 | 0147 | 37 |
| 14 |  | 1121 | 1034 | 0933 | 0825 | 0714 | 0607 | 0505 | 0412 | 0330 | 0260 | 36 |
| 15 |  | 1209 | 1168 | 1103 | 1020 | 0923 | 0819 | 0712 | 0606 | 0507 | 0415 | 35 |
| 16 |  | 1188 | 1202 | 1189 | 1149 | 1088 | 1008 | 0914 | 0813 | 0709 | 0606 | 34 |
| 17 |  | 1068 | 1132 | 1171 | 1184 | 1171 | 1133 | 1074 | 0997 | 0906 | 0808 | 33 |
| 18 |  | 0880 | 0976 | 1057 | 1118 | 1156 | 1169 | 1156 | 1120 | 1062 | 0987 | 32 |
| 19 |  | 0666 | 0774 | 0877 | 0970 | 1048 | 1107 | 1144 | 1156 | 1144 | 1109 | 31 |
| 20 |  | 0463 | 0564 | 0670 | 0775 | 0875 | 0956 | 1041 | 1098 | 1134 | 1146 | 30 |
| 21 |  | 0297 | 0379 | 0471 | 0570 | 0673 | 0776 | 0874 | 0962 | 1035 | 1091 | 29 |
| 22 |  | 0176 | 0235 | 0306 | 0387 | 0478 | 0575 | 0676 | 0777 | 0873 | 0959 | 28 |
| 23 |  | 0096 | 0135 | 0183 | 0243 | 0313 | 0394 | 0484 | 0580 | 0679 | 0778 | 27 |
| 24 |  | 0049 | 0071 | 0102 | 0141 | 0190 | 0249 | 0319 | 0400 | 0489 | 0584 | 26 |
| 25 |  | 0023 | 0035 | 0052 | 0075 | 0106 | 0146 | 0195 | 0255 | 0325 | 0405 | 25 |
| 26 |  | 0010 | 0016 | 0025 | 0037 | 0055 | 0079 | 0110 | 0150 | 0200 | 0259 | 24 |
| 27 |  | 0004 | 0007 | 0011 | 0017 | 0026 | 0039 | 0058 | 0082 | 0113 | 0154 | 23 |
| 28 |  | 0001 | 0003 | 0004 | 0007 | 0012 | 0018 | 0028 | 0041 | 0060 | 0084 | 22 |
| 29 |  | 0000 | 0001 | 0002 | 0003 | 0005 | 0008 | 0012 | 0019 | 0029 | 0043 | 21 |
| 30 |  | 0000 | 0000 | 0001 | 0001 | 0002 | 0003 | 0005 | 0008 | 0013 | 0020 | 20 |
| 31 |  | 0000 | 0000 | 0000 | 0000 | 0001 | 0001 | 0002 | 0003 | 0005 | 0009 | 19 |
| 32 |  | 0000 | 0000 | 0000 | 0000 | 0000 | 0000 | 0001 | 0001 | 0002 | 0003 | 18 |
| 33 |  | 0000 | 0000 | 0000 | 0000 | 0000 | 0000 | 0000 | 0000 | 0001 | 0001 | 17 |
|  | p | 69 | 68 | 67 | 66 | 65 | 64 | 63 | 62 | 61 | 60 | x |

| x | p | 41 | 42 | 43 | 44 | 45 | 46 | 47 | 48 | 49 | 50 | x |
|---|---|----|----|----|----|----|----|----|----|----|----|---|
| 8 |  | 0001 | 0001 | 0000 | 0000 | 0000 | 0000 | 0000 | 0000 | 0000 | 0000 | 42 |
| 9 |  | 0003 | 0002 | 0001 | 0001 | 0000 | 0000 | 0000 | 0000 | 0000 | 0000 | 41 |
| 10 |  | 0009 | 0006 | 0004 | 0002 | 0002 | 0001 | 0001 | 0000 | 0000 | 0000 | 40 |
| 11 |  | 0024 | 0016 | 0011 | 0007 | 0005 | 0003 | 0002 | 0001 | 0001 | 0000 | 39 |
| 12 |  | 0054 | 0037 | 0025 | 0018 | 0012 | 0008 | 0005 | 0003 | 0002 | 0001 | 38 |
| 13 |  | 0109 | 0079 | 0057 | 0040 | 0027 | 0018 | 0012 | 0008 | 0005 | 0003 | 37 |
| 14 |  | 0200 | 0152 | 0113 | 0082 | 0059 | 0041 | 0029 | 0019 | 0013 | 0008 | 36 |
| 15 |  | 0334 | 0264 | 0204 | 0155 | 0116 | 0085 | 0061 | 0043 | 0030 | 0020 | 35 |
| 16 |  | 0508 | 0418 | 0337 | 0267 | 0207 | 0158 | 0118 | 0086 | 0062 | 0044 | 34 |
| 17 |  | 0706 | 0605 | 0508 | 0419 | 0339 | 0269 | 0209 | 0159 | 0119 | 0087 | 33 |

(*Continued*)

Table I (Continued)

**n = 50**

| x | 41 | 42 | 43 | 44 | 45 | 46 | 47 | 48 | 49 | 50 | x |
|---|----|----|----|----|----|----|----|----|----|----|---|
| 18 | 0899 | 0803 | 0703 | 0604 | 0508 | 0420 | 0340 | 0270 | 0210 | 0160 | 32 |
| 19 | 1053 | 0979 | 0893 | 0799 | 0700 | 0602 | 0507 | 0419 | 0340 | 0270 | 31 |
| 20 | 1134 | 1099 | 1044 | 0973 | 0888 | 0795 | 0697 | 0600 | 0506 | 0419 | 30 |
| 21 | 1126 | 1137 | 1126 | 1092 | 1030 | 0967 | 0884 | 0791 | 0695 | 0598 | 29 |
| 22 | 1031 | 1086 | 1119 | 1131 | 1119 | 1086 | 1033 | 0963 | 0880 | 0788 | 28 |
| 23 | 0872 | 0957 | 1028 | 1082 | 1115 | 1126 | 1115 | 1082 | 1029 | 0960 | 27 |
| 24 | 0682 | 0780 | 0872 | 0956 | 1026 | 1079 | 1112 | 1124 | 1112 | 1080 | 26 |
| 25 | 0493 | 0587 | 0684 | 0781 | 0873 | 0956 | 1026 | 1079 | 1112 | 1123 | 25 |
| 26 | 0329 | 0409 | 0497 | 0590 | 0687 | 0783 | 0875 | 0957 | 1027 | 1080 | 24 |
| 27 | 0203 | 0263 | 0333 | 0412 | 0500 | 0593 | 0690 | 0786 | 0877 | 0960 | 23 |
| 28 | 0116 | 0157 | 0206 | 0266 | 0336 | 0415 | 0502 | 0596 | 0692 | 0788 | 22 |
| 29 | 0061 | 0086 | 0118 | 0159 | 0208 | 0268 | 0338 | 0417 | 0504 | 0598 | 21 |
| 30 | 0030 | 0044 | 0062 | 0087 | 0119 | 0160 | 0210 | 0270 | 0339 | 0419 | 20 |
| 31 | 0013 | 0020 | 0030 | 0044 | 0063 | 0088 | 0120 | 0161 | 0210 | 0270 | 19 |
| 32 | 0006 | 0009 | 0014 | 0021 | 0031 | 0044 | 0063 | 0088 | 0120 | 0160 | 18 |
| 33 | 0002 | 0003 | 0006 | 0009 | 0014 | 0021 | 0031 | 0044 | 0063 | 0087 | 17 |
| 34 | 0001 | 0001 | 0002 | 0003 | 0006 | 0009 | 0014 | 0020 | 0030 | 0044 | 16 |
| 35 | 0000 | 0000 | 0001 | 0001 | 0002 | 0003 | 0005 | 0009 | 0013 | 0020 | 15 |
| 36 | 0000 | 0000 | 0000 | 0000 | 0001 | 0001 | 0002 | 0003 | 0005 | 0006 | 14 |
| 37 | 0000 | 0000 | 0000 | 0000 | 0000 | 0000 | 0001 | 0001 | 0002 | 0003 | 13 |
| 38 | 0000 | 0000 | 0000 | 0000 | 0000 | 0000 | 0000 | 0000 | 0001 | 0001 | 12 |
| | 59 | 58 | 57 | 56 | 55 | 54 | 53 | 52 | 51 | 50 | p |

**n = 100**

| x | 01 | 02 | 03 | 04 | 05 | 06 | 07 | 08 | 09 | 10 | x |
|---|----|----|----|----|----|----|----|----|----|----|---|
| 10 | 0000 | 0000 | 0007 | 0046 | 0167 | 0399 | 0712 | 1024 | 1243 | 1319 | 90 |
| 11 | 0000 | 0000 | 0002 | 0016 | 0072 | 0209 | 0439 | 0728 | 1006 | 1199 | 89 |
| 12 | 0000 | 0000 | 0000 | 0005 | 0028 | 0099 | 0245 | 0470 | 0738 | 0988 | 88 |
| 13 | 0000 | 0000 | 0000 | 0001 | 0010 | 0043 | 0125 | 0276 | 0494 | 0743 | 87 |
| 14 | 0000 | 0000 | 0000 | 0000 | 0003 | 0017 | 0058 | 0149 | 0304 | 0513 | 86 |
| 15 | 0000 | 0000 | 0000 | 0000 | 0001 | 0006 | 0025 | 0074 | 0172 | 0327 | 85 |
| 16 | 0000 | 0000 | 0000 | 0000 | 0000 | 0002 | 0010 | 0034 | 0090 | 0193 | 84 |
| 17 | 0000 | 0000 | 0000 | 0000 | 0000 | 0001 | 0004 | 0015 | 0044 | 0106 | 83 |
| 18 | 0000 | 0000 | 0000 | 0000 | 0000 | 0000 | 0001 | 0006 | 0020 | 0054 | 82 |
| 19 | 0000 | 0000 | 0000 | 0000 | 0000 | 0000 | 0000 | 0002 | 0009 | 0026 | 81 |
| 20 | 0000 | 0000 | 0000 | 0000 | 0000 | 0000 | 0000 | 0001 | 0003 | 0012 | 80 |
| 21 | 0000 | 0000 | 0000 | 0000 | 0000 | 0000 | 0000 | 0000 | 0001 | 0005 | 79 |
| 22 | 0000 | 0000 | 0000 | 0000 | 0000 | 0000 | 0000 | 0000 | 0000 | 0002 | 78 |
| 23 | 0000 | 0000 | 0000 | 0000 | 0000 | 0000 | 0000 | 0000 | 0000 | 0001 | 77 |
| | 99 | 98 | 97 | 96 | 95 | 94 | 93 | 92 | 91 | 90 | p |

| x | 11 | 12 | 13 | 14 | 15 | 16 | 17 | 18 | 19 | 20 | x |
|---|----|----|----|----|----|----|----|----|----|----|---|
| 1 | 0001 | 0000 | 0000 | 0000 | 0000 | 0000 | 0000 | 0000 | 0000 | 0000 | 99 |
| 2 | 0007 | 0003 | 0001 | 0000 | 0000 | 0000 | 0000 | 0000 | 0000 | 0000 | 98 |
| 3 | 0027 | 0012 | 0005 | 0002 | 0001 | 0000 | 0000 | 0000 | 0000 | 0000 | 97 |
| 4 | 0080 | 0038 | 0018 | 0008 | 0003 | 0001 | 0000 | 0000 | 0000 | 0000 | 96 |
| 5 | 0189 | 0100 | 0050 | 0024 | 0011 | 0005 | 0002 | 0001 | 0000 | 0000 | 95 |
| 6 | 0369 | 0215 | 0119 | 0063 | 0031 | 0015 | 0007 | 0003 | 0001 | 0001 | 94 |
| 7 | 0613 | 0394 | 0238 | 0137 | 0075 | 0039 | 0020 | 0009 | 0004 | 0002 | 93 |
| 8 | 0881 | 0625 | 0414 | 0259 | 0153 | 0086 | 0047 | 0024 | 0012 | 0006 | 92 |
| 9 | 1112 | 0871 | 0632 | 0430 | 0276 | 0168 | 0098 | 0054 | 0029 | 0015 | 91 |
| 10 | 1251 | 1080 | 0860 | 0637 | 0444 | 0292 | 0182 | 0108 | 0062 | 0034 | 90 |
| 11 | 1265 | 1205 | 1051 | 0849 | 0640 | 0454 | 0305 | 0194 | 0118 | 0069 | 89 |
| 12 | 1160 | 1219 | 1165 | 1025 | 0838 | 0642 | 0463 | 0316 | 0206 | 0128 | 88 |
| 13 | 0970 | 1125 | 1179 | 1130 | 1001 | 0827 | 0642 | 0470 | 0327 | 0216 | 87 |
| 14 | 0745 | 0954 | 1094 | 1143 | 1098 | 0979 | 0817 | 0641 | 0476 | 0335 | 86 |
| 15 | 0528 | 0745 | 0938 | 1067 | 1111 | 1070 | 0960 | 0807 | 0640 | 0481 | 85 |
| 16 | 0347 | 0540 | 0744 | 0922 | 1041 | 1082 | 1044 | 0941 | 0798 | 0638 | 84 |
| 17 | 0213 | 0364 | 0549 | 0742 | 0908 | 1019 | 1057 | 1021 | 0924 | 0789 | 83 |
| 18 | 0121 | 0229 | 0379 | 0557 | 0739 | 0895 | 0998 | 1033 | 1000 | 0909 | 82 |
| 19 | 0064 | 0135 | 0244 | 0391 | 0563 | 0736 | 0882 | 0979 | 1012 | 0981 | 81 |
| 20 | 0032 | 0074 | 0148 | 0258 | 0402 | 0567 | 0732 | 0870 | 0962 | 0993 | 80 |
| | 89 | 88 | 87 | 86 | 85 | 84 | 83 | 82 | 81 | 80 | p |

**n = 100**

| x | 01 | 02 | 03 | 04 | 05 | 06 | 07 | 08 | 09 | 10 | x |
|---|----|----|----|----|----|----|----|----|----|----|---|
| 0 | 3660 | 1326 | 0476 | 0169 | 0059 | 0021 | 0007 | 0002 | 0001 | 0000 | 100 |
| 1 | 3697 | 2707 | 1471 | 0703 | 0312 | 0131 | 0053 | 0021 | 0008 | 0003 | 99 |
| 2 | 1849 | 2734 | 2252 | 1450 | 0812 | 0414 | 0198 | 0090 | 0039 | 0016 | 98 |
| 3 | 0610 | 1823 | 2275 | 1973 | 1396 | 0864 | 0486 | 0254 | 0125 | 0059 | 97 |
| 4 | 0149 | 0902 | 1706 | 1994 | 1781 | 1338 | 0888 | 0536 | 0301 | 0159 | 96 |
| 5 | 0029 | 0353 | 1013 | 1595 | 1800 | 1639 | 1283 | 0895 | 0571 | 0339 | 95 |
| 6 | 0005 | 0114 | 0496 | 1052 | 1500 | 1657 | 1529 | 1233 | 0895 | 0596 | 94 |
| 7 | 0001 | 0031 | 0206 | 0589 | 1060 | 1420 | 1545 | 1440 | 1188 | 0889 | 93 |
| 8 | 0000 | 0007 | 0074 | 0285 | 0649 | 1054 | 1302 | 1455 | 1366 | 1148 | 92 |
| 9 | 0000 | 0002 | 0023 | 0121 | 0349 | 0687 | 1040 | 1293 | 1381 | 1304 | 91 |
| | 99 | 98 | 97 | 96 | 95 | 94 | 93 | 92 | 91 | 90 | p |

(Continued)

A–16

# Table I (Continued)

## n = 100

| p \ x | 21 | 22 | 23 | 24 | 25 | 26 | 27 | 28 | 29 | 30 | x |
|---|---|---|---|---|---|---|---|---|---|---|---|
| 22 | 0931 | 0959 | 0932 | 0858 | 0749 | 0623 | 0495 | 0376 | 0273 | 0190 | 78 |
| 23 | 0839 | 0917 | 0944 | 0919 | 0847 | 0743 | 0621 | 0495 | 0378 | 0277 | 77 |
| 24 | 0716 | 0830 | 0905 | 0931 | 0906 | 0837 | 0736 | 0618 | 0496 | 0380 | 76 |
| 25 | 0578 | 0712 | 0822 | 0893 | 0918 | 0894 | 0828 | 0731 | 0615 | 0496 | 75 |
| 26 | 0444 | 0579 | 0708 | 0814 | 0883 | 0906 | 0883 | 0819 | 0725 | 0613 | 74 |
| 27 | 0323 | 0448 | 0580 | 0704 | 0806 | 0873 | 0896 | 0873 | 0812 | 0720 | 73 |
| 28 | 0224 | 0329 | 0451 | 0580 | 0701 | 0799 | 0864 | 0886 | 0864 | 0804 | 72 |
| 29 | 0148 | 0231 | 0335 | 0455 | 0580 | 0697 | 0793 | 0855 | 0876 | 0856 | 71 |
| 30 | 0093 | 0154 | 0237 | 0340 | 0458 | 0580 | 0694 | 0787 | 0847 | 0868 | 70 |
| 31 | 0056 | 0098 | 0160 | 0242 | 0344 | 0460 | 0580 | 0691 | 0781 | 0840 | 69 |
| 32 | 0032 | 0060 | 0103 | 0165 | 0248 | 0349 | 0462 | 0579 | 0688 | 0776 | 68 |
| 33 | 0018 | 0035 | 0063 | 0107 | 0170 | 0252 | 0352 | 0464 | 0579 | 0685 | 67 |
| 34 | 0009 | 0019 | 0037 | 0067 | 0112 | 0175 | 0257 | 0356 | 0466 | 0579 | 66 |
| 35 | 0005 | 0010 | 0021 | 0040 | 0070 | 0116 | 0179 | 0261 | 0359 | 0468 | 65 |
| 36 | 0002 | 0005 | 0011 | 0023 | 0042 | 0073 | 0120 | 0183 | 0265 | 0362 | 64 |
| 37 | 0001 | 0003 | 0006 | 0012 | 0024 | 0045 | 0077 | 0123 | 0187 | 0268 | 63 |
| 38 | 0000 | 0001 | 0003 | 0006 | 0013 | 0026 | 0047 | 0079 | 0127 | 0191 | 62 |
| 39 | 0000 | 0001 | 0001 | 0003 | 0007 | 0015 | 0028 | 0049 | 0082 | 0130 | 61 |
| 40 | 0000 | 0000 | 0001 | 0002 | 0003 | 0008 | 0016 | 0029 | 0051 | 0085 | 60 |
| 41 | 0000 | 0000 | 0000 | 0001 | 0002 | 0004 | 0008 | 0017 | 0031 | 0053 | 59 |
| 42 | 0000 | 0000 | 0000 | 0000 | 0001 | 0002 | 0004 | 0009 | 0018 | 0032 | 58 |
| 43 | 0000 | 0000 | 0000 | 0000 | 0000 | 0001 | 0002 | 0005 | 0010 | 0019 | 57 |
| 44 | 0000 | 0000 | 0000 | 0000 | 0000 | 0000 | 0001 | 0002 | 0005 | 0010 | 56 |
| 45 | 0000 | 0000 | 0000 | 0000 | 0000 | 0000 | 0000 | 0001 | 0003 | 0005 | 55 |
| 46 | 0000 | 0000 | 0000 | 0000 | 0000 | 0000 | 0000 | 0001 | 0001 | 0003 | 54 |
| 47 | 0000 | 0000 | 0000 | 0000 | 0000 | 0000 | 0000 | 0000 | 0001 | 0001 | 53 |
| 48 | 0000 | 0000 | 0000 | 0000 | 0000 | 0000 | 0000 | 0000 | 0000 | 0001 | 52 |
| p | 79 | 78 | 77 | 76 | 75 | 74 | 73 | 72 | 71 | 70 | x |

| p \ x | 31 | 32 | 33 | 34 | 35 | 36 | 37 | 38 | 39 | 40 | x |
|---|---|---|---|---|---|---|---|---|---|---|---|
| 15 | 0001 | 0001 | 0000 | 0000 | 0000 | 0000 | 0000 | 0000 | 0000 | 0000 | 85 |
| 16 | 0003 | 0001 | 0001 | 0000 | 0000 | 0000 | 0000 | 0000 | 0000 | 0000 | 84 |
| 17 | 0006 | 0003 | 0002 | 0001 | 0000 | 0000 | 0000 | 0000 | 0000 | 0000 | 83 |
| 18 | 0013 | 0007 | 0004 | 0002 | 0001 | 0000 | 0000 | 0000 | 0000 | 0000 | 82 |
| 19 | 0025 | 0014 | 0008 | 0004 | 0002 | 0001 | 0000 | 0000 | 0000 | 0000 | 81 |

*(Continued)*

## n = 100

| p \ x | 11 | 12 | 13 | 14 | 15 | 16 | 17 | 18 | 19 | 20 | x |
|---|---|---|---|---|---|---|---|---|---|---|---|
| 21 | 0015 | 0039 | 0084 | 0160 | 0270 | 0412 | 0571 | 0728 | 0859 | 0946 | 79 |
| 22 | 0007 | 0019 | 0045 | 0094 | 0171 | 0282 | 0420 | 0574 | 0724 | 0849 | 78 |
| 23 | 0003 | 0009 | 0023 | 0052 | 0103 | 0182 | 0292 | 0427 | 0576 | 0720 | 77 |
| 24 | 0001 | 0004 | 0011 | 0027 | 0058 | 0111 | 0192 | 0301 | 0433 | 0577 | 76 |
| 25 | 0000 | 0002 | 0005 | 0013 | 0031 | 0064 | 0119 | 0201 | 0309 | 0439 | 75 |
| 26 | 0000 | 0001 | 0002 | 0006 | 0016 | 0035 | 0071 | 0127 | 0209 | 0317 | 74 |
| 27 | 0000 | 0000 | 0001 | 0003 | 0008 | 0018 | 0040 | 0076 | 0134 | 0217 | 73 |
| 28 | 0000 | 0000 | 0000 | 0001 | 0004 | 0009 | 0021 | 0044 | 0082 | 0141 | 72 |
| 29 | 0000 | 0000 | 0000 | 0000 | 0002 | 0004 | 0011 | 0024 | 0048 | 0088 | 71 |
| 30 | 0000 | 0000 | 0000 | 0000 | 0001 | 0002 | 0005 | 0012 | 0027 | 0052 | 70 |
| 31 | 0000 | 0000 | 0000 | 0000 | 0000 | 0001 | 0002 | 0006 | 0014 | 0029 | 69 |
| 32 | 0000 | 0000 | 0000 | 0000 | 0000 | 0000 | 0001 | 0003 | 0007 | 0016 | 68 |
| 33 | 0000 | 0000 | 0000 | 0000 | 0000 | 0000 | 0000 | 0001 | 0003 | 0008 | 67 |
| 34 | 0000 | 0000 | 0000 | 0000 | 0000 | 0000 | 0000 | 0001 | 0002 | 0004 | 66 |
| 35 | 0000 | 0000 | 0000 | 0000 | 0000 | 0000 | 0000 | 0000 | 0001 | 0002 | 65 |
| 36 | 0000 | 0000 | 0000 | 0000 | 0000 | 0000 | 0000 | 0000 | 0000 | 0001 | 64 |
| p | 89 | 88 | 87 | 86 | 85 | 84 | 83 | 82 | 81 | 80 | x |

| p \ x | 21 | 22 | 23 | 24 | 25 | 26 | 27 | 28 | 29 | 30 | x |
|---|---|---|---|---|---|---|---|---|---|---|---|
| 7 | 0001 | 0000 | 0000 | 0000 | 0000 | 0000 | 0000 | 0000 | 0000 | 0000 | 93 |
| 8 | 0003 | 0001 | 0000 | 0000 | 0000 | 0000 | 0000 | 0000 | 0000 | 0000 | 92 |
| 9 | 0007 | 0003 | 0002 | 0001 | 0000 | 0000 | 0000 | 0000 | 0000 | 0000 | 91 |
| 10 | 0018 | 0009 | 0004 | 0002 | 0001 | 0000 | 0000 | 0000 | 0000 | 0000 | 90 |
| 11 | 0038 | 0021 | 0011 | 0005 | 0003 | 0001 | 0001 | 0000 | 0000 | 0000 | 89 |
| 12 | 0076 | 0043 | 0024 | 0012 | 0006 | 0003 | 0001 | 0001 | 0000 | 0000 | 88 |
| 13 | 0136 | 0082 | 0048 | 0027 | 0014 | 0007 | 0004 | 0002 | 0001 | 0000 | 87 |
| 14 | 0225 | 0144 | 0089 | 0052 | 0030 | 0016 | 0009 | 0004 | 0002 | 0001 | 86 |
| 15 | 0343 | 0233 | 0152 | 0095 | 0057 | 0033 | 0018 | 0010 | 0005 | 0002 | 85 |
| 16 | 0484 | 0350 | 0241 | 0159 | 0100 | 0061 | 0035 | 0020 | 0011 | 0006 | 84 |
| 17 | 0636 | 0487 | 0356 | 0248 | 0165 | 0106 | 0065 | 0038 | 0022 | 0012 | 83 |
| 18 | 0780 | 0634 | 0490 | 0361 | 0254 | 0171 | 0111 | 0069 | 0041 | 0024 | 82 |
| 19 | 0895 | 0772 | 0631 | 0492 | 0365 | 0259 | 0177 | 0115 | 0072 | 0044 | 81 |
| 20 | 0963 | 0881 | 0764 | 0629 | 0493 | 0369 | 0264 | 0182 | 0120 | 0076 | 80 |
| 21 | 0975 | 0947 | 0869 | 0756 | 0626 | 0494 | 0373 | 0269 | 0186 | 0124 | 79 |

*(Continued)*

# Table I (Continued)

## n = 100

| x | 31 | 32 | 33 | 34 | 35 | 36 | 37 | 38 | 39 | 40 | x |
|---|----|----|----|----|----|----|----|----|----|----|---|
| 20 | 0046 | 0027 | 0015 | 0008 | 0004 | 0002 | 0001 | 0001 | 0000 | 0000 | 80 |
| 21 | 0079 | 0049 | 0029 | 0016 | 0009 | 0005 | 0002 | 0001 | 0000 | 0000 | 79 |
| 22 | 0127 | 0082 | 0051 | 0030 | 0017 | 0009 | 0005 | 0003 | 0001 | 0001 | 78 |
| 23 | 0194 | 0131 | 0085 | 0053 | 0032 | 0018 | 0010 | 0006 | 0003 | 0001 | 77 |
| 24 | 0280 | 0198 | 0134 | 0088 | 0055 | 0033 | 0019 | 0011 | 0006 | 0003 | 76 |
| 25 | 0382 | 0283 | 0201 | 0137 | 0090 | 0057 | 0035 | 0020 | 0012 | 0006 | 75 |
| 26 | 0496 | 0384 | 0286 | 0204 | 0140 | 0092 | 0059 | 0036 | 0021 | 0012 | 74 |
| 27 | 0610 | 0495 | 0386 | 0288 | 0207 | 0143 | 0095 | 0060 | 0037 | 0022 | 73 |
| 28 | 0715 | 0608 | 0495 | 0387 | 0290 | 0209 | 0145 | 0097 | 0062 | 0038 | 72 |
| 29 | 0797 | 0710 | 0605 | 0495 | 0388 | 0292 | 0211 | 0147 | 0098 | 0063 | 71 |
| 30 | 0848 | 0791 | 0706 | 0603 | 0494 | 0389 | 0294 | 0213 | 0149 | 0100 | 70 |
| 31 | 0860 | 0840 | 0785 | 0702 | 0601 | 0494 | 0389 | 0295 | 0215 | 0151 | 69 |
| 32 | 0833 | 0853 | 0833 | 0779 | 0698 | 0599 | 0493 | 0390 | 0296 | 0217 | 68 |
| 33 | 0771 | 0827 | 0846 | 0827 | 0774 | 0694 | 0597 | 0493 | 0390 | 0297 | 67 |
| 34 | 0683 | 0767 | 0821 | 0840 | 0821 | 0769 | 0691 | 0595 | 0492 | 0391 | 66 |
| 35 | 0578 | 0680 | 0763 | 0816 | 0834 | 0816 | 0765 | 0688 | 0593 | 0491 | 65 |
| 36 | 0469 | 0578 | 0678 | 0759 | 0811 | 0829 | 0811 | 0761 | 0685 | 0591 | 64 |
| 37 | 0365 | 0471 | 0578 | 0676 | 0755 | 0806 | 0824 | 0807 | 0757 | 0682 | 63 |
| 38 | 0272 | 0367 | 0472 | 0577 | 0674 | 0752 | 0802 | 0820 | 0803 | 0754 | 62 |
| 39 | 0194 | 0275 | 0369 | 0473 | 0577 | 0672 | 0749 | 0799 | 0816 | 0799 | 61 |
| 40 | 0133 | 0197 | 0277 | 0372 | 0474 | 0577 | 0671 | 0746 | 0795 | 0812 | 60 |
| 41 | 0087 | 0136 | 0200 | 0280 | 0373 | 0475 | 0577 | 0670 | 0744 | 0792 | 59 |
| 42 | 0055 | 0090 | 0138 | 0203 | 0282 | 0375 | 0476 | 0576 | 0668 | 0742 | 58 |
| 43 | 0033 | 0057 | 0092 | 0141 | 0205 | 0285 | 0377 | 0477 | 0576 | 0667 | 57 |
| 44 | 0019 | 0035 | 0059 | 0094 | 0143 | 0207 | 0287 | 0378 | 0477 | 0576 | 56 |
| 45 | 0011 | 0020 | 0036 | 0060 | 0096 | 0145 | 0210 | 0289 | 0380 | 0478 | 55 |
| 46 | 0006 | 0011 | 0021 | 0037 | 0062 | 0098 | 0147 | 0211 | 0290 | 0381 | 54 |
| 47 | 0003 | 0006 | 0012 | 0022 | 0039 | 0063 | 0099 | 0149 | 0213 | 0292 | 53 |
| 48 | 0001 | 0003 | 0007 | 0012 | 0023 | 0039 | 0064 | 0101 | 0151 | 0215 | 52 |
| 49 | 0001 | 0002 | 0003 | 0007 | 0013 | 0023 | 0040 | 0066 | 0102 | 0152 | 51 |
| 50 | 0000 | 0001 | 0002 | 0004 | 0007 | 0013 | 0024 | 0041 | 0067 | 0103 | 50 |
| 51 | 0000 | 0000 | 0001 | 0002 | 0004 | 0007 | 0014 | 0025 | 0042 | 0068 | 49 |
| 52 | 0000 | 0000 | 0000 | 0001 | 0002 | 0004 | 0008 | 0014 | 0025 | 0042 | 48 |
| 53 | 0000 | 0000 | 0000 | 0000 | 0001 | 0002 | 0004 | 0008 | 0015 | 0026 | 47 |
| 54 | 0000 | 0000 | 0000 | 0000 | 0000 | 0001 | 0002 | 0004 | 0008 | 0015 | 46 |
| p | 69 | 68 | 67 | 66 | 65 | 64 | 63 | 62 | 61 | 60 | |

## n = 100

| x | 31 | 32 | 33 | 34 | 35 | 36 | 37 | 38 | 39 | 40 | x |
|---|----|----|----|----|----|----|----|----|----|----|---|
| 55 | 0000 | 0000 | 0000 | 0000 | 0000 | 0000 | 0001 | 0002 | 0004 | 0008 | 45 |
| 56 | 0000 | 0000 | 0000 | 0000 | 0000 | 0000 | 0000 | 0001 | 0002 | 0004 | 44 |
| 57 | 0000 | 0000 | 0000 | 0000 | 0000 | 0000 | 0000 | 0000 | 0001 | 0002 | 43 |
| 58 | 0000 | 0000 | 0000 | 0000 | 0000 | 0000 | 0000 | 0000 | 0001 | 0001 | 42 |
| 59 | 0000 | 0000 | 0000 | 0000 | 0000 | 0000 | 0000 | 0000 | 0000 | 0001 | 41 |
| p | 69 | 68 | 67 | 66 | 65 | 64 | 63 | 62 | 61 | 60 | |

| x | 41 | 42 | 43 | 44 | 45 | 46 | 47 | 48 | 49 | 50 | x |
|---|----|----|----|----|----|----|----|----|----|----|---|
| 23 | 0001 | 0000 | 0000 | 0000 | 0000 | 0000 | 0000 | 0000 | 0000 | 0000 | 77 |
| 24 | 0002 | 0001 | 0000 | 0000 | 0000 | 0000 | 0000 | 0000 | 0000 | 0000 | 76 |
| 25 | 0003 | 0002 | 0001 | 0000 | 0000 | 0000 | 0000 | 0000 | 0000 | 0000 | 75 |
| 26 | 0007 | 0003 | 0002 | 0001 | 0000 | 0000 | 0000 | 0000 | 0000 | 0000 | 74 |
| 27 | 0013 | 0007 | 0004 | 0002 | 0001 | 0000 | 0000 | 0000 | 0000 | 0000 | 73 |
| 28 | 0023 | 0013 | 0007 | 0004 | 0002 | 0001 | 0000 | 0000 | 0000 | 0000 | 72 |
| 29 | 0039 | 0024 | 0014 | 0008 | 0004 | 0002 | 0001 | 0000 | 0000 | 0000 | 71 |
| 30 | 0065 | 0040 | 0024 | 0014 | 0008 | 0004 | 0002 | 0001 | 0000 | 0000 | 70 |
| 31 | 0102 | 0066 | 0041 | 0025 | 0014 | 0008 | 0004 | 0002 | 0001 | 0001 | 69 |
| 32 | 0152 | 0103 | 0067 | 0042 | 0025 | 0015 | 0008 | 0004 | 0002 | 0001 | 68 |
| 33 | 0218 | 0154 | 0104 | 0068 | 0043 | 0026 | 0015 | 0008 | 0004 | 0002 | 67 |
| 34 | 0298 | 0219 | 0155 | 0105 | 0069 | 0043 | 0026 | 0015 | 0009 | 0005 | 66 |
| 35 | 0391 | 0299 | 0220 | 0156 | 0106 | 0069 | 0044 | 0026 | 0015 | 0009 | 65 |
| 36 | 0491 | 0391 | 0300 | 0221 | 0157 | 0107 | 0070 | 0044 | 0027 | 0016 | 64 |
| 37 | 0590 | 0490 | 0391 | 0300 | 0222 | 0157 | 0107 | 0070 | 0044 | 0027 | 63 |
| 38 | 0680 | 0588 | 0489 | 0391 | 0301 | 0222 | 0158 | 0108 | 0071 | 0045 | 62 |
| 39 | 0751 | 0677 | 0587 | 0489 | 0391 | 0301 | 0223 | 0158 | 0108 | 0071 | 61 |
| 40 | 0796 | 0748 | 0675 | 0586 | 0488 | 0391 | 0301 | 0223 | 0159 | 0108 | 60 |
| 41 | 0809 | 0793 | 0745 | 0673 | 0584 | 0487 | 0391 | 0301 | 0223 | 0159 | 59 |
| 42 | 0790 | 0806 | 0790 | 0743 | 0672 | 0583 | 0487 | 0390 | 0301 | 0223 | 58 |
| 43 | 0740 | 0787 | 0804 | 0788 | 0741 | 0670 | 0582 | 0486 | 0390 | 0301 | 57 |
| 44 | 0666 | 0739 | 0785 | 0802 | 0786 | 0739 | 0669 | 0581 | 0485 | 0390 | 56 |
| 45 | 0576 | 0666 | 0737 | 0784 | 0800 | 0784 | 0738 | 0668 | 0580 | 0485 | 55 |
| 46 | 0479 | 0576 | 0665 | 0736 | 0782 | 0798 | 0783 | 0737 | 0667 | 0580 | 54 |
| 47 | 0382 | 0480 | 0576 | 0665 | 0736 | 0781 | 0797 | 0781 | 0736 | 0666 | 53 |
| 48 | 0293 | 0383 | 0480 | 0577 | 0665 | 0735 | 0781 | 0797 | 0781 | 0735 | 52 |
| 49 | 0216 | 0295 | 0384 | 0481 | 0577 | 0664 | 0735 | 0780 | 0796 | 0780 | 51 |
| 50 | 0153 | 0218 | 0296 | 0385 | 0482 | 0577 | 0665 | 0735 | 0780 | 0796 | 50 |
| 51 | 0104 | 0155 | 0219 | 0297 | 0386 | 0482 | 0578 | 0665 | 0735 | 0780 | 49 |
| 52 | 0068 | 0105 | 0156 | 0220 | 0298 | 0387 | 0483 | 0578 | 0665 | 0735 | 48 |
| p | 59 | 58 | 57 | 56 | 55 | 54 | 53 | 52 | 51 | 50 | |

(Continued)

# Table I (Continued)

n = 100

| x | p | 41 | 42 | 43 | 44 | 45 | 46 | 47 | 48 | 49 | 50 | |
|---|---|----|----|----|----|----|----|----|----|----|----|---|
| 53 | | 0043 | 0069 | 0106 | 0156 | 0221 | 0299 | 0388 | 0483 | 0579 | 0666 | 47 |
| 54 | | 0026 | 0044 | 0070 | 0107 | 0157 | 0221 | 0299 | 0388 | 0484 | 0580 | 46 |
| 55 | | 0015 | 0026 | 0044 | 0070 | 0108 | 0157 | 0222 | 0300 | 0389 | 0485 | 45 |
| 56 | | 0008 | 0015 | 0027 | 0044 | 0071 | 0108 | 0158 | 0222 | 0300 | 0390 | 44 |
| 57 | | 0005 | 0009 | 0016 | 0027 | 0045 | 0071 | 0108 | 0158 | 0223 | 0301 | 43 |
| 58 | | 0002 | 0005 | 0009 | 0016 | 0027 | 0045 | 0071 | 0108 | 0159 | 0223 | 42 |
| 59 | | 0001 | 0002 | 0005 | 0009 | 0016 | 0027 | 0045 | 0071 | 0109 | 0159 | 41 |
| 60 | | 0001 | 0001 | 0002 | 0005 | 0009 | 0016 | 0027 | 0045 | 0071 | 0108 | 40 |
| 61 | | 0000 | 0001 | 0001 | 0002 | 0005 | 0009 | 0016 | 0027 | 0045 | 0071 | 39 |
| 62 | | 0000 | 0000 | 0001 | 0001 | 0002 | 0005 | 0009 | 0016 | 0027 | 0045 | 38 |
| 63 | | 0000 | 0000 | 0000 | 0001 | 0001 | 0002 | 0005 | 0009 | 0016 | 0027 | 37 |
| 64 | | 0000 | 0000 | 0000 | 0000 | 0001 | 0001 | 0002 | 0005 | 0009 | 0016 | 36 |
| 65 | | 0000 | 0000 | 0000 | 0000 | 0000 | 0001 | 0001 | 0002 | 0005 | 0009 | 35 |
| 66 | | 0000 | 0000 | 0000 | 0000 | 0000 | 0000 | 0001 | 0001 | 0002 | 0005 | 34 |
| 67 | | 0000 | 0000 | 0000 | 0000 | 0000 | 0000 | 0000 | 0001 | 0001 | 0002 | 33 |
| 68 | | 0000 | 0000 | 0000 | 0000 | 0000 | 0000 | 0000 | 0000 | 0001 | 0001 | 32 |
| 69 | | 0000 | 0000 | 0000 | 0000 | 0000 | 0000 | 0000 | 0000 | 0000 | 0001 | 31 |
| | | 59 | 58 | 57 | 56 | 55 | 54 | 53 | 52 | 51 | 50 | x |
| | | | | | | | | | | | p | |

# Table II  Poisson Distribution

From *Handbook of Probability and Statistics* by R. S. Burington and D. C. May, Jr. Copyright 1953 by McGraw-Hill, Inc. Used with permission of McGraw-Hill Book Company.

The following table gives the probability of exactly x successes, for various values of λ, as defined by the Poisson mass function.

$$P(x) = \frac{e^{-\lambda}\lambda^x}{x!}$$

*Examples:* If λ = 1.5, then P(2) = 0.2510, P(3) = 0.1255.

## Poisson Probabilities

λ

| x | 0.1 | 0.2 | 0.3 | 0.4 | 0.5 | 0.6 | 0.7 | 0.8 | 0.9 | 1.0 |
|---|-----|-----|-----|-----|-----|-----|-----|-----|-----|-----|
| 0 | 9048 | 8187 | 7408 | 6703 | 6065 | 5488 | 4966 | 4493 | 4066 | 3679 |
| 1 | 0905 | 1637 | 2222 | 2681 | 3033 | 3293 | 3476 | 3595 | 3659 | 3679 |
| 2 | 0045 | 0164 | 0333 | 0536 | 0758 | 0988 | 1217 | 1438 | 1647 | 1839 |
| 3 | 0002 | 0011 | 0033 | 0072 | 0126 | 0198 | 0284 | 0383 | 0494 | 0613 |
| 4 | 0000 | 0001 | 0002 | 0007 | 0016 | 0030 | 0050 | 0077 | 0111 | 0153 |
| 5 | 0000 | 0000 | 0000 | 0001 | 0002 | 0004 | 0007 | 0012 | 0020 | 0031 |
| 6 | 0000 | 0000 | 0000 | 0000 | 0000 | 0000 | 0001 | 0002 | 0003 | 0005 |
| 7 | 0000 | 0000 | 0000 | 0000 | 0000 | 0000 | 0000 | 0000 | 0000 | 0001 |

λ

| x | 1.1 | 1.2 | 1.3 | 1.4 | 1.5 | 1.6 | 1.7 | 1.8 | 1.9 | 2.0 |
|---|-----|-----|-----|-----|-----|-----|-----|-----|-----|-----|
| 0 | 3329 | 3012 | 2725 | 2466 | 2231 | 2019 | 1827 | 1653 | 1496 | 1353 |
| 1 | 3662 | 3614 | 3543 | 3452 | 3347 | 3230 | 3106 | 2975 | 2842 | 2707 |
| 2 | 2014 | 2169 | 2303 | 2417 | 2510 | 2584 | 2640 | 2678 | 2700 | 2707 |
| 3 | 0738 | 0867 | 0998 | 1128 | 1255 | 1378 | 1496 | 1607 | 1710 | 1804 |
| 4 | 0203 | 0260 | 0324 | 0395 | 0471 | 0551 | 0636 | 0723 | 0812 | 0902 |
| 5 | 0045 | 0062 | 0084 | 0111 | 0141 | 0176 | 0216 | 0260 | 0309 | 0361 |
| 6 | 0008 | 0012 | 0018 | 0026 | 0035 | 0047 | 0061 | 0078 | 0098 | 0120 |
| 7 | 0001 | 0002 | 0003 | 0005 | 0008 | 0011 | 0015 | 0020 | 0027 | 0034 |
| 8 | 0000 | 0000 | 0001 | 0001 | 0001 | 0002 | 0003 | 0005 | 0006 | 0009 |
| 9 | 0000 | 0000 | 0000 | 0000 | 0000 | 0000 | 0001 | 0001 | 0001 | 0002 |

(*Continued*)

Table II (Continued)

λ

| x | 4.1 | 4.2 | 4.3 | 4.4 | 4.5 | 4.6 | 4.7 | 4.8 | 4.9 | 5.0 |
|---|---|---|---|---|---|---|---|---|---|---|
| 0 | .0166 | .0150 | .0136 | .0123 | .0111 | .0101 | .0091 | .0082 | .0074 | .0067 |
| 1 | .0679 | .0630 | .0583 | .0540 | .0500 | .0462 | .0427 | .0395 | .0365 | .0337 |
| 2 | .1393 | .1323 | .1254 | .1188 | .1125 | .1063 | .1005 | .0948 | .0894 | .0842 |
| 3 | .1904 | .1852 | .1798 | .1743 | .1687 | .1631 | .1574 | .1517 | .1460 | .1404 |
| 4 | .1951 | .1944 | .1933 | .1917 | .1898 | .1875 | .1849 | .1820 | .1789 | .1755 |
| 5 | .1600 | .1633 | .1662 | .1687 | .1708 | .1725 | .1738 | .1747 | .1753 | .1755 |
| 6 | .1093 | .1143 | .1191 | .1237 | .1281 | .1323 | .1362 | .1398 | .1432 | .1462 |
| 7 | .0640 | .0686 | .0732 | .0778 | .0824 | .0869 | .0914 | .0959 | .1002 | .1044 |
| 8 | .0328 | .0360 | .0393 | .0428 | .0463 | .0500 | .0537 | .0575 | .0614 | .0653 |
| 9 | .0150 | .0168 | .0188 | .0209 | .0232 | .0255 | .0280 | .0307 | .0334 | .0363 |
| 10 | .0061 | .0071 | .0081 | .0092 | .0104 | .0118 | .0132 | .0147 | .0164 | .0181 |
| 11 | .0023 | .0027 | .0032 | .0037 | .0043 | .0049 | .0056 | .0064 | .0073 | .0082 |
| 12 | .0008 | .0009 | .0011 | .0014 | .0016 | .0019 | .0022 | .0026 | .0030 | .0034 |
| 13 | .0002 | .0003 | .0004 | .0005 | .0006 | .0007 | .0008 | .0009 | .0011 | .0013 |
| 14 | .0001 | .0001 | .0001 | .0001 | .0002 | .0002 | .0003 | .0003 | .0004 | .0005 |
| 15 | .0000 | .0000 | .0000 | .0000 | .0001 | .0001 | .0001 | .0001 | .0001 | .0002 |

λ

| x | 5.1 | 5.2 | 5.3 | 5.4 | 5.5 | 5.6 | 5.7 | 5.8 | 5.9 | 6.0 |
|---|---|---|---|---|---|---|---|---|---|---|
| 0 | .0061 | .0055 | .0050 | .0045 | .0041 | .0037 | .0033 | .0030 | .0027 | .0025 |
| 1 | .0311 | .0287 | .0265 | .0244 | .0225 | .0207 | .0191 | .0176 | .0162 | .0149 |
| 2 | .0793 | .0746 | .0701 | .0659 | .0618 | .0580 | .0544 | .0509 | .0477 | .0446 |
| 3 | .1348 | .1293 | .1239 | .1185 | .1133 | .1082 | .1033 | .0985 | .0938 | .0892 |
| 4 | .1719 | .1681 | .1641 | .1600 | .1558 | .1515 | .1472 | .1428 | .1383 | .1339 |
| 5 | .1753 | .1748 | .1740 | .1728 | .1714 | .1697 | .1678 | .1656 | .1632 | .1606 |
| 6 | .1490 | .1515 | .1537 | .1555 | .1571 | .1584 | .1594 | .1601 | .1605 | .1606 |
| 7 | .1086 | .1125 | .1163 | .1200 | .1234 | .1267 | .1298 | .1326 | .1353 | .1377 |
| 8 | .0692 | .0731 | .0771 | .0810 | .0849 | .0887 | .0925 | .0962 | .0998 | .1033 |
| 9 | .0392 | .0423 | .0454 | .0486 | .0519 | .0552 | .0586 | .0620 | .0654 | .0688 |
| 10 | .0200 | .0220 | .0241 | .0262 | .0285 | .0309 | .0334 | .0359 | .0386 | .0413 |
| 11 | .0093 | .0104 | .0116 | .0129 | .0143 | .0157 | .0173 | .0190 | .0207 | .0225 |
| 12 | .0039 | .0045 | .0051 | .0058 | .0065 | .0073 | .0082 | .0092 | .0102 | .0113 |
| 13 | .0015 | .0018 | .0021 | .0024 | .0028 | .0032 | .0036 | .0041 | .0046 | .0052 |
| 14 | .0006 | .0007 | .0008 | .0009 | .0011 | .0013 | .0015 | .0017 | .0019 | .0022 |
| 15 | .0002 | .0002 | .0003 | .0003 | .0004 | .0005 | .0006 | .0007 | .0008 | .0009 |
| 16 | .0001 | .0001 | .0001 | .0001 | .0001 | .0002 | .0002 | .0002 | .0003 | .0003 |
| 17 | .0000 | .0000 | .0000 | .0000 | .0001 | .0001 | .0001 | .0001 | .0001 | .0001 |

λ

| x | 2.1 | 2.2 | 2.3 | 2.4 | 2.5 | 2.6 | 2.7 | 2.8 | 2.9 | 3.0 |
|---|---|---|---|---|---|---|---|---|---|---|
| 0 | .1225 | .1108 | .1003 | .0907 | .0821 | .0743 | .0672 | .0608 | .0550 | .0498 |
| 1 | .2572 | .2438 | .2306 | .2177 | .2052 | .1931 | .1815 | .1703 | .1596 | .1494 |
| 2 | .2700 | .2681 | .2652 | .2613 | .2565 | .2510 | .2450 | .2384 | .2314 | .2240 |
| 3 | .1890 | .1966 | .2033 | .2090 | .2138 | .2176 | .2205 | .2225 | .2237 | .2240 |
| 4 | .0992 | .1082 | .1169 | .1254 | .1336 | .1414 | .1488 | .1557 | .1622 | .1680 |
| 5 | .0417 | .0476 | .0538 | .0602 | .0668 | .0735 | .0804 | .0872 | .0940 | .1008 |
| 6 | .0146 | .0174 | .0206 | .0241 | .0278 | .0319 | .0362 | .0407 | .0455 | .0504 |
| 7 | .0044 | .0055 | .0068 | .0083 | .0099 | .0118 | .0139 | .0163 | .0188 | .0216 |
| 8 | .0011 | .0015 | .0019 | .0025 | .0031 | .0038 | .0047 | .0057 | .0068 | .0081 |
| 9 | .0003 | .0004 | .0005 | .0007 | .0009 | .0011 | .0014 | .0018 | .0022 | .0027 |
| 10 | .0001 | .0001 | .0001 | .0002 | .0002 | .0003 | .0004 | .0005 | .0006 | .0008 |
| 11 | .0000 | .0000 | .0000 | .0000 | .0000 | .0001 | .0001 | .0001 | .0002 | .0002 |
| 12 | .0000 | .0000 | .0000 | .0000 | .0000 | .0000 | .0000 | .0000 | .0000 | .0001 |

λ

| x | 3.1 | 3.2 | 3.3 | 3.4 | 3.5 | 3.6 | 3.7 | 3.8 | 3.9 | 4.0 |
|---|---|---|---|---|---|---|---|---|---|---|
| 0 | .0450 | .0408 | .0369 | .0334 | .0302 | .0273 | .0247 | .0224 | .0202 | .0183 |
| 1 | .1397 | .1304 | .1217 | .1135 | .1057 | .0984 | .0915 | .0850 | .0789 | .0733 |
| 2 | .2165 | .2087 | .2008 | .1929 | .1850 | .1771 | .1692 | .1615 | .1539 | .1465 |
| 3 | .2237 | .2226 | .2209 | .2186 | .2158 | .2125 | .2087 | .2046 | .2001 | .1954 |
| 4 | .1734 | .1781 | .1823 | .1858 | .1888 | .1912 | .1931 | .1944 | .1951 | .1954 |
| 5 | .1075 | .1140 | .1203 | .1264 | .1322 | .1377 | .1429 | .1477 | .1522 | .1563 |
| 6 | .0555 | .0608 | .0662 | .0716 | .0771 | .0826 | .0881 | .0936 | .0989 | .1042 |
| 7 | .0246 | .0278 | .0312 | .0348 | .0385 | .0425 | .0466 | .0508 | .0551 | .0595 |
| 8 | .0095 | .0111 | .0129 | .0148 | .0169 | .0191 | .0215 | .0241 | .0269 | .0298 |
| 9 | .0033 | .0040 | .0047 | .0056 | .0066 | .0076 | .0089 | .0102 | .0116 | .0132 |
| 10 | .0010 | .0013 | .0016 | .0019 | .0023 | .0028 | .0033 | .0039 | .0045 | .0053 |
| 11 | .0003 | .0004 | .0005 | .0006 | .0007 | .0009 | .0011 | .0013 | .0016 | .0019 |
| 12 | .0001 | .0001 | .0001 | .0002 | .0002 | .0003 | .0003 | .0004 | .0005 | .0006 |
| 13 | .0000 | .0000 | .0000 | .0000 | .0001 | .0001 | .0001 | .0001 | .0002 | .0002 |
| 14 | .0000 | .0000 | .0000 | .0000 | .0000 | .0000 | .0000 | .0000 | .0000 | .0001 |

(Continued)

# Table II (Continued)

λ

| x | 7.1 | 7.2 | 7.3 | 7.4 | 7.5 | 7.6 | 7.7 | 7.8 | 7.9 | 8.0 |
|---|---|---|---|---|---|---|---|---|---|---|
| 10 | .0740 | .0770 | .0800 | .0829 | .0858 | .0887 | .0914 | .0941 | .0967 | .0993 |
| 11 | .0478 | .0504 | .0531 | .0558 | .0585 | .0613 | .0640 | .0667 | .0695 | .0722 |
| 12 | .0283 | .0303 | .0323 | .0344 | .0366 | .0388 | .0411 | .0434 | .0457 | .0481 |
| 13 | .0154 | .0168 | .0181 | .0196 | .0211 | .0227 | .0243 | .0260 | .0278 | .0296 |
| 14 | .0078 | .0086 | .0095 | .0104 | .0113 | .0123 | .0134 | .0145 | .0157 | .0169 |
| 15 | .0037 | .0041 | .0046 | .0051 | .0057 | .0062 | .0069 | .0075 | .0083 | .0090 |
| 16 | .0016 | .0019 | .0021 | .0024 | .0026 | .0030 | .0033 | .0037 | .0041 | .0045 |
| 17 | .0007 | .0008 | .0009 | .0010 | .0012 | .0013 | .0015 | .0017 | .0019 | .0021 |
| 18 | .0003 | .0003 | .0004 | .0004 | .0005 | .0006 | .0006 | .0007 | .0008 | .0009 |
| 19 | .0001 | .0001 | .0001 | .0002 | .0002 | .0002 | .0003 | .0003 | .0003 | .0004 |
| 20 | .0000 | .0000 | .0001 | .0001 | .0001 | .0001 | .0001 | .0001 | .0001 | .0002 |
| 21 | .0000 | .0000 | .0000 | .0000 | .0000 | .0000 | .0000 | .0000 | .0001 | .0001 |

λ

| x | 8.1 | 8.2 | 8.3 | 8.4 | 8.5 | 8.6 | 8.7 | 8.8 | 8.9 | 9.0 |
|---|---|---|---|---|---|---|---|---|---|---|
| 0 | .0003 | .0003 | .0002 | .0002 | .0002 | .0002 | .0002 | .0002 | .0001 | .0001 |
| 1 | .0025 | .0023 | .0021 | .0019 | .0017 | .0016 | .0014 | .0013 | .0012 | .0011 |
| 2 | .0100 | .0092 | .0086 | .0079 | .0074 | .0068 | .0063 | .0058 | .0054 | .0050 |
| 3 | .0269 | .0252 | .0237 | .0222 | .0208 | .0195 | .0183 | .0171 | .0160 | .0150 |
| 4 | .0544 | .0517 | .0491 | .0466 | .0443 | .0420 | .0398 | .0377 | .0357 | .0337 |
| 5 | .0882 | .0849 | .0816 | .0784 | .0752 | .0722 | .0692 | .0663 | .0635 | .0607 |
| 6 | .1191 | .1160 | .1128 | .1097 | .1066 | .1034 | .1003 | .0972 | .0941 | .0911 |
| 7 | .1378 | .1358 | .1338 | .1317 | .1294 | .1271 | .1247 | .1222 | .1197 | .1171 |
| 8 | .1395 | .1392 | .1388 | .1382 | .1375 | .1366 | .1356 | .1344 | .1332 | .1318 |
| 9 | .1256 | .1269 | .1280 | .1290 | .1299 | .1306 | .1311 | .1315 | .1317 | .1318 |
| 10 | .1017 | .1040 | .1063 | .1084 | .1104 | .1123 | .1140 | .1157 | .1172 | .1186 |
| 11 | .0749 | .0776 | .0802 | .0828 | .0853 | .0878 | .0902 | .0925 | .0948 | .0970 |
| 12 | .0505 | .0530 | .0555 | .0579 | .0604 | .0629 | .0654 | .0679 | .0703 | .0728 |
| 13 | .0315 | .0334 | .0354 | .0374 | .0395 | .0416 | .0438 | .0459 | .0481 | .0504 |
| 14 | .0182 | .0196 | .0210 | .0225 | .0240 | .0256 | .0272 | .0289 | .0306 | .0324 |
| 15 | .0098 | .0107 | .0116 | .0126 | .0136 | .0147 | .0158 | .0169 | .0182 | .0194 |
| 16 | .0050 | .0055 | .0060 | .0066 | .0072 | .0079 | .0086 | .0093 | .0101 | .0109 |
| 17 | .0024 | .0026 | .0029 | .0033 | .0036 | .0040 | .0044 | .0048 | .0053 | .0058 |
| 18 | .0011 | .0012 | .0014 | .0015 | .0017 | .0019 | .0021 | .0024 | .0026 | .0029 |
| 19 | .0005 | .0005 | .0006 | .0007 | .0008 | .0009 | .0010 | .0011 | .0012 | .0014 |
| 20 | .0002 | .0002 | .0002 | .0003 | .0003 | .0004 | .0004 | .0005 | .0005 | .0006 |
| 21 | .0001 | .0001 | .0001 | .0001 | .0001 | .0002 | .0002 | .0002 | .0002 | .0003 |
| 22 | .0000 | .0000 | .0000 | .0000 | .0001 | .0001 | .0001 | .0001 | .0001 | .0001 |

λ

| x | 6.1 | 6.2 | 6.3 | 6.4 | 6.5 | 6.6 | 6.7 | 6.8 | 6.9 | 7.0 |
|---|---|---|---|---|---|---|---|---|---|---|
| 0 | .0022 | .0020 | .0018 | .0017 | .0015 | .0014 | .0012 | .0011 | .0010 | .0009 |
| 1 | .0137 | .0126 | .0116 | .0106 | .0098 | .0090 | .0082 | .0076 | .0070 | .0064 |
| 2 | .0417 | .0390 | .0364 | .0340 | .0318 | .0296 | .0276 | .0258 | .0240 | .0223 |
| 3 | .0848 | .0806 | .0765 | .0726 | .0688 | .0652 | .0617 | .0584 | .0552 | .0521 |
| 4 | .1294 | .1249 | .1205 | .1162 | .1118 | .1076 | .1034 | .0992 | .0952 | .0912 |
| 5 | .1579 | .1549 | .1519 | .1487 | .1454 | .1420 | .1385 | .1349 | .1314 | .1277 |
| 6 | .1605 | .1601 | .1595 | .1586 | .1575 | .1562 | .1546 | .1529 | .1511 | .1490 |
| 7 | .1399 | .1418 | .1435 | .1450 | .1462 | .1472 | .1480 | .1486 | .1489 | .1490 |
| 8 | .1066 | .1099 | .1130 | .1160 | .1188 | .1215 | .1240 | .1263 | .1284 | .1304 |
| 9 | .0723 | .0757 | .0791 | .0825 | .0858 | .0891 | .0923 | .0954 | .0985 | .1014 |
| 10 | .0441 | .0469 | .0498 | .0528 | .0558 | .0588 | .0618 | .0649 | .0679 | .0710 |
| 11 | .0245 | .0265 | .0285 | .0307 | .0330 | .0353 | .0377 | .0401 | .0426 | .0452 |
| 12 | .0124 | .0137 | .0150 | .0164 | .0179 | .0194 | .0210 | .0227 | .0245 | .0264 |
| 13 | .0058 | .0065 | .0073 | .0081 | .0089 | .0098 | .0108 | .0119 | .0130 | .0142 |
| 14 | .0025 | .0029 | .0033 | .0037 | .0041 | .0046 | .0052 | .0058 | .0064 | .0071 |
| 15 | .0010 | .0012 | .0014 | .0016 | .0018 | .0020 | .0023 | .0026 | .0029 | .0033 |
| 16 | .0004 | .0005 | .0005 | .0006 | .0007 | .0008 | .0010 | .0011 | .0013 | .0014 |
| 17 | .0001 | .0002 | .0002 | .0002 | .0003 | .0003 | .0004 | .0004 | .0005 | .0006 |
| 18 | .0000 | .0001 | .0001 | .0001 | .0001 | .0001 | .0001 | .0002 | .0002 | .0002 |
| 19 | .0000 | .0000 | .0000 | .0000 | .0000 | .0000 | .0000 | .0001 | .0001 | .0001 |

λ

| x | 7.1 | 7.2 | 7.3 | 7.4 | 7.5 | 7.6 | 7.7 | 7.8 | 7.9 | 8.0 |
|---|---|---|---|---|---|---|---|---|---|---|
| 0 | .0008 | .0007 | .0007 | .0006 | .0006 | .0005 | .0005 | .0004 | .0004 | .0003 |
| 1 | .0059 | .0054 | .0049 | .0045 | .0041 | .0038 | .0035 | .0032 | .0029 | .0027 |
| 2 | .0208 | .0194 | .0180 | .0167 | .0156 | .0145 | .0134 | .0125 | .0116 | .0107 |
| 3 | .0492 | .0464 | .0438 | .0413 | .0389 | .0366 | .0345 | .0324 | .0305 | .0286 |
| 4 | .0874 | .0836 | .0799 | .0764 | .0729 | .0696 | .0663 | .0632 | .0602 | .0573 |
| 5 | .1241 | .1204 | .1167 | .1130 | .1094 | .1057 | .1021 | .0986 | .0951 | .0916 |
| 6 | .1468 | .1445 | .1420 | .1394 | .1367 | .1339 | .1311 | .1282 | .1252 | .1221 |
| 7 | .1489 | .1486 | .1481 | .1474 | .1465 | .1454 | .1442 | .1428 | .1413 | .1396 |
| 8 | .1321 | .1337 | .1351 | .1363 | .1373 | .1382 | .1388 | .1392 | .1395 | .1396 |
| 9 | .1042 | .1070 | .1096 | .1121 | .1144 | .1167 | .1187 | .1207 | .1224 | .1241 |

*(Continued)*

# Table II (Continued)

$\lambda$

| x | 11 | 12 | 13 | 14 | 15 | 16 | 17 | 18 | 19 | 20 |
|---|----|----|----|----|----|----|----|----|----|----|
| 5 | .0224 | .0127 | .0070 | .0037 | .0019 | .0010 | .0005 | .0002 | .0001 | .0001 |
| 6 | .0411 | .0255 | .0152 | .0087 | .0048 | .0026 | .0014 | .0007 | .0004 | .0002 |
| 7 | .0646 | .0437 | .0281 | .0174 | .0104 | .0060 | .0034 | .0018 | .0010 | .0005 |
| 8 | .0888 | .0655 | .0457 | .0304 | .0194 | .0120 | .0072 | .0042 | .0024 | .0013 |
| 9 | .1085 | .0874 | .0661 | .0473 | .0324 | .0213 | .0135 | .0083 | .0050 | .0029 |
| 10 | .1194 | .1048 | .0859 | .0663 | .0486 | .0341 | .0230 | .0150 | .0095 | .0058 |
| 11 | .1194 | .1144 | .1015 | .0844 | .0663 | .0496 | .0355 | .0245 | .0164 | .0106 |
| 12 | .1094 | .1144 | .1099 | .0984 | .0829 | .0661 | .0504 | .0368 | .0259 | .0176 |
| 13 | .0926 | .1056 | .1099 | .1060 | .0956 | .0814 | .0658 | .0509 | .0378 | .0271 |
| 14 | .0728 | .0905 | .1021 | .1060 | .1024 | .0930 | .0800 | .0655 | .0514 | .0387 |
| 15 | .0534 | .0724 | .0885 | .0989 | .1024 | .0992 | .0906 | .0786 | .0650 | .0516 |
| 16 | .0367 | .0543 | .0719 | .0866 | .0960 | .0992 | .0963 | .0884 | .0772 | .0646 |
| 17 | .0237 | .0383 | .0550 | .0713 | .0847 | .0934 | .0963 | .0936 | .0863 | .0760 |
| 18 | .0145 | .0256 | .0397 | .0554 | .0706 | .0830 | .0909 | .0936 | .0911 | .0844 |
| 19 | .0084 | .0161 | .0272 | .0409 | .0557 | .0699 | .0814 | .0887 | .0911 | .0888 |
| 20 | .0046 | .0097 | .0177 | .0286 | .0418 | .0559 | .0692 | .0798 | .0866 | .0888 |
| 21 | .0024 | .0055 | .0109 | .0191 | .0299 | .0426 | .0560 | .0684 | .0783 | .0846 |
| 22 | .0012 | .0030 | .0065 | .0121 | .0204 | .0310 | .0433 | .0560 | .0676 | .0769 |
| 23 | .0006 | .0016 | .0037 | .0074 | .0133 | .0216 | .0320 | .0438 | .0559 | .0669 |
| 24 | .0003 | .0008 | .0020 | .0043 | .0083 | .0144 | .0226 | .0328 | .0442 | .0557 |
| 25 | .0001 | .0004 | .0010 | .0024 | .0050 | .0092 | .0154 | .0237 | .0336 | .0446 |
| 26 | .0000 | .0002 | .0005 | .0013 | .0029 | .0057 | .0101 | .0164 | .0246 | .0343 |
| 27 | .0000 | .0001 | .0002 | .0007 | .0016 | .0034 | .0063 | .0109 | .0173 | .0254 |
| 28 | .0000 | .0000 | .0001 | .0003 | .0009 | .0019 | .0038 | .0070 | .0117 | .0181 |
| 29 | .0000 | .0000 | .0001 | .0002 | .0004 | .0011 | .0023 | .0044 | .0077 | .0125 |
| 30 | .0000 | .0000 | .0000 | .0001 | .0002 | .0006 | .0013 | .0026 | .0049 | .0083 |
| 31 | .0000 | .0000 | .0000 | .0000 | .0001 | .0003 | .0007 | .0015 | .0030 | .0054 |
| 32 | .0000 | .0000 | .0000 | .0000 | .0001 | .0001 | .0004 | .0009 | .0018 | .0034 |
| 33 | .0000 | .0000 | .0000 | .0000 | .0000 | .0001 | .0002 | .0005 | .0010 | .0020 |
| 34 | .0000 | .0000 | .0000 | .0000 | .0000 | .0000 | .0001 | .0002 | .0005 | .0012 |
| 35 | .0000 | .0000 | .0000 | .0000 | .0000 | .0000 | .0000 | .0001 | .0003 | .0007 |
| 36 | .0000 | .0000 | .0000 | .0000 | .0000 | .0000 | .0000 | .0001 | .0002 | .0004 |
| 37 | .0000 | .0000 | .0000 | .0000 | .0000 | .0000 | .0000 | .0000 | .0001 | .0002 |
| 38 | .0000 | .0000 | .0000 | .0000 | .0000 | .0000 | .0000 | .0000 | .0000 | .0001 |
| 39 | .0000 | .0000 | .0000 | .0000 | .0000 | .0000 | .0000 | .0000 | .0000 | .0001 |

$\lambda$

| x | 9.1 | 9.2 | 9.3 | 9.4 | 9.5 | 9.6 | 9.7 | 9.8 | 9.9 | 10 |
|---|----|----|----|----|----|----|----|----|----|----|
| 0 | .0001 | .0001 | .0001 | .0001 | .0001 | .0001 | .0001 | .0001 | .0001 | .0000 |
| 1 | .0010 | .0009 | .0009 | .0008 | .0007 | .0007 | .0006 | .0005 | .0005 | .0005 |
| 2 | .0046 | .0043 | .0040 | .0037 | .0034 | .0031 | .0029 | .0027 | .0025 | .0023 |
| 3 | .0140 | .0131 | .0123 | .0115 | .0107 | .0100 | .0093 | .0087 | .0081 | .0076 |
| 4 | .0319 | .0302 | .0285 | .0269 | .0254 | .0240 | .0226 | .0213 | .0201 | .0189 |
| 5 | .0581 | .0555 | .0530 | .0506 | .0483 | .0460 | .0439 | .0418 | .0398 | .0378 |
| 6 | .0881 | .0851 | .0822 | .0793 | .0764 | .0736 | .0709 | .0682 | .0656 | .0631 |
| 7 | .1145 | .1118 | .1091 | .1064 | .1037 | .1010 | .0982 | .0955 | .0928 | .0901 |
| 8 | .1302 | .1286 | .1269 | .1251 | .1232 | .1212 | .1191 | .1170 | .1148 | .1126 |
| 9 | .1317 | .1315 | .1311 | .1306 | .1300 | .1293 | .1284 | .1274 | .1263 | .1251 |
| 10 | .1198 | .1210 | .1219 | .1228 | .1235 | .1241 | .1245 | .1249 | .1250 | .1251 |
| 11 | .0991 | .1012 | .1031 | .1049 | .1067 | .1083 | .1098 | .1112 | .1125 | .1137 |
| 12 | .0752 | .0776 | .0799 | .0822 | .0844 | .0866 | .0888 | .0908 | .0928 | .0948 |
| 13 | .0526 | .0549 | .0572 | .0594 | .0617 | .0640 | .0662 | .0685 | .0707 | .0729 |
| 14 | .0342 | .0361 | .0380 | .0399 | .0419 | .0439 | .0459 | .0479 | .0500 | .0521 |
| 15 | .0208 | .0221 | .0235 | .0250 | .0265 | .0281 | .0297 | .0313 | .0330 | .0347 |
| 16 | .0118 | .0127 | .0137 | .0147 | .0157 | .0168 | .0180 | .0192 | .0204 | .0217 |
| 17 | .0063 | .0069 | .0075 | .0081 | .0088 | .0095 | .0103 | .0111 | .0119 | .0128 |
| 18 | .0032 | .0035 | .0039 | .0042 | .0046 | .0051 | .0055 | .0060 | .0065 | .0071 |
| 19 | .0015 | .0017 | .0019 | .0021 | .0023 | .0026 | .0028 | .0031 | .0034 | .0037 |
| 20 | .0007 | .0008 | .0009 | .0010 | .0011 | .0012 | .0014 | .0015 | .0017 | .0019 |
| 21 | .0003 | .0003 | .0004 | .0004 | .0005 | .0006 | .0006 | .0007 | .0008 | .0009 |
| 22 | .0001 | .0001 | .0002 | .0002 | .0002 | .0002 | .0003 | .0003 | .0004 | .0004 |
| 23 | .0000 | .0001 | .0001 | .0001 | .0001 | .0001 | .0001 | .0001 | .0002 | .0002 |
| 24 | .0000 | .0000 | .0000 | .0000 | .0000 | .0000 | .0000 | .0001 | .0001 | .0001 |

$\lambda$

| x | 11 | 12 | 13 | 14 | 15 | 16 | 17 | 18 | 19 | 20 |
|---|----|----|----|----|----|----|----|----|----|----|
| 0 | .0000 | .0000 | .0000 | .0000 | .0000 | .0000 | .0000 | .0000 | .0000 | .0000 |
| 1 | .0002 | .0001 | .0000 | .0000 | .0000 | .0000 | .0000 | .0000 | .0000 | .0000 |
| 2 | .0010 | .0004 | .0002 | .0001 | .0000 | .0000 | .0000 | .0000 | .0000 | .0000 |
| 3 | .0037 | .0018 | .0008 | .0004 | .0002 | .0001 | .0001 | .0000 | .0000 | .0000 |
| 4 | .0102 | .0053 | .0027 | .0013 | .0006 | .0003 | .0001 | .0001 | .0000 | .0000 |

*(Continued)*

**Table III**  Cumulative Normal Distribution

$$F(z) = \int_{-\infty}^{z} \frac{1}{\sqrt{2\pi}}\, e^{-z^2/2}\, dz \quad \textit{Example:}$$

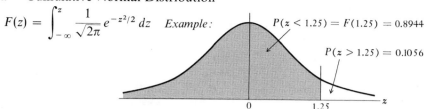

$P(z < 1.25) = F(1.25) = 0.8944$

$P(z > 1.25) = 0.1056$

| z | .00 | .01 | .02 | .03 | .04 | .05 | .06 | .07 | .08 | .09 |
|---|-----|-----|-----|-----|-----|-----|-----|-----|-----|-----|
| .0 | .5000 | .5040 | .5080 | .5120 | .5160 | .5199 | .5239 | .5279 | .5319 | .5359 |
| .1 | .5398 | .5438 | .5478 | .5517 | .5557 | .5596 | .5636 | .5675 | .5714 | .5753 |
| .2 | .5793 | .5832 | .5871 | .5910 | .5948 | .5987 | .6026 | .6064 | .6103 | .6141 |
| .3 | .6179 | .6217 | .6255 | .6293 | .6331 | .6368 | .6406 | .6443 | .6480 | .6517 |
| .4 | .6554 | .6591 | .6628 | .6664 | .6700 | .6736 | .6772 | .6808 | .6844 | .6879 |
| .5 | .6915 | .6950 | .6985 | .7019 | .7054 | .7088 | .7123 | .7157 | .7190 | .7224 |
| .6 | .7257 | .7291 | .7324 | .7357 | .7389 | .7422 | .7454 | .7486 | .7517 | .7549 |
| .7 | .7580 | .7611 | .7642 | .7673 | .7704 | .7734 | .7764 | .7794 | .7823 | .7852 |
| .8 | .7881 | .7910 | .7939 | .7967 | .7995 | .8023 | .8051 | .8078 | .8106 | .8133 |
| .9 | .8159 | .8186 | .8212 | .8238 | .8264 | .8289 | .8315 | .8340 | .8365 | .8389 |
| 1.0 | .8413 | .8438 | .8461 | .8485 | .8508 | .8531 | .8554 | .8577 | .8599 | .8621 |
| 1.1 | .8643 | .8665 | .8686 | .8708 | .8729 | .8749 | .8770 | .8790 | .8810 | .8830 |
| 1.2 | .8849 | .8869 | .8888 | .8907 | .8925 | .8944 | .8962 | .8980 | .8997 | .9015 |
| 1.3 | .9032 | .9049 | .9066 | .9082 | .9099 | .9115 | .9131 | .9147 | .9162 | .9177 |
| 1.4 | .9192 | .9207 | .9222 | .9236 | .9251 | .9265 | .9279 | .9292 | .9306 | .9319 |
| 1.5 | .9332 | .9345 | .9357 | .9370 | .9382 | .9394 | .9406 | .9418 | .9429 | .9441 |
| 1.6 | .9452 | .9463 | .9474 | .9484 | .9495 | .9505 | .9515 | .9525 | .9535 | .9545 |
| 1.7 | .9554 | .9564 | .9573 | .9582 | .9591 | .9599 | .9608 | .9616 | .9625 | .9633 |
| 1.8 | .9641 | .9649 | .9656 | .9664 | .9671 | .9678 | .9686 | .9693 | .9699 | .9706 |
| 1.9 | .9713 | .9719 | .9726 | .9732 | .9738 | .9744 | .9750 | .9756 | .9761 | .9767 |
| 2.0 | .9772 | .9778 | .9783 | .9788 | .9793 | .9798 | .9803 | .9808 | .9812 | .9817 |
| 2.1 | .9821 | .9826 | .9830 | .9834 | .9838 | .9842 | .9846 | .9850 | .9854 | .9857 |
| 2.2 | .9861 | .9864 | .9868 | .9871 | .9875 | .9878 | .9881 | .9884 | .9887 | .9890 |
| 2.3 | .9893 | .9896 | .9898 | .9901 | .9904 | .9906 | .9909 | .9911 | .9913 | .9916 |
| 2.4 | .9918 | .9920 | .9922 | .9925 | .9927 | .9929 | .9931 | .9932 | .9934 | .9936 |
| 2.5 | .9938 | .9940 | .9941 | .9943 | .9945 | .9946 | .9948 | .9949 | .9951 | .9952 |
| 2.6 | .9953 | .9955 | .9956 | .9957 | .9959 | .9960 | .9961 | .9962 | .9963 | .9964 |
| 2.7 | .9965 | .9966 | .9967 | .9968 | .9969 | .9970 | .9971 | .9972 | .9973 | .9974 |
| 2.8 | .9974 | .9975 | .9976 | .9977 | .9977 | .9978 | .9979 | .9979 | .9980 | .9981 |
| 2.9 | .9981 | .9982 | .9982 | .9983 | .9984 | .9984 | .9985 | .9985 | .9986 | .9986 |
| 3.0 | .9987 | .9987 | .9987 | .9988 | .9988 | .9989 | .9989 | .9989 | .9990 | .9990 |
| 3.1 | .9990 | .9991 | .9991 | .9991 | .9992 | .9992 | .9992 | .9992 | .9993 | .9993 |
| 3.2 | .9993 | .9993 | .9994 | .9994 | .9994 | .9994 | .9994 | .9995 | .9995 | .9995 |
| 3.3 | .9995 | .9995 | .9995 | .9996 | .9996 | .9996 | .9996 | .9996 | .9996 | .9997 |
| 3.4 | .9997 | .9997 | .9997 | .9997 | .9997 | .9997 | .9997 | .9997 | .9997 | .9998 |

**Table IV**  Cumulative Chi-Square Distribution

$$F(\chi^2) = \int_0^{\chi^2} \frac{x^{(v-2)/2}\,e^{-x/2}}{2^{v/2}[(v-2)/2]!}\,dx$$

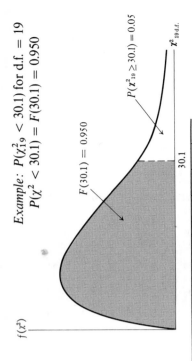

Example: $P(\chi^2_{19} < 30.1)$ for d.f. $= 19$
$P(\chi^2 < 30.1) = F(30.1) = 0.950$

$F(30.1) = 0.950$

$P(\chi^2_{19} \geq 30.1) = 0.05$

$\chi^2_{19\,\text{d.f.}}$

30.1

| $v$ \ $F$ | .005 | .010 | .025 | .050 | .100 | .250 | .500 | .750 | .900 | .950 | .975 | .990 | .995 |
|---|---|---|---|---|---|---|---|---|---|---|---|---|---|
| 1 | 0³393 | 0³157 | 0³982 | 0³393 | .0158 | .102 | .455 | 1.32 | 2.71 | 3.84 | 5.02 | 6.63 | 7.88 |
| 2 | .0100 | .0201 | .0506 | .103 | .211 | .575 | 1.39 | 2.77 | 4.61 | 5.99 | 7.38 | 9.21 | 10.6 |
| 3 | .0717 | .115 | .216 | .352 | .584 | 1.21 | 2.37 | 4.11 | 6.25 | 7.81 | 9.35 | 11.3 | 12.8 |
| 4 | .207 | .297 | .484 | .711 | 1.06 | 1.92 | 3.36 | 5.39 | 7.78 | 9.49 | 11.1 | 13.3 | 14.9 |
| 5 | .412 | .554 | .831 | 1.15 | 1.61 | 2.67 | 4.35 | 6.63 | 9.24 | 11.1 | 12.8 | 15.1 | 16.7 |
| 6 | .676 | .872 | 1.24 | 1.64 | 2.20 | 3.45 | 5.35 | 7.84 | 10.6 | 12.6 | 14.4 | 16.8 | 18.5 |
| 7 | .989 | 1.24 | 1.69 | 2.17 | 2.83 | 4.25 | 6.35 | 9.04 | 12.0 | 14.1 | 16.0 | 18.5 | 20.3 |
| 8 | 1.34 | 1.65 | 2.18 | 2.73 | 3.49 | 5.07 | 7.34 | 10.2 | 13.4 | 15.5 | 17.5 | 20.1 | 22.0 |
| 9 | 1.73 | 2.09 | 2.70 | 3.33 | 4.17 | 5.90 | 8.34 | 11.4 | 14.7 | 16.9 | 19.0 | 21.7 | 23.6 |
| 10 | 2.16 | 2.56 | 3.25 | 3.94 | 4.87 | 6.74 | 9.34 | 12.5 | 16.0 | 18.3 | 20.5 | 23.2 | 25.2 |
| 11 | 2.60 | 3.05 | 3.82 | 4.57 | 5.58 | 7.58 | 10.3 | 13.7 | 17.3 | 19.7 | 21.9 | 24.7 | 26.8 |
| 12 | 3.07 | 3.57 | 4.40 | 5.23 | 6.30 | 8.44 | 11.3 | 14.8 | 18.5 | 21.0 | 23.3 | 26.2 | 28.3 |
| 13 | 3.57 | 4.11 | 5.01 | 5.89 | 7.04 | 9.30 | 12.3 | 16.0 | 19.8 | 22.4 | 24.7 | 27.7 | 29.8 |
| 14 | 4.07 | 4.66 | 5.63 | 6.57 | 7.79 | 10.2 | 13.3 | 17.1 | 21.1 | 23.7 | 26.1 | 29.1 | 31.3 |
| 15 | 4.60 | 5.23 | 6.26 | 7.26 | 8.55 | 11.0 | 14.3 | 18.2 | 22.3 | 25.0 | 27.5 | 30.6 | 32.8 |
| 16 | 5.14 | 5.81 | 6.91 | 7.96 | 9.31 | 11.9 | 15.3 | 19.4 | 23.5 | 26.3 | 28.8 | 32.0 | 34.3 |
| 17 | 5.70 | 6.41 | 7.56 | 8.67 | 10.1 | 12.8 | 16.3 | 20.5 | 24.8 | 27.6 | 30.2 | 33.4 | 35.7 |
| 18 | 6.26 | 7.01 | 8.23 | 9.39 | 10.9 | 13.7 | 17.3 | 21.6 | 26.0 | 28.9 | 31.5 | 34.8 | 37.2 |
| 19 | 6.84 | 7.63 | 8.91 | 10.1 | 11.7 | 14.6 | 18.3 | 22.7 | 27.2 | 30.1 | 32.9 | 36.2 | 38.6 |
| 20 | 7.43 | 8.26 | 9.59 | 10.9 | 12.4 | 15.5 | 19.3 | 23.8 | 28.4 | 31.4 | 34.2 | 37.6 | 40.0 |
| 21 | 8.03 | 8.90 | 10.3 | 11.6 | 13.2 | 16.3 | 20.3 | 24.9 | 29.6 | 32.7 | 35.5 | 38.9 | 41.4 |
| 22 | 8.64 | 9.54 | 11.0 | 12.3 | 14.0 | 17.2 | 21.3 | 26.0 | 30.8 | 33.9 | 36.8 | 40.3 | 42.8 |
| 23 | 9.26 | 10.2 | 11.7 | 13.1 | 14.8 | 18.1 | 22.3 | 27.1 | 32.0 | 35.2 | 38.1 | 41.6 | 44.2 |
| 24 | 9.89 | 10.9 | 12.4 | 13.8 | 15.7 | 19.0 | 23.3 | 28.2 | 33.2 | 36.4 | 39.4 | 43.0 | 45.6 |
| 25 | 10.5 | 11.5 | 13.1 | 14.6 | 16.5 | 19.9 | 24.3 | 29.3 | 34.4 | 37.7 | 40.6 | 44.3 | 46.9 |
| 26 | 11.2 | 12.2 | 13.8 | 15.4 | 17.3 | 20.8 | 25.3 | 30.4 | 35.6 | 38.9 | 41.9 | 45.6 | 48.3 |
| 27 | 11.8 | 12.9 | 14.6 | 16.2 | 18.1 | 21.7 | 26.3 | 31.5 | 36.7 | 40.1 | 43.2 | 47.0 | 49.6 |
| 28 | 12.5 | 13.6 | 15.3 | 16.9 | 18.9 | 22.7 | 27.3 | 32.6 | 37.9 | 41.3 | 44.5 | 48.3 | 51.0 |
| 29 | 13.1 | 14.3 | 16.0 | 17.7 | 19.8 | 23.6 | 28.3 | 33.7 | 39.1 | 42.6 | 45.7 | 49.6 | 52.3 |
| 30 | 13.8 | 15.0 | 16.8 | 18.5 | 20.6 | 24.5 | 29.3 | 34.8 | 40.3 | 43.8 | 47.0 | 50.9 | 53.7 |

This table is abridged from "Tables of percentage points of the incomplete beta function and of the chi-square distribution," *Biometrika*, Vol. 32 (1941). It is here published with the kind permission of its author, Catherine M. Thompson, and the editor of *Biometrika*.

**Table V** Values of $F(T)$ where $T$ has the exponential distribution $f(T) = \lambda e^{-\lambda T}$

*Example:* If $\lambda = \frac{1}{6}$, the probability of observing a value less than $T = 9$ is found by $F(T)$ for $\lambda T = \frac{1}{6}(9) = 1.5$; $P(T < 9) = 0.777$.

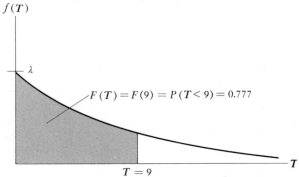

| $\lambda T$ | $F(T)$ | $\lambda T$ | $F(T)$ | $\lambda T$ | $F(T)$ | $\lambda T$ | $F(T)$ |
|------|-------|------|-------|------|--------|------|---------|
| 0.0 | 0.000 | 2.5 | 0.918 | 5.0 | 0.9933 | 7.5 | 0.99945 |
| 0.1 | 0.095 | 2.6 | 0.926 | 5.1 | 0.9939 | 7.6 | 0.99950 |
| 0.2 | 0.181 | 2.7 | 0.933 | 5.2 | 0.9945 | 7.7 | 0.99955 |
| 0.3 | 0.259 | 2.8 | 0.939 | 5.3 | 0.9950 | 7.8 | 0.99959 |
| 0.4 | 0.330 | 2.9 | 0.945 | 5.4 | 0.9955 | 7.9 | 0.99963 |
| | | | | | | | |
| 0.5 | 0.393 | 3.0 | 0.950 | 5.5 | 0.9959 | 8.0 | 0.99966 |
| 0.6 | 0.451 | 3.1 | 0.955 | 5.6 | 0.9963 | 8.1 | 0.99970 |
| 0.7 | 0.503 | 3.2 | 0.959 | 5.7 | 0.9967 | 8.2 | 0.99972 |
| 0.8 | 0.551 | 3.3 | 0.963 | 5.8 | 0.9970 | 8.3 | 0.99975 |
| 0.9 | 0.593 | 3.4 | 0.967 | 5.9 | 0.9973 | 8.4 | 0.99978 |
| | | | | | | | |
| 1.0 | 0.632 | 3.5 | 0.970 | 6.0 | 0.9975 | 8.5 | 0.99980 |
| 1.1 | 0.667 | 3.6 | 0.973 | 6.1 | 0.9978 | 8.6 | 0.99982 |
| 1.2 | 0.699 | 3.7 | 0.975 | 6.2 | 0.9980 | 8.7 | 0.99983 |
| 1.3 | 0.727 | 3.8 | 0.978 | 6.3 | 0.9982 | 8.8 | 0.99985 |
| 1.4 | 0.753 | 3.9 | 0.980 | 6.4 | 0.9983 | 8.9 | 0.99986 |
| | | | | | | | |
| 1.5 | 0.777 | 4.0 | 0.982 | 6.5 | 0.9985 | 9.0 | 0.99989 |
| 1.6 | 0.798 | 4.1 | 0.983 | 6.6 | 0.9986 | 9.1 | 0.99989 |
| 1.7 | 0.817 | 4.2 | 0.985 | 6.7 | 0.9988 | 9.2 | 0.99990 |
| 1.8 | 0.835 | 4.3 | 0.986 | 6.8 | 0.9989 | 9.3 | 0.99991 |
| 1.9 | 0.850 | 4.4 | 0.988 | 6.9 | 0.9990 | 9.4 | 0.99992 |
| | | | | | | | |
| 2.0 | 0.865 | 4.5 | 0.989 | 7.0 | 0.9991 | 9.5 | 0.99992 |
| 2.1 | 0.878 | 4.6 | 0.990 | 7.1 | 0.9992 | 9.6 | 0.99993 |
| 2.2 | 0.889 | 4.7 | 0.991 | 7.2 | 0.9993 | 9.7 | 0.99994 |
| 2.3 | 0.900 | 4.8 | 0.992 | 7.3 | 0.9993 | 9.8 | 0.99994 |
| 2.4 | 0.909 | 4.9 | 0.993 | 7.4 | 0.9993 | 9.9 | 0.99995 |

**Table VI**  Random Digits

From RAND Corporation, *A Million Random Digits*. By permission.

| | | | | | | | | | |
|---|---|---|---|---|---|---|---|---|---|
| 07018 | 31172 | 12572 | 23968 | 55216 | 85366 | 56223 | 09300 | 94564 | 18172 |
| 52444 | 65625 | 97918 | 46794 | 62370 | 59344 | 20149 | 17596 | 51669 | 47429 |
| 72161 | 57299 | 87521 | 44351 | 99981 | 55008 | 93371 | 60620 | 66662 | 27036 |
| 17918 | 75071 | 91057 | 46829 | 47992 | 26797 | 64423 | 42379 | 91676 | 75127 |
| 13623 | 76165 | 43195 | 50205 | 75736 | 77473 | 07268 | 31330 | 07337 | 55901 |
| | | | | | | | | | |
| 27426 | 97534 | 89707 | 97453 | 90836 | 78967 | 00704 | 85734 | 21776 | 85764 |
| 96039 | 21338 | 88169 | 69530 | 53300 | 29895 | 71507 | 28517 | 77761 | 17244 |
| 68282 | 98888 | 25545 | 69406 | 29470 | 46476 | 54562 | 79373 | 72993 | 98998 |
| 54262 | 21477 | 33097 | 48125 | 92982 | 98382 | 11265 | 25366 | 06636 | 25349 |
| 66290 | 27544 | 72780 | 91384 | 47296 | 54892 | 59168 | 83951 | 91075 | 04724 |
| | | | | | | | | | |
| 53348 | 39044 | 04072 | 62210 | 01209 | 43999 | 54952 | 68699 | 31912 | 09317 |
| 34482 | 42758 | 40128 | 48436 | 30254 | 50029 | 19016 | 56837 | 05206 | 33851 |
| 99268 | 98715 | 07545 | 27317 | 52459 | 75366 | 43688 | 27460 | 65145 | 65429 |
| 95342 | 97178 | 10401 | 31615 | 95784 | 77026 | 33087 | 65961 | 10056 | 72834 |
| 38556 | 60373 | 77935 | 64608 | 28949 | 94764 | 45312 | 71171 | 15400 | 72182 |
| | | | | | | | | | |
| 39159 | 04795 | 51163 | 84475 | 60722 | 35268 | 05044 | 56420 | 39214 | 89822 |
| 41786 | 18169 | 96649 | 92406 | 42773 | 23672 | 37333 | 85734 | 99886 | 81200 |
| 95627 | 30768 | 30607 | 89023 | 60730 | 31519 | 53462 | 90489 | 81693 | 17849 |
| 98738 | 15548 | 42263 | 79489 | 85118 | 97073 | 01574 | 57310 | 59375 | 54417 |
| 75214 | 61575 | 27805 | 21930 | 94726 | 39454 | 19616 | 72239 | 93791 | 22610 |
| | | | | | | | | | |
| 73904 | 89123 | 19271 | 15792 | 72675 | 62175 | 48746 | 56084 | 54029 | 22296 |
| 33329 | 08896 | 94662 | 05781 | 59187 | 53284 | 28024 | 45421 | 37956 | 14252 |
| 66364 | 94799 | 62211 | 37539 | 80172 | 43269 | 91133 | 05562 | 82385 | 91760 |
| 68349 | 16984 | 86532 | 96186 | 53893 | 48268 | 82821 | 19526 | 63257 | 14288 |
| 19193 | 99621 | 66899 | 12351 | 72438 | 99839 | 24228 | 32079 | 53517 | 18558 |
| | | | | | | | | | |
| 09237 | 23489 | 19172 | 80439 | 76263 | 98918 | 59330 | 20121 | 89779 | 58862 |
| 11941 | 77008 | 27646 | 82072 | 28048 | 41589 | 70883 | 72035 | 81800 | 50296 |
| 60430 | 25875 | 26446 | 25738 | 32962 | 24266 | 26814 | 01194 | 48587 | 93319 |
| 79023 | 26895 | 65304 | 34978 | 43053 | 28951 | 22676 | 05303 | 39725 | 60054 |
| 11337 | 74487 | 83196 | 61939 | 05045 | 20405 | 69324 | 80823 | 20905 | 68727 |
| | | | | | | | | | |
| 09773 | 36773 | 21247 | 54735 | 68996 | 16937 | 18134 | 51873 | 10973 | 77090 |
| 63279 | 85087 | 94186 | 67793 | 18178 | 82224 | 17069 | 87880 | 54945 | 73489 |
| 34968 | 76028 | 54285 | 90845 | 35464 | 68076 | 15868 | 70063 | 26794 | 81386 |
| 99696 | 78454 | 21700 | 12301 | 88832 | 96796 | 59341 | 16136 | 01803 | 17537 |
| 55282 | 61051 | 97260 | 89829 | 69121 | 86547 | 62195 | 72492 | 33536 | 60137 |

**Table VII**  Cumulative $t$-Distribution

$$F(t) = \int_{-\infty}^{t} \frac{\left(\frac{v-1}{2}\right)!}{\left(\frac{v-2}{2}\right)! \sqrt{\pi n}\left(1 + \frac{t^2}{v}\right)^{(v+1)/2}} \, dt$$

$F(t) = P(t_{19} < 2.093) = 0.975$

0.025

*Example:*  $n = 20, v = 19$

| $v$ \ $F$ | .75 | .90 | .95 | .975 | .99 | .995 | .9995 |
|---|---|---|---|---|---|---|---|
| 1 | 1.000 | 3.078 | 6.314 | 12.706 | 31.821 | 63.657 | 636.619 |
| 2 | .816 | 1.886 | 2.920 | 4.303 | 6.965 | 9.925 | 31.598 |
| 3 | .765 | 1.638 | 2.353 | 3.182 | 4.541 | 5.841 | 12.941 |
| 4 | .741 | 1.533 | 2.132 | 2.776 | 3.747 | 4.604 | 8.610 |
| 5 | .727 | 1.476 | 2.015 | 2.571 | 3.365 | 4.032 | 6.859 |
| 6 | .718 | 1.440 | 1.943 | 2.447 | 3.143 | 3.707 | 5.959 |
| 7 | .711 | 1.415 | 1.895 | 2.365 | 2.998 | 3.499 | 5.405 |
| 8 | .706 | 1.397 | 1.860 | 2.306 | 2.896 | 3.355 | 5.041 |
| 9 | .703 | 1.383 | 1.833 | 2.262 | 2.821 | 3.250 | 4.781 |
| 10 | .700 | 1.372 | 1.812 | 2.228 | 2.764 | 3.169 | 4.587 |
| 11 | .697 | 1.363 | 1.796 | 2.201 | 2.718 | 3.106 | 4.437 |
| 12 | .695 | 1.356 | 1.782 | 2.179 | 2.681 | 3.055 | 4.318 |
| 13 | .694 | 1.350 | 1.771 | 2.160 | 2.650 | 3.012 | 4.221 |
| 14 | .692 | 1.345 | 1.761 | 2.145 | 2.624 | 2.977 | 4.140 |
| 15 | .691 | 1.341 | 1.753 | 2.131 | 2.602 | 2.947 | 4.073 |
| 16 | .690 | 1.337 | 1.746 | 2.120 | 2.583 | 2.921 | 4.015 |
| 17 | .689 | 1.333 | 1.740 | 2.110 | 2.567 | 2.898 | 3.965 |
| 18 | .688 | 1.330 | 1.734 | 2.101 | 2.552 | .2878 | 3.922 |
| 19 | .688 | 1.328 | 1.729 | 2.093 | 2.539 | 2.861 | 3.883 |
| 20 | .687 | 1.325 | 1.725 | 2.086 | 2.528 | 2.845 | 3.850 |
| 21 | .686 | 1.323 | 1.721 | 2.080 | 2.518 | 2.831 | 3.819 |
| 22 | .686 | 1.321 | 1.717 | 2.074 | 2.508 | 2.819 | 3.792 |
| 23 | .685 | 1.319 | 1.714 | 2.069 | 2.500 | 2.807 | 3.767 |
| 24 | .685 | 1.318 | 1.711 | 2.064 | 2.492 | 2.797 | 3.745 |
| 25 | .684 | 1.316 | 1.708 | 2.060 | 2.485 | 2.787 | 3.725 |
| 26 | .684 | 1.315 | 1.706 | 2.056 | 2.479 | 2.779 | 3.707 |
| 27 | .684 | 1.314 | 1.703 | 2.052 | 2.473 | 2.771 | 3.690 |
| 28 | .683 | 1.313 | 1.701 | 2.048 | 2.467 | 2.763 | 3.674 |
| 29 | .683 | 1.311 | 1.699 | 2.045 | 2.462 | 2.756 | 3.659 |
| 30 | .683 | 1.310 | 1.697 | 2.042 | 2.457 | 2.750 | 3.646 |
| 40 | .681 | 1.303 | 1.684 | 2.021 | 2.423 | 2.704 | 3.551 |
| 60 | .679 | 1.296 | 1.671 | 2.000 | 2.390 | 2.660 | 3.460 |
| 120 | .677 | 1.289 | 1.658 | 1.980 | 2.358 | 2.617 | 3.373 |
| ∞ | .674 | 1.282 | 1.645 | 1.960 | 2.326 | 2.576 | 3.291 |

This table is abridged from the "Statistical Tables" of R. A. Fisher and Frank Yates published by Oliver & Boyd, Ltd., Edinburgh and London, 1938.  It is here published with the kind permission of the authors and their publishers.

**Table VIII** a)  Critical Values of the *F*-Distribution ($\alpha = 0.05$)

Tables VI(a) and (b) from M. Merrington and C. M. Thompson, "Tables of percentage points of the inverted beta (F) distribution." *Biometrica*, Vol. 33 (1943) by permission of the *Biometrica* Trustees.

The following table gives the critical values of the *F* distribution for $\alpha = 0.05$. This probability represents the area exceeding the value of $F_{0.05, \nu_1, \nu_2}$, as shown by the shaded area in the figure below.

*Examples:* If $\nu_1 = 15$ (representing the greater mean square), and $\nu_2 = 20$, then the critical value for $\alpha = 0.05$ is 2.20.

$$P(F > 2.20) = 0.05,$$
$$P(F < 2.20) = 0.95.$$

VALUES OF $F_{0.05, \nu_1, \nu_2}$

$\nu_1 = $ Degrees of freedom for numerator

| $\nu_2$ | 1 | 2 | 3 | 4 | 5 | 6 | 7 | 8 | 9 | 10 | 12 | 15 | 20 | 24 | 30 | 40 | 60 | 120 | ∞ |
|---|---|---|---|---|---|---|---|---|---|---|---|---|---|---|---|---|---|---|---|
| 1 | 161 | 200 | 216 | 225 | 230 | 234 | 237 | 239 | 241 | 242 | 244 | 246 | 248 | 249 | 250 | 251 | 252 | 253 | 254 |
| 2 | 18.5 | 19.0 | 19.2 | 19.2 | 19.3 | 19.3 | 19.4 | 19.4 | 19.4 | 19.4 | 19.4 | 19.4 | 19.4 | 19.5 | 19.5 | 19.5 | 19.5 | 19.5 | 19.5 |
| 3 | 10.1 | 9.55 | 9.28 | 9.12 | 9.01 | 8.94 | 8.89 | 8.85 | 8.81 | 8.79 | 8.74 | 8.70 | 8.66 | 8.64 | 8.62 | 8.59 | 8.57 | 8.55 | 8.53 |
| 4 | 7.71 | 6.94 | 6.59 | 6.39 | 6.26 | 6.16 | 6.09 | 6.04 | 6.00 | 5.96 | 5.91 | 5.86 | 5.80 | 5.77 | 5.75 | 5.72 | 5.69 | 5.66 | 5.63 |
| 5 | 6.61 | 5.79 | 5.41 | 5.19 | 5.05 | 4.95 | 4.88 | 4.82 | 4.77 | 4.74 | 4.68 | 4.62 | 4.56 | 4.53 | 4.50 | 4.46 | 4.43 | 4.40 | 4.37 |
| 6 | 5.99 | 5.14 | 4.76 | 4.53 | 4.39 | 4.28 | 4.21 | 4.15 | 4.10 | 4.06 | 4.00 | 3.94 | 3.87 | 3.84 | 3.81 | 3.77 | 3.74 | 3.70 | 3.67 |
| 7 | 5.59 | 4.74 | 4.35 | 4.12 | 3.97 | 3.87 | 3.79 | 3.73 | 3.68 | 3.64 | 3.57 | 3.51 | 3.44 | 3.41 | 3.38 | 3.34 | 3.30 | 3.27 | 3.23 |
| 8 | 5.32 | 4.46 | 4.07 | 3.84 | 3.69 | 3.58 | 3.50 | 3.44 | 3.39 | 3.35 | 3.28 | 3.22 | 3.15 | 3.12 | 3.08 | 3.04 | 3.01 | 2.97 | 2.93 |
| 9 | 5.12 | 4.26 | 3.86 | 3.63 | 3.48 | 3.37 | 3.29 | 3.23 | 3.18 | 3.14 | 3.07 | 3.01 | 2.94 | 2.90 | 2.86 | 2.83 | 2.79 | 2.75 | 2.71 |
| 10 | 4.96 | 4.10 | 3.71 | 3.48 | 3.33 | 3.22 | 3.14 | 3.07 | 3.02 | 2.98 | 2.91 | 2.85 | 2.77 | 2.74 | 2.70 | 2.66 | 2.62 | 2.58 | 2.54 |

$\nu_2 = $ Degrees of freedom for denominator

$v_2$ = Degrees of freedom for denominator

| | | | | | | | | | | | | | | | | | | | |
|---|---|---|---|---|---|---|---|---|---|---|---|---|---|---|---|---|---|---|---|
| 11 | 4.84 | 3.98 | 3.59 | 3.36 | 3.20 | 3.09 | 3.01 | 2.95 | 2.90 | 2.85 | 2.79 | 2.72 | 2.65 | 2.61 | 2.57 | 2.53 | 2.49 | 2.45 | 2.40 |
| 12 | 4.75 | 3.89 | 3.49 | 3.26 | 3.11 | 3.00 | 2.91 | 2.85 | 2.80 | 2.75 | 2.69 | 2.62 | 2.54 | 2.51 | 2.47 | 2.43 | 2.38 | 2.34 | 2.30 |
| 13 | 4.67 | 3.81 | 3.41 | 3.18 | 3.03 | 2.92 | 2.83 | 2.77 | 2.71 | 2.67 | 2.60 | 2.53 | 2.46 | 2.42 | 2.38 | 2.34 | 2.30 | 2.25 | 2.21 |
| 14 | 4.60 | 3.74 | 3.34 | 3.11 | 2.96 | 2.85 | 2.76 | 2.70 | 2.65 | 2.60 | 2.53 | 2.46 | 2.39 | 2.35 | 2.31 | 2.27 | 2.22 | 2.18 | 2.13 |
| 15 | 4.54 | 3.68 | 3.29 | 3.06 | 2.90 | 2.79 | 2.71 | 2.64 | 2.59 | 2.54 | 2.48 | 2.40 | 2.33 | 2.29 | 2.25 | 2.20 | 2.16 | 2.11 | 2.07 |
| 16 | 4.49 | 3.63 | 3.24 | 3.01 | 2.85 | 2.74 | 2.66 | 2.59 | 2.54 | 2.49 | 2.42 | 2.35 | 2.28 | 2.24 | 2.19 | 2.15 | 2.11 | 2.06 | 2.01 |
| 17 | 4.45 | 3.59 | 3.20 | 2.96 | 2.81 | 2.70 | 2.61 | 2.55 | 2.49 | 2.45 | 2.38 | 2.31 | 2.23 | 2.19 | 2.15 | 2.10 | 2.06 | 2.01 | 1.96 |
| 18 | 4.41 | 3.55 | 3.16 | 2.93 | 2.77 | 2.66 | 2.58 | 2.51 | 2.46 | 2.41 | 2.34 | 2.27 | 2.19 | 2.15 | 2.11 | 2.06 | 2.02 | 1.97 | 1.92 |
| 19 | 4.38 | 3.52 | 3.13 | 2.90 | 2.74 | 2.63 | 2.54 | 2.48 | 2.42 | 2.38 | 2.31 | 2.23 | 2.16 | 2.11 | 2.07 | 2.03 | 1.98 | 1.93 | 1.88 |
| 20 | 4.35 | 3.49 | 3.10 | 2.87 | 2.71 | 2.60 | 2.51 | 2.45 | 2.39 | 2.35 | 2.28 | 2.20 | 2.12 | 2.08 | 2.04 | 1.99 | 1.95 | 1.90 | 1.84 |
| 21 | 4.32 | 3.47 | 3.07 | 2.84 | 2.68 | 2.57 | 2.49 | 2.42 | 2.37 | 2.32 | 2.25 | 2.18 | 2.10 | 2.05 | 2.01 | 1.96 | 1.92 | 1.87 | 1.81 |
| 22 | 4.30 | 3.44 | 3.05 | 2.82 | 2.66 | 2.55 | 2.46 | 2.40 | 2.34 | 2.30 | 2.23 | 2.15 | 2.07 | 2.03 | 1.98 | 1.94 | 1.89 | 1.84 | 1.78 |
| 23 | 4.28 | 3.42 | 3.03 | 2.80 | 2.64 | 2.53 | 2.44 | 2.37 | 2.32 | 2.27 | 2.20 | 2.13 | 2.05 | 2.01 | 1.96 | 1.91 | 1.86 | 1.81 | 1.76 |
| 24 | 4.26 | 3.40 | 3.01 | 2.78 | 2.62 | 2.51 | 2.42 | 2.36 | 2.30 | 2.25 | 2.18 | 2.11 | 2.03 | 1.98 | 1.94 | 1.89 | 1.84 | 1.79 | 1.73 |
| 25 | 4.24 | 3.39 | 2.99 | 2.76 | 2.60 | 2.49 | 2.40 | 2.34 | 2.28 | 2.24 | 2.16 | 2.09 | 2.01 | 1.96 | 1.92 | 1.87 | 1.82 | 1.77 | 1.71 |
| 30 | 4.17 | 3.32 | 2.92 | 2.69 | 2.53 | 2.42 | 2.33 | 2.27 | 2.21 | 2.16 | 2.09 | 2.01 | 1.93 | 1.89 | 1.84 | 1.79 | 1.74 | 1.68 | 1.62 |
| 40 | 4.08 | 3.23 | 2.84 | 2.61 | 2.45 | 2.34 | 2.25 | 2.18 | 2.12 | 2.08 | 2.00 | 1.92 | 1.84 | 1.79 | 1.74 | 1.69 | 1.64 | 1.58 | 1.51 |
| 60 | 4.00 | 3.15 | 2.76 | 2.53 | 2.37 | 2.25 | 2.17 | 2.10 | 2.04 | 1.99 | 1.92 | 1.84 | 1.75 | 1.70 | 1.65 | 1.59 | 1.53 | 1.47 | 1.39 |
| 120 | 3.92 | 3.07 | 2.68 | 2.45 | 2.29 | 2.18 | 2.09 | 2.02 | 1.96 | 1.91 | 1.83 | 1.75 | 1.66 | 1.61 | 1.55 | 1.50 | 1.43 | 1.35 | 1.25 |
| $\infty$ | 3.84 | 3.00 | 2.60 | 2.37 | 2.21 | 2.10 | 2.01 | 1.94 | 1.88 | 1.83 | 1.75 | 1.67 | 1.57 | 1.52 | 1.46 | 1.39 | 1.32 | 1.22 | 1.00 |

**Table VIII** b)   Critical Values of the **F**-Distribution ($\alpha = 0.01$)

The following table gives the critical values of the **F** distribution for $\alpha = 0.01$. This probability represents the area exceeding the value of $F_{0.01, \nu_1, \nu_2}$, as shown by the shaded area in the figure below.

*Examples:* If $\nu_1 = 15$ (representing the greater mean square), and $\nu_2 = 20$, then the critical value for $\alpha = 0.01$ is 3.09.

$$P(F > 3.09) = 0.01,$$
$$P(F < 3.09) = 0.99.$$

VALUES OF $F_{0.01, \nu_1, \nu_2}$

$\nu_1$ = Degrees of freedom for numerator

| $\nu_2$ | 1 | 2 | 3 | 4 | 5 | 6 | 7 | 8 | 9 | 10 | 12 | 15 | 20 | 24 | 30 | 40 | 60 | 120 | $\infty$ |
|---|---|---|---|---|---|---|---|---|---|---|---|---|---|---|---|---|---|---|---|
| 1 | 4,052 | 5,000 | 5,403 | 5,625 | 5,764 | 5,859 | 5,928 | 5,982 | 6,023 | 6,056 | 6,106 | 6,157 | 6,209 | 6,235 | 6,261 | 6,287 | 6,313 | 6,339 | 6,366 |
| 2 | 98.5 | 99.0 | 99.2 | 99.2 | 99.3 | 99.3 | 99.4 | 99.4 | 99.4 | 99.4 | 99.4 | 99.4 | 99.4 | 99.5 | 99.5 | 99.5 | 99.5 | 99.5 | 99.5 |
| 3 | 34.1 | 30.8 | 29.5 | 28.7 | 28.2 | 27.9 | 27.7 | 27.5 | 27.3 | 27.2 | 27.1 | 26.9 | 26.7 | 26.6 | 26.5 | 26.4 | 26.3 | 26.2 | 26.1 |
| 4 | 21.2 | 18.0 | 16.7 | 16.0 | 15.5 | 15.2 | 15.0 | 14.8 | 14.7 | 14.5 | 14.4 | 14.2 | 14.0 | 13.9 | 13.8 | 13.7 | 13.7 | 13.6 | 13.5 |
| 5 | 16.3 | 13.3 | 12.1 | 11.4 | 11.0 | 10.7 | 10.5 | 10.3 | 10.2 | 10.1 | 9.89 | 9.72 | 9.55 | 9.47 | 9.38 | 9.29 | 9.20 | 9.11 | 9.02 |
| 6 | 13.7 | 10.9 | 9.78 | 9.15 | 8.75 | 8.47 | 8.26 | 8.10 | 7.98 | 7.87 | 7.72 | 7.56 | 7.40 | 7.31 | 7.23 | 7.14 | 7.06 | 6.97 | 6.88 |
| 7 | 12.2 | 9.55 | 8.45 | 7.85 | 7.46 | 7.19 | 6.99 | 6.84 | 6.72 | 6.62 | 6.47 | 6.31 | 6.16 | 6.07 | 5.99 | 5.91 | 5.82 | 5.74 | 5.65 |
| 8 | 11.3 | 8.65 | 7.59 | 7.01 | 6.63 | 6.37 | 6.18 | 6.03 | 5.91 | 5.81 | 5.67 | 5.52 | 5.36 | 5.28 | 5.20 | 5.12 | 5.03 | 4.95 | 4.86 |
| 9 | 10.6 | 8.02 | 6.99 | 6.42 | 6.06 | 5.80 | 5.61 | 5.47 | 5.35 | 5.26 | 5.11 | 4.96 | 4.81 | 4.73 | 4.65 | 4.57 | 4.48 | 4.40 | 4.31 |
| 10 | 10.0 | 7.56 | 6.55 | 5.99 | 5.64 | 5.39 | 5.20 | 5.06 | 4.94 | 4.85 | 4.71 | 4.56 | 4.41 | 4.33 | 4.25 | 4.17 | 4.08 | 4.00 | 3.91 |

$\nu_2$ = Degrees of freedom for denominator

$v_2$ = Degrees of freedom for denominator

| $v_2$ | | | | | | | | | | | | | | | | | | | |
|---|---|---|---|---|---|---|---|---|---|---|---|---|---|---|---|---|---|---|---|
| 11 | 9.65 | 7.21 | 6.22 | 5.67 | 5.32 | 5.07 | 4.89 | 4.74 | 4.63 | 4.54 | 4.40 | 4.25 | 4.10 | 4.02 | 3.94 | 3.86 | 3.78 | 3.69 | 3.60 |
| 12 | 9.33 | 6.93 | 5.95 | 5.41 | 5.06 | 4.82 | 4.64 | 4.50 | 4.39 | 4.30 | 4.16 | 4.01 | 3.86 | 3.78 | 3.70 | 3.62 | 3.54 | 3.45 | 3.36 |
| 13 | 9.07 | 6.70 | 5.74 | 5.21 | 4.86 | 4.62 | 4.44 | 4.30 | 4.19 | 4.10 | 3.96 | 3.82 | 3.66 | 3.59 | 3.51 | 3.43 | 3.34 | 3.25 | 3.17 |
| 14 | 8.86 | 6.51 | 5.56 | 5.04 | 4.70 | 4.46 | 4.28 | 4.14 | 4.03 | 3.94 | 3.80 | 3.66 | 3.51 | 3.43 | 3.35 | 3.27 | 3.18 | 3.09 | 3.00 |
| 15 | 8.68 | 6.36 | 5.42 | 4.89 | 4.56 | 4.32 | 4.14 | 4.00 | 3.89 | 3.80 | 3.67 | 3.52 | 3.37 | 3.29 | 3.21 | 3.13 | 3.05 | 2.96 | 2.87 |
| 16 | 8.53 | 6.23 | 5.29 | 4.77 | 4.44 | 4.20 | 4.03 | 3.89 | 3.78 | 3.69 | 3.55 | 3.41 | 3.26 | 3.18 | 3.10 | 3.02 | 2.93 | 2.84 | 2.75 |
| 17 | 8.40 | 6.11 | 5.19 | 4.67 | 4.34 | 4.10 | 3.93 | 3.79 | 3.68 | 3.59 | 3.46 | 3.31 | 3.16 | 3.08 | 3.00 | 2.92 | 2.83 | 2.75 | 2.65 |
| 18 | 8.29 | 6.01 | 5.09 | 4.58 | 4.25 | 4.01 | 3.84 | 3.71 | 3.60 | 3.51 | 3.37 | 3.23 | 3.08 | 3.00 | 2.92 | 2.84 | 2.75 | 2.66 | 2.57 |
| 19 | 8.19 | 5.93 | 5.01 | 4.50 | 4.17 | 3.94 | 3.77 | 3.63 | 3.52 | 3.43 | 3.30 | 3.15 | 3.00 | 2.92 | 2.84 | 2.76 | 2.67 | 2.58 | 2.49 |
| 20 | 8.10 | 5.85 | 4.94 | 4.43 | 4.10 | 3.87 | 3.70 | 3.56 | 3.46 | 3.37 | 3.23 | 3.09 | 2.94 | 2.86 | 2.78 | 2.69 | 2.61 | 2.52 | 2.42 |
| 21 | 8.02 | 5.78 | 4.87 | 4.37 | 4.04 | 3.81 | 3.64 | 3.51 | 3.40 | 3.31 | 3.17 | 3.03 | 2.88 | 2.80 | 2.72 | 2.64 | 2.55 | 2.46 | 2.36 |
| 22 | 7.95 | 5.72 | 4.82 | 4.31 | 3.99 | 3.76 | 3.59 | 3.45 | 3.35 | 3.26 | 3.12 | 2.98 | 2.83 | 2.75 | 2.67 | 2.58 | 2.50 | 2.40 | 2.31 |
| 23 | 7.88 | 5.66 | 4.76 | 4.26 | 3.94 | 3.71 | 3.54 | 3.41 | 3.30 | 3.21 | 3.07 | 2.93 | 2.78 | 2.70 | 2.62 | 2.54 | 2.45 | 2.35 | 2.26 |
| 24 | 7.82 | 5.61 | 4.72 | 4.22 | 3.90 | 3.67 | 3.50 | 3.36 | 3.26 | 3.17 | 3.03 | 2.89 | 2.74 | 2.66 | 2.58 | 2.49 | 2.40 | 2.31 | 2.21 |
| 25 | 7.77 | 5.57 | 4.68 | 4.18 | 3.86 | 3.63 | 3.46 | 3.32 | 3.22 | 3.13 | 2.99 | 2.85 | 2.70 | 2.62 | 2.53 | 2.45 | 2.36 | 2.27 | 2.17 |
| 30 | 7.56 | 5.39 | 4.51 | 4.02 | 3.70 | 3.47 | 3.30 | 3.17 | 3.07 | 2.98 | 2.84 | 2.70 | 2.55 | 2.47 | 2.39 | 2.30 | 2.21 | 2.11 | 2.01 |
| 40 | 7.31 | 5.18 | 4.31 | 3.83 | 3.51 | 3.29 | 3.12 | 2.99 | 2.89 | 2.80 | 2.66 | 2.52 | 2.37 | 2.29 | 2.20 | 2.11 | 2.02 | 1.92 | 1.80 |
| 60 | 7.08 | 4.98 | 4.13 | 3.65 | 3.34 | 3.12 | 2.95 | 2.82 | 2.72 | 2.63 | 2.50 | 2.35 | 2.20 | 2.12 | 2.03 | 1.94 | 1.84 | 1.73 | 1.60 |
| 120 | 6.85 | 4.79 | 3.95 | 3.48 | 3.17 | 2.96 | 2.79 | 2.66 | 2.56 | 2.47 | 2.34 | 2.19 | 2.03 | 1.95 | 1.86 | 1.76 | 1.66 | 1.53 | 1.38 |
| $\infty$ | 6.63 | 4.61 | 3.78 | 3.32 | 3.02 | 2.80 | 2.64 | 2.51 | 2.41 | 2.32 | 2.18 | 2.04 | 1.88 | 1.79 | 1.70 | 1.59 | 1.47 | 1.32 | 1.00 |

A–31

# Answers to Odd-Numbered Problems

## CHAPTER 1

5. a) $\mu = 0.8$     7. a) 4;     b) 2.45;     c) 4;     d) 7

9. $\mu = 8.5$;   $\sigma = 3.91$

11. $\mu = 6.5$;   $\sigma^2 = 3.25$     b) $\pm 1\sigma = 66.6\%$; $\pm 2\sigma = 96.6\%$     13. a) 10.25;     b) 5.36

15. b) Mean = \$19,384.62, mode = 21,000;     c) $\sigma^2 = 65,260,540$

17. a) 5;     b) 8.33;     c) $0 < S_K < 1$; slight positive skewness

19. $\mu = 16.6$, range = 29, mode = 14 = median, $Q_1 = 11.5$, $Q_3 = 18.75$

25. b) $\sigma = 2.27$;     c) $\sigma = 3.34$

27. a) range = \$3,500;     b) \$13,100;     c) 995;     d) 68.75%, 100%

31. Median = 20,238 (by interpolation); $Q_1 = 13,950$, $Q_3 = 25,625$

## CHAPTER 2

1. a) 2/5;     b) 3/5          3. 1/64

5. 1/7

9. $P(\text{Sum} < 6) = 0.40$;     c) 0.60

11. a) $P(A_1 \cap S_2) = 1/52$;     b) $P(A_1 \cup S_2) = 16/52$
    c) Indep.     d) Answers all the same

13. a) 0.21,  0.12,  0.209;     b) 0.21;     c) 0.38          15. 0.50

19. a) 0.06;     b) 0.84;     c) 2/3;     d) Not independent

21. a) 2/20,  12/20,  6/20

23. 7/10

25. a) 3/8;     b) 4/5

27. 0.625

29. a) 0.0043;     b) 0.2097

31. a) 40,320;     b) 336;     c) 56

33. $(_{14}C_3)(_{10}C_5)/(_{24}C_8)$

37. a) 1/4;     b) 1/2;     c) 1/3

41. 0.6

43. b) 1/3

47. 7/102

**CHAPTER 3**

3. b) 10.25;   c) $E[x^2] = 137.25, V[x] = 32.19$

5. b) 6.0;   c) 1.732

7. $E[x] = E[y] = 20,000; V[x] = 200,000,000, V[y] = 66,666,667$

9. a) $E[x \cdot y] = 50, E[x + 2y] = 25, E[13 - 2x] = 3$
   b) $V[x - y] = 34, V[x + 2y] = 109, V(13 - 2x) = 36$
   c) $C(x, y) = 0$

13. a) $P_x(1) = 7/20, P_x(5) = 6/20, P_x(10) = 7/20;$
    $P_y(1) = 6/20, P_y(2) = 7/20, P_y(3) = 7/20$
    b) $E[x] = 107/20, V[x] = 14.2275, E[x + y] = 7.4$
    c) Not independent

15. a) 3.0;   b) 3.0

17. $E[x] = 10¢$, so return $= \$25.00$

19. $E[y^2] = p, V[y^2] = p(1 - p)$

**CHAPTER 4**

1. 0.9590

3. 0.1536

5. b) 0.0086;   c) $\mu = 3.2, \sigma^2 = 1.92$

7. a) 0.1382;   b) 0.9533;   c) $E[x] = 2.4$

9. 300

11. a) 0.3125;   b) 0.6562;   c) 0.6562;   d) 0.7813;   e) $\mu = 3, \sigma^2 = 1.50$

13. a) $\binom{5}{3}\binom{5}{3}\Big/\binom{10}{6}$;   b) $\binom{5}{4}\binom{5}{2}\Big/\binom{10}{6} + \binom{5}{5}\binom{5}{1}\Big/\binom{10}{6}$

15. 0.5518

17. 12/27

19. a) $\binom{3}{1}\left(\frac{3}{10}\right)^1\left(\frac{7}{10}\right)^2 + \binom{3}{2}\left(\frac{3}{10}\right)^2\left(\frac{7}{10}\right)^1 + \binom{3}{3}\left(\frac{3}{10}\right)^3\left(\frac{7}{10}\right)^0$
    b) $\binom{3}{1}\binom{7}{2}\Big/\binom{10}{3} + \binom{3}{2}\binom{7}{1}\Big/\binom{10}{3} + \binom{3}{3}\binom{7}{0}\Big/\binom{10}{3}$

21. Binomial 0.0005; Poisson $= 0.0012$;
    $V[\text{Binomial}] = 0.736, V[\text{Poisson}] = \lambda = 0.8$

23. 0.4060

25. a) 29%;   b) 0.525;   c) 0.256

27. a) 0.023

29. a) 7 games;   b) 5 games

## CHAPTER 5

1. c) 3/4
3. a) 2.33, $-1.96$
   b) $-2.33, 2.05$
5. $P(z < -2.33) = 0.01$
7. 0.1359
9. a) 0.0539;  b) 0.6423
11. a) They are the same;  b) 0.5;  c) $z_i = 2.25, z_p = -2.4$ pole more unusual
13. $n = 27$
15. b) $\mu = \frac{1}{3}, \sigma^2 = \frac{1}{9}$;  c) 0.865, 0.950
17. a) 0.2148;  b) 0.135;  c) 0.206
19. $\mu = 2/3, \sigma^2 = 1/18$
21. c) $\frac{1}{2}, F(1) = \frac{1}{4}$;  d) 4/3, $V[x] = 2/9$
23. Reject lot if it has 16 or more defectives.
25. a) 0.2148, 0.0916;  b) 0.0003, 0.8647

## CHAPTER 6

5. b) 0.4219
7. $\bar{x} = 700, s = 49.6$
9. $\bar{x} = 2.5, s_x = \sqrt{0.5789}$
11. a) 0.8944, 0.6915, 0.5468
    b) 0.9987, 0.8413, 0.9544
15. $P(x \geqslant 11) = 0.0228, P(\bar{x} \geqslant 11) = 0.00$
17. b) $\sigma_{\bar{x}} = 8.70$
19. a) 1.812;  b) 2.764;  c) 1.725;  d) 2.528;  e) 4.541
21. a) $0.05 > P(t \geqslant 2) > 0.025; 0.01 > P(t < -3) > 0.005;$ between 0.95 and 0.98;
    b) 0.0, 0.0, 1.0.
23. 0.05, between 0.05 and 0.025
29. $\bar{x} = 122.75$, sampling error and grouping error
31. a) 105;  b) 2.58;  c) More suspicious;  d) $R = 6$
33. a) $\mu = 6.0, \sigma = 2,516$;
    b) Mean $= 5.99$; CLT says $E[\bar{x}] = 6.0$ and $\sqrt{V[x]} = 2.76/\sqrt{5}$.
39. $x = 96, s = 15.1$
    a) $t = 1.854$ with 3 d.f. so $0.10 \geqslant P(t > 1.854) \geqslant 0.05$;
       $t = 2.12$ with 3 d.f. so $0.10 \geqslant P(t \geqslant 2.12) \geqslant 0.05$
    b) $z = 2.0, \quad P(z \geqslant 2) = 0.0228; \quad z = 0.80, \quad P(z \geqslant 0.80) = 0.2119$

## CHAPTER 7

7. $21.01 \leqslant \mu \leqslant 27.59$

9. $75.23 \leqslant \mu \leqslant 92.765$

11. $0.37 \leqslant p \leqslant 0.43$

13. $0.304 \leqslant p \leqslant 0.496$

15. $n = 1537$

17. $n = 166$

21. $0.00000231 \leqslant \sigma^2 \leqslant 0.00000856$

23. If $V[\hat{\theta}] = 0$

25. a) $0.0013$;     b) $1.99 \leqslant \bar{x} \leqslant 2.51$

27. $z = 2.00$: cannot have 99% confidence, but do have 97.72%

29. $7219.4 \leqslant \mu \leqslant 8780.6$

31. $N = 13{,}260$

33. $N = 662$

## CHAPTER 8

3. b) $z = 2.0$; reject $H_0$. Process not working correctly

5. $t = -2.50$; average income is $< \$10{,}000$

7. $\bar{x} = 6, s = $     $2.0, t = -3.464$. Less than 8 meals eaten beyond one mile.

9. $\bar{x} = 21, s = 4.1, t = -2.195$;     b) $p < .05$

11. $t = -2$; reject $H_0$

13. $P(1 \text{ or less}) = 0.0371$

15. $z = 4.08$; more than $\frac{2}{3}$ large fish

17. $z = 0.924$, Larger do not survive

21. $\chi^2 = 8.0$; do not reject $H_0$

23. Probability of sample this extreme is $0.9 \geqslant p \geqslant 0.1$, so do not reject $H_0$.

25. $z = 1.645$ gives $n = 271$

27. a) If $\mu_0 = 310, \beta = 0.9082$     b) If $\mu_0 = 330, \beta = 0.2514$

29. a) $\alpha = 0.50, \beta = 0.75$;     b) $\alpha = 0.75, \beta = 0.625$

31. a) $0.414, 0.207$;     b) $x \geqslant 19; x \leqslant 8$ or $x \geqslant 18$

33. a) $\alpha = 0.0114, \beta = 0.6171$     b) $\$5.70$ and $\$123.42$
    c) $\$41.02$     d) $x \geqslant 5$ is best

35. $\chi^2 = 1.921$ with 3 d.f. Do not reject $H_0$.

## CHAPTER 9

7. $0.229$

9 a) $E[\text{Air}] = 3400$;     $E[\text{Sea}] = 3000$
   b) $P(B_1 \mid R) = 0.64$
   c) Ship by sea

11. a) If $P_1 > \frac{2}{3}$, choose $A_1$
    b) $P(B_1 \mid x) = 0.615$; this is less than $\frac{2}{3}$, so choose $A_2$.
13. a) Expected value $= \$10$; fly.
    b) Worth an additional $50 over part (a).
15. EVSI $= \$1.90$;    ENGS $= \$0.90$
17. a) $U(-5) = -10, U(-10) = -30, U(20) = 30, U(50) = 50, U(100) = 60$
19. $\mu = 0.0578, \sigma^2 = 0.0264$
21. a) 124.2;    b) $\mu_1 = 118.57, \sigma_1^2 = 7.143$;
    c) 120.81;    d) 120, 118.57

## CHAPTER 10

1. $s_{\bar{x}_1 - \bar{x}_2} = 0.344, z = 4.33$; reject $H_0$.
3. $t = 2.0$; $A$ sells more.
5. $s_{\bar{x}_1 - \bar{x}_2} = 6.79,   t = 2.21$
7. $t =    2.19 > t_{(0.05,4)} = 2.13$, so sales are better in $B$.
9. $t = 3.0$; switch to Capt. Joe.
13. Probability $= 0.3748$, so no difference.
17. $F = 0.60$, there is no difference.
19. $\chi^2 = 7.66$ with 2 d.f.; reject $H_0$.
21. $t = 4.07$; new one is superior.
23. $F = 3.64$; no difference
27. $p = 0.3438$ (two-sided); $p = 0.1719$ (one-sided)

## CHAPTER 11

3. For $x = 50, \hat{y} = 2820$
5. b) $\hat{y} = -27{,}224 + 0.126x$;    c) 10,576
7. $\hat{y} = 68.6 - 2.7x$
11. a) 8 maids;    b) $-5, -7, +2$
13. a) $\hat{y} = 3.33 + (1/3)x$;    b) $833;    c) $s_e = 2.236$
15. a) $\hat{y} = 75 - 5x$;    b) $r^2 = 0.65$;    c) 45
17. a) $\hat{y} = 2.2 + 0.4x$;    b) $r^2 = 0.933$;    c) $s_e = \sqrt{0.2}$;    d) $+1, +1.2, -0.2$
21. a) 0;    b) 1;    c) $\sqrt{1/3}$
25. a) $t = 2.25 > t_{(0.05,\ 9)} = 1.833$;    b) 0.64
27. $t = 4.0$, reject $H_0$
29. a) $r^2 = 0.11, 11\%$;    b) $t = 1.72 < t_{(0.05,\ 24)}$; fail to reject.
    c) $F = 2.96$

31. b) $\hat{y} = 1750 + 6.5x$;    c) $6300; $1750;    d) $3979 \leqslant E[y] \leqslant 6021$
    e) $2496 \leqslant y_i \leqslant 7504$;    f) $t = 4.04 > t_{(0.05,\ 3)}$; reject $H_0$.
    g) 0.919;    h) $t = 4.04$

35. $\hat{b} = \sum(y - \bar{y})(x - \bar{x})^3 / \sum(x - \bar{x})^6$

37. SSE greater, so $r^2$ less and $s_e$ more than in OLS.

41. $1.12 < b_3 < 3.88$

45. b) 18.125, 20.575;    d) $s_e = 3$;    c) $t = 2.45$;    e) $0.225 \leqslant \beta \leqslant 2.225$
    f) SSR = 54, MSR = 9, $F = 6.0$;    g) $r = 0.301$;    h) $t = 2.45$;    i) 9.09%

47. c) $\hat{y} = -1.187 + 2.757x'$, $e_i = -0.570, -0.121, 0.328, 0.224, 0.673, -0.533$;
    d) $r = 0.917$; $r' = 0.964$

49. $\log \hat{y} = -0.2861 + x(0.1735)$;    b) $e_i = 0.03, -0.05, -0.01, 0.05, -0.01, 0.02, 0.03$

## CHAPTER 12

5. a) $R^2 = 0.68$;    $r^2 = 0.333$

7. a) True;    b) False! True if $R_y \cdot x_2, x_3 = 1$.
   c) True;    d) True;
   e) False! Only true if $r_{x_1 x_2} = 0$; in general should be ($\geqslant$).

9. a) $R^2 = 0.8$;    b) $r^2_{yx_3 \cdot x_1 x_2} = 80/140 = 0.571$
   c) $F = 26.67 = 80/3 = $ MSR/MSE $> F_{(01.4,\ 19)} = 4.94$; relation is significant

11. a) $t_1 = 1.5, t_2 = 2.5, t_3 = -5$
    b) $x_3$ most and $x_1$ least important
    c) $\Delta y = \sum b_i \Delta x_i = +60$.

13. a) $r^2_{yx_4,\ x_1 x_2 x_3} = 0.15$
    b) $F = 2.37 < 4.12 = F_{(4.7)}$; relation not significant.

19. $r_s = 0.66$,    $\tau = 0.42$

23. a) $\partial y / \partial x_2 \mid_{x_1,\ x_3 \text{constant}} = 6$
    b) $t_1 = 3.3$,    $t_2 = 1.2$,    $t_3 = -5$;    $x_3$ most significant
    c) SSE increases and $r^2$ decreases; $a, b_1$ and $b_3$ change; change in $s_e$ indeterminate because
       SSE increases, but $n - m - 1$ increases by one also.

25. a) Heteroscedasticity;    b) Autocorrelation

29. a) $x_1$ has higher $t$-value, 28.3 to 10.4
    b) 8.6 since $\hat{y} = 211.4$
    c) 2 and 9 since $n = 12$
    d) $R^2 \times 100 = 99.0\%$
    e) No evidence
    f) Slight negative autocorrelation
    g) No, since $r_{x_1 x_2}$ near zero
    h) Accuracy better at 10 and 12 since these values are closer to mean (see 11.9)

## CHAPTER 13

3. a) $b = \$300$;    b) $y_{1963} = \$9200$ so \$200 below trend;
   <br>$\phantom{b) y}{}_{x=9}$

5. b) $\hat{y} = 330 + 16.25x$
   c) $\hat{y} = 330 + 16.08x$
   d) 476.25, 474.72

7. a) $\hat{y} = (5.715)(0.682)^{(0.298)x}$
   b) $1/\hat{y} = 0.1842 + (0.0722)(0.1638)^x$
   c) About the same

9. —, 20, 20.33, 20.67, 21.33, 22.0, 22.67, 23.67, —

11. a) $\hat{y} = 50 + 5.5x$; origin mid-1968; $x$ units of one year
    b) $\hat{y}_{1970} = 61$; $MA_{1970} = 61\frac{1}{3}$

13. $\hat{y} = 300 + 40x$, $\hat{y}_{(42)} = 1980$, April output $= 1980(0.80) = 1584$.

17. 1st $Q = \$3.9$ mil, 2nd $Q = \$2.7$ mil

19. a) $S = 76.7$;    b) Monthly average $= 20,000$;   March $= \$18,000$.

21. 1965, \$3.68;   1973, \$3.21

25. 1.9% growth

27. \$500 increase

31. a) $_LP_{67} = 100$, $_LP_{68} = 120$, $_LP_{69} = 120$, $_LP_{70} = 160$
    b) $_PP_{67} = 100$, $_PP_{68} = 120$, $_PP_{69} = 120$, $_PP_{70} = 160$

33. a) 150;    b) \$1160 in 1960, so had more in 1950.

37. $I = 1.105$, or index of 110.5.

39. $\hat{y} = 37.5 + (6/12)(x + 20) = 47.5 + 0.5x$

43. a) $\hat{y} = 706(2.39)^x$                    c) $\hat{y} = -38.46 + 738.46(5.33)^x$
    b) $\hat{y} = 23,016$                          d) Modified exponential gives better fit

45. a) $\hat{y} = 90,000 + 5000x$, $x$ in quarters, $x = 0$ at 1st quarter 1975
    b) 1st, 125,000; 2nd, 115,000; 3rd, 109,000; 4th, 138,000
    c) 175,500

47. a) $S_{\text{Mar.}} = 105.0$; $S_{\text{Apr.}} = 115.5$; $S_{\text{Aug.}} = 75.6$
    b) April $= 11.55$; August $= 7.56$

49. a) $\hat{y} = 4.028 + 0.139x$;    b) $r = 0.14$;    c) 1.9%
    d) $\hat{y}_{73} = 6.8$ above trend; $\hat{y}_{74} = 7.3$ below trend
    e) No! Annual data;    f) $MA_{67} = 4.67$, $MA_{70} = 8.67$

# Index